Pathways in Mathematics

Series Editors

Takayuki Hibi, Department of Pure and Applied Mathematics, Osaka University, Suita, Osaka, Japan

Wolfgang König, Weierstraß-Institut, Berlin, Germany

Johannes Zimmer, Fakultät für Mathematik, Technische Universität München, Garching, Germany

Birkhäuser

Each “Pathways in Mathematics” book offers a roadmap to a currently well developing mathematical research field and is a first-hand information and inspiration for further study, aimed both at students and researchers. It is written in an educational style, i.e., in a way that is accessible for advanced undergraduate and graduate students. It also serves as an introduction to and survey of the field for researchers who want to be quickly informed about the state of the art. The point of departure is typically a bachelor/masters level background, from which the reader is expeditiously guided to the frontiers. This is achieved by focusing on ideas and concepts underlying the development of the subject while keeping technicalities to a minimum. Each volume contains an extensive annotated bibliography as well as a discussion of open problems and future research directions as recommendations for starting new projects. Titles from this series are indexed by Scopus.

More information about this series at http://www.springer.com/series/15133

Joël Bellaïche

The Eigenbook

Eigenvarieties, families of Galois representations, p-adic L-functions

Birkhäuser

Joël Bellaïche
Department of Mathematics
Brandeis University
Waltham, MA, USA

ISSN 2367-3451 ISSN 2367-346X (electronic)
Pathways in Mathematics
ISBN 978-3-030-77265-9 ISBN 978-3-030-77263-5 (eBook)
https://doi.org/10.1007/978-3-030-77263-5

Mathematics Subject Classification: 11F11, 11F33, 11F80, 11Rxx, 11Sxx

This book is published under the imprint Birkhäuser, www.birkhauser-science.com, by the registered company Springer Nature Switzerland AG.
The registered company address is: Gewerbestrasse 11, 6330 Cham, Switzerland

Contents

1 Introduction ... 1

Part I The 'Eigen' Construction

2 Eigenalgebras ... 7
2.1 A Reminder on the Ring of Endomorphisms of a Module ... 7
2.2 Construction of Eigenalgebras ... 8
2.3 First Properties ... 9
2.4 Behavior Under Base Change ... 12
2.5 Eigenalgebras Over a Field ... 13
2.5.1 Structure of the Scheme Spec $\mathcal{T}$ and of the $\mathcal{T}$-Module M ... 13
2.5.2 System of Eigenvalues, Eigenspaces and Generalized Eigenspaces ... 13
2.5.3 Systems of Eigenvalues and Points of Spec $\mathcal{T}$ ... 14
2.6 The Fundamental Example of Hecke Operators Acting on a Space of Modular Forms ... 16
2.6.1 Complex Modular Forms and Diamond Operators ... 16
2.6.2 General Theory of Hecke Operators ... 18
2.6.3 Hecke Operators on Modular Forms ... 19
2.6.4 A Brief Reminder of Atkin–Lehner–Li's Theory (Without Proofs) ... 20
2.6.5 Hecke Eigenalgebra Constructed on Spaces of Complex Modular Forms ... 24
2.6.6 Galois Representations Attached to Eigenforms ... 28
2.6.7 Reminder on Pseudorepresentations ... 28
2.6.8 Pseudorepresentations and Eigenalgebra ... 29
2.7 Eigenalgebras Over Discrete Valuation Rings ... 30
2.7.1 Closed Points and Irreducible Components of Spec $\mathcal{T}$ ... 31
2.7.2 Reduction of Characters ... 32

2.7.3 The Case of a Complete Discrete Valuation Ring......... 34
2.7.4 A Simple Application: Deligne–Serre's Lemma.......... 35
2.7.5 The Theory of Congruences.............................. 36
2.8 Modular Forms with Integral Coefficients.......................... 41
2.8.1 The Specialization Morphism for Hecke Algebras of Modular Forms.............................. 41
2.8.2 An Application to Galois Representations............... 42
2.9 A Comparison Theorem.. 44
2.10 Notes and References.. 45

3 Eigenvarieties.. 47
3.1 Non-archimedean Fredholm's Theory.............................. 48
3.1.1 General Notions.. 49
3.1.2 Compact Operators.. 49
3.1.3 Orthonormalizable and Potentially Orthonormalizable Banach Modules..................... 50
3.1.4 Serre's Sufficient Condition for Being Orthonormalizable.. 52
3.1.5 Fredholm's Determinant of a Compact Endomorphism.. 53
3.1.6 Property (Pr).. 56
3.1.7 Extension of Scalars... 57
3.2 Everywhere Convergent Formal Series and Riesz's Theory....... 58
3.2.1 The ν-Valuation and ν-Dominant Polynomials........... 59
3.2.2 Euclidean Division.. 60
3.2.3 Good Zeros.. 62
3.2.4 A Piece of Resultant Theory.............................. 62
3.2.5 Riesz's Theory... 64
3.3 Adapted Pairs.. 66
3.3.1 Strongly ν-Dominant Polynomials....................... 66
3.3.2 A Canonical Factorization of Everywhere Convergent Power Series................................. 67
3.3.3 Adapted Pairs.. 69
3.4 Submodules of Bounded Slope...................................... 73
3.5 Links.. 75
3.6 The Eigenvariety Machine... 76
3.6.1 Eigenvariety Data.. 76
3.6.2 Construction of the Eigenvariety.......................... 77
3.7 Properties of Eigenvarieties.. 80
3.8 A Comparison Theorem for Eigenvarieties........................ 83
3.8.1 Classical Structures... 83
3.8.2 A Reducedness Criterion.................................... 85
3.8.3 A Comparison Theorem..................................... 85

3.9 A Simple Generalization: The Eigenvariety Machine for Complexes ... 89
3.9.1 Data for a Cohomological Eigenvariety ... 90
3.9.2 Construction of the Cohomological Eigenvariety ... 90
3.9.3 Properties of the Cohomological Eigenvariety ... 91
3.9.4 Classical Structures and an Application to Reducedness ... 92
3.10 Notes and References ... 93

Part II Modular Symbols and L-Functions

4 Abstract Modular Symbols ... 97
4.1 The Notion of Modular Symbols ... 97
4.2 Action of the Hecke Operators on Modular Symbols ... 99
4.3 Reminder on Cohomology with Local Coefficients ... 100
4.3.1 Local Systems ... 100
4.3.2 Local Systems and Representations of the Fundamental Group ... 101
4.3.3 Singular Simplices ... 102
4.3.4 Homology with Local Coefficients ... 102
4.3.5 Cohomology with Local Coefficients ... 103
4.3.6 Relative Homology and Relative Cohomology ... 104
4.3.7 Formal Duality Between Homology and Cohomology ... 104
4.3.8 Cohomology with Compact Support and Interior Cohomology ... 105
4.3.9 Cup-Products and Cap-Products ... 107
4.3.10 Poincaré Duality ... 108
4.3.11 Singular Cohomology and Sheaf Cohomology ... 109
4.4 Modular Symbols and Cohomology ... 110
4.4.1 Right Action on the Cohomology ... 111
4.4.2 Modular Symbols and Relative Cohomology ... 111
4.4.3 Modular Symbols and Cohomology with Compact Support ... 112
4.4.4 Pairings on Modular Symbols ... 115
4.5 Notes and References ... 118

5 Classical Modular Symbols, Modular Forms, L-functions ... 119
5.1 On a Certain Monoid and Some of Its Modules ... 119
5.1.1 The Monoid S ... 119
5.1.2 The S-modules $\mathcal{P}_k$ and $\mathcal{V}_k$ ... 120
5.2 Classical Modular Symbols ... 122
5.2.1 Definition ... 122
5.2.2 The Standard Pairing on Classical Modular Symbols ... 122
5.2.3 Adjoint of Hecke Operators for the Standard Pairing ... 123

5.3 Classical Modular Symbols and Modular Forms ... 125
5.3.1 Modular Forms and Real Classical Modular Symbols ... 125
5.3.2 Modular Forms and Complex Classical Modular Symbols ... 129
5.3.3 The Involution ι, and How to Get Rid of the Complex Conjugation ... 133
5.3.4 The Endomorphism W_N and the Corrected Scalar Product ... 136
5.3.5 Boundary Modular Symbols and Eisenstein Series ... 139
5.3.6 Summary ... 142
5.4 Applications of Classical Modular Symbols to L-functions and Congruences ... 144
5.4.1 Reminder About L-functions ... 144
5.4.2 Modular Symbols and L-functions ... 147
5.4.3 Scalar Product and Congruences ... 149
5.5 Notes and References ... 150

6 Rigid Analytic Modular Symbols and p-Adic L-functions ... 153
6.1 Rigid Analytic Functions and Distributions ... 153
6.1.1 Some Modules of Sequences and Their Dual ... 153
6.1.2 Modules of Functions over $\mathbb{Z}_p$ ... 155
6.1.3 Modules of Convergent Distributions ... 158
6.2 Overconvergent Functions and Distributions ... 159
6.2.1 Semi-normic Modules and Fréchet Modules ... 159
6.2.2 Modules of Overconvergent Functions and Distributions ... 162
6.2.3 Integration of Functions Against Distributions ... 163
6.2.4 Order of Growth of a Distribution ... 163
6.3 The Weight Space ... 168
6.3.1 Definition and Description of the Weight Space ... 168
6.3.2 Local Analyticity of Characters ... 170
6.3.3 Some Remarkable Elements in the Weight Space ... 171
6.3.4 The Functions $\log_p^{[k]}$ on the Weight Space ... 172
6.3.5 The Iwasawa Algebra and the Weight Space ... 172
6.4 The Monoid $S_0(p)$ and Its Actions on Overconvergent Distributions ... 173
6.4.1 The Monoid $S_0(p)$ ... 173
6.4.2 Actions of $S_0(p)$ on Functions and Distributions ... 174
6.4.3 The Module of Locally Constant Polynomials and Its Dual ... 175
6.4.4 The Fundamental Exact Sequence for Overconvergent Functions ... 176
6.4.5 The Fundamental Exact Sequence for Overconvergent Distributions ... 179
6.5 Rigid Analytic and Overconvergent Modular Symbols ... 179

6.5.1 Definitions and Compactness of U_p 179
6.5.2 Space of Overconvergent Modular Symbols of Finite Slope 182
6.5.3 Computation of an H_0 184
6.5.4 The Fundamental Exact Sequence for Modular Symbols 185
6.5.5 Stevens's Control Theorem 186
6.6 The Mellin Transform 188
6.6.1 The Real Mellin Transform 188
6.6.2 The p-Adic Mellin Transform 188
6.6.3 Properties of the p-Adic Mellin Transform 189
6.6.4 The Mellin Transform over a Banach Algebra 191
6.7 Applications to the p-Adic L-functions of Non-critical Slope Modular Forms 191
6.7.1 Refinements 191
6.7.2 Construction of the p-Adic L-functions 193
6.7.3 Computation of the p-Adic L-functions at Special Characters 195
6.8 Notes and References 199

Part III The Eigencurve and its p-Adic L-Functions

7 The Eigencurve of Modular Symbols 203
7.1 Construction of the Eigencurve Using Rigid Analytic Modular Symbols 203
7.1.1 Overconvergent Modular Symbols Over an Admissible Open Affinoid of the Weight Space 204
7.1.2 The Restriction Theorem 206
7.1.3 The Specialization Theorem 209
7.1.4 Construction 213
7.2 Comparison with the Coleman-Mazur Eigencurve 214
7.2.1 The Coleman-Mazur Full Eigencurve $\mathcal{C}$ 214
7.2.2 The Cuspidal Eigencurve 216
7.2.3 Applications of Chenevier's Comparison Theorem 217
7.3 Points of the Eigencurve 220
7.3.1 Interpretations of the Points as Systems of Eigenvalues of Overconvergent Modular Symbols 220
7.3.2 Very Classical Points 221
7.3.3 Classical Points 221
7.3.4 Hida Classical Points 222
7.4 The Family of Galois Representations Carried by the Eigencurve 225
7.4.1 Construction of the Family of Galois Representations 225
7.4.2 Local Properties at $l \neq p$ of the Family of Galois Representations 227

7.4.3 Local Properties at p of the Family of Galois Representations ... 230
7.5 The Ordinary Locus ... 230
7.6 Local Geometry of the Eigencurve ... 231
7.6.1 Clean Neighborhoods ... 231
7.6.2 Étaleness of the Eigencurve at Non-critical Slope Classical Points ... 235
7.6.3 Geometry of the Eigencurve at Critical Slope Very Classical Points ... 236
7.6.4 Critical Slope Eigenforms and Points on $\mathcal{C}^{\pm}$ ... 242
7.6.5 Complements on the Geometry of the Eigencurve at Classical Points ... 243
7.7 Global Properties of the Eigencurve ... 245
7.7.1 Integrality of Fredholm Determinants and Integral Models of the Eigencurves ... 245
7.7.2 Valuative Criterion of Properness ... 246
7.7.3 Open Questions ... 246
7.8 Notes and References ... 246

8 *p*-Adic *L*-Functions on the Eigencurve ... 249
8.1 Good Points and p-Adic L-Functions ... 249
8.1.1 Good Points on the Eigencurve ... 249
8.1.2 The p-Adic L-Function of a Good Point ... 251
8.1.3 Companion Points and p-Adic L-Functions ... 251
8.2 p-Adic L-Functions of an Overconvergent Eigenform ... 252
8.2.1 Definition ... 252
8.2.2 Classical Cuspidal Eigenforms of Non-critical Slope ... 253
8.2.3 Ordinary Eisenstein Eigenforms ... 254
8.2.4 Classical Eigenforms of Critical Slope ... 255
8.2.5 Classical Eigenforms of Weight 1 ... 257
8.3 The 2-Variable p-Adic L-Function ... 257
8.4 Notes and References ... 261

9 The Adjoint *p*-Adic *L*-Function and the Ramification Locus of the Eigencurve ... 263
9.1 The L-Ideal of a Scalar Product ... 264
9.1.1 The Noether Different of T/R ... 264
9.1.2 Duality ... 267
9.1.3 The L-Ideal of a Scalar Product ... 268
9.2 Kim's Scalar Product ... 270
9.2.1 A Bilinear Product on the Space of Overconvergent Modular Symbols of Weight k ... 270
9.2.2 Interpolation of Those Scalar Products ... 273
9.3 The Cuspidal Eigencurve, the Interior Cohomological Eigencurves, and Their Good Points ... 276

9.4 Construction of the Adjoint p-Adic L-Function on the Cuspidal Eigencurve ... 278
9.5 Relation Between the Adjoint p-Adic L-Function and the Classical Adjoint L-Function ... 283
9.5.1 Scalar Product and Refinements ... 283
9.5.2 Adjoint L-Function and Peterson's Product ... 285
9.5.3 p-Adic and Classical Adjoint L-Function ... 286

10 Solutions and Hints to Exercises ... 287

Bibliography ... 311

Chapter 1
Introduction

The aim of this book is to give a gentle but complete introduction to the two interrelated theory of p-adic families of modular forms and of p-adic L-functions of modular forms. These theories and their generalizations to automorphic forms for group of higher ranks are now of fundamental importance in number theory. To our knowledge, this is the first textbook containing this material, which now is to be found scattered in many articles published during the last 40 years. Here we present a self-contained exposition of the theory leading to the construction of the eigencurves, and of the families of p-adic L-functions and adjoint p-adic L-functions on them.

The book is intended both for beginners in the field (graduate students and established researchers in other fields alike) and for researchers working in the field of p-adic families of automorphic forms and p-adic L-functions, who want a solid foundation, in one place, for further work in the theory. Actually, I wrote it in a large part for myself, to help me prepare for my current work on p-adic L-functions and p-adic families of automorphic forms, with a view toward the Beilinson-Bloch-Kato conjectures. In particular, while the focus is for families of modular forms, the main tools used are presented in an abstract way in view of their applications to higher-dimensional situation (e.g. the *eigenvariety machine*, or the theory of the L-*ideal* used to construct adjoint p-adic L-functions).

The prerequisites for this book includes a familiarity with basic commutative algebra, algebraic geometry (the language of schemes), and algebraic number theory. In addition, a basic knowledge of the theory of classical modular forms, and the Galois representation attached to them (for example, the content of a book like [56]) is needed, as well as the very basics in the theory of rigid analytic geometry. Even so, we take care to recall the definitions and results we need in these theories, with precise references when we don't provide proofs.

Let us describe the content of this book.

In Chap. 2 we introduce the very simple idea which is central to the construction of eigenvariety: when we have a module M together with a family of commuting

J. Bellaïche, *The Eigenbook*, Pathways in Mathematics,
https://doi.org/10.1007/978-3-030-77263-5_1

linear endomorphisms of it (such as a module of modular forms with the Hecke operators), we consider the subring generated by the operators in End(M); since this ring is commutative, its spectrum is defined, and we focus our attention on the geometry of this scheme. The Chap. 2 tries to help the reader to develop an intuition of this basic idea, with the analysis of many special cases and examples.

Chapter 3 presents a version at large of this idea, the *eigenvariety machine*. It is largely inspired from the very readable article by K. Buzzard of the same title. The main change is that we offer a completely self-contained presentation, proving the results of p-adic functional analysis that we need (due to Serre, Lazard, Coleman). We also give a more general and more precise construction (removing some unnecessary hypotheses, and describing a more convenient admissible covering, which will be of much help for the rest of the book), and complete it with many results due to Chenevier, including the extremely important Chenevier's comparisons theorem.

Chapters 4, 5, and 6 are devoted to the theory of modular symbols, and their connection to modular forms. They are central tools in our study: they will be the elements of the modules to which we will apply the methods of Chaps. 2 and 3, and they will also be the matrix of which p-adic L-functions are extracted. Chapter 4 introduces the abstract theory of modular symbols, including their cohomological interpretation. This requires quite a bit of algebraic topology, that we completely recall with definition and proofs, for the sake of self-containment. Chapter 5 exposes the classical theory of modular symbols, as developed by Manin, Shokurov, Amice-Velu, and Mazur-Tate-Teitelbaum, and their close, but subtle connections to classical modular forms. Chapter 6 develops the theory, initiated by Stevens, and developed by Pollack-Stevens and the author, of rigid analytic modular symbols. Chapter 6 recalls the method of Stevens to construct the p-adic L-function of classical cuspidal non-critical slope modular form.

In the last three chapters we use the tools developed earlier to construct and study the eigencurve, and the families of p-adic L-function it carries. Chapter 7 gives the construction of the eigencurve using families of rigid analytic modular symbols. It also compares this eigencurve to the eigencurve constructed by Coleman and Mazur using overconvergent modular forms. The family of Galois representations on the eigencurve is introduced, and used to study the local geometry of the eigencurve at most interesting points, following earlier work of the author and Chenevier. In Chap. 8, we construct the a p-adic family of L-function on the eigencurve, locally in the neighborhoods of *good* points. Here a *good* point is a technical condition satisfied by all classical points and many others. At the same times, we define the p-adic L-functions of classical modular forms that were not susceptible to the method of Stevens or Mazur-Tate-Teitelbaum, and when possible, we compute them. In Chap. 9, we construct, globally on the eigencurve, an ideal, the L-ideal, and, locally near good points, a function that generates it, the adjoint p-adic L-function; this function of characterized by the property of interpolating, at any classical point, the near-central value of the archimedean adjoint L-function of the modular form corresponding to that point. It has also the property that its zero, and their order, are precisely related to the geometry of the eigencurve.

This books started with the notes I wrote for a graduate course given at Brandeis. I want to thank the auditors of this course, John Bergdall, Yu Fang, Dawn Nelson, Dipramit Majumdar, Anna Medvedovsky, Yurong Zhan. My students John Bergdall and Tarakaram Gollamudi also read part of this books and suggested improvements, and I want to thank them for that.

This book would never have been possible without Gaëtan Chenevier. I learned most of the theory either from him or with him, and I want to thank him warmly here. Also I benefited from conversations with too many mathematicians to name them all, but I want to thank particularly Glenn Stevens, Robert Pollack, Kevin Buzzard, Shin Hattori and Samit Dasgupta.

Part I
The 'Eigen' Construction

Chapter 2
Eigenalgebras

Except for two sections (Sects. 2.6 and 2.8) concerning modular forms, intended as motivations for and illustrations of the general theory, this chapter is purely algebraic. We want to explain a very simple, even trivial, construction that has played an immense role in the arithmetic theory of automorphic forms during the last 40 years. This construction attaches to a family of commuting operators acting on some space or module an algebraic object, called the *eigenalgebra*, that parameterizes the systems of eigenvalues for those operators appearing in the said space or module.

In all this chapter, R is a commutative noetherian ring.

2.1 A Reminder on the Ring of Endomorphisms of a Module

Let M be a **finite** R**-module**. We shall denote by $\mathrm{End}_R(M)$ the R-algebra of R-linear endomorphisms $\phi : M \to M$. If R' is a commutative R-algebra, and $M' = M \otimes_R R'$, there is a natural morphism of R-algebras $\mathrm{End}_R(M) \to \mathrm{End}_{R'}(M')$ sending $\phi : M \to M$ to $\phi \otimes \mathrm{Id}_{R'} : M' \to M'$. Hence there is a natural morphism of R'-algebras

$$\mathrm{End}_R(M) \otimes_R R' \to \mathrm{End}_{R'}(M'). \tag{2.1.1}$$

When R' is a fraction ring of R, and more generally when R' is R-flat, this morphism is an isomorphism. In other words, **the formation of** $\mathrm{End}_R(M)$ **commutes with localization**, and more generally, flat base change.

J. Bellaïche, *The Eigenbook*, Pathways in Mathematics,
https://doi.org/10.1007/978-3-030-77263-5_2

Exercise 2.1.1 Show more generally that if M is a finitely presented R-module, N any R-module, and $R \to R'$ is flat, then the morphism $\mathrm{Hom}_R(M, N) \otimes_R R' \to \mathrm{Hom}_{R'}(M', N')$ where $M' = M \otimes_R R'$ and $N' = N \otimes_R R'$ is an isomorphism.

Exercise 2.1.2 Give an example of R-algebra R' and finite R-module M where the morphism (2.1.1) is not an isomorphism.

We now assume that in addition of being a finite R-module, M **is flat**, or what amounts to the same, projective, or locally free. Recall that the *rank* of M is the locally constant function $\mathrm{Spec}\, R \to \mathbb{N}$ sending $x \in \mathrm{Spec}\, R$ to the dimension over $k(x)$ of $M \otimes_R k(x)$, where $k(x)$ is the residue field of the point x. When this function is constant (which is always the case when $\mathrm{Spec}\, R$ is connected, for example when R is local), we also call *rank* of M its value.

Since the formation of $\mathrm{End}_R(M)$ commutes with localizations, many properties enjoyed by $\mathrm{End}_R(M)$ in the case where M is free (in which case $\mathrm{End}_R(M)$ is just a matrix algebra $M_d(R)$ if M is of rank d) are still true in the flat case. For example, if M is flat, the natural morphism (2.1.1)

$$\mathrm{End}_R(M) \otimes_R R' \to \mathrm{End}_{R'}(M')$$

is an isomorphism for all R'-algebras R. To see this, note that it suffices to check that the morphism (2.1.1) is an isomorphism after localization at every prime ideal $\mathfrak{p}'$ of R', and thus after localization of R as well at the prime ideal $\mathfrak{p}$ of R below $\mathfrak{p}'$. Hence we can assume that R is local, so that M is free, in which case the result is clear by the description of $\mathrm{End}_R(M)$ as an algebra of square matrices. More generally, when M and N are finite free, $\mathrm{Hom}_R(M, N) \otimes_R R' \to \mathrm{Hom}_{R'}(M', N')$ is an isomorphism for all R'-algebras R.

Similarly, we can define the characteristic polynomial $P_\phi(X) \in R[X]$ of an endomorphism $\phi \in \mathrm{End}_R(M)$ of a finite flat module M by gluing the definitions $P_\phi(X) = \det(\phi - X\mathrm{Id})$ in the free case. The Cayley–Hamilton theorem $P_\phi(\phi) = 0 \in \mathrm{End}_R(M)$ holds since it holds locally. Observe that $P_\phi(X)$ is not necessarily monic in general, but is monic of degree d when M has constant rank d. In particular, $P_\phi(X)$ is monic if $\mathrm{Spec}\, R$ is connected.

Exercise 2.1.3 Show that the formation of $P_\phi(X)$ commutes with arbitrary base change, that is that if $R \to R'$ is any map, and $\phi' = \phi \otimes \mathrm{Id}_{R'} \in \mathrm{End}_{R'}(M')$, then $P_{\phi'}(X)$ is the image of $P_\phi(X)$ in $R'[X]$.

2.2 Construction of Eigenalgebras

Let M be a finite flat R-module (or equivalently, finite locally free, or finite projective). In the applications, elements of M will be modular forms, or automorphic forms, or families thereof, over R. Finally, we suppose given a commutative ring $\mathcal{H}$ and a morphism of rings $\psi : \mathcal{H} \to \mathrm{End}_R(M)$.

To those data $(R, M, \mathcal{H}, \psi)$, we attach the sub-$R$-module $\mathcal{T} = \mathcal{T}(R, M, \mathcal{H}, \psi)$ of $\mathrm{End}_R(M)$ generated by the image $\psi(\mathcal{H})$. It is clear that $\mathcal{T}$ is a sub-algebra of $\mathrm{End}_R(M)$.

Equivalently, $\mathcal{T}$ is the quotient of $\mathcal{H} \otimes_{\mathbb{Z}} R$ that acts faithfully on M, that is to say, the quotient of $\mathcal{H} \otimes_{\mathbb{Z}} R$ by the ideal annihilator of M.

Definition 2.2.1 The R-algebra $\mathcal{T}$ is called the *eigenalgebra* of $\mathcal{H}$ acting on the module M.

Another version of the same beginning is simpler and more direct: we are given R, M as above and we attach to a family of commuting endomorphisms $(T_i)_{i \in I}$ (infinite in general) in $\mathrm{End}_R(M)$, the R-subalgebra $\mathcal{T}$ they generate. This is equivalent to the preceding situation, since we can take for $\mathcal{H}$ the ring of polynomials $\mathbb{Z}[(X_i)_{i \in I}]$ with independent variables X_i, and for ψ the map that sends X_i on T_i, giving the same $\mathcal{T}$. Conversely, the situation with $\mathcal{H}$ and ψ can be converted into the situation with the T_i's by choosing a family of generators $(X_i)_{i \in I}$ of $\mathcal{H}$ as a $\mathbb{Z}$-algebra and setting $T_i = \psi(X_i)$.

The aim of this section is to explain the meaning and properties of this simple construction.

2.3 First Properties

We keep assuming that M is a finite flat R-module. By construction $\mathcal{T}$ is a commutative R-algebra, and M has naturally a structure of $\mathcal{T}$-module. We record some obvious properties.

Lemma 2.3.1 *As an R-module, $\mathcal{T}$ is finite and torsion-free.*[1] *As a ring, $\mathcal{T}$ is noetherian. The module M is finite over $\mathcal{T}$.*

Proof The first assertion is obvious since $\mathrm{End}_R(M)$ is finite and R is noetherian. Moreover, $\mathrm{End}_R(M)$ is torsion-free since M is, and so is its submodule $\mathcal{T}$. That $\mathcal{T}$ is noetherian follows form the first assertion and Hilbert's theorem. And the last assertion is clear since M is already finite as an R-module. □

Exercise 2.3.2 Assume that R and $\mathcal{T}$ are domains. Show that M is torsion-free as a $\mathcal{T}$-module.

Proposition 2.3.3 *We assume that M is a faithful (finite flat) R-module. Then, the map $\mathrm{Spec}\,\mathcal{T} \to \mathrm{Spec}\,R$ is surjective and closed. It maps every irreducible component of $\mathrm{Spec}\,\mathcal{T}$ surjectively on an irreducible component of $\mathrm{Spec}\,R$. In*

---

[1] Remember that in an R-module M, an element m is said to be *torsion* if there exists an $x \in R$, not a zero-divisor, such that $xm = 0$. The module M is *torsion* if all its elements are torsion, and is *torsion-free* if only 0 is. A flat module M is torsion-free.

particular if Spec R *is equidimensional*[2] *of dimension* n*, then* Spec $\mathcal{T}$ *is also equidimensional of dimension* n*.*

Let us note that if M is a finite flat R-module, the condition M faithful is equivalent, in the case Spec R connected, to $M \neq 0$.

Proof The hypothesis tells us that the map $R \to \mathrm{End}_R(M)$ is injective, hence also the map $R \to \mathcal{T}$. Since $\mathcal{T}$ is also finite over R, the first sentence is well-known: see [59, Prop. 4.15] for a proof. For the second, it suffices to prove that the intersection with R of every minimal prime ideal of $\mathcal{T}$ is a minimal prime ideal of R, and it follows from [59, Cor 4.18]. The assertion about equidimensionality is then clear. □

Proposition 2.3.4 *If* R *has no embedded prime,*[3] *then* $\mathcal{T}$ *has no embedded prime either.*

Proof Let Q be an associated prime of $\mathcal{T}$, and set $P = Q \cap R$. Hence P is a prime ideal of R which kills a non-zero element of $\mathcal{T}$. Since $\mathcal{T}$ is torsion-free over R, every element of P is a divisor of 0 in R, and by [59, cor. 3.2, page 90], P is an associated prime of R. Since R has no embedded prime, P is therefore a minimal prime ideal of R, and since $\mathcal{T}$ is finite over R, Q is a minimal prime ideal of $\mathcal{T}$ (cf. [59, Cor 4.18]). Hence Q is not an embedded prime. □

The finiteness of $\mathcal{T}$ over R implies that it is integral over R, which can be made more precise by the following lemma.

Lemma 2.3.5 *Assume that* M *has constant rank* d *over* R*. Every element of* $\mathcal{T}$ *is killed by a monic polynomial of degree* d *with coefficients in* R*.*

Proof Every element of $\mathcal{T}$ is killed by its characteristic polynomial, according to the Cayley–Hamilton theorem. □

Exercise 2.3.6 Assume that M is a finite flat faithful R-module

1. Suppose given a projective R-submodule A of M that is $\mathcal{H}$-stable. Then we can also define the eigenalgebra of $\mathcal{H}$ acting on A, and denote it by $\mathcal{T}_A$. Construct a natural surjective morphism of algebras $\mathcal{T} \to \mathcal{T}_A$. Show that if M/A is torsion, then this map is an isomorphism.
2. Suppose given two $\mathcal{H}$-stable submodules A and B of M such that $A \cap B = 0$. Show that the map $\mathcal{T} \to \mathcal{T}_A \times \mathcal{T}_B$ is neither injective nor surjective in general.
3. Same hypotheses as in 2. and assume that $M/(A \oplus B)$ is torsion. Show that the map $\mathcal{T} \to \mathcal{T}_A \times \mathcal{T}_B$ is injective.

---

[2] A scheme X is *equidimensional of dimension* n if all its irreducible components have Krull dimension n.

[3] Remember that a prime ideal in a ring R is an *associated* prime if it is the annihilator of some non-zero element of R; associated primes which are minimal among associated primes are called *isolated* primes, and those who are not are called *embedded* primes.

4. Same hypothesis as in 2. and assume there is a $T \in \mathcal{H}$ that acts by multiplication by a scalar a on A and by a scalar b and B, and that $b-a$ is invertible in R. Show that the map $\mathcal{T} \to \mathcal{T}_A \times \mathcal{T}_B$ is surjective.

Exercise 2.3.7 Let $R, M, \mathcal{H}, \psi$ be as above. We write $M^\vee$ for the dual module $\mathrm{Hom}_R(M, R)$. There is a natural map $\psi^\vee : \mathcal{H} \to \mathrm{End}_R(M^\vee)$ defined by $\psi^\vee(h) = {}^t\psi(h)$ where ^{t} denotes the transpose of a map. We denote by $\mathcal{T}$ and $\mathcal{T}^\vee$ the eigenalgebras of $\mathcal{H}$ acting on M and $M^\vee$.

1. Show that $\mathcal{T}$ is canonically isomorphic to $\mathcal{T}^\vee$.
2. Show that if R is a field, and $\mathcal{H}$ is generated by one element, then $M^\vee \simeq M$ as $\mathcal{H}$-modules and as $\mathcal{T}$-modules.
3. Show that the same results may fail to hold if either R is not a field, or R is a field (even $R = \mathbb{C}$) but $\mathcal{H}$ is not generated by a single element.

Can we say more about the abstract structure of $\mathcal{T}$ than just its being finite and torsion-free over R? Or on the contrary, is any finite and torsion-free R-algebra S an eigenalgebra $\mathcal{T}$ for some $M, \mathcal{H}, \psi$. The question is clearly equivalent to the following:

Question 2.3.8 Is any finite and torsion-free R-algebra S a sub-algebra of some $\mathrm{End}_R(M)$ for some finite flat R-module M?

The answer obviously depends on the nature of the noetherian ring R.

When R is a Dedekind domain, the answer is yes. Indeed, any finite torsion-free R-algebra S is also a flat R-module, so we can take $M = S$ and see S as a sub-algebra of $\mathrm{End}_R(M)$ by left-multiplication.

A Dedekind domain is regular of dimension 1. The answer to Question 2.3.8 is also yes when R is a regular ring of dimension 2. Indeed the bi-dual of any finite torsion-free module over R is projective by a result of Serre (see [110, Lemme 6]), so if we define M as the bi-dual of S seen as an R-module, then S embeds naturally in $\mathrm{End}_R(M)$.

However, it seems[4] that Question 2.3.8 is still open for regular rings R of dimension ≥ 3. For general non-regular domains, the answer is no: there may exist a torsion-free R-algebra S that cannot be a $\mathcal{T}$. Here is an example:[5] $R = k[[x^2, xy, y^2]]$ and $S = k[[x, y]]$ for k a field.

Exercise 2.3.9

1. If I is an ideal of R, observe that $R \oplus \epsilon I$ is a sub-R-algebra of $R[\epsilon]/(\epsilon^2)$. Show that all those algebras $R \oplus \epsilon I$ are eigenalgebras $\mathcal{T}$. Deduce that there are examples of eigenalgebras that are not flat over R, and not reduced.
2. Is there an example of eigenalgebra $\mathcal{T}$ that is reduced and non-flat over R?

[4] I thank Kevin Buzzard from this information, who got it from Mel Hochster of Michigan University.

[5] Also due to Mel Hochster.

2.4 Behavior Under Base Change

We keep assuming that M is finite flat R-module.

Let us investigate the behavior of the construction of eigenalgebras with respect to base change. Let $R \to R'$ be a morphism of noetherian rings, and define $M' = M \otimes_R R'$, and let $\psi' : \mathcal{H} \to \mathrm{End}_R(M) \to \mathrm{End}_{R'}(M')$ be the obvious composition. We call $\mathcal{T}'$ the R'-eigenalgebra of $\mathcal{H}'$ acting on M'.

Through the isomorphism $\mathrm{End}_R(M) \otimes_R R' \to \mathrm{End}_{R'}(M')$, the image of $\mathcal{T} \otimes_R R'$ in $\mathrm{End}_{R'}(M')$ is precisely $\mathcal{T}'$. Therefore, there is a natural surjective map of R'-algebras

$$\mathcal{T} \otimes_R R' \to \mathcal{T}'. \tag{2.4.1}$$

Proposition 2.4.1 *The kernel of the base change map (2.4.1) is a nilpotent ideal. This map is an isomorphism when R' is R-flat.*

Proof We first prove the second assertion. By definition, the map $\mathcal{T} \to \mathrm{End}_R(M)$ is injective. By flatness of R', so is the map $\mathcal{T} \otimes_R R' \to \mathrm{End}_{R'}(M')$. In other words $\mathcal{T} \otimes_R R'$ acts faithfully on M'. Since $\mathcal{T}'$ is the quotient of $\mathcal{H} \otimes R'$ that acts faithfully on M', we have $\mathcal{T}' = \mathcal{T} \otimes_R R'$.

For the first assumption, let us assume first that the map $R \to R'$ is surjective, of kernel I. We may assume in addition that M is free, of rank d. Let $\tilde{\phi}$ in $\mathcal{T} \otimes_R R' = \mathcal{T}/I\mathcal{T}$ be an element whose image $\phi' \in \mathcal{T}' \subset \mathrm{End}_{R'}(M')$ is 0. Let ϕ in $\mathcal{T}$ be an element that lifts $\tilde{\phi}$. We thus have $0 = \phi' = \phi \otimes 1 \in \mathrm{End}_{R'}(M')$. The characteristic polynomial of ϕ' is X^d. Since the formation of characteristic polynomial commutes with base change (see Exercise 2.1.3), we deduce that the characteristic polynomial of ϕ belongs to $X^d + IR[X]$. By Cayley–Hamilton, we thus have $\phi^d \in I\mathrm{End}_R(M)$, and so $\tilde{\phi}^d = 0$. Since $\mathcal{T} \otimes_R R'$ is Noetherian, it follows that the kernel of $\mathcal{T} \otimes_R R' \to \mathcal{T}'$ is nilpotent.

In general, any morphism $R \to R'$ may be factorized at $R \to R'' \to R'$ with $R \to R''$ flat and $R'' \to R'$ surjective. By the second assumption, the map $\mathcal{T} \otimes_R R'' \to \mathcal{T}''$ is an isomorphism and by the first in the surjective case, the map $\mathcal{T} \otimes_R R' = \mathcal{T}'' \otimes_{R''} R' \to \mathcal{T}'$ has nilpotent kernel. □

Exercise 2.4.2 Let $R = \mathbb{Z}_p$, $M = R^2$, $T \in \mathrm{End}_R(M)$ given by the matrix $\begin{pmatrix} 1 & p \\ 0 & 1 \end{pmatrix}$. Describe $\mathcal{T}$ in this case. Set $R' = \mathbb{F}_p$. Describe $\mathcal{T}'$ and the morphism $\mathcal{T} \otimes \mathbb{F}_p \to \mathcal{T}'$ in this case.

Proposition 2.4.3 *If M is flat as a $\mathcal{T}$-module, then the map (2.4.1) is an isomorphism for any R-algebra R'. In particular, this holds if $\mathcal{T}$ is étale over R.*

Proof By assumption, M is flat as a $\mathcal{T}$-module; by construction it is also faithful, and we know that it is finite. Therefore M is a faithfully flat $\mathcal{T}$-module. It follows that $M \otimes_R R'$ is faithfully flat over $\mathcal{T} \otimes_R R'$, and in particular, is faithful. This implies that the morphism (2.4.1) is injective, hence an isomorphism.

For the second assertion, we just use that if M is a finite $\mathcal{T}$-module and $\mathcal{T}$ an étale finite R-algebra, then M is flat over $\mathcal{T}$ if and only if it is flat over R. This is true because flatness of a module over a ring may be checked on the strict henselianizations of this ring at all prime ideals, and those are the same for $\mathcal{T}$ and R. □

2.5 Eigenalgebras Over a Field

In this section we assume that R **is a field** k. We assume that M is a finite-dimensional k-vector space.

2.5.1 Structure of the Scheme Spec $\mathcal{T}$ and of the $\mathcal{T}$-Module M

Since $\mathcal{T}$ is a finite algebra over k, it is an artinian semi-local ring: $\mathcal{T}$ has only finitely many prime ideals, and those prime ideals are maximal, say $\mathfrak{m}_1, \dots, \mathfrak{m}_l$. The algebra $\mathcal{T}$ is canonically isomorphic to the product of the local artinian k-algebras $\mathcal{T}_{\mathfrak{m}_i}$. We shall set $M_{\mathfrak{m}_i} = M \otimes_{\mathcal{T}} \mathcal{T}_{\mathfrak{m}_i}$ so that we have a decomposition $M = \oplus_{i=1}^{l} M_{\mathfrak{m}_i}$. This decomposition is stable by the action of $\mathcal{T}$, each $\mathcal{T}_{\mathfrak{m}_i}$ acting by 0 on the summands other than $M_{\mathfrak{m}_i}$. Since $\mathcal{T}$ acts faithfully on M, $\mathcal{T}_{\mathfrak{m}_i}$ acts faithfully on $M_{\mathfrak{m}_i}$ and in particular those subspace are non-zero. (See e.g. [59, §2.4] if any of those results is not clear)

If we write $M[I]$ for the subspace of M of elements killed by an ideal I of $\mathcal{T}$, then $M_{\mathfrak{m}_i}$ can be canonically identified with $M[\mathfrak{m}_i^\infty] := \cup_{n \in \mathbb{N}} M[\mathfrak{m}_i^n]$ which since M is noetherian is the same as $M[\mathfrak{m}_i^n]$ for n large enough. Indeed, $M[\mathfrak{m}_i^\infty] = \oplus_j M_{\mathfrak{m}_j}[\mathfrak{m}_i^\infty]$ and $M_{\mathfrak{m}_j}[\mathfrak{m}_i^\infty] = 0$ if $j \neq i$ since $M_{\mathfrak{m}_j}$ has a finite composition series with factors $R/\mathfrak{m}_j$ on which for any n some elements of $\mathfrak{m}_i^n$ acts in an invertible way, and $M_{\mathfrak{m}_i}[\mathfrak{m}_i^\infty] = M_{\mathfrak{m}_i}$ since $M_{\mathfrak{m}_i}$ has a finite composition series with factors $R/\mathfrak{m}_i$. In particular, $M[\mathfrak{m}_i^\infty]$ is non-zero, from which we easily deduce by descending induction that $M[\mathfrak{m}_i]$ is non-zero.

2.5.2 System of Eigenvalues, Eigenspaces and Generalized Eigenspaces

We recall here some basic definitions from linear algebra:

Definition 2.5.1 A vector $v \in M$ is a common *eigenvector* (resp. *generalized eigenvector*) for $\mathcal{H}$ if for every $T \in \mathcal{H}$, there exists a scalar $\chi(T) \in k$ (resp. and an integer n) such that $\psi(T)v = \chi(T)v$ (resp. $(\psi(T) - \chi(T)\mathrm{Id})^n v = 0$).

Note that if $v \neq 0$ is an eigenvector or a generalized eigenvector, the scalar $\chi(T)$, called the *eigenvalue* or *generalized eigenvalue*, is well-determined, and the map $T \mapsto \chi(T)$, $\mathcal{H} \to k$ is a character (that is a morphism of rings); we then say that v is an *eigenvector* (resp. *generalized eigenvector*) *for* the character χ. By convention, we shall say that 0 is an eigenvector for any character $\chi : \mathcal{H} \to k$.

Definition 2.5.2 If $\chi : \mathcal{H} \to k$ is a character, the set of $v \in M$ that are eigenvectors (resp. generalized eigenvectors) for χ is a vector space called the *eigenspace of* χ and denoted $M[\chi]$ (resp. the *generalized eigenspace* of χ denoted $M_{(\chi)}$).

Definition 2.5.3 A character $\chi : \mathcal{H} \to k$ (or what amounts to the same, $\mathcal{H}\otimes k \to k$) is said to be a *system of eigenvalues appearing in M* if $M[\chi] \neq 0$, or equivalently $M_{(\chi)} \neq 0$.

Exercise 2.5.4 If k is algebraically closed, show that we have a decomposition $M = \oplus_\chi M_{(\chi)}$ where χ runs in the finite set of all systems of eigenvalues of $\mathcal{H}$ appearing in M.

Exercise 2.5.5 Let k' be an extension of k and $M' = M \otimes k'$. Let $\chi : \mathcal{H} \to k$ be a character and $\chi' = \chi \otimes 1 : \mathcal{H} \otimes k' \to k'$. Show that $M[\chi] \otimes k' = M'[\chi']$ and $M_{(\chi)} \otimes k' = M'_{(\chi')}$.

Exercise 2.5.6 Prove that $M[\chi] \neq 0$ if and only if $M_{(\chi)} \neq 0$, as asserted in Definition 2.5.3.

Exercise 2.5.7 Let $0 \to K \to M \to N \to 0$ be an exact sequence of k-vector spaces with actions of $\mathcal{H}$, and let $\chi : \mathcal{H} \to k$ be a character. Show that the sequence $0 \to K_{(\chi)} \to M_{(\chi)} \to N_{(\chi)} \to 0$ is exact, but not necessarily the sequence $0 \to K[\chi] \to M[\chi] \to N[\chi] \to 0$.

Exercise 2.5.8 Show that $M^\vee_{(\chi)}$ is isomorphic as $\mathcal{H}$-module with $(M^\vee)_{(\chi)}$ where $M^\vee$ is defined as in Exercise 2.3.7, but that in general we do not have $\dim M[\chi] = \dim M^\vee[\chi]$

2.5.3 Systems of Eigenvalues and Points of Spec $\mathcal{T}$

Theorem 2.5.9 *Let $\chi : \mathcal{H} \to k$ be a character. Then χ is a system of eigenvalues appearing in M if and only if χ factors as a character $\mathcal{T} \to k$. Moroever, if this is the case, then denoting $\mathfrak{m}$ the kernel of $\chi : \mathcal{T} \to k$ one has*

$$M[\chi] = M[\mathfrak{m}] \quad \textit{and} \quad M_{(\chi)} = M_{\mathfrak{m}}.$$

Proof If χ is a system of eigenvalues appearing in M, and v a non-zero eigenvector for χ, the relation $\psi(T)v = \chi(T)v$ shows that $\chi(T)$ depends only of $\psi(T)$, that is that χ factors through $\mathcal{T}$.

Conversely, let $\chi : \mathcal{T} \to k$ be a character, and let $\mathfrak{m}$ be its kernel, which is a maximal ideal of $\mathcal{T}$. We see χ as a character $\chi : \mathcal{H} \to k$, by precomposition with ψ. The ideal $\mathfrak{m}$ of $\mathcal{T}$ is generated, as a k-vector space, by the elements $\psi(T) - \chi(T)1_{\mathcal{T}}$ for $T \in \mathcal{H}$ (indeed, an element m of $\mathfrak{m}$ can be written as a finite sum $m = \sum \lambda_r \psi(T_r)$ for $T_r \in \mathcal{H}$, and since $\chi(m) = 0$, we get $\sum \lambda_r \chi(T_r) = 0$, so $m = \sum \lambda_r(\psi(T_r) - \chi(T_r)1_{\mathcal{T}})$.) It follows that $M[\chi] = M[\mathfrak{m}]$ and that $M_{(\chi)} = M_{\mathfrak{m}}$. Moreover, since $M_{\mathfrak{m}} \neq 0$ (cf. Sect. 2.5.1), $M_{(\chi)} \neq 0$ and χ is a system of eigenvalues appearing in M. □

We can now describe the k'-points of the eigenalgebra Spec $\mathcal{T}$, for any extension k' of k:

Corollary 2.5.10 *For any field k' containing k, we have a natural bijection between $(\mathrm{Spec}\,\mathcal{T})(k')$ and the set of systems of eigenvalues of $\mathcal{H}$ that appear in $M \otimes_k k'$.*

Proof When $k' = k$, the natural bijection is implicit in the preceding theorem: a system of eigenvalues of $\mathcal{H}$ appearing in M corresponds bijectively to a character $\mathcal{T} \to k$, that is a point of $(\mathrm{Spec}\,\mathcal{T})(k)$. In general, the same result applied to k' gives a bijection between the set of systems of eigenvalues of $\mathcal{H}$ appearing in $M \otimes_k k'$ and $(\mathrm{Spec}\,\mathcal{T}')(k')$, where $\mathcal{T}'$ is the eigenalgebra of $\mathcal{H}$ acting on $M \otimes_k k'$, but $\mathrm{Spec}\,\mathcal{T}'(k') = \mathrm{Spec}\,\mathcal{T}(k')$ by Proposition 2.4.1. □

Remark 2.5.11 This result explains the name *eigenalgebra*.

Corollary 2.5.12 *Let $\bar{k}$ be an algebraic closure of k, and write $G_k = Aut(\bar{k}/k)$. There is a natural bijection between Spec $\mathcal{T}$ and the set of G_k-orbits of characters $\chi : \mathcal{H} \to \bar{k}$ appearing in $M \otimes \bar{k}$.*

Proof If k is algebraically closed, since $\mathcal{T}$ is finite over k, there is a natural bijection between Spec $\mathcal{T}$ and $(\mathrm{Spec}\,\mathcal{T})(k)$. In the case of a general field k, the result follows because Spec $\mathcal{T}$ is in natural bijection with $\mathcal{T}(\bar{k})^{G_k}$. □

Of course, if k is perfect, then G_k is just the absolute Galois group of k.

Corollary 2.5.13 *The algebra $\mathcal{T}$ is étale over k if and only if $\mathcal{H}$ acts semi-simply on $M \otimes_k \bar{k}$. If this holds, one has $\dim_k \mathcal{T} \leq \dim_k M$.*

Proof Let $\mathcal{T}_{\bar{k}}$ be the eigenalgebra generated by $\mathcal{H}$ on $M \otimes_k \bar{k}$. By Proposition 2.4.1, $\mathcal{T}_{\bar{k}} = \mathcal{T} \otimes_k \bar{k}$, hence $\mathcal{T}_{\bar{k}}$ is étale over $\bar{k}$ if and only if $\mathcal{T}$ is étale over k. So we may assume that k is algebraically closed.

Then, $\mathcal{T}$ is étale if and only if for every maximal ideal $\mathfrak{m}$ of $\mathcal{T}$, $\mathcal{T}_{\mathfrak{m}} = k$. The later condition is equivalent to $M[\mathfrak{m}] = M_{\mathfrak{m}}$ for every maximal ideal $\mathfrak{m}$ of $\mathcal{T}$ (see Sect. 2.5.1) hence using Theorem 2.5.9, to $M[\chi] = M_{(\chi)}$ for every system of eigenvalues appearing in M, which (using Exercise 2.5.4) is equivalent to $\mathcal{H}$ acting semi-simply on M. Moreover, if this holds, $\mathcal{T} = \prod_{\mathfrak{m}} \mathcal{T}_m = k^r$ where r is the number of systems of eigenvalues appearing in M, hence clearly $\dim_k \mathcal{T} = r \leq \dim_k M$. □

Remember that $\mathcal{T}$ is étale over k is equivalent to $\mathcal{T}$ being a finite product of fields that are finite separable extensions of k. Remember also that when k is perfect, $\mathcal{H}$ acts semi-simply on $M \otimes_k \bar{k}$ if and only if it acts semi-simply on M.

Remark 2.5.14 When $\mathcal{H}$ does not act semi-simply on $M \otimes_k \bar{k}$, the dimension of $\mathcal{T}$ may be larger than the dimension of M. An old result of Schur (see [106] and [73] for a simple proof) states that the maximal possible dimension of $\mathcal{T}$ is $1 + \lfloor (\dim_k M)^2/4 \rfloor$.

Exercise 2.5.15 Show that there exist commutative subalgebras of $\mathrm{End}_k(M)$ of that dimension.

2.6 The Fundamental Example of Hecke Operators Acting on a Space of Modular Forms

The motivating example of the theory above is the action of Hecke operators on spaces of modular forms. We assume that the reader is familiar with the basic theory of modular forms as exposed in classical textbooks, e.g. [93, 113], or [56], but we nevertheless recall the definitions and main results that we will use.

2.6.1 *Complex Modular Forms and Diamond Operators*

We recall the standard action of $\mathrm{GL}_2^+(\mathbb{Q})$ (the + indicates matrices with positive determinant) on the Poincaré upper half-plane $\mathcal{H}$:

$$\gamma \cdot z = (az + b)/(cz + d) \text{ for } \gamma = \begin{pmatrix} a & b \\ c & d \end{pmatrix}, \quad z \in \mathcal{H}, \tag{2.6.1}$$

and the standard right-action of weight k on the space of functions on $\mathcal{H}$:

$$f_{|k\gamma}(z) = (\det \gamma)^{k-1}(cz + d)^{-k} f(\gamma \cdot z). \tag{2.6.2}$$

Let $k \geq 0$ be an integer and let Γ be a congruence subgroup of $\mathrm{SL}_2(\mathbb{Z})$, that is a subgroup containing all matrices congruent to the identity matrix mod N for a certain integer $N \geq 1$. A *modular form of weight k and level* Γ is an holomorphic function on $\mathcal{H}$ invariant by Γ for that action, which is *holomorphic at all cusps of* $\mathcal{H}/\Gamma$, a condition whose meaning we now recall.

The group $\mathrm{SL}_2(\mathbb{Z})$ acts on $\mathbb{P}^1(\mathbb{Q})$ on the left, transitively, by the formula (2.6.1). The *cusps of* Γ are the element of $\mathbb{P}^1(\mathbb{Q})^\Gamma$. We call *infinity* the cusp which contains the point at infinity of $\mathbb{P}^1(\mathbb{Q})$. Let us call M the smallest positive integer such that

$\begin{pmatrix} 1 & M \\ 0 & 1 \end{pmatrix} \in \Gamma$. Note that for $\Gamma \supset \Gamma_1(N)$ we have $M = 1$, and this is the only case we shall have to consider in the sequel. An holomorphic function f satisfying (2.6.2) is *holomorphic at infinity* if it can be written $f(z) = \sum_{m=0}^{\infty} a_n q^{n/M}$ for $z \in \mathcal{H}$, $q = e^{2i\pi z}$. The a_n's are called the *the coefficients* of f (at infinity), and a_0 is the *value at infinity* of f. If c is any cusp, then c can be written $\gamma \cdot \infty$ for $\gamma \in \mathrm{SL}_2(\mathbb{Z})$, and we say that f is *holomorphic at the cusp* c if $f_{|_k}(\gamma)$ is holomorphic at infinity, that is can be written $\sum_{n=0} a_n q^{n/M}$. It is easily seen that this condition, and a_0, depends only on the cusp c, not of the element γ such that $\gamma \cdot \infty = c$. The coefficient a_0 is called *the value of* f *at* c.

A modular form is *cuspidal* if it vanishes at all the cusps. We shall denote by $M_k(\Gamma)$ (resp. $S_k(\Gamma)$) the complex vector space of modular forms (resp. cuspidal modular forms) of weight k and level Γ.

The fundamental examples of congruence subgroups are, for $N \geq 1$ an integer, the subgroup $\Gamma_0(N)$ of matrices which are upper-triangular modulo N, and its normal subgroup $\Gamma_1(N)$ of matrices which are unipotent modulo N. Actually any congruence subgroup is conjugate to another one which contains $\Gamma_1(N)$ for some N, so these congruences subgroups are in some sense universal.

Exercise 2.6.1 Prove the last sentence.

Let us identify the quotient $\Gamma_0(N)/\Gamma_1(N)$ with $(\mathbb{Z}/N\mathbb{Z})^*$ by sending a matrix in $\Gamma_0(N)$ to the reduction modulo N of its upper-left coefficient. Then the spaces $M_k(\Gamma_1(N))$ and $S_k(\Gamma_1(N))$ have a natural action of the group $(\mathbb{Z}/N\mathbb{Z})^* = \Gamma_0(N)/\Gamma_1(N)$. We denote by $\langle a \rangle$ the action of $a \in (\mathbb{Z}/N\mathbb{Z})^*$, and call $\langle a \rangle$ a *diamond operator*. For $\epsilon : (\mathbb{Z}/\mathbb{N}\mathbb{Z})^* \to \mathbb{C}^*$ any character, which we shall call a *nebentypus* in this context, we denote by $M_k(\Gamma_1(N), \epsilon)$ the common eigenspace in $M_k(\Gamma_1(N))$ for the diamond operators with system of eigenvalues ϵ, and similarly for S_k. We obviously have

$$M_k(\Gamma_1(N)) = \oplus_\epsilon M_k(\Gamma_1(N), \epsilon)$$

where ϵ runs among the set of nebentypus, and

$$M_k(\Gamma_1(N), 1) = M_k(\Gamma_0(N)).$$

Similar results hold for S_k.

Finally, we call $\mathcal{E}_k(\Gamma_1(N))$ the submodule generated by the Eisenstein series (specifically by the forms $E_{k,\chi,\psi,t}$ as in Proposition 2.6.12 below), and $\mathcal{E}_k(\Gamma_1(N).\epsilon)$ the common eigenspace with system of eigenvalues ϵ. We have

$$M_k(\Gamma_1(N), \epsilon) = S_k(\Gamma_1(N), \epsilon) \oplus \mathcal{E}_k(\Gamma_1(N), \epsilon).$$

2.6.2 General Theory of Hecke Operators

We suppose given a monoid Σ, an abelian group V on which Σ acts, say on the right, and a subgroup Γ of Σ. The general theory of Hecke operators is the construction, under a technical finiteness condition, of certain operators (endomorphisms of abelian groups) on the subgroup V^Γ of Γ-invariants in V, coming from Σ. Those operators are called the *Hecke operators*.

To get an intuition of what is behind this construction, it is useful to look at a simpler special case: let us assume for a minute that Σ is a group and that Γ is a *normal* subgroup of Σ. Then it is clear that the action of Σ on V preserves the subgroup V^Γ, and since Γ acts trivially on V^Γ, we get an action of the quotient group $\Gamma\backslash\Sigma$ on V^Γ. Hence the elements of $\Gamma\backslash\Sigma$ defines operators on V^Γ, and these are the Hecke operators in this special case. Elements of $\Gamma\backslash\Sigma$ can be seen as right Γ-cosets Γs, or equivalently, since Γ is normal, as left Γ-cosets $s\Gamma$, or also as double Γ-cosets $\Gamma s\Gamma$. It turns out that it is the latter point of view which is more prone to generalization.

Back to the general case, let us make the following technical hypothesis:

$$\forall s \in \Sigma,\ \Gamma s\Gamma \text{ is the union of \textit{finitely many} left } \Gamma\text{-cosets } \Gamma s_i. \tag{2.6.3}$$

Definition 2.6.2 Given a double Γ-coset $\Gamma s\Gamma$ in Σ such that $\Gamma s\Gamma = \coprod_i \Gamma s_i$ (disjoint finite union), the *Hecke operator* $[\Gamma s\Gamma]$ on V^Γ is the endomorphism $v \mapsto v_{|[\Gamma s\Gamma]} := \sum_i v_{|s_i}$ of V^Γ.

Exercise 2.6.3 Prove that the definition of $[\Gamma s\Gamma]$ is correct by showing that $\sum_i v_{|s_i}$ is independent of the choice of the s_i's and belongs to V^Γ.

Consider the free abelian group $\mathcal{H}(\Sigma,\Gamma)$ of $\mathbb{Z}$-valued functions with finite support on the set of double Γ-cosets $\Gamma\backslash\Sigma/\Gamma$, or equivalently, the free abelian group on the set of the symbols $[\Gamma s\Gamma]$ of Hecke operators. This abelian group is naturally isomorphic, in view of hypothesis (2.6.3), to the subgroup of Γ-invariants in the abelian group with right Σ-action of functions on $\Gamma\backslash\Sigma$ with finite support. Thus $\mathcal{H}(\Sigma,\Gamma)$ gets an action of Hecke operators $[\Gamma s\Gamma]$, which by additivity defines a *multiplication map* $\mathcal{H}(\Sigma,\Gamma)\times\mathcal{H}(\Sigma,\Gamma)\to\mathcal{H}(\Sigma,\Gamma)$ which makes of $\mathcal{H}(\Sigma,\Gamma)$ a ring, not necessarily commutative, but with unity $[\Gamma 1\Gamma] = [\Gamma]$. The ring $\mathcal{H}(\Sigma,\Gamma)$ is known as the *(abstract) Hecke ring*. Its elements are linear combinations of the symbols $[\Gamma s\Gamma]$ representing Hecke operators.

By additivity again, the Hecke operators on V^Γ gives V^Γ the structure of a right-$\mathcal{H}(\Sigma,\Gamma)$-module. The map $V \mapsto V^\Gamma$ defines, as is easily seen, a functor from the category of right Σ-modules to the category of right-$\mathcal{H}(\Sigma,\Gamma)$-modules. When V is an R-module, and the action of Σ is by R-linear operators, then obviously the Hecke operators are R-linear on V^Γ.

2.6.3 Hecke Operators on Modular Forms

The general theory above applies when $\Sigma = \mathrm{GL}_2^+(\mathbb{Q})$, and Γ a congruence subgroup contained in Σ, for in this case the condition (2.6.3) is true and not very hard to check (cf. e.g. [113, §3].) Thus if V is defined as the $\mathbb{C}$-vector space of holomorphic function $\mathcal{H} \to \mathbb{C}$, given a weight $k \in \mathbb{Z}$, V has a right action of Σ given by (2.6.2), and modular forms are those elements of V^Γ that satisfy certain condition at cusp. The space V^Γ thus inherits an action of Hecke operators $[\Gamma s \Gamma]$ for $s \in \mathrm{GL}_2^+(\mathbb{Q})$, and it is not hard to see that these operators leave stable the subspace $M_k(\Gamma)$ and $S_k(\Gamma)$

For instance, for $\Gamma = \Gamma_1(N)$, the operator $[\Gamma_1(N) g \Gamma_1(N)]$ when g is a matrix in $\Gamma_0(N)$ whose upper-left coefficient is a is the diamond operator $\langle a \rangle$ defined above on $M_k(\Gamma_1(N))$. This follows from the definition since in this case $\Gamma_1(N) g \Gamma_1(N) = \Gamma_1(N) g$.

Of fundamental importance are the operators T_l defined, for l a prime number, by

$$T_l = [\Gamma_1(N) \begin{pmatrix} l & 0 \\ 0 & 1 \end{pmatrix} \Gamma_1(N)]. \tag{2.6.4}$$

They are called *Hecke operators at* ℓ, and they commute with each other and with the diamond operators. In particular they stabilize the spaces $M_k(\Gamma_1(N), \epsilon)$. They also stabilize the subspaces of cuspidal forms $S_k(\Gamma_1(N))$ and $S_k(\Gamma_1(N), \epsilon)$, and of Eisenstein series $\mathcal{E}_k(\Gamma_1(N))$ and $\mathcal{E}_k(\Gamma_1(N), \epsilon)$.

More generally (cf. [93]) we define, for every integer n, by induction on the number of prime factors of n, the Hecke operators T_n by the formula

$$T_n = T_{l_1^{m_1}} \dots T_{l_r^{m_r}} \text{ if } n = l_1^{m_1} \dots l_r^{m_r} \tag{2.6.5}$$

where the l_i are distinct primes and the m_i are positive integers (in particular $T_1 = \mathrm{Id}$), and

$$T_{l^{m+1}} = T_l T_{l^m} - l^{k-1} \langle l \rangle T_{l^{m-1}} \tag{2.6.6}$$

for l a prime, $m \geq 1$. Thus the Hecke operators T_n are polynomials in the Hecke operators T_l when l is prime, and in the diamond operators, and thus commute with each other, and stabilize all the subspaces stabilized by the T_l and diamond operators.

To emphasize the difference of behavior of T_l when l divides N or not, we shall use the notation U_l instead of T_l when l divides N. One has, as is easily seen using the definition,

$$T_l f = \sum_{a=0}^{l-1} f_{|_k \begin{pmatrix} 1 & a \\ 0 & l \end{pmatrix}} + \langle l \rangle f_{|_k \begin{pmatrix} l & 0 \\ 0 & 1 \end{pmatrix}} \text{ when } l \nmid N \tag{2.6.7}$$

$$U_l f = \sum_{a=0}^{l-1} f_{|_k \begin{pmatrix} 1 & a \\ 0 & l \end{pmatrix}} \text{ when } l \mid N \tag{2.6.8}$$

When $N' \mid N$, $\Gamma_1(N) \subset \Gamma_1(N')$ so we have $M_k(\Gamma_1(N')) \subset M_k(\Gamma_1(N))$. When we consider $M_k(\Gamma_1(N'))$ as a space of modular forms on its own, it gets its own Hecke operators T_l for $l \nmid N'$ and U_l for $l \mid N'$ and also $\langle a \rangle$ for a in $\mathbb{Z}$, a coprime to N'. When we consider $M_k(\Gamma_1(N'))$ as a subspace of $M_k(\Gamma_1(N))$, the formulas above show that it is stable by the T_l, $l \nmid N$, by the U_l, $l \mid N'$ and by the diamond operators $\langle a \rangle$, $(a, N) = 1$, and that each of these operators induces the operator of the same name on $M_k(\Gamma_1(N))$. However, when l is a prime dividing N but not N', the operator T_l on $M_k(\Gamma_1(N))$ does not necessarily stabilize $M_k(\Gamma_1(N'))$.

We also recall the action of T_l and U_l on the coefficients at ∞ of a modular form $f \in M_k(\Gamma_1(N))$, which is easily deduced from (2.6.7) and (2.6.8).

$$T_l f = \sum_{n=0}^{\infty} a_{nl}(f) q^n + l^{k-2} \sum_{n=0}^{\infty} a_n(\langle l \rangle f) q^{ln} \text{ when } l \nmid N \tag{2.6.9}$$

$$(T_l f =)\ U_l f = \sum_{n=0}^{\infty} a_{ln}(f) q^n \text{ when } l \mid N. \tag{2.6.10}$$

Exercise 2.6.4 Prove the last equality.

In particular, we have $a_1(T_l f) = a_l(f)$ for every prime l, from which it is easy to deduce that for all modular forms $f \in M_k(\Gamma_1(N)$, and all integers $n \geq 1$,

$$a_1(T_n f) = a_n(f). \tag{2.6.11}$$

2.6.4 A Brief Reminder of Atkin–Lehner–Li's Theory (Without Proofs)

We fix an integer $N \geq 1$. We shall denote by $\mathcal{H}$ (or by $\mathcal{H}(N)$ when this precision is useful) the polynomial ring over $\mathbb{Z}$ in infinitely many variables with names T_l for $l \nmid N$, U_l for $l \mid N$ and $\langle a \rangle$ for $a \in (\mathbb{Z}/N\mathbb{Z})^*$. We let $\mathcal{H}$ acts on $M_k(\Gamma_1(N))$ by letting each variable acts by the Hecke operator of the same name. This action stabilizes $S_k(\Gamma_1(N))$ and $\mathcal{E}_k(\Gamma_1(N))$. We shall denote by $\mathcal{H}_0$ (or by $\mathcal{H}_0(N)$) the subring generated by the variables T_l for $l \nmid N$ and $\langle a \rangle$ for $a \in (\mathbb{Z}/N\mathbb{Z})^*$. So $\mathcal{H}_0$ acts on $M_k(\Gamma_1(N))$ as well, with the advantage that its action stabilizes all the subspaces $M_k(\Gamma_1(N'))$ for $N' \mid N$ (cf. the discussion at the end of the preceding subsection).

Definition 2.6.5 Let $\lambda : \mathcal{H}_0 \to \mathbb{C}$ be a system of eigenvalues appearing in $M_k(\Gamma_1(N))$. We shall say that λ is *new* if λ does not appear (cf. Definition 2.5.3) in any $M_k(\Gamma_1(N'))$ for N' a proper divisor of N. We shall say that λ is *old* if it is not new.

We shall need an obvious refinement of that definition: if l is a prime factor of N, we shall say that λ is *new at l* if it does not appear in $M_k(\Gamma_1(N/l))$. Obviously a system of eigenvalues is new if and only if it is new at every prime factors of N.

The first fundamental result of the Atkin–Lehner–Li theory is the multiplicity-one theorem:

Theorem 2.6.6 *A system of eigenvalues $\lambda : \mathcal{H}_0 \to \mathbb{C}$ appearing in $M_k(\Gamma_1(N))$ is new if and only if the eigenspace $M_k(\Gamma_1(N))[\lambda]$ has dimension* 1. *Moreover, when λ is new, there is a unique form $f \in M_k(\Gamma_1(N))[\lambda]$ which is normalized (that is, its coefficient a_1 is* 1*).*

The form f of the theorem is called the *newform* of the system of eigenvalues λ. Since all the Hecke operators commute, we see that f is an eigenform for $\mathcal{H}$, not only for $\mathcal{H}_0$.

Definition 2.6.7 The *system of eigenvalues of E_2* is the morphism $\mathcal{H}_0 \to \mathbb{C}$ that sends T_l for $l \nmid N$ to $1 + l$, and all the $\langle a \rangle$ to 1.

Theorem 2.6.8 *Assume that λ is a system of eigenvalues for $\mathcal{H}_0$ appearing in $M_k(\Gamma_1(N))$, different from the system of eigenvalues of E_2.*

There is a divisor N_0 of N such that for every divisor N' of N, λ appears in $M_k(\Gamma_1(N'))$ if and only if N_0 divides N'. For such an N', the dimension of $M_k(\Gamma_1(N'))[\lambda]$ is $\sigma(N'/N_0)$ where $\sigma(n)$ is the number of divisors of n.

In particular, the space $M_k(\Gamma_1(N_0))[\lambda]$ has dimension 1. *If $f(z)$ is a generator of that space, then for any N' such that $N_0 \mid N' \mid N$, a basis of $M_k(\Gamma_1(N'))[\lambda]$ is given by the forms $f(dz)$, $d \mid (N'/N_0)$.*

Definition 2.6.9 We call N_0 the *minimal level* of λ.

Remark 2.6.10 Obviously λ is *new* if and only if $N = N_0$. In general, any system of eigenvalues for $\mathcal{H}_0$ appearing in $M_k(\Gamma_1(N))$ can be considered new when seen as a system of eigenvalues appearing in $M_k(\Gamma_1(N_0))$. More precisely, the above theorem says that $M_k(\Gamma_1(N_0))[\lambda]$ has dimension 1, where the eigenspace is for the algebra $\mathcal{H}_0(N)$. The algebra $\mathcal{H}_0(N_0)$ may be bigger that $\mathcal{H}_0(N)$; nevertheless since all the Hecke operators commute, they stabilize $M_k(\Gamma_1(N_0))[\lambda]$ which, being of dimension 1, is therefore also an eigenspace for $\mathcal{H}_0(N_0)$ of system of eigenvalues some well-defined extension $\tilde{\lambda}$ of λ to $\mathcal{H}_0(N_0)$, and this system of eigenvalues is new.

Remark 2.6.11 By definition, the operator T_l, $l \nmid N$ acts on $M_k(\Gamma_1(N))[\lambda]$ by the scalar $\lambda(T_l)$. It is also possible to describe the action of certain of the operators U_l for $l \mid N$. Let f be, as in Theorem 2.6.8, a generator of $M_k(\Gamma_1(N_0))[\lambda]$. Then an easy computation using (2.6.7) and (2.6.8) (or alternatively the description of the action of the Hecke operators on coefficients) gives for d any positive divisor of N/N_0:

$$U_l(f(dz)) = f(\frac{d}{l} z) \text{ if } l \mid d \tag{2.6.12}$$

$$U_l(f(dz)) = (T_l f)(dz) - l^{k-1}\langle l\rangle(f(dlz)) \text{ if } l \nmid dN_0 \tag{2.6.13}$$

In the last formula, $T_l f$ is to be understood as the action of the operator T_l of $M_k(\Gamma_1(N_0))$ on f. In other words, $T_l f = \tilde{\lambda}(T_l) f$ where $\tilde{\lambda}$ is defined as in the preceding remark.

Let us briefly indicate where the reader can find a proof of Theorems 2.6.6 and 2.6.8. All the results above are well known and due to Atkin and Lehner in the case of cuspidal forms for $\Gamma_0(N)$, and were extended by them and Li to the case of forms for $\Gamma_1(N)$. Modern treatments can be found in the books [93] and [56]. For $\mathcal{E}_{k+2}(\Gamma_1(N))$ the theorem follows easily from the explicit description of all Eisenstein series that can be found in the chapter 7 of [93] or the chapter 4 (and §5.2) of [56]. We recall this description.

Let χ and ψ be two primitive Dirichlet characters of conductors L and R. We assume that $\chi(-1)\psi(-1) = (-1)^k$. Let

$$E_{k,\chi,\psi}(q) = c_0 + \sum_{m\geq 1} q^m \sum_{n|m} (\psi(n)\chi(m/n)n^{k-1}) \tag{2.6.14}$$

where $c_0 = 0$ if $L > 1$ and $c_0 = -B_{k,\psi}/2k$ if $L = 1$. If t is a positive integer, let $E_{k,\chi,\psi,t}(q) = E_{k,\chi,\psi}(q^t)$ except in the case $k = 2$, $\chi = \psi = 1$, where one sets $E_{2,1,1,t} = E_{2,1,1}(q) - tE_{2,1,1}(q^t)$.

Proposition 2.6.12 *The series $E_{k,\chi,\psi,t}(q)$ are modular forms of level $\Gamma_1(N)$ and nebentypus $\chi\psi$ for all positive integers L, R, t such that $LRt \mid N$ and all primitive Dirichlet character χ of conductor L and ψ of conductor R, satisfying $\chi(-1)\psi(-1) = -1$ (and $t > 1$ in the case $k = 2$, $\chi = \psi = 1$) and moreover they form a basis of the space $\mathcal{E}_k(\Gamma_1(N))$.*

For every prime l not dividing N, we have

$$T_l E_{k,\chi,\psi,t} = (\chi(l) + \psi(l)l^{k-1})E_{k,\chi,\psi,t}.$$

Proof See Miyake ([93, chapter 7]) for the computations leading to those results, Stein ([119, Theorems 5.8, 5.9, 5.10]) for the results stated exactly as here. □

From this description it follows easily that

Corollary 2.6.13 *The Eisenstein series in $M_{k+2}(\Gamma_1(N))$ that are new are exactly: the* normal *new Eisenstein series $E_{k+2,\chi,\psi}$ with $N = LR$ (excepted of course $E_{2,1,1}$ which is not even a modular form); the* exceptional *new Eisenstein series $E_{2,1,1,l}$, that we shall denote simply by $E_{2,l}$ when $N = l$ is prime.*

The reader may check as an exercise that all the statements of Atkin–Lehner–Li's theory recalled above hold for $\mathcal{E}_{k+2}(\Gamma_1(N))$ (and thus for $M_{k+2}(\Gamma_1(N))$) for a system of eigenvalues λ that is different of the one of E_2. When λ is the system of eigenvalues of E_2, then the minimal level of λ is not well-defined anymore: all prime factors l of N are minimal elements of the set of divisors N' of N such that λ appears in $M_2(\Gamma_1(N'))$.

Let us also note the following important corollary:

Corollary 2.6.14 *The algebra $\mathcal{H}_0$ acts semi-simply on $M_k(\Gamma_1(N))$.*

Proof We prove separately that $\mathcal{H}_0$ acts semi-simply on $S_k(\Gamma_1(N))$ and on $\mathcal{E}_k(\Gamma_1(N))$. On $\mathcal{E}_k(\Gamma_1(N))$, Proposition 2.6.12 provides a basis of eigenforms for $\mathcal{H}_0$, proving the semi-simplicity. On $S_k(\Gamma_1(N))$, there exists a natural Hermitian product, the Peterson inner product (see (5.3.1) below), for which the adjoint of the Hecke operator T_l, $l \nmid N$ is $\langle l \rangle T_\ell$ and the adjoint of $\langle a \rangle$ is $\langle a^{-1} \rangle$. It follows that all operators in $\mathcal{H}_0$ commute with their adjoints, hence are diagonalizable, and $\mathcal{H}_0$ acts semi-simply on $S_k(\Gamma_1(N))$. □

Exercise 2.6.15 Let p be a prime number not dividing N. In this exercise we are interested in the space of Eisenstein series $\mathcal{E}_k(\Gamma_1(N) \cap \Gamma_0(p))$. This is the subspace of $\mathcal{E}_k(\Gamma_1(Np))$ on which the diamond operators $\langle a \rangle$ for $a \in (\mathbb{Z}/p\mathbb{Z})^* \subset (\mathbb{Z}/Np\mathbb{Z})^*$ act trivially. The operator U_p acts on that space. On questions a. to e. we assume that $k \neq 2$.

1. Show that a basis of $\mathcal{E}_k(\Gamma_1(N) \cap \Gamma_0(p))$ is the set of Eisenstein series $E_{k,\chi,\psi,t}(z)$ and $E_{k,\chi,\psi,tp}(z)$ where χ, ψ are Dirichlet characters of conductors L, R respectively, and $LRt \mid N$. In particular, all these forms are old at p, and the dimension of $\mathcal{E}_k(\Gamma_1(N) \cap \Gamma_0(p))$ is twice the dimension of $\mathcal{E}_k(\Gamma_1(N))$.
2. Show that the Eisenstein series
$$E_{k,\chi,\psi,t,\mathrm{ord}} := E_{k,\chi,\psi,t}(z) - \psi(p)p^{k-1}E_{k,\chi,\psi,tp}(z)$$
and
$$E_{k,\chi,\psi,t,\mathrm{crit}} := E_{k,\chi,\psi,t}(z) - \chi(p)E_{k,\chi,\psi,tp}(z)$$
also form a basis of $\mathcal{E}_k(\Gamma_1(N) \cap \Gamma_0(p))$ (where (χ, ψ, t) runs in the same set as in 1.).
3. Show that $U_p E_{k,\chi,\psi,t,\mathrm{ord}} = \chi(p) E_{k,\chi,\psi,t,\mathrm{ord}}$ and $U_p E_{k,\chi,\psi,t,\mathrm{crit}} = \psi(p)p^{k-1} E_{k,\chi,\psi,t,\mathrm{crit}}$.

4. Show that
$$\mathcal{E}_k(\Gamma_1(N)\cap\Gamma_0(p)) = \mathcal{E}_k(\Gamma_1(N)\cap\Gamma_0(p))_{\text{ord}} \oplus \mathcal{E}_k(\Gamma_1(N)\cap\Gamma_0(p))_{\text{crit}},$$
where these two subspaces are the subspaces generated by the $E_{k,\chi,\psi,t,\text{ord}}$, $E_{k,\chi,\psi,t,\text{crit}}$ respectively, and the eigenvalue of U_p on the first are roots of unity, and on the second space are roots of unity times p^{k-1}.
5. Show that if $g \in \mathcal{E}_k(\Gamma_1(N)\cap\Gamma_0(p))_{\text{crit}}$, g vanishes at every cusp in the $\Gamma_1(N)$-class of $\{\infty\}$. Conversely, if $g \in \mathcal{E}_k(\Gamma_1(N)\cap\Gamma_0(p))_{\text{crit}}$ and g vanishes at every cusp in the $\Gamma_1(N)$-class of $\{\infty\}$, then $g \in \mathcal{E}_k(\Gamma_1(N)\cap\Gamma_0(p))_{\text{crit}}$.
6. How should the above be modified for $k = 2$?

2.6.5 Hecke Eigenalgebra Constructed on Spaces of Complex Modular Forms

For a choice of a space $M \subset M_k(\Gamma_1(N))$ which is $\mathcal{H}$-stable, we define $\mathcal{T}_0$ as the eigenalgebra of $\mathcal{H}_0$ acting on M and $\mathcal{T}$ as the eigenalgebra of $\mathcal{H}$ acting on M.

The great advantage of $\mathcal{T}_0$ is that it is semisimple. Therefore $\mathcal{T}_0$ is a product of a certain number r of copies of $\mathbb{C}$, one for each system of eigenvalues appearing in M (see Corollary 2.5.13). The problem is that in general we do not have multiplicity 1: if χ is a system of eigenvalues $\mathcal{H}_0 \to \mathcal{T}_0 \to \mathbb{C}$, $M[\chi]$ may have dimension greater than 1, and, what is worse, depending on χ. Actually, we know that the dimension of that space is $\sigma(N/N_0)$ where N_0 is the minimal level of χ, at least when χ is not the system of eigenvalues of E_2. So M is not, in general, a free module over $\mathcal{T}_0$.

The operators U_p, $p \mid N$, acting on the space of forms of level N, are not semi-simple in general (but see Exercise 2.6.21 for a discussion of when they are). Therefore the algebra $\mathcal{T}$ is not semi-simple in general, that is it may have nilpotent elements. However, we shall see that the multiplicity one principle holds and that the structure of the $\mathcal{T}$-modules M and $M^\vee = \operatorname{Hom}_{\mathbb{C}}(M, \mathbb{C})$ are very simple.

For the latter, there is a simple standard argument.

Proposition 2.6.16 *Assume $k > 0$. The pairing $\mathcal{T} \times M \to \mathbb{C}$, $(T, f) \mapsto \langle T, f\rangle = a_1(Tf)$ is a perfect $\mathcal{T}$-equivariant pairing.*

Proof That the pairing $\langle T, f\rangle$ is $\mathcal{T}$-equivariant means that for all $T' \in \mathcal{T}$, we have $\langle T'T, f\rangle = \langle T, T'f\rangle$, and this is obvious since $\mathcal{T}$ is commutative.

If $f \in M$ is such that $\langle T, f\rangle = 0$ for all $T \in \mathcal{T}$, then $a_1(T_n f) = 0$ for every integer $n \geq 1$, so $a_n(f) = 0$ for every $n \geq 1$ by (2.6.11) and f is a constant. Since the non-zero constant modular forms are of weight 0, our hypothesis implies $f = 0$.

If $T \in \mathcal{T}$ is such that $\langle T, f\rangle = 0$ for every $f \in M$, then for any given $f \in M$ we have $a_1(TT_n f) = 0$, so $a_n(Tf) = 0$ for all $n \geq 1$ by (2.6.11), hence $Tf = 0$ by the same argument as above. Since this is true for all f, and $\mathcal{T}$ acts faithfully on M, $T = 0$. Hence the pairing is perfect. □

For $k = 0$, $M = M_k(\Gamma_1(N)) = \mathcal{E}_k(\Gamma_1(N)) = \mathbb{C}$, and $\mathcal{T} = \mathbb{C}$, but in this case the pairing $(T, f) \mapsto a_1(Tf)$ is 0.

Corollary 2.6.17 *$M^\vee$ is free of rank one over $\mathcal{T}$.*

Proof This follows from the above proposition if $k > 0$, and this is trivial if $k = 0$ since in this case either $M = M_0(\Gamma_1(N)) = \mathcal{E}_0(\Gamma_1(N))$, M is of dimension 1, and $\mathcal{T} = \mathbb{C}$, or $M = 0$ and $\mathcal{T}$ is the zero ring. □

Corollary 2.6.18 *The multiplicity one principle holds, that is for every character $\chi : \mathcal{H} \to \mathcal{T} \to \mathbb{C}$, we have* $\dim M[\chi] = 1$.

Proof Let $\mathfrak{m}$ be the maximal ideal of $\mathcal{T}$ corresponding to χ. Since $M^\vee \simeq \mathcal{T}$, one has $M \simeq \mathcal{T}^\vee$ and $M[\chi] = M[\mathfrak{m}] \simeq \mathcal{T}^\vee[\mathfrak{m}] = (\mathcal{T}/\mathfrak{m}\mathcal{T})^\vee = \mathbb{C}$. □

The next theorem needs the full force of the Atkin–Lehner theory.

Theorem 2.6.19 *Let M be either $S_k(\Gamma_1(N))$, $M_k(\Gamma_1(N))$ or $\mathcal{E}_k(\Gamma_1(N))$. Then M is free of rank one as a $\mathcal{T}$-module.*

Proof Write $M = \oplus_\lambda M[\lambda]$ when λ runs among the finite number of systems of eigenvalues of $\mathcal{H}_0$ that appear in M. Let $\mathcal{T}_\lambda$ be the eigenalgebra attached to the action of $\mathcal{H}$ on $M[\lambda]$. Then $\mathcal{T} = \prod_\lambda \mathcal{T}_\lambda$ and the action of $\mathcal{T}$ on M is the product of the action of $\mathcal{T}_\lambda$ over $M[\lambda]$ (cf. Exercise 2.3.6). Hence it is enough to prove that $M[\lambda]$ is free of rank one over $\mathcal{T}_\lambda$. By the corollary above, we know that $\dim M[\lambda] = \dim \mathcal{T}_\lambda$. Therefore it suffices to prove that $M[\lambda]$ is generated by one element over $\mathcal{T}_\lambda$.

Assume first that λ is not the system of eigenvalues of E_2. We use Atkin–Lehner's theory: let N_0 be the minimal level of the character λ of $\mathcal{H}_0$, and f a generator of $M(N_0)[\lambda]$. Then $f\left(\frac{N}{N_0}z\right)$ generates $M(N_0)[\lambda]$ under $\mathcal{H}$ since for $d \mid N/N_0$, writing $N/(N_0 d) = l_1^{a_1} \dots l_m^{a_m}$, we have $f(dz) = U_{l_1}^{a_1} \dots U_{l_m}^{a_m} f\left(\frac{N}{N_0}z\right)$ by (2.6.12) and these forms $f(dz)$ generate $M[\lambda]$ as a vector space (Theorem 2.6.8).

The case where λ is the system of eigenvalues of E_2 deserves a special treatment. In this case the forms $E_{2,d}(z) = E_2(z) - dE_2(dz)$ for $d \mid N$, $d \neq 1$, form a basis of $M[\lambda]$. For a prime $l \mid N$, we have, as easily follows from (2.6.12) and (2.6.13)

$$U_l E_{2,l} = E_{2,l} \tag{2.6.15}$$

$$U_l E_{2,d} = E_{2,l} + l E_{2,d/l} \text{ if } l \mid d, l \neq d \tag{2.6.16}$$

$$U_l E_{2,d} = E_{2,l} + (l+1)E_{2,d} - l E_{2,dl} \text{ if } l \nmid d \tag{2.6.17}$$

Choose a prime factor l of N. For $d \mid N$, $d = d'l^n$ with $l \nmid d'$ and some integer $n \geq 1$, and we have

$$U_{l^k} E_{2,d'l^n} = l^k E_{2,d'l^{n-k}} + \frac{l^k - 1}{l - 1} E_{2,l}, \text{ for } k = 0, 1, \dots, n-1$$

by an easy induction using (2.6.16). Applying U_l once more, we get (using (2.6.15 and (2.6.16))

$$U_{l^n} E_{2,l^n} = \frac{l^n - 1}{l - 1} E_{2,l} \tag{2.6.18}$$

$$U_{l^n} E_{2,d'l^n} = l^n E_{2,d'} + \frac{l^n - 1}{l - 1} E_{2,l} \text{ if } d' \neq 1 \tag{2.6.19}$$

In the case $d' \neq 1$, let us apply once more U_l, to get, using (2.6.17)

$$U_{l^{n+1}} E_{2,d'l^n} = l^n (l+1) E_{d'} - l^{n+1} E_{2,d'l} + \frac{l^{n+1} - 1}{l - 1} E_{2,l}.$$

We claim that the forms $F_{d'l^i}$, for $i = 0, \ldots, n$, all belong to the module generated by F_d under U_l. To prove the claim, consider first the case $d' = 1$. In this case, the square matrix expressing the n vectors $U_{l^k} F_{l^n}$ for $k = 0, 1, \ldots, n-1$ for $k = 0, \ldots, n-1$ in terms of the n independent vectors $F_{l^{n-k}}$ is

$$\begin{pmatrix} 1 & & & & \\ & l & & & \\ & & \ddots & & \\ & & & l^{n-2} & \\ 0 & 1 & \ldots & \frac{l^{n-2}-1}{l-1} & \frac{l^n-1}{l-1} \end{pmatrix}.$$

As this matrix is invertible, the claim follows in the case $d' = 1$. When $d' \neq 1$ the square matrix of size $n+2$ expressing the family of $n+2$ vectors $U_{l^k} F_{d'l^n}$ for $k = 0, 1, \ldots, n, n+1$ in term of the family of $n+2$ vectors which consists in $F_{d'l^{n-k}}$ for $k = 0, 1, \ldots, n$ and F_l is

$$\begin{pmatrix} 1 & & & & & \\ & l & & & & \\ & & \ddots & & & \\ & & & l^{n-1} & & -l^{n+1} \\ & & & & l^n & l^n(l+1) \\ 0 & 1 & \frac{l^2-1}{l-1} & \ldots & \frac{l^n-1}{l-1} & \frac{l^{n+1}-1}{l-1} \end{pmatrix}.$$

By replacing the last column R_{n+2} by $R_{n+2} - (l+1)R_{n+1} - l^2 R_n$, we get the equivalent triangular matrix

$$\begin{pmatrix} 1 & & & & & & \\ & l & & & & & \\ & & \ddots & & & & \\ & & & l^{n-1} & & & \\ & & & & l^n & & \\ 0 & 1 & \frac{l^2-1}{l-1} & \cdots & \frac{l^n-1}{l-1} & \frac{l^{n+1}-l^n-l^2-1}{l-1} \end{pmatrix}.$$

The denominator $l^{n+1} - l^n - l^2 - 1$ of the last coefficient never vanishes for l prime and $n \geq 1$, since it is congruent to $1 \pmod{l}$. The matrix is therefore invertible and the claim follows in the case $d = 1$.

By induction on the number of prime factors of N, we deduce that all F_d for $d \mid N$ are in the $\mathcal{H}$-module generated by F_N, hence F_N generates the $\mathcal{H}$-module $M[\lambda]$.

□

Corollary 2.6.20 *Let M be as in the above theorem. We have $M \simeq M^\vee \simeq \mathcal{T}$ as an $\mathcal{H}$-module. The eigenalgebra $\mathcal{T}$ is a Gorenstein $\mathbb{C}$-algebra.*

The fact that $\mathcal{T}$ is Gorenstein is a serious restriction on its possible structure. For example $\mathbb{C}[X]/X^2$ or $\mathbb{C}[X,Y]/(X^2,Y^2)$ are Gorenstein, but $\mathbb{C}[X,Y]/(X^2,Y^2,XY)$ is not. For an introduction to Gorenstein rings, see [59, Chapter 21].

Exercise 2.6.21 Let λ be a system of eigenvalues of $\mathcal{H}_0$ different of the system of E_2 that appears on $M_k(\Gamma_1(N))$ and has minimal level N_0. Let f be a generator of the one-dimensional vector space $M_k(\Gamma_1(N_0))[\lambda]$.

For l a prime dividing N, denote by a_l and ϵ_l the eigenvalues of T_l and $\langle l \rangle$ on f in the case $l \nmid N_0$; denote by u_l the eigenvalue of U_l on f in the case $l \mid N_0$.

1. First assume that $N/N_0 = l^{m_l}$ for $m_l \in \mathbb{N}$. Show that:
   (i) If $l \nmid N_0$, U_l acts semi-simply on $M[\lambda]$ if and only if either $m_l = 0$, or $m_l = 1, 2$ and the equation $X^2 - a_l X + l^{k-1}\epsilon_l = 0$ has distinct roots.
   (ii) If $l \mid N_0$, U_l acts semi-simply on $M[\lambda]$ if and only if either $m_l = 0$, or $m_l = 1$ and $u_l \neq 0$.
2. In general, show that $\mathcal{H}$ acts semi-simply on $M[\lambda]$ if and only if, for all prime factors l on N/N_0, with m_l defined so that l^{m_l} is the maximal power of l that divides N/N_0, the same condition for the semi-simplicity of U_l given above holds.

2.6.6 Galois Representations Attached to Eigenforms

Instead of working over $\mathbb{C}$, we can work rationally. Namely, if K is a subfield of $\mathbb{C}$, we write $M_k(\Gamma_1(N), K)$ for the K-subspace of $M_k(\Gamma_1(N))$ of forms that have a q-expansion at ∞ with coefficients in K, and we define similarly $S_k(\Gamma_1(N), K)$, etc. Those spaces are stable by the Hecke operators, and, provided that K contains the image of ϵ in the case of a space of modular forms with Nebentypus ϵ, their formations commute with base change $K \subset K'$ for subfields of $\mathbb{C}$ (see [113]). Hence we can define more generally, if K is any field of characteristic 0, $M_k(\Gamma_1(N), K)$ as $M_k(\Gamma_1(N)) \otimes_{\mathbb{Q}} K$ and $S_k(\Gamma_1(N), K)$ as $S_k(\Gamma_1(N)) \otimes_{\mathbb{Q}} K$.

Let us call M_K any of these K-vector spaces, and M the corresponding $\mathbb{C}$-vector space. Then M_K is stable by $\mathcal{H}$ and $M_K \otimes_K \mathbb{C} = M$. Hence we can define K-algebras $\mathcal{T}_{0,K}$ and $\mathcal{T}_K$ using M_K instead of M, and we have $\mathcal{T}_{0,K} \otimes_K \mathbb{C} = \mathcal{T}_0$, $\mathcal{T}_K \otimes_K \mathbb{C} = \mathcal{T}$. Hence we see easily by descent that the results we proved above also holds over K (semi-simplicity of $\mathcal{T}_{0,K}$, freeness of rank one of M_K over $\mathcal{T}_K$, Gorensteinness of $\mathcal{T}_K, \ldots$)

Let us assume that $k \geq 1$ and that p is a prime number.

Theorem 2.6.22 *Let K be any finite extension of $\mathbb{Q}_p$, and f a normalized eigenform (for $\mathcal{H}_0$) in $M_k(\Gamma_1(N), K)$. There exists a unique semi-simple continuous Galois representation $\rho_f : G_{\mathbb{Q},Np} \to \mathrm{GL}_2(\bar{\mathbb{Q}}_p)$ such that for all prime number l not dividing Np, $\operatorname{tr} \rho_f(\mathrm{Frob}_l) = a_l$. Here $G_{\mathbb{Q},Np}$ is the Galois group of the maximal extension of $\mathbb{Q}$ unramified outside Np and Frob_l is any element in the conjugacy class of Frobenius at l.*

Moreover, we have $\det \rho_f = \omega_p^{k-1}\epsilon$, where $\omega_p : G_{\mathbb{Q},Np} \to \mathbb{Z}_p^*$ is the p-adic cyclotomic character, and ϵ is seen as a character of $G_{\mathbb{Q},Np}$ by composition with the morphism $G_{\mathbb{Q},Np} \to (\mathbb{Z}/N\mathbb{Z})^*$ defined by the action of $G_{\mathbb{Q},Np}$ over N-roots of unity in $\bar{\mathbb{Q}}$.

The case $k = 2$ is due to Eichler and Shimura. A modern reference is [56] ([50] also contains a useful sketch). The case $k > 2$ is due to Deligne [52]. The case $k = 1$ is due to Deligne and Serre [54].

2.6.7 Reminder on Pseudorepresentations

This subsection is a brief reminder of Chenevier's theory of pseudorepresentations (the most general version of the various theories of pseudorepresentations or pseudocharacters) in dimension 2. For the theory in general dimension d, see [36].

Definition 2.6.23 A (two-dimensional) *pseudorepresentation* of a group Π with values in a commutative ring A is a pair of maps $\tau : \Pi \to A$, $\delta : \Pi \to A$, such that

(1) δ is a group homomorphism from Π to A^*.
(2) τ is a central function from Π to A.

(3) $\tau(1) = 2$.
(4) $\tau(xy) + \delta(y)\tau(xy^{-1}) = \tau(x)\tau(y)$ for all $x, y \in \Pi$.

If Π is a topological group, A a topological ring, one says that the pseudo-representation (τ, δ) is *continuous* if τ and δ are.

If 2 is invertible in A, δ can be recovered from τ by the formula $\delta(x) = \frac{\tau(x)^2 - \tau(x^2)}{2}$. In this case, τ is a pseudocharacter in the sense of Rouquier.[6]

If ρ is any representation $\Pi \to \mathrm{GL}_2(A)$, then it is easy to check that $(\operatorname{tr} \rho, \det \rho)$ is a pseudo-representation of dimension 2.

Conversely, one has:

Theorem 2.6.24

(i) *If (τ, δ) is a pseudorepresentation of Π with values in a field L, then there exists a finite extension L' of L and a semi-simple representation $\rho : \Pi \to \mathrm{GL}_2(L')$ such that $\tau = \operatorname{tr} \rho$, $\delta = \det \rho$. Moreover ρ is unique up to isomorphism.*
(ii) *Let A be an henselian local ring of maximal ideal $\mathfrak{m}$ and residue field L, and (τ, δ) be a pseudorepresentation of Π with values in A. We assume that there exists an absolutely irreducible representation $\bar{\rho} : \Pi \to \mathrm{GL}_2(L)$ such that $\tau \equiv \operatorname{tr} \bar{\rho} \pmod{\pi}$ and $\delta \equiv \det \bar{\rho} \pmod{\pi}$. Then there exists a representation $\rho : \Pi \to \mathrm{GL}_2(A)$ such that $\tau = \operatorname{tr} \rho$, $\delta = \det \rho$. Moreover ρ is unique up to isomorphism.*
(iii) *In the situation of (i) or (ii), if (τ, δ) is continuous, then the representation ρ is continuous, provided that L is a local field in case (i), or a complete discrete valuation ring in case (ii).*

These results (and their generalization in any dimension) are due to Chenevier: see [36, Theorem 2.12] for (i), [36, Theorem 2.22] for (ii); for (iii) see [13, §1.5.5]. In characteristic greater than 2 (or greater than the dimension in general), (i) and (ii) were previously known by works of Taylor [121], Nyssen [95] and Rouquier [104].

2.6.8 *Pseudorepresentations and Eigenalgebra*

Theorem 2.6.25 *Let K be any finite extension of $\mathbb{Q}_p$. There exists a unique continuous pseudorepresentation of dimension 2*

$$(\tau, \delta) : G_{\mathbb{Q}, Np} \to \mathcal{T}_{0,K}$$

such that $\tau(\mathrm{Frob}_l) = T_l$ for all $l \nmid Np$. Here $\mathcal{T}_{0,K}$ is provided with its natural topology as a finite K-vector space.

[6] For the definition of a pseudocharacter, see [104]. For an introduction to pseudocharacters, see [10].

Moreover one has $\tau(c) = 0$, *where* c *is any complex conjugation in* $G_{\mathbb{Q},Np}$, *and* $\delta(\mathrm{Frob}_l) = l^{k-1}\langle l\rangle$ *for all* $l \nmid Np$.

Proof The key remark is that it is enough to prove the corollary for K replaced by a finite extension K' of K. Indeed, note that $\mathcal{T}_{0,K}$ is a closed subspace in $\mathcal{T}_{0,K} = \mathcal{T}_{0,K} \otimes_K K'$. So if we have a pseudorepresentation $(\tau, \delta) : G_{\mathbb{Q},Np} \to \mathcal{T}_{0,K'}$, satisfying the condition of the theorem, then τ and δ send a dense subset of $G_{\mathbb{Q},Np}$ (namely the set consisting of the Frob_l for $l \nmid Np$ and their conjugates) into $\mathcal{T}_{0,K}$ (because $T_l \in \mathcal{T}_{0,K}$ and $\langle l\rangle \in \mathcal{T}_{0,K}$). Therefore, τ and δ have image in $\mathcal{T}_{0,K}$ and they define a pseudorepresentation on $\mathcal{T}_{0,K}$ which obviously satisfies the required property.

Now if K is large enough, $\mathcal{T}_{0,K} = K^r$, where every factor corresponds to a system of eigenvalues of $\mathcal{H}_0$ appearing in $M_k(\Gamma_1(N), K)$. Let $\chi_i : \mathcal{H}_0 \to \mathcal{T}_{0,K} = K^r \xrightarrow{\text{projection on the } i\text{-th component}} K$ be one of those systems. There exists an eigenform f_i in $M_k(\Gamma_1(N), K)$ such that $\psi(T)f_i = \chi_i(T)f_i$ for every $T \in \mathcal{H}_0$. Theorem 2.6.22 attaches to f_i a continuous Galois representation $\rho_i : G_{\mathbb{Q},Np} \to \mathrm{GL}_2(K)$ such that $\mathrm{tr}\,\rho_i(\mathrm{Frob}_l) = \chi_i(T_l)$. Therefore, the product $\rho = \prod_{i=1}^r \rho_i : G_{\mathbb{Q},Np} \to \mathrm{GL}_2(K^r) = \mathrm{GL}_2(\mathcal{T}_K)$ is a representation whose trace $\tau := \mathrm{tr}\,\rho$ and determinant $\delta := \det \rho$ is a pseudorepresentation satisfying the required properties.

The continuity of τ and its values at Frob_l for $l \nmid Np$ completely determines τ by Chebotarev, which in turns completely determine δ since 2 is invertible in $\mathcal{T}_K$, so the uniqueness follows from Theorem 2.6.24(i). □

2.7 Eigenalgebras Over Discrete Valuation Rings

The study of Hecke algebras over a discrete valuation ring R is important to get a better understanding of the general case and is fundamental for the applications to number theory. Hecke algebras over a DVR are the framework in which the proofs of the Taniyama–Weil conjecture, of the Serre conjecture, of most cases of the Fontaine–Mazur were developed. Actually, in these applications, the discrete valuation ring R is also complete, and this hypothesis simplifies somewhat the theory (cf. Sect. 2.7.3). However, the theory for a general DVR is only slightly harder, and we expose it first, in Sects. 2.7.1 and 2.7.2. After a brief discussion of the Deligne–Serre's lemma, which in this point of view is just a simple consequence of the general structure theory of Hecke algebra, we give an exposition of the theory of congruences over a DVR.

We now fix some notations and terminology for all this section: R is a discrete valuation ring. So R is a domain, is principal and local, and has exactly two prime ideals, the maximal ideal $\mathfrak{m}$ and the minimal ideal (0). As usual, we refer to the two corresponding points of $\mathrm{Spec}\, R$ as the *special* point and the *generic* point. A uniformizer of R is chosen and denoted by π. We call $k = R/\mathfrak{m}$ the residue field and K the fraction field of R. As in Sect. 2.2, M is a projective module of finite type over R, that is a free module of a module of finite rank since R is principal. We write $M_K := M \otimes_R K$ and $M_k := M \otimes_R k = M/\mathfrak{m}M$. We suppose given a commutative ring $\mathcal{H}$ and a map $\psi : \mathcal{H} \to \mathrm{End}_R(M)$, and we write $\mathcal{T}$, $\mathcal{T}_K$ and $\mathcal{T}_k$ the Hecke algebras constructed on M, M_K and M_k.

2.7.1 Closed Points and Irreducible Components of Spec $\mathcal{T}$

In this section, we describe and interpret as systems of eigenvalues the points of Spec $\mathcal{T}$, in other words the prime ideals of $\mathcal{T}$. We shall distinguish between the closed points of Spec $\mathcal{T}$, which are the maximal ideals of $\mathcal{T}$, and the non-closed points, which are the prime ideals of $\mathcal{T}$ that are not maximal.

Proposition 2.7.1 *The closed immersion Spec $\mathcal{T}_k \hookrightarrow$ Spec $\mathcal{T}$ induced by the natural surjective map $\mathcal{T} \to \mathcal{T}_k$ defines an homeomorphism of Spec $\mathcal{T}_k$ onto the special fiber of Spec $\mathcal{T} \to$ Spec R, or equivalently, onto the set of closed points of Spec $\mathcal{T}$.*

The morphism Spec $\mathcal{T}_K \to$ Spec $\mathcal{T}$ induced by the natural map $\mathcal{T} \to \mathcal{T}_K$ is an homeomorphism of Spec $\mathcal{T}_K$ onto the generic fiber of Spec $\mathcal{T} \to$ Spec R, or equivalently, onto the set of non-closed points of Spec $\mathcal{T}$. Further, the map sending a point to its closure realizes a bijection between that set of non-closed points in Spec $\mathcal{T}$ and the set of irreducible components of Spec $\mathcal{T}$.

Proof The algebra $\mathcal{T}$ is finite, hence integral, over R, so the so-called *incomparability of prime ideals* apply (cf. [59, Cor 4.18]): If $\mathfrak{p} \subset \mathfrak{p}'$ are two disjoint primes in $\mathcal{T}$, then $\mathfrak{p} \cap R \neq \mathfrak{p}' \cap R$. Since $\mathfrak{p} \cap R$ and $\mathfrak{p}' \cap R$ are obviously primes of R, this means that $\mathfrak{p} \cap R = (0)$ and $\mathfrak{p} \cap R' = \mathfrak{m}$. It follows that the prime ideals $\mathfrak{p}$ of $\mathcal{T}$ such that $\mathfrak{p} \cap R = (0)$ (resp. $\mathfrak{p} \cap R = \mathfrak{m}$) are minimal prime ideals (resp. maximal ideals) in $\mathcal{T}$. On the other hand, since R is a discrete valuation ring, $\mathcal{T}$ is free, hence flat, over R, and the going-down lemma holds (cf. [59, Lemma 10.11]): for every prime ideal $\mathfrak{p}'$ of $\mathcal{T}$ such that $\mathfrak{p}' \cap R = \mathfrak{m}$, there exists a prime $\mathfrak{p}$ of $\mathcal{T}$, contained in $\mathfrak{p}'$, such that $\mathfrak{p} \cap R = (0)$. It follows that no maximal ideal of $\mathcal{T}$ is a minimal prime ideal. Thus we have the following equivalences: $\mathfrak{p} \cap R = (0)$ if and only if $\mathfrak{p}$ is a minimal prime ideal of $\mathcal{T}$; $\mathfrak{p} \cap R = \mathfrak{m}$ if and only if $\mathfrak{p}$ is a maximal ideal of $\mathcal{T}$.

Translated into geometric terms, this gives the first assertion in each of the paragraph of the proposition. The second sentence follows immediately since by Proposition 2.4.1 the points of Spec $\mathcal{T}_k$ (resp Spec $\mathcal{T}_K$) are the same as the points of the special fiber (resp. of the generic fiber) of Spec $\mathcal{T} \to$ Spec R. For the last sentence, it is enough to recall that the irreducible components are the closed subsets of Spec $\mathcal{T}$ corresponding to the minimal prime ideals. □

One can reformulate the proposition using the simple concept of *multivalued map*. If A and B are two sets, by a *multivalued map from A to B*, denoted $\tilde{f} : A \rightsquigarrow B$, we shall mean a map f in the ordinary sense from A to the set of non-empty parts of B. We shall say that the multivalued map $\tilde{f}$ is surjective if $\cup_{a\in A}\tilde{f}(a) = B$, and we shall denote such a map $\tilde{f} : A \rightsquigarrow\!\!\twoheadrightarrow B$.

The proposition allows one to define a multivalued *specialization* map, $\widetilde{\mathrm{sp}}$, from the set of non-closed points of Spec $\mathcal{T}$ to the set of closed points of Spec $\mathcal{T}$. Indeed, to every non-closed point of Spec $\mathcal{T}$, we attach the set of closed points of the irreducible component it belongs to. For example, in the following picture, Spec $\mathcal{T}$ has four irreducible components (namely from bottom to top the two straight lines, the oval, and the ugly curve) hence four non-closed points. It has also four closed points, but the multivalued map from the non-closed points to the closed point is not a bijection. Instead, it sends the two non-closed points corresponding to the two straight lines to the same closed point, the non-closed point corresponding to the oval to a set of two closed points, and the non-closed point corresponding to the ugly curve to one closed point.

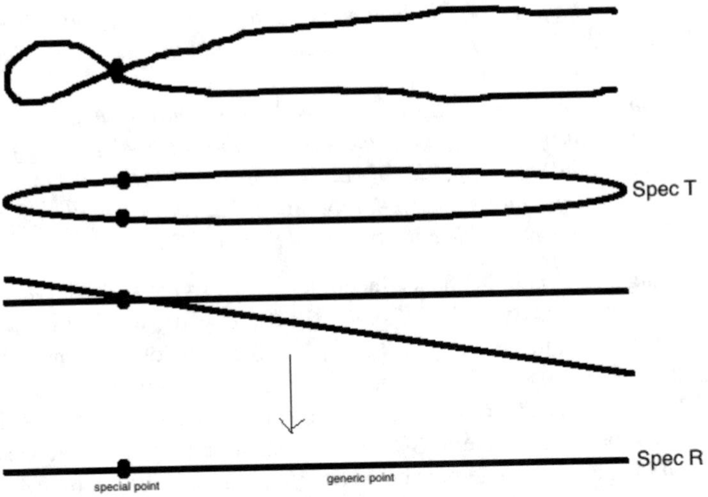

2.7.2 Reduction of Characters

Let K' be a finite extension of K and $\chi : \mathcal{T} \to K'$ a character (that is, a morphism of R-algebras). We are going to define a finite sets of characters $\bar{\chi}_1, \ldots, \bar{\chi}_r$ from $\mathcal{T}$ to $k_1, \ldots, k_r$, where the k_i are algebraic extensions of k which depend only on K', not on χ.

Let R' be any sub R-algebra of K', containing $\chi(\mathcal{T})$, and integral over R. Two examples of such sub-algebras are $\chi(\mathcal{T})$ and the integral closure of R in K', and all other such algebras are contained between those two. By the Krull–Akizuki theorem ([59, Theorem 11.13]), R' is a noetherian dimension 1 domain and it has only finitely many ideals containing $\mathfrak{m}$. The prime ideals of R' are therefore the minimal prime ideal (0), which of course lies above the ideal (0) of R, and its maximal ideals which (as is easily seen using [59, Cor 4.18] using that R' is integral over R, as in the proof of Proposition 2.7.1) lies over the maximal ideal $\mathfrak{m}$ of R. Hence R' has only finitely many prime ideals $\mathfrak{m}_1, \dots, \mathfrak{m}_r$, and they satisfy $\mathfrak{m}_i \cap R = \mathfrak{m}$. Let us call, for $i = 1, \dots, r$, $k_i = R'/\mathfrak{m}_i$; this is an algebraic extension of k since R' is integral over R. For $i = 1, \dots r$, we define a character $\bar{\chi}_i : \mathcal{T} \to k_i$ by reducing $\chi : \mathcal{T} \to R'$ modulo $\mathfrak{m}_i$.

Definition 2.7.2 The characters $\bar{\chi}_i : \mathcal{T} \to k_i$ for $i = 1, \dots, r$ are called the *reductions of the character* $\chi : \mathcal{T} \to K'$ *along* R'.

Remark 2.7.3 In the case $K' = K$, taking $R' = R$ is the only choice since R is integrally closed. In this case, R' has only one maximal ideal $\mathfrak{m}_1 = \mathfrak{m}$, and χ has only one reduction $\bar{\chi}_1$ that we denote simply by $\bar{\chi}$.

With R' as above, one can group all the $\bar{\chi}_i$'s into one character $\bar{\chi} : \mathcal{T} \to (R'/\mathfrak{m}R')^{\mathrm{red}}$, the reduction mod $\mathfrak{m}R'$ of χ, since $R'/\mathfrak{m}R'$ is just the product of the k_i. This is particularly convenient to express the (obvious) functoriality of the construction of the reductions of a character, which is as follows: *if K_1 and K_2 are two finite extensions of K, $\sigma : K_1 \to K_2$ a K-morphism, $\chi_i : \mathcal{T} \to K_i$ for $i = 1, 2$ two characters such that $\chi_2 = \sigma \circ \chi_1$, R_i for $i = 1, 2$ two R-subalgebras of K_i containing $\chi_i(\mathcal{T})$ and integral over R, such that $\sigma(R_1) = R_2$ and $\bar{\chi}_i : \mathcal{T} \to (R_i/\mathfrak{m}R_i)^{\mathrm{red}}$ the reduction of χ_i along R_i for $i = 1, 2$, then one has*

$$\bar{\chi}_2 = \bar{\sigma} \circ \bar{\chi}_1,$$

where $\bar{\sigma} : (R_1/\mathfrak{m}R_1)^{\mathrm{red}} \to (R_2/\mathfrak{m}R_2)^{\mathrm{red}}$ is the morphism induced by $\sigma : R_1 \to R_2$.

This functoriality property shows that the reductions of a character χ along $\chi(R)$ are *universal* amongst all reductions modulo an algebra R'. We can use these reductions to define a natural *reduction multivalued map*

$$\widetilde{\mathrm{red}} : \{\text{characters } \mathcal{T} \to \bar{K}\}/G_K \rightsquigarrow \{\text{characters } \mathcal{T} \to \bar{k}\}/G_k \ .$$

Here, as in Sect. 2.5.3 we have denoted by $\bar{K}$ and $\bar{k}$ some algebraic closures of K and k respectively, and we have set $G_K = \mathrm{Aut}(\bar{K}/K)$ and $G_k = \mathrm{Aut}(\bar{k}/k)$. One proceeds as follows: if $\chi : \mathcal{T} \to \bar{K}$ is a character, let $R' = \chi(\mathcal{T})$. If $\mathfrak{m}_1, \dots, \mathfrak{m}_r$ are the maximal ideals of R, and $k_i = R'/\mathfrak{m}_i$, the reduction $\bar{\chi}_i : \mathcal{T} \to k_i$ can be seen as a character $\bar{\chi}_i : \mathcal{T} \to \bar{k}$ by choosing a k-embedding $k_i \hookrightarrow \bar{k}$, which is then well-defined up to the action of G_k. Hence a well-defined multivalued map

$$\{\text{characters } \mathcal{T} \to \bar{K}\} \rightsquigarrow \{\text{characters } \mathcal{T} \to \bar{k}\}/G_k$$

which sends χ to the set of the $\bar{\chi}_i$, for $i = 1, \dots, r$. One checks easily that this multivalued map factors through $\{\text{characters } \mathcal{T} \to \bar{K}\}/G_K$, defining $\widetilde{\mathrm{red}}$.

Theorem 2.7.4 *One has the following natural commutative diagram of sets, where the horizontal arrows are bijection, and the vertical arrows are surjective-multivalued maps:*

$$\begin{array}{ccccc}
\{\textit{non-closed points of Spec}\,\mathcal{T}\} & \xrightarrow{\simeq} & \{\textit{points of Spec}\,\mathcal{T}_K\} & \xrightarrow{\simeq} & \{\textit{characters}\,\mathcal{T}\to\bar{K}\}/G_K \\
\big\downarrow \widetilde{sp} & & & & \big\downarrow \widetilde{red} \\
\{\textit{closed points of Spec}\,\mathcal{T}\} & \xrightarrow{\simeq} & \{\textit{points of Spec}\,\mathcal{T}_k\} & \xrightarrow{\simeq} & \{\textit{characters}\,\mathcal{T}\to\bar{k}\}/G_k
\end{array}$$

Proof The morphisms have been constructed above. To check the commutativity of the diagram, start with a character $\chi : \mathcal{T} \to \bar{K}$. Then $\mathfrak{p} = \ker\chi$ is a minimal prime ideal of $\mathcal{T}$, and is the non-closed point of Spec $\mathcal{T}$ corresponding to χ. Since $\chi(\mathcal{T}) \simeq \mathcal{T}/\mathfrak{p}$, the maximal ideals of $\mathcal{T}$ containing $\mathfrak{p}$ are the maximal ideals $\mathfrak{m}_1, \ldots, \mathfrak{m}_r$ of $\chi(\mathcal{T})$, and those ideals are the kernel of the reduced character $\bar{\chi}_1, \ldots, \bar{\chi}_r$. Since by definitions $\widetilde{\mathrm{sp}}(\mathfrak{p}) = \{\mathfrak{m}_1, \ldots, \mathfrak{m}_r\}$, and $\widetilde{\mathrm{red}}(\mathfrak{p}) = \{\bar{\chi}_1, \ldots, \bar{\chi}_r\}$, the commutativity of the diagram is proved. The surjectivity of $\widetilde{\mathrm{red}}$ then follows from the surjectivity of $\widetilde{\mathrm{sp}}$. □

2.7.3 The Case of a Complete Discrete Valuation Ring

When R is a **complete** discrete valuation ring the situation becomes simpler.

Proposition 2.7.5 *If R is complete, every irreducible component of Spec $\mathcal{T}$ contains exactly one closed point. Moreover, the map from the set of closed points of Spec $\mathcal{T}$ to its set of connected components, which to a closed point attaches the connected component where it belongs, is a bijection.*

Proof Since $\mathcal{T}$ is finite over R local complete, by Eisenbud [59, Cor. 7.6], one has $\mathcal{T} = \prod_{i=1}^{r} \mathcal{T}_{\mathfrak{m}_i}$ where $\mathfrak{m}_1, \ldots, \mathfrak{m}_r$ are the maximal ideals of $\mathcal{T}$. This means that Spec $\mathcal{T}$ is the disjoint union of the schemes Spec $\mathcal{T}_{\mathfrak{m}_i}$, which are connected since $\mathcal{T}_{\mathfrak{m}_i}$ is local. Hence the Spec $\mathcal{T}_{\mathfrak{m}_i}$ are the connected components of Spec $\mathcal{T}$, and they obviously contain exactly one closed point, namely $\mathfrak{m}_i$. Since connected components contain at least one irreducible component, hence at least one closed point by Proposition 2.7.1, the second assertion follows. But an irreducible component, being contained in a connected component, cannot contain more than one closed point, and the first assertion follows as well. □

The consequences for our general picture are as follows. First, when R is complete, the multivalued maps $\widetilde{\mathrm{sp}}$ and $\widetilde{\mathrm{red}}$ becomes ordinary single valued maps and we thus denote them sp and red. Similarly, a character $\chi : \mathcal{T} \to K'$ has only one reduction along any subalgebra R' of K containing $\chi(\mathcal{T})$ and integral over R.

Second, the set of closed (resp. non-closed) points being identified with the set of connected (resp. irreducible) components of Spec $\mathcal{T}$, we get a new description of the map sp, as the map $\widetilde{\text{incl}}$ that sends an irreducible component of Spec $\mathcal{T}$ to the connected component that contains it. We summarize this information in the following corollary of Theorem 2.7.4.

Corollary 2.7.6 *If R is complete, one has the following natural commutative diagram of sets, where the horizontal arrows are bijections, and the vertical arrows are surjective maps:*

$$\begin{array}{ccccc}
\{\text{irreducible comp. of Spec}\,\mathcal{T}\} & \xrightarrow{\simeq} & \text{Spec}\,\mathcal{T}_K & \xrightarrow{\simeq} & \{\text{characters}\,\mathcal{T} \to \bar{K}\}/G_K \\
\downarrow \widetilde{\text{incl}} & & & & \downarrow \text{red} \\
\{\text{connected comp. of Spec}\,\mathcal{T}\} & \xrightarrow{\simeq} & \text{Spec}\,\mathcal{T}_k & \xrightarrow{\simeq} & \{\text{characters}\,\mathcal{T} \to \bar{k}\}/G_k
\end{array}$$

Exercise 2.7.7 Assume that R is a complete DVR.

1. Show that there is a connected component in Spec $\mathcal{T}$ which is not irreducible if and only if there are two systems of eigenvalues $\mathcal{T} \to \bar{K}$, not in the same G_K-orbit, that have the same reduction.
2. Assume that $\mathcal{T}_K$ is étale over K. Show that Spec $\mathcal{T}$ is not étale over Spec R if and only if there are two distinct systems of eigenvalues $\mathcal{T} \to \bar{K}$ that have the same reduction.

Exercise 2.7.8 Let $R = \mathbb{Z}_p$, $M = \mathbb{Z}_p^2$, and $\mathcal{H} = \mathbb{Z}_p[T]$ with $\psi(T) = \begin{pmatrix} 0 & \pi^a \\ \pi^b & 0 \end{pmatrix}$, for some $a, b \in \mathbb{N}$. Compute $\mathcal{T}$ in this case. Describe prime and maximal ideals of $\mathcal{T}$. When is $\mathcal{T}$ regular ? when is $\mathcal{T}$ irreducible ? When is $\mathcal{T}$ connected ? When is $\mathcal{T}$ étale over R ? When is M free over $\mathcal{T}$?

2.7.4 A Simple Application: Deligne–Serre's Lemma

It is the very useful following simple result. If $m, m' \in M$, we say that $m \equiv m' \pmod{\mathfrak{m}}$ if $m - m' \in \mathfrak{m}M$.

Lemma 2.7.9 (Deligne–Serre) *Assume there is an $m \in M$, $m \not\equiv 0 \pmod{\mathfrak{m}}$ such that for every $T \in \mathcal{H}$, we have $\psi(T)m \equiv \alpha(T)m \pmod{\mathfrak{m}}$ for some $\alpha(T) \in R$. Then there is a finite extension K' of K, such that if R' is the integral closure of R in K', there is a vector $m' \in M \otimes_R R'$ which is a common eigenvector for $\mathcal{H}$ and whose system of eigenvalues χ satisfies $\chi(T) \equiv \alpha(T) \pmod{\mathfrak{m}'}$ for every $T \in \mathcal{H}$.*

In other words, if we have an eigenvector $\pmod{\mathfrak{m}}$ then one can lift its eigenvalues $\pmod{\mathfrak{m}}$ (but maybe not the eigenvector itself) into true eigenvalues (after possibly extending R to R').

The proof is actually contained in what we have said above: it follows from the relation $\psi(T)m \equiv \alpha(T)m \pmod{m}$ that $\alpha(T) \pmod{m}$ depends only of $\psi(T)$ and hence is a character $\mathcal{T} \to k$. We have seen that those characters can be lifted into characters $\mathcal{T} \to R' \subset K'$ for a suitable finite extension K' of $\mathfrak{m}$.

Actually the need to replace K by a finite extension K' follows from the fact that point of Spec $\mathcal{T}_K$ may not be defined over K, but only on a finite extension. The same proof gives the following variant, which is sometimes useful:

Lemma 2.7.10 (Variant of Deligne–Serre's Lemma) *Assume there is an $m \in M$, $m \not\equiv 0 \pmod{m}$ such that for every $T \in \mathcal{H}$, we have $\psi(T)m \equiv \alpha(T)m \pmod{m}$ for some $\alpha(T) \in R$. Also assume that all points of $\mathcal{T}_K$ are defined over K. Then there is a vector $m' \in M$ which is a common eigenvector for $\mathcal{H}$ and whose system of eigenvalues χ satisfies $\chi(T) \equiv \alpha(T) \pmod{\mathfrak{m}}$ for every $T \in \mathcal{H}$.*

The variant of Deligne–Serre's lemma implies the classical version, since there is always a finite extension K' of K such that all points of $\mathcal{T}_{K'}$ are defined over K', and applying the variant to K' gives the classical Deligne–Serre's lemma. The variant has the advantage it gives some control on what extension K' is needed, if any.

Exercise 2.7.11 It is often said that the Deligne–Serre's lemma does not hold modulo $\mathfrak{m}^2$, or $\mathfrak{m}^c$ for $c > 1$. While technically this is not true for the classical version (any congruence mod $\mathfrak{m}$ becomes a congruence mod $\mathfrak{m}'^c$ if K is replaced by an extension of index of ramification at least c), this is true for the variant of the Deligne–Serre's lemma.

Indeed, prove that a character $\mathcal{T} \to R/\mathfrak{m}^2$ needs not be liftable to a character $\mathcal{T} \to R$ even if every point of Spec $\mathcal{T}_K$ is defined over K.

2.7.5 *The Theory of Congruences*

Congruences Between Two Submodules

As usual, M is a finite projective module over R. We write $M_K = M \otimes K$ and we suppose given a decomposition $M_K = A \oplus B$ of K-vector spaces. We write p_A and p_B for the first and second projections of M_K on the factors of that decomposition, and we define $M_A = p_A(M)$, $M_B = p_B(M)$. We thus have exact sequences

$$0 \longrightarrow M \cap B \longrightarrow M \xrightarrow{p_A} M_A \longrightarrow 0,$$

$$0 \longrightarrow M \cap A \longrightarrow M \xrightarrow{p_B} M_B \longrightarrow 0.$$

In this situation, we define the *congruence module*

$$C = M/((M \cap A) \oplus (M \cap B)).$$

Exercise 2.7.12 Let $M = \mathbb{Z}_2^2$, and A (resp. B) be the $\mathbb{Q}_2$-subspace of $M_{\mathbb{Q}}$ generated by $(1, 1)$ (resp. $(1, -1)$). What is C in this case?

Exercise 2.7.13

1. Show that C is a finite torsion module.
2. Show that the map p_A identifies $M \cap A$ with a submodule of M_A, and show that $C = M_A/(M \cap A)$. By symmetry, $C = M_B/(M \cap B)$.
3. Show that (p_A, p_B) identifies M with a sub-module of $M_A \oplus M_B$. Show that $C = (M_A \oplus M_B)/M$.

Definition 2.7.14 In this situation, C is called the *module of congruences*. Its annihilator is called the *ideal of congruences*.

To explain the name, we shall relate C to actual congruences between elements of A and B. For $c \geq 1$, $f, g \in M$, we shall write $f \equiv g \pmod{\pi^c}$ if $f - g \in \pi^c M$. We define a *congruence in M modulo π^c between A and B* as the data of $f \in M \cap A$, $g \in M \cap B$, such that $f \equiv g \pmod{\pi^c}$ and $f \not\equiv 0 \pmod{\pi}$ (which is the same as $g \not\equiv 0 \pmod{\pi}$).

Proposition 2.7.15 *There exists a congruence in M between A and B modulo π^c if and only if C contains a submodule isomorphic to R/π^c.*

Proof Let x be a generator of a sub-module of $C = M/(M \cap A \oplus M \cap B)$ isomorphic to R/π^c. Then we have $\pi^c x \in M \cap A \oplus M \cap B$, so we can write $\pi^c x = f - g$ with $f \in M \cap A$ and $g \in M \cap B$, and we also have $\pi^{c-1} f \notin (M \cap A) \oplus (M \cap B)$ which implies that $f \notin \pi M$ (otherwise, we would have $f = \pi f'$, $g = \pi g'$, with $f' \in M \cap A$, $g' \in M \cap B$ and $\pi^{c-1} x = f' - g'$.) Thus $f \equiv g \pmod{\pi^c}$ while $f \not\equiv 0$.

Conversely, if f, g define a congruence in M between A and B then $f - g = \pi^c x$ for some $x \in M$. Now if $\pi^{c-1} x = f' - g'$ with $f' \in M \cap A$, $g' \in M \cap B$, then $\pi f' - \pi g' = f - g$ which implies $f = \pi f'$, which is absurd. So $\pi^{c-1} x \notin M \cap A \oplus M \cap B$. This shows that x generates a module isomorphic to R/π^c in C. □

In other words, the ideal of congruences is the maximal ideal modulo which there are congruences between A and B.

Congruences in Presence of a Bilinear Product

There is a situation, which arises in applications, where it is easy to compute the congruence module. It is the situation where there is a *bilinear product*, on M: $M \times M \to R$, $(x, y) \mapsto \langle x, y \rangle$, which is *non-degenerate* (or *perfect*), which means that the maps $p : x \mapsto (y \mapsto \langle x, y \rangle)$ and $q : y \mapsto (y \mapsto \langle x, y \rangle)$ are isomorphisms of M onto M^*.

Proposition 2.7.16 *Assume that M has a non-degenerate bilinear product as above such that $\langle A, B \rangle = 0$ (in other words, $A^\perp = B$). Then there is an isomorphism*

$C = (M \cap A)^*/p(M \cap A)$. *In particular,* $C = 0$ *if and only if the restriction of the bilinear product to* $M \cap A$ *is still non-degenerate.*

Proof We consider the composition $r : M \xrightarrow{p} M^* \to (M \cap A)^* \to (M \cap A)^*/p(M \cap A)$, where the second morphism is the restriction map (which is surjective since $M \cap A$ is a direct summand of M). The morphism r is surjective as the composition of three surjective morphisms. By definition, an element $m \in M$ is in $\ker r$ if and only if there exist an element $a \in M \cap A$ such that $\langle m, a' \rangle = \langle a, a' \rangle$. This is equivalent to $m - a \in A^{\perp} = B$, but since $m - a \in M$, this is also equivalent to $m - a \in M \cap B$. Therefore $\ker r = M \cap A \oplus M \cap B$, and the results follows. □

Exercise 2.7.17 Assume that M has a non-degenerate bilinear product. Let $f \in M$, $f \not\equiv 0 \pmod{M}$, and $\langle f, f \rangle \neq 0$. Let $c \in \mathbb{N}$. Show that there exists $g \in M$, $\langle f, g \rangle = 0$ and $g \equiv f \pmod{\mathfrak{m}^c}$ if and only if $\langle f, f \rangle \in \mathfrak{m}^c$.

Congruences and Eigenalgebras

We place ourselves in the situation of the beginning Sect. 2.7.5: M is a finite free R-module such that $M_K = A \oplus B$. We also assume that we are in a situation which gives rise to an eigenalgebra: we have a commutative ring $\mathcal{H}$, a morphism $\psi : \mathcal{H} \to \mathrm{End}_R(M)$. We assume the following compatibility between those data: $\psi(H)$ stabilizes A and B.

A natural question in this context is weither there are congruences between **eigenvectors** in A and B, not simply vectors, or better if there are congruences between **system of eigenvalues** appearing in A and B. To discuss those questions, let us introduce some terminology: For any submodule N of M stable by $\psi(\mathcal{H})$, let us call $\mathcal{T}_N$ sub-algebra of $\mathrm{End}_R(N)$ generated by $\psi(\mathcal{H})$. So in particular $\mathcal{T}_M = \mathcal{T}$.

Note that by Exercise 2.3.6, $\mathcal{T}_{M \cap A} = \mathcal{T}_{M_A}$ and $\mathcal{T}_{M \cap B} = \mathcal{T}_{M_B}$. We call those algebras $\mathcal{T}_A$ and $\mathcal{T}_B$ for simplicity. Also, by the same exercise, the natural maps $\mathcal{T} \to \mathcal{T}_A$ and $\mathcal{T} \to \mathcal{T}_B$ are surjective, while their product $\mathcal{T} \to \mathcal{T}_A \times \mathcal{T}_B$ is injective.

Definition 2.7.18

(a) We say that there is a *congruence modulo* π^c *between eigenvectors of* A *and* B if there exist $f \in M \cap A$, $g \in M \cap B$ both eigenvectors for $\psi(\mathcal{H})$ (equivalently: for $\mathcal{T}$) such that $f \equiv g \pmod{\pi^c}$ and $f \not\equiv 0 \pmod{\pi}$.
(b) We say that there is a *congruence modulo* π^c *between system of eigenvalues of* A *and* B if there exists characters $\chi_A : \mathcal{T} \to \mathcal{T}_A \to R$ and $\chi_B : \mathcal{T} \to \mathcal{T}_B \to R$ such that $\chi_A \equiv \chi_B \pmod{\pi^c}$.
(c) We say that there is an *eigencongruence modulo* π^c *between* A *and* B if there is a character $\chi : \mathcal{T} \to R/\pi^c$ that factors both through $\mathcal{T}_A$ and $\mathcal{T}_B$.

Obviously, (a) implies (b) implies (c). However, those inclusions are strict, even if we assume that all points of $\mathrm{Spec}\, \mathcal{T}_K$ are defined over K.

Example 2.7.19 For an example of situation where (b) holds but not (a), let $R = \mathbb{Z}_p$, $K = \mathbb{Q}_p$ $M = \mathbb{Z}_p^2$, $M_K = \mathbb{Q}_p^2$, $A = \mathbb{Q}_p e_1$, $B = \mathbb{Q}_p e_2$ where (e_1, e_2) is the standard basis of $\mathbb{Z}_p^2$, and let $T \in \mathrm{End}_R(M)$ be the matrix $\begin{pmatrix} 0 & 0 \\ 0 & p \end{pmatrix}$. Then e_1 and e_2 are eigenvectors for T of eigenvalues 0 and p, so there is a congruence modulo (p) between systems of eigenvalues appearing in A and B. Yet there is no congruences between A and B, as $M \cap A = \mathbb{Z}_p e_1$, $M \cap B = \mathbb{Z}_p e_2$, so $M = (M \cap A) \oplus (M \cap B)$ and the module of congruence C is trivial. Note that in this situation the eigenalgebra $\mathcal{T}$ is $\mathbb{Z}_p[X]/X(X-p)$, X acting on M by T. This algebra $\mathcal{T}$ is naturally isomorphic as a $\mathbb{Z}_p$-algebra to $\{(a, b) \in \mathbb{Z}_p^2, a \equiv b \pmod{p}\}$, the isomorphism sendind $P(X)$ to $(P(0), P(p))$. The algebras $\mathcal{T}_A$ and $\mathcal{T}_B$ are just $\mathbb{Z}_p$ and the map $\mathcal{T} \mapsto \mathcal{T}_A \times \mathcal{T}_B$ is just the obvious inclusion $\mathcal{T} \subset \mathbb{Z}_p^2$ (obvious in terms of the second description of $\mathcal{T}$).

Exercise 2.7.20 Give an example of situation where (c) holds but not (b), even if all points of Spec $\mathcal{T}_K$ are defined over K.

Exercise 2.7.21 Show that when all points Spec $\mathcal{T}_K$ are defined over K, any eigencongruence modulo $\mathfrak{m}$ between A and B comes form a congruence between system of eigenvalues appearing in A and B. (N.B. we are talking here of congruence modulo $\mathfrak{m}$, not $\mathfrak{m}^c$.)

Among the three notions of congruences that we define, the notion (b) of congruences between system of eigenvalues is the most natural and useful. Unfortunately, it is also the more difficult to detect. Let us just mention the following easy but weak result:

Exercise 2.7.22 Assume that $\dim_K A = 1$ and that all points Spec $\mathcal{T}_K$ are defined over K. If the module of congruence C is not trivial, show that there is a congruence modulo $\mathfrak{m}$ between systems of eigenvalues appearing in A and B. Show that the converse is false.

It is relatively easy, however, to detect eigencongruences. But for that we need a better tool than the ideal of congruence, whose definition does not take into account the action of $\mathcal{H}$ on M.

Definition 2.7.23 The *ideal of fusion* F is the conductor of the morphism $\mathcal{T} \to \mathcal{T}_A \times \mathcal{T}_B$.

Let us recall that the *conductor* of an injective morphism of rings $S \to S'$ is the set of $x \in S$ such that $xS' \subset S$, that is the annihilator of the S-module S'/S. It is an ideal of S, and also an ideal of S'. Actually, it is easy to see that the conductor is the largest S-ideal which is also an S'-ideal (or which is the same, the largest S'-ideal which is contained in S).

Proposition 2.7.24 *Let F_A and F_B be the image of F in $\mathcal{T}_A$ and $\mathcal{T}_B$. Then we have natural isomorphisms of R-algebras $\mathcal{T}_A/F_A \simeq \mathcal{T}/F \simeq \mathcal{T}_B/F_B$.*

Proof Since F is an ideal of $\mathcal{T}_A \times \mathcal{T}_B$, we have $F = F_A \times F_B$. The map $\mathcal{T} \to \mathcal{T}_A \to \mathcal{T}_A/F_A$ is obviously surjective; let us call I its kernel. We obviously have $F \subset I$. Let $x \in I$, whose image in $\mathcal{T}_A \times \mathcal{T}_B$ is (x_A, x_B). We have $x_A \in F_A$.

Therefore there is f in F, whose image (f_A, f_B) in $\mathcal{T}_A \times \mathcal{T}_B$ is such that $f_A = x_A$. Then $x - f = (0, x_B - f_B)$ in $\mathcal{T}_A \times \mathcal{T}_B$. Therefore $x - f \in F$ (the claim here is that any element in $\mathcal{T}$ whose image in $\mathcal{T}_A$ is 0 is in F. For if we have such an element $(0, y) \in \mathcal{T}$, then for any $(u, v) \in \mathcal{T}_A \times \mathcal{T}_B$, v is the image of some element $z \in \mathcal{T}$, and $(0, y)(u, v) = (0, y)z$ is in $\mathcal{T}$ as a product of two elements in $\mathcal{T}$, which shows that $(0, u) \in F$). Hence $x \in F$, so finally $I = F$. Therefore we have an isomorphism $\mathcal{T}/F \simeq \mathcal{T}_{M_A}/F_A$. The other isomorphism is symmetric. □

What is interesting here is that we get an isomorphism $\mathcal{T}_A/F_A \simeq \mathcal{T}_B/F_B$ that can be interpreted as eigencongruences:

Theorem 2.7.25 *Let c be an integer. Then there is an eigencongruence modulo $\mathfrak{m}^c$ between A and B if and only if there is a surjective map of R-algebras $\mathcal{T}/F \mapsto R/\mathfrak{m}^c$.*

Proof An eigencongruence modulo $\mathfrak{m}^c$ is a surjective map $\chi : \mathcal{T} \to R/\mathfrak{m}^c$ that factors both through $\mathcal{T}_A$ and $\mathcal{T}_B$. Let $x \in F$, and (x_A, x_B) its image in $\mathcal{T}_A \times \mathcal{T}_B$. Then $(x_A, 0)$ and $(0, x_B)$ are in $\mathcal{T}$, since they are $x(1, 0)$ and $x(0, 1)$, and x is in F. But $\chi(x_A, 0) = \psi_B(0) = 0$ and similarly $\chi(0, x_B) = 0$, so $\chi(x) = 0$. We have just shown $F \subset \ker \chi$, so χ factors into a map $\mathcal{T}/F \to R/\mathfrak{m}^c$, necessarily surjective.

Conversely, a map $\chi : \mathcal{T} \to \mathcal{T}/F \to R/\mathfrak{m}^c$ factors through both $\mathcal{T}_A$ and $\mathcal{T}_B$ by the above proposition. □

What is the relation between the ideal of fusion and the module of congruence? Let $I_\mathcal{T}$ be the annihilator of the module of congruence $C = M/(M \cap A \oplus M \cap B)$ in $\mathcal{T}$ (not to be confused with the ideal of congruence I defined as the annihilator of C in R. We have $I_\mathcal{T} \cap R = I$.)

Theorem 2.7.26 *We have an inclusion $F \subset I_\mathcal{T}$. If there exists a non-degenerate bilinear product $\mathcal{T} \times M \to R$ satisfying $\langle TT', f\rangle = \langle T', Tf\rangle$ for all $T, T' \in \mathcal{T}$, $f \in M$, then we have $F = I_\mathcal{T}$.*

Proof By definition, $I_\mathcal{T}$ is the set of $T \in \mathcal{T}$ such that $T((M \cap A) \oplus (M \cap B))$ is in M. So $I_\mathcal{T} = \{T \in \mathcal{T}, T(\mathcal{T}_A \times \mathcal{T}_B)M \subset M\}$ since clearly, $(M \cap A) \oplus (M \cap B)$ is generated by M as a $(\mathcal{T}_A \times \mathcal{T}_B)$-module. So

$$I_\mathcal{T} = \{T \in \mathcal{T}, T(\mathcal{T}_A \times \mathcal{T}_B) \subset \mathcal{O}_M\}$$

where $\mathcal{O}_M = \{(T_A, T_B) \in \mathcal{T}_A \times \mathcal{T}_B, (T_A, T_B)M \subset M\}$. On the other hand,

$$F = \{T \in \mathcal{T}, T(\mathcal{T}_A \times \mathcal{T}_B) \subset \mathcal{T}\}$$

Therefore, the inclusion $F \subset I_\mathcal{T}$ follows from the trivial inclusion $T \subset \mathcal{O}_M$, which proves the first half of the theorem.

Similarly, $F = I_\mathcal{T}$ would follow from $T = \mathcal{O}_M$ which we shall prove assuming the existence of a bilinear product as in the statement. First, because of the perfect pairing $\dim_K \mathcal{T}_K = \dim_K M_K$, and similarly $\dim_K \mathcal{T}_A = \dim_K A$ and $\dim_K \mathcal{T}_B = \dim_K B$ because the pairing induces a perfect pairing between $\mathcal{T}_A \otimes K$ and A, and

$\mathcal{T}_B \otimes K$ and B. It follows that the map $\mathcal{T} \hookrightarrow \mathcal{T}_A \times \mathcal{T}_B$ becomes an isomorphism after tensorizing by K. The pairing extends to a pairing $\mathcal{T}_K \times M_K \to K$, and the non-degeneracy of the pairing means that for $T \in \mathcal{T}_K$, we have $T \in \mathcal{T}$ if and only if $\langle T, f \rangle \in R$ for all $f \in M$. If $T \in \mathcal{O}_M \in \mathcal{T}_A \times \mathcal{T}_B \subset \mathcal{T}_K$, and $f \in M$, we have $\langle T, f \rangle = \langle 1, Tf \rangle$ which is in R since $Tf \in M$ by definition of $\mathcal{O}_M$. Hence $T \in \mathcal{T}$ and $\mathcal{O}_M = \mathcal{T}$. □

Exercise 2.7.27 Prove that if we release the hypothesis that the pairing is non-degenerate in the above theorem, but we assume instead that it has discriminant of valuation $c \in \mathbb{N}$, then $I_{\mathcal{T}}/F$ is killed by π^c.

2.8 Modular Forms with Integral Coefficients

We use the same notation as in Sect. 2.6.4: N is an integer, and $\mathcal{H}$ is the polynomial algebra over $\mathbb{Z}$ generated by indeterminates named T_l (for every $l \nmid N$), U_l (for every $l \mid N$), and $\langle a \rangle$ (for every $a \in (\mathbb{Z}/N\mathbb{Z})^*$). The same without the U_l's is called $\mathcal{H}_0$.

Let $w \geq 0$ be an integer (the weight). Recall that the $\mathbb{Z}$-submodule $M_w(\Gamma_1(N), \mathbb{Z})$ of $M_w(\Gamma_1(N), \mathbb{Q})$ (same statement with S_w) with integral coefficients is a lattice. So we can extend the preceding definitions by setting $M_w(\Gamma_1(N), R) = M_w(\Gamma_1(N), \mathbb{Z}) \otimes_{\mathbb{Z}} R$ for any commutative ring R. The same holds for S_w.

We now fix a prime number p, a finite extension K of $\mathbb{Q}_p$, of ring of integers R. As above, $\mathfrak{m}$ is the maximal ideal of R, and $k = R/m$. We consider the module $S_w(\Gamma_1(N), R)$ and the eigenalgebras $\mathcal{T}$, $\mathcal{T}_0$ of $\mathcal{H}$, $\mathcal{H}_0$ acting on this module. Similarly, we define the eigenalgebras $\mathcal{T}_K$, $\mathcal{T}_{0,K}$, and $\mathcal{T}_k$, $\mathcal{T}_{0,k}$ for the action of $\mathcal{H}$, $\mathcal{H}_0$ on $S_w(\Gamma_1(N), K)$ and $S_w(\Gamma_1(N), k)$.

We let $\mathcal{T}_k$ be the eigenalgebras generated by all Hecke operators on $S_w(\Gamma_1(N), k)$.

2.8.1 *The Specialization Morphism for Hecke Algebras of Modular Forms*

Proposition 2.8.1 *The natural surjective morphism $\mathcal{T} \otimes_R k \to \mathcal{T}_k$ is an isomorphism, and the module $S_w(\Gamma_1(N), R)^\vee$ is free over $\mathcal{T}$.*

Proof Let n be the dimension of $S_w(\Gamma_1(N), K)$. Since $S_w(\Gamma_1(N), R)$ is finite free over R, this is also the dimension of $S_w(\Gamma_1(N), k)$. By Proposition 2.6.16 $\dim \mathcal{T}_K = n$. Since $\mathcal{T}$ is finite flat over R, $\dim \mathcal{T} \otimes k = n$. The same argument as in Proposition 2.6.16 and its corollary shows that $\dim \mathcal{T}_k = \dim S_w(\Gamma_1(N), k) = n$ (Indeed, the only facts used in Proposition 2.6.16 are the fact that an element of $S_w(\Gamma_1(N), k)$ is determined by its q-expansion, which is a tautology with our definition of that space, and the fact that $S_w(\Gamma_1(N), k)$ contains no non-zero element with constant q-expansion, which is clear since the q-expansion of an element of that space has 0 constant term.) It follows by equality of dimensions that the surjective map $\mathcal{T}_R \otimes_R k \to \mathcal{T}_k$ is an isomorphism. Then the pairing $(t, f) \mapsto a_1(tf)$, $\mathcal{T}_R \times S_w(\Gamma_1(N), R)^\vee$ is perfect because its generic and special fibers are perfect. □

Exercise 2.8.2 Show that both assertions of the theorem may be false when $\mathcal{T}$ and $\mathcal{T}_K$ are defined using the full space of modular forms $M_w(\Gamma_1(N), K)$.

Exercise 2.8.3 Let $R = \mathbb{Z}_p$, $w = 12p$ an integer, and define $\mathcal{T}^p$ (resp. $\mathcal{T}_k^p$) as the Hecke algebra of all the Hecke operators **excepted** T_p acting on $S_w(\mathrm{SL}_2(\mathbb{Z}), R)$ (resp. on $S_w(\mathrm{SL}_2(\mathbb{Z}), k)$). The specialization map $\mathcal{T}^p \otimes_R k \to \mathcal{T}_k^p$ is not an isomorphism.

Exercise 2.8.4 When is the map $\mathcal{T}_0 \otimes_R k \to \mathcal{T}_k$ an isomorphism?

2.8.2 An Application to Galois Representations

Theorem 2.8.5 *There exists a unique continuous two-dimensional pseudorepresentation (τ, δ) of $G_{\mathbb{Q},Np}$ with values in $\mathcal{T}_0$, of dimension 2, such that $\tau(\mathrm{Frob}_l) = T_l$ for all $l \nmid Np$. We also have $\tau(c) = 0$ if c is any complex conjugation and $\delta(\mathrm{Frob}_l) = l^{k-1}\langle l\rangle$ for all $l \nmid Np$.*

Proof We have already seen (Theorem 2.6.25) how to construct a continuous pseudorepresentation $(\tau, \delta) : G_{\mathbb{Q},Np} \to \mathcal{T}_{0,K} = \mathcal{T}_0 \otimes_R K$ satisfying the required properties, and since $\mathcal{T}_0$ is closed in $\mathcal{T}_{0,K}$ and $\tau(\mathrm{Frob}_l)$, $\delta(\mathrm{Frob}_l) \in \mathcal{T}_0$, this pseudo representation takes values in $\mathcal{T}_0$. □

Thus we have glued all the Galois representations attached to eigenforms in $M_k(\Gamma_1(N))$ in one unique pseudorepresentation with values in the algebra $\mathcal{T}_{0,R}$.

Let us give one application.

Corollary 2.8.6 *Let $n \geq 1$ be an integer, $f \in M_k(\Gamma_1(N), \epsilon, R)$ be a normalized form such that for every prime l not dividing Np, $T_l f \equiv \chi(T_l) f \pmod{\mathfrak{m}^n}$ where $\chi(T)$ is some element in $R/\mathfrak{m}^n$. There exists a continuous pseudorepresentation (τ_f, δ_f) of $G_{\mathbb{Q},Np}$ with values in $R/\mathfrak{m}^n$ such that $\tau_f(\mathrm{Frob}_l) = \chi(T_l)$ in $R/\mathfrak{m}^n$ and $\delta_f(\mathrm{Frob}_l) = l^{k-1}\epsilon(l)$.*

So we can attach Galois pseudorepresentations not only to true eigenforms, but more generally to eigenforms modulo $\mathfrak{m}^n$. Of course the pseudorepresentations we get takes values in $R/\mathfrak{m}^n$ instead of R.

Had we not the above theorem, the obvious method to prove the result of the corollary would be to replace f by a true eigenform g with eigenvalues congruent to the $\chi(T_l)$ modulo $\mathfrak{m}^n$. Then the would reduce the Eichler–Shimura–Deligne representations ρ_g modulo $\mathfrak{m}^n$ and take its trace and determinant. The Deligne–

Serre's lemma tells us that we can find such a g when $n = 1$, but in general we cannot. So this approach fails. Yet with what we have done, the proof of the corollary is trivial:

Proof Since f is normalized, the element $\chi(T_l)$ of $R/\mathfrak{m}^n$ depends only on $T_l \in \mathcal{T}_0$, hence χ extends to a character $\mathcal{T}_0 \to R/\mathfrak{m}^n$. Composing $\tau : G_{\mathbb{Q},Np} \to \mathcal{T}_0$ of the above theorem with χ gives the result. □

For any maximal ideals $\mathfrak{m}_i$ of $\mathcal{T}_0$, the quotient $k_i := \mathcal{T}_0/\mathfrak{m}_i$ is a finite field, so the pseudorepresentation $\bar{\tau}_{\mathfrak{m}_i}, \bar{\delta}_{\mathfrak{m}_i} : G_{\mathbb{Q},Np} \to \mathcal{T}_0 \to \mathcal{T}_0/\mathfrak{m}_i = k_i$ is the pseudo-representation of a unique semi-simple representation $\bar{\rho}_{\mathfrak{m}_i} : G_{\mathbb{Q}_p} : G_{\mathbb{Q},Np} \to \mathrm{GL}_2(k_i')$ by Theorem 2.6.24, where k_i' is a finite extension of k_i. Actually $\bar{\rho}_{\mathfrak{m}_i}$ is defined over k_i as the Brauer group of a finite field is trivial.

We call the maximal ideal $\mathfrak{m}_i$ of $\mathcal{T}_0$ (or their connected component in Spec $\mathcal{T}_0$) *non-Eisenstein* if $\bar{\rho}_{\mathfrak{m}_i}$ is absolutely irreducible, and *Eisenstein* otherwise.

Theorem 2.8.7 *Let* Spec $\mathcal{T}_{0,\mathfrak{m}_i}$ *be a non-Eisenstein component. There exists a unique continuous Galois representation* $\rho_{\mathfrak{m}_i} : G_{\mathbb{Q},Np} \to \mathrm{GL}_2(\mathcal{T}_{0,\mathfrak{m}_i})$ *such that* $\mathrm{tr}\,\rho_{\mathfrak{m}_i}(\mathrm{Frob}_l) = T_l$ *in* $\mathcal{T}_{0,m_i}$.

Proof The pseudorepresentation $(\tau_{\mathfrak{m}_i}, \delta_{\mathfrak{m}_i})$ of $G_{\mathbb{Q},Np}$ with value in the complete local ring $\mathcal{T}_{0,\mathfrak{m}_i}$ obtained by post-composing (τ, δ) with the map $\mathcal{T}_0 \to \mathcal{T}_{0,\mathfrak{m}_i}$, satisfies the hypothesis of Theorem 2.6.24(i) because $\mathfrak{m}_i$ is non-Eisenstein. Therefore, there exists a representation $\rho_{\mathfrak{m}_i}$ such that $\mathrm{tr}\,\rho_{\mathfrak{m}_i}(\mathrm{Frob}_l) = T_l$. The continuity of this representation follows from Theorem 2.6.24(iii).

For the uniqueness, it suffices to note that if $\rho\mathfrak{m}_i, 1$ and $\rho_{\mathfrak{m}_i,2}$ both satisfy the conditions of the statement, then using Chebotarev, $\mathrm{tr}\,\rho_{\mathfrak{m}_i,1} = \tau_{\mathfrak{m}_i} = \mathrm{tr}\,\rho_{\mathfrak{m}_i,2}$ on $G_{\mathbb{Q},Np}$. Thus $2\det\rho_{\mathfrak{m}_i,1} = 2\det\rho_{m_i,2}$ using the formula $2\det(x) = \mathrm{tr}\,(x)^2 - tr(x^2)$ for x a $(2, 2)$-matrix), and since $\mathcal{T}_{\mathfrak{m}_i}$ is flat over $\mathbb{Z}_p$, 2 is not a divisor of zero in $\mathcal{T}_{\mathfrak{m}_i}$ and $\det\rho_{\mathfrak{m}_i,1} = \det\rho_{\mathfrak{m}_i,2}$. Therefore, $\rho_{\mathfrak{m}_i,1}$ and $\rho_{\mathfrak{m}_i,2}$ are isomorphic by Theorem 2.6.24(ii). □

The fact that we have a true representation over $\mathcal{T}_{0,\mathfrak{m}_i}$ (assume this component non-Eisenstein) allows us to make a comparison with Mazur's deformation theory. Let $\mathcal{R}_{\mathfrak{m}_i}$ be the universal deformation ring of the residual representation $\bar{\rho}_{\mathfrak{m}_i} : G_{\mathbb{Q},Np} \to \mathrm{GL}_2(k_i)$. Since $\mathcal{T}_{0,\mathfrak{m}_i}$ is a coefficient ring (that is a complete local noetherian of residue field k_i), and since $\rho_{\mathfrak{m}_i}$ obviously deforms $\bar{\rho}_{\mathfrak{m}_i}$, there is a morphism of coefficient rings

$$\mathcal{R}_{\mathfrak{m}_i} \to \mathcal{T}_{0,\mathfrak{m}_i}$$

which sends the universal deformation representation on $\mathcal{R}_{\mathfrak{m}_i}$ to $\rho_{\mathfrak{m}_i}$.

Exercise 2.8.8 Show that the map $\mathcal{R}_{\mathfrak{m}_i} \to \mathcal{T}_{0,\mathfrak{m}_i}$ is surjective.

The remarkable idea of Wiles leading to the proof of the Shimura–Taniyama–Weil conjecture in the semi-stable case (hence of Fermat's last theorem) was a method to determine the kernel of this map in simple cases (weight $k = 2$, and

$\rho_{\mathfrak{m}_i}$ of a very special type). This method has then been generalized to many more cases, leading to the proof of the Shimura–Taniyama–Weil conjecture in general, and even of the Fontaine–Mazur conjecture in many cases.

2.9 A Comparison Theorem

In this section, we only assume that R is noetherian and reduced. As usual, M is a finite flat R-module, $\mathcal{H}$ is a commutative ring and $\psi : \mathcal{H} \to \mathrm{End}_R(M)$ a morphism of rings. For every point $x \in \mathrm{Spec}\,(R)$, of field $k(x)$, we call $M_x = M \otimes_R k(x)$ and $\mathcal{T}_x$ the subalgebra generated by $\psi(\mathcal{H})$ in $\mathrm{End}_{k(x)}(M_x)$.

Proposition 2.9.1 *Assume that for x in a Zariski-dense subset Z in Spec R, $\mathcal{H}$ acts semi-simply on M_x. Then $\mathcal{T}$ is reduced.*

Proof The hypothesis implies that for $x \in Z$, $\mathcal{T}_x$ is reduced.

Let $t \in \mathcal{T}$ such that $t^n = 0$ for some $n \geq 1$. Let t_x be the image of t by the natural map $\mathrm{End}_R(M) \to \mathrm{End}_{k(x)}(M_x)$. We have $t_x^n = 0$, so for $x \in Z$, we get $t_x = 0$.

If M is free over R, that means that every matrix element of $t \in \mathrm{End}_R(M)$ is 0 at every $x \in Z$. Since Z is dense and R is reduced, every matrix element of t is 0, and $t = 0$.

In the general case, M is projective, so $N = M \oplus Q$ say is free. Apply the result in the free case to the endomorphism t' of N such that $t'_{|M} = t$, $t'_{|Q} = 0$. □

Now let us assume that M and M' are two finite projective R-modules with action ψ and ψ' of $\mathcal{H}$, and let $\mathcal{T}$ and $\mathcal{T}'$ be there associated algebras.

Theorem 2.9.2 *Let us assume that for every x in a Zariski-dense subset Z of R, there is an $\mathcal{H}$-injection $M'^{ss}_x \subset M^{ss}_x$. Then there is a canonical surjection $\mathcal{T}^{red} \to \mathcal{T}'^{red}$ (that is, a closed immersion Spec $\mathcal{T}' \subset$ Spec $\mathcal{T}$). Moreover, for every $x \in$ Spec R, there is an $\mathcal{H}$-injection $(M'_x)^{ss} \to (M_x)^{ss}$*

Proof Let $t \in \mathcal{H} \otimes R$. Let $P_t(X)$, $P'_t(X)$ in $R[X]$ be the characteristic polynomials of $\psi(t)$ and $\psi'(t)$. If $x \in \mathrm{Spec}\,R$, let $P_{t,x}(X)$ and $P'_{t,x}(X)$ be the images of those polynomial in $k(x)[X]$. They are also the characteristic polynomials of $(\psi \otimes 1)(t)$ and $(\psi' \otimes 1)(t)$ in $\mathrm{End}_{k(x)}(M_x)$, $\mathrm{End}_{k(x)}(M'_x)$. By hypothesis, for $x \in Z$, $P'_{t,x}$ divides $P_{t,x}$ in $k(x)[X]$ for $x \in Z$. As Z is dense, P_t divides P'_t in $R[X]$, from which it follows that $P_{t,x} \mid P'_{t,x}$ for every $x \in \mathrm{Spec}\,R$. By elementary representation theory, this proves the second assertion.

For the first assertion, we can work locally and assume that M is free of rank d. Let us call J (resp. J') the kernel J of $\psi : \mathcal{H} \otimes_R \to \mathrm{End}_R(M)$ (resp. $\psi' : \mathcal{H} \otimes R \to \mathrm{End}_R(M')$). If $t \in J$, $\psi(t) = 0$ so $P_t = X^d$. So $P'_t \mid X^d$, and by Cayley–Hamilton $\psi(t')^d = 0$, so $\psi(t') = 0$. Hence $J \subset \sqrt{J'}$ and the theorem follows. □

Corollary 2.9.3 *Let us assume that for every x in a Zariski-dense subset Z of R, M_x is a semi-simple $\mathcal{H}$-module, and there is an $\mathcal{H}$-isomorphism $M'_x \simeq M_x$. Then* *there is a canonical isomorphism* $\mathcal{T} = \mathcal{T}'$. *Moreover, for every* $x \in \operatorname{Spec} R$, *there is an* $\mathcal{H}$*-isomorphism* $(M'_x)^{ss} \simeq (M_x)^{ss}$.

2.10 Notes and References

In a sense, the idea, in the context of Hecke operators acting on spaces or modules of modular forms, of considering the sub-algebra of the endomorphism algebra of that space or module generated by those operators—that we call the *eigenalgebra* here—is so tautological that it is impossible to trace back its origin: as soon as we consider operators acting on a space or module, we implicitly or explicitly consider the subalgebra they generate. Yet the realization that this *eigenalgebra* was an important object, even perhaps the central object on the theory, that the algebraic properties of this commutative algebra (which by Grothendieck's theory of schemes can be reformulated into geometric properties of its spectrum) were a way to express and study the properties of eigenforms, this realization I say was long to come, and a fundamental progress when it did.

Over a field (a subfield of $\mathbb{C}$), this idea certainly occurred to Shimura,[7] who defines $\mathcal{T}$ and prove some elementary properties of these algebras in his book [113]. He for example raised the following question: is the eigenalgebra $\mathcal{T}$ generated by all the Hecke operators acting on the space $S_2(\Gamma_0(N), \mathbb{Q})$ isomorphic to the algebra of isogenies of the modular Jacobian $J_0(N)$. This question was solved affirmatively by Ribet [102] in 1975.

Over a ring like the ring of integers of a number field, or a localization or completion of such a ring, the study of the eigenalgebra is intimately related to the study of the arithmetic property of modular forms. The eigenalgebra appears in [54], in the second proof of the famous Deligne–Serre's lemma, but it seems that Mazur was the first to study and use the algebraico-geometric properties of those eigenalgebras, in its very influential 1978s paper [90], to prove deep arithmetic results about modular forms and elliptic curves. Most of the important progresses in the arithmetic theory of modular forms ever since have used and developed that very idea (often named Hecke algebras). To mention a few of them: uses of the Hecke algebras to study congruences between modular forms [68, 69, 103], use of big Hecke Algebra (Eigenalgebra over Iwasawa's algebra) by Hida to construct families of p-adic modular forms, use of those Hecke algebra to prove the main conjectures [91, 126], use of eigenalgebras (and their fine algebraico-geometric properties) in the proof of Fermat's last theorem [122, 127], and use of new kind of "big Hecke algebras" by Coleman [38] to construct non-ordinary families of modular forms, and more generally by Coleman–Mazur [40] and other to construct the *eigencurve* or *eigenvarieties* (called *variétés de Hecke* in French).

[7] Perhaps even earlier. Eichler? Selberg?

In this chapter, we have tried to expose the formal aspects of the construction of eigenalgebras, relying on the work of the aforementioned mathematicians and many other. We now try to give a proper attribution to most of the results that are not standard commutative algebra.

In Sects. 2.1, 2.2, and 2.3, almost all results are elementary. Exercise 2.3.9 is due to Buzzard. Proposition 2.3.3 is due to Chenevier [34], and Proposition 2.3.4 is also due to him [35].

Proposition 2.4.1 is due to Chenevier [34], at lest in the case of a surjective map $R \to R'$.

The results of Sect. 2.5 are essentially linear algebras, and commutative linear algebra.

Most of the results of Sect. 2.6 are due to Shimura [113] though the emphasis on the notion of Gorensteinness in this context seems due to Mazur. To our knowledge, Theorem 2.6.19 is new.

As for eigenalgebras over a base ring which is a D.V.R. (or more generally a Dedekind domain), as we said they appear in [54] and [90]. The interpretation of those maximal and prime ideals as eigenforms, etc. given in Sect. 2.7.1, which is again elementary commutative algebra, is sketched in a few sentences in [54]. Each subsequent author stated the part he needed, until all those things became completely standard at the time of the proof of Fermat's Last Theorem—The survey [50] contains an exposition of that theory, which also mention the link with Deligne–Serre's lemma. The theory of congruences exposed in Sect. 2.7.5 is essentially due to Hida (e.g. [68, 69]) and Ribet (e.g. [103]). A good survey is [60].

The results of Sect. 2.9 are due to Chenevier [35].

Chapter 3
Eigenvarieties

In this chapter, we explain the construction of *eigenvarieties* from suitable families of Banach modules with action of a commutative algebra, and prove their general properties. This is a version at large and in a rigid analytic setting of the construction of eigenalgebras of the previous chapter.

In this endeavor, our work is greatly facilitated by the now classical exposition, formalization and generalization of Mazur and Coleman's method by Buzzard, cf. [29], as well as the treatment of similar questions in Chenevier's thesis (cf. [34] and [35]). The main difference with those treatments is that we offer a self-contained all-in-one-place exposition of this material, from the first principles in non-archimedean functional analysis up to the hard comparison theorems for eigenvarieties of Chenevier. Our only prerequisite beyond elementary mathematics are some familiarity with Tate's rigid analytic geometry, at the level of [27, Chapters 7 and 8] (other good references, sufficient for this chapter, include [48, §1 and §2] or the first half of [26]). Even so, rigid analytic geometry is not used before Sect. 3.3.3, that is roughly for the first half of this chapter. We also give some simplifications and generalizations of the methods and results in [29].

Compared with the setting of a construction of an eigenalgebra of the preceding chapter (a base ring R, a finite flat R-module M, a morphism $\psi : \mathcal{H} \to \mathrm{End}_R(M)$ where $\mathcal{H}$ is some commutative ring), the setting for eigenvarieties is enlarged in two directions. The base ring R, or rather its spectrum $\mathrm{Spec}\, R$ is replaced by a rigid analytic variety $\mathcal{W}$, which will not be an affinoid in general. Hence it would be natural to replace M by a coherent locally-free sheaf over $\mathcal{W}$, but here comes a second, more important, enlargement: we want to replace the finite module M by a non-finite Banach module. To avoid talking of sheaf of Banach modules (which we could do, as in [40] or [35]), we simply define our data as in [29] by the giving, for every affinoid subdomain $W = \mathrm{Sp}\, R$ of $\mathcal{W}$, of a Banach-module M_W over R (satisfying some technical condition of orthonormability), provided with an action of the ring $\mathcal{H}$. The modules M_W for various W are required to satisfy some compatibility relations, that are explained in Sect. 3.5. To deal with the new

J. Bellaïche, *The Eigenbook*, Pathways in Mathematics,
https://doi.org/10.1007/978-3-030-77263-5_3

infiniteness of our modules, we introduce a crucial hypothesis: that our ring $\mathcal{H}$ contains a privileged element $U_p \in \mathcal{H}$ that acts compactly on the various Banach modules M_W.

From those data, our aim is to construct in a canonical way an eigenvariety $\mathcal{E}$, that is a rigid analytic variety together with a locally finite map $\kappa : \mathcal{E} \to \mathcal{W}$, and a morphism of rings $\psi : \mathcal{H} \to \mathcal{O}(\mathcal{E})$ allowing us to see elements of $\mathcal{H}$ as functions on $\mathcal{E}$. The **eigen**variety should have the fundamental property that its points z (over a given point $w = \kappa(z) \in \mathcal{W}$) classify the systems of $\mathcal{H}$-**eigen**values appearing in the fiber M_w of the modules M_W at w that are *of finite slope*, that is for which U_p has a non-zero eigenvalue (the set of such systems do not depend of the chosen affinoid W of $\mathcal{W}$). The bijection should be naturally given as follows: to a point z should correspond the system of eigenvalues $T \mapsto \psi(T)(z)$, for any $T \in \mathcal{H}$.

In order to construct the eigenvariety, we glue some local pieces that are rigid analytic spectrum of eigenalgebras constructed as in the preceding chapter. More precisely, if $W = \operatorname{Sp} R$ is an admissible affinoid subdomain of $\mathcal{W}$, and ν is a real number, it is often, though not always, possible to define in a natural way a finite flat direct summand R-submodule $M_W^{\leq \nu}$ of M_W which is morally the "slope less or equal to ν" part of M_W, that is the submodule of vectors on which U_p acts with generalized eigenvalues of valuation less or equal than ν. When this is possible, we say that W and ν are *adapted.* This critical notion is explained in Sect. 3.3, which also proves that there are enough, in a precise technical sense, pairs (W, ν) that are adapted. In Sect. 3.4, we give the definition and basic properties of the modules $M_W^{\leq \nu}$.

In Sect. 3.6, we explain our variant of Buzzard's eigenvariety machine of [29], constructing the eigenvariety by gluing the spectrum of eigenalgebras $\mathcal{T}_{W,\nu}$ defined by the action of $\mathcal{H}$ on $M_W^{\leq \nu}$ for adapted pairs (W, ν).

In Sect. 3.7 we list the most important properties of the eigenvarieties, due to Coleman–Mazur [40] or Chenevier [34], in particular we explain why their points parametrize system of eigenvalues of finite slope.

In Sect. 3.8 we discuss two very useful results of Chenevier (cf. [35]) allowing to compare two eigenvarieties, or to prove that an eigenvariety is reduced.

Finally, in the last Sect. 3.9, we give a slight generalization of the eigenvariety machine, which takes as input, for every affinoid W, instead of a single Banach module M_W with action of a commutative ring $\mathcal{H}$, a finite complex $(M_{W,i})$ of such modules.

3.1 Non-archimedean Fredholm's Theory

In this section, we give a self-contained treatment of the theory of the Fredholm's determinant of compact operators over Banach algebras. Our treatment is inspired by the one in [29], himself inspired by Coleman [38], and by the paper [108]. We weaken the noetherian hypothesis that is used in [29].

3.1.1 General Notions

Let R be a (commutative, with unity) $\mathbb{Q}_p$-algebra with a norm $|\ |$ extending the standard p-adic norm on $\mathbb{Q}_p$. We recall that this means that $|\ |$ is a map $R \to \mathbb{R}_+$, satisfying, for every $\lambda \in \mathbb{Q}_p$, $x, y \in R$, $|\lambda 1_R| = |\lambda|$, $|x+y| \leq \max(|x|, |y|)$, $|xy| \leq |x||y|$, and $|x| = 0$ if and only if $x = 0$. We assume moreover that R is a *Banach* algebra, that is complete for the topology defined by $|\ |$.

Exercise 3.1.1 An element x in R^* is called *multiplicative* for $|\ |$ if $|x||x^{-1}| = 1$. Show that x is *multiplicative* if and only if for all $y \in R$, $|xy| = |x||y|$. Show that elements of $\mathbb{Q}_p^*$ are multiplicative in R.

A *norm* on an R-module M is a map, usually also denoted $|\ |$, from M to $\mathbb{R}_+$, such that for every $x \in R$, $m, n \in M$, we have $|m+n| \leq \max(|m|, |n|)$, $|xm| \leq |x||m|$, and $|m| = 0$ if and only if $m = 0$. An R-module M with a norm is a *Banach R-module* if it is complete for the topology defined by the norm. Two norms $|\ |_1$ and $|\ |_2$ on M are *equivalent* if there exists positive real constants $c < C$ such that $c|m|_1 \leq |m|_2 \leq C|m|_1$ for every $m \in M$.

Exercise 3.1.2 Show that two norms on M are equivalent if and only if they define the same topology.

A morphism $\phi : M \to N$ between two Banach R-modules is continuous if and only if it is *bounded*, that is if there exists a constant $C > 0$ such that $|\phi(m)| \leq C|m|$ for every m in M. In this case, the norm of ϕ, denoted $|\phi|$, is defined as the smallest constant C satisfying this condition. One evidently has $|\phi'\phi| \leq |\phi'||\phi|$ when $\phi : M \to N$ and $\phi' : N \to K$ are continuous morphisms of Banach R-modules. We denote by $\mathrm{Hom}_R(M, N)$ the R-module of continuous morphisms of R-modules from M to N. It is a Banach R-module for the norm we just defined. When $M = N$ we write $\mathrm{End}_R(M)$ for $\mathrm{Hom}_R(M, M)$.

The Open Mapping Theorem states that every continuous surjective map between Banach R-modules is open. It is proven exactly as in the case of real Banach spaces, using the Baire category theorem, see Wikipedia or any standard textbook of functional analysis—or see [28, Théorème 1, Chapitre 1, §3.3] for a general proof valid in our case. The open mapping theorem implies that if $f : M \to N$ is a continuous surjective map of R-modules, there exists a constant $c > 0$ such that for every n in N we can find $m \in M$ with $f(m) = n$ and $|m| \leq c|n|$.

3.1.2 Compact Operators

Definition 3.1.3 We say that a morphism of R-modules $\phi : M \to N$ is *of finite rank* if its image is contained in a finite (i.e. finitely generated) submodule of N. We say that a continuous morphism $\phi : M \to N$ between Banach R-modules is *compact* (or *completely continuous*) if it is in the closure of the module of morphisms of finite rank in $\mathrm{Hom}_R(M, N)$.

It follows from the definition that the set of compact morphisms is a closed submodule of $\mathrm{Hom}_R(M, N)$. In particular the sum of two compact morphisms is compact.

Lemma 3.1.4 *Let M, N, K be three Banach R-modules and $\phi : M \to N$ a compact morphism. If $f : N \to K$ is continuous, then $f \circ \phi$ is compact. Similarly if $f : K \to M$ is continuous, then $\phi \circ f$ is compact.*

Proof It is enough to prove the same result with 'compact' replaced by 'finite rank'. If ϕ is of finite rank, $\phi(M)$ is contained in a finite submodule N_0 of N, and for $f : N \to K$, $(f \circ \phi)(M)$ is contained in $f(N_0)$ which is finite as the image of a finite module, and so $f \circ \phi$ is of finite rank. For $f : K \to M$, $(\phi \circ f)(K) \subset \phi(M) \subset N_0$ and $\phi \circ f$ is also of finite rank. □

3.1.3 Orthonormalizable and Potentially Orthonormalizable Banach Modules

Definition 3.1.5 Let M be a Banach R-module, whose norm is also denoted $|\ |$. An *orthonormal basis* for the module M is a family $(e_i)_{i \in I}$ of elements of M such that $|e_i| = 1$ for every $i \in I$, and such that for every element $m \in M$ there exists a unique sequence $(a_i)_{i \in I}$ of elements of R converging to zero[1] such that $m = \sum_{i \in I} a_i e_i$ and $|m| = \sup_{i \in I} |a_i|$.

Definition 3.1.6 A Banach module M is *orthonormalizable* if it has an orthonormal basis. It is *potentially orthonormalizable* if there exists a norm $|\ |'$ on M equivalent to $|\ |$ such that $(M, |\ |')$ is orthonormalizable.

Example 3.1.7 The Banach R-module $c_I(R)$ of sequences $(r_i)_{i \in I}$ of elements of R converging to 0, normed with the sup norm, is orthonormalizable. There is even a canonical orthonormal basis $(e_i)_{i \in I}$ defined by setting e_i the sequence whose i-th term is 1 and all others 0. A Banach R-module M is orthonormalizable (resp. potentially orthonormalizable) if and only if it is isometric (resp. isomorphic) to $c_I(R)$ for some set I.

Let M and N be two Banach R-modules with orthonormal bases $(e_i)_{i \in I}$ and $(f_j)_{j \in J}$ respectively. If $\phi \in \mathrm{Hom}_R(M, N)$ we can write, for every i, $\phi(e_i) = \sum_{j \in J} a_{i,j} f_j$, where $(a_{i,j})_{j \in J}$ is a sequence of elements of R converging to 0. It is clear that $|\phi| = \sup_{i \in I, j \in J} |a_{i,j}|$. Conversely every matrix $(a_{i,j})_{i \in I, j \in J}$ of elements of R, whose rows $(a_{i,j})_{j \in J}$ go to zero for every $i \in I$, and whose coefficients are bounded, defines an element $\phi \in \mathrm{Hom}_R(M, N)$.

[1] Let m_i, $i \in I$, be elements in M. We say that $(m_i)_{i \in I}$ *converges to* m if for every $\epsilon > 0$ there exists a finite subset S of I such that $|m_i - m| < \epsilon$ for every $i \in I - S$. When the sequence $(m_i)_{i \in I}$ converges to 0, it is easy to see that the sequence $(\sum_{i \in J} m_i)_{J \in F(I)}$ (where $F(I)$ is the set of finite subsets of I) converges to some limit in M. We denote this limit by $\sum_{i \in I} m_i$.

If M is an orthonormalizable module with basis $(e_i)_{i \in I}$, and $S \subset I$ a finite subset, we denote by M_S the finite free sub-module of M generated by the e_s, $s \in S$, and by $\pi_S : M \to M$ the projection of M onto M_S sending e_i to e_i if $i \in S$, and e_i to 0 if $i \notin S$.

In all the rest of Sect. 3.1, we shall make the following hypothesis on the $\mathbb{Q}_p$-Banach algebra A:

Hypothesis 3.1.8 *Every finite-type sub-module of an orthonormalizable A-module is closed.*

Lemma 3.1.9 *If A is noetherian, Hypothesis 3.1.8 is satisfied.*

Proof Let M be an ON A-module, P a finite submodule of M, provided with restriction of the norm of M. Let $f : A^r \to P$ be a surjective morphism. Let Q be the quotient $A^r / \ker f$, provided with the quotient norm, which is complete by [27, Prop. 3.7.2]. The module Q is algebraically isomorphic to P, and the isomorphism $f : Q \to P$ is continuous.

Choose an ON basis $(e_i)_{i \in I}$. We claim that for $S \subset I$ finite and large enough, π_S is injective on P. Indeed, assume that P is generated by $p_1, \dots, p_r$, with $p_k = \sum_{i \in I} a_{k,i} e_i$. For any k, the ideal generated by all the $a_{k,i}$ has a finite family of generators, that all are linear combinations of finitely many $a_{k,i}$. Let S be the finite set of all the indices i of those $a_{k,i}$. Then clearly π_S is injective on P.

Choosing S as above, we see that the map $P \to \pi_S(P)$ is a continuous bijection, where $\pi_S(P)$ is provided with the norm restricted from A^S. Since $\pi_S(P)$ is closed in A^S by [27, Prop 3.1.3/3], it is complete, and by the open mapping theorem, the composition $\pi_S \circ f : Q \to P \to \pi_S(P)$ is an homeomorphism. It follows that f is also an homeomorphism, hence P is complete and thus closed in M. □

However, Hypothesis 3.1.8 is not satisfied by every $\mathbb{Q}_p$-Banach algebra, as seen in the following exercise:

Exercise 3.1.10 Find a Banach $\mathbb{Q}_p$-algebra A with a non-closed principal ideal.

Nevertheless, Hypothesis 3.1.8 is still satisfied by many non-noetherian Banach algebras A.

Exercise 3.1.11 Let A be a Banach $\mathbb{Q}_p$-algebra whose norm is an absolute value (that is satisfies $|xy| = |x||y|$ for every $x, y \in A$), and such that $|x| \leq |y| \Leftrightarrow x \in Ay$. (These hypotheses are equivalent to $(A, |\,|)$ being a complete valuation ring of height one containing $\mathbb{Q}_p$. It is easy to find examples of such A of arbitrary countable rational rank, and those with rational rank > 1 are non-noetherian). Show that A satisfies Hypothesis 3.1.8.

Lemma 3.1.12 *Let M be an orthonormalizable module with basis $(e_i)_{i \in I}$ and P a finite sub-module. For any $\epsilon > 0$, there exists a finite subset S of I such that for every $p \in P$, $|\pi_S(p) - p| \leq \epsilon |p|$.*

Proof Since P is finite, there exists a surjective continuous morphism $\pi : A^r \to P$. By Hypothesis 3.1.8, P is complete for the norm induced from M, hence by the open mapping theorem, there is a constant $c > 0$ such that for every $p \in P$, there exists $m \in A^r$ such that $\pi(m) = p$ and $|m| \leq c|p|$.

If $e_1, \ldots, e_r$ is an orthonormal basis of A^r, there exists a finite subset S of I such that $|\pi_S(\pi(e_i)) - \pi(e_i)| \leq \epsilon/c$ for $i = 1, \ldots, r$. If $m \in A^r$, one can write $m = \sum_{i=1}^r a_i e_i$, and $|\pi_S(\pi(m)) - \pi(m)| = |\sum a_i(\pi_S(\pi(e_i)) - \pi(e_i))| \leq \max(|a_i|)\epsilon/c = \epsilon|m|/c$.

Now let $p \in P$, and choose $m \in A^r$ such that $\pi(m) = p$ and $|m| \leq c|p|$. Then $|\pi_S(p) - p| = |\pi_S(\pi(m)) - \pi(m)| \leq \epsilon|m|/c \leq \epsilon|p|$. □

Proposition 3.1.13 *With notations as above, a morphism $\phi \in \mathrm{Hom}_R(M, N)$ is compact if and only if the sequence $(\sup_{i \in I} |a_{i,j}|)_{j \in J}$ converges to 0. (In other words, ϕ is compact if and only if the rows $(a_{i,j})_{j \in J}$ go to zero* uniformly in *i.)*

Proof If $\sup_{i \in I} |a_{i,j}|$ converges to 0, then the sequence $(\pi_S \circ \phi)_{S \subset I,\ S \text{ finite}}$ converges to ϕ in $\mathrm{Hom}_R(M, M)$ and therefore ϕ is compact.

Conversely assume that ϕ is compact and let $\epsilon > 0$. By definition there exists a finite rank operator ϕ' such that $|\phi - \phi'| < \epsilon$. Since ϕ' has finite rank, there exists by Lemma 3.1.12 a finite set $S \subset J$ such that $|\pi_S \circ \phi' - \phi'| < \epsilon$. Thus $|\pi_S \circ \phi - \phi| = |(\pi_S \circ \phi - \pi_S \circ \phi') + (\pi_S \circ \phi' - \phi') + (\phi' - \phi)| \leq \max(\epsilon, \epsilon, \epsilon) = \epsilon$. This means that $\sup_{i \in I} |a_{i,j}| < \epsilon$ for $j \notin S$, which shows that $\sup_{i \in I} |a_{i,j}|$ converges to 0. □

Let us record an important technical result that we have obtained while proving Proposition 3.1.13:

Scholium 3.1.14 *Let M be a potentially orthonormalizable Banach module and N an orthonormalizable Banach module with basis $(f_j)_{j \in J}$. If $\phi \in \mathrm{Hom}_R(M, N)$ is compact, then the sequence of operators $(\pi_S \phi)_{S \subset J,\ S\text{finite}}$ converges in operator norm to ϕ.*

3.1.4 Serre's Sufficient Condition for Being Orthonormalizable

In this section, we shall need the following hypothesis about R:

Hypothesis 3.1.15 *Among the elements $r \in R$ such that $|r| < 1$, there is one of largest norm which is multiplicative (see Exercise 3.1.1).*

We choose such an element π. It is then clear that the set of non-zero norms $|R^*|$ is the discrete subgroup $|\pi|^{\mathbb{Z}}$ of the multiplicative group of positive real numbers. We denote by R^0 the set of elements r of R such that $|r| \leq 1$, which is a subring of R, and by $\tilde{R}$ the quotient ring $R^0/\pi R^0$. For M a Banach R-module we define $M^0 = \{m \in M, |m| \leq 1\}$, which is an R^0-submodule of M, and $\tilde{M} = M^0/\pi M^0$, which is a $\tilde{R}$-module.

Lemma 3.1.16 *Assume Hypothesis 3.1.15. Let M be a Banach R-module such that $|M| \subset |R|$. Let $(e_i)_{i \in I}$ be a family of elements of M^0, and for $i \in I$ denote by $\tilde{e}_i$ the image of e_i in $\tilde{M}$. Then $(e_i)_{i \in I}$ is an orthonormal basis of M if and only if $(\tilde{e}_i)_{i \in I}$ is* *a basis (in the algebraic sense) of $\tilde{M}$. In particular, M is orthonormalizable if and only if $\tilde{M}$ is free over $\tilde{R}$.*

Proof We note that since π is multiplicative, $|\pi m| = |\pi||m|$ for every $m \in M$: the proof is the same as in Exercise 3.1.1.

Assume that $(\tilde{e}_i)$ is a basis of $\tilde{M}$. If $m \in M^0$, denote by $\tilde{m}$ the image of m in $\tilde{M}$. We can write $\tilde{m} = \sum \alpha_i \tilde{e}_i$ with the α_i in $\tilde{R}$, almost all 0. Choosing lifts a_i^1 of the α_i in R_0, we have $m - \sum a_i^1 e_i = \pi m_1$ with $m_1 \in M^0$. Applying the same result to m_1, we get $m - \sum a_i^2 e_i = \pi^2 m_2$ with $m_2 \in M^0$, and by induction $m - \sum a_i^n e_n = \pi^n m_n$ with $m_n \in M_0$. By construction, the sequence $(a_i^n)_{n \in \mathbb{N}}$ satisfies $|a_i^n - a_i^{n+1}| \leq |\pi|^n$, hence is Cauchy, and therefore converges to an element $a_i \in R^0$ for every $i \in I$. One has $m = \sum a_i e_i$. If $|m| = 1$, then some a_i^1 has norm 1, and so does a_i, and thus $|m| = \sup_i |a_i|$. By replacing m by $\pi^n m$ for the n such that $|m| = |\pi|^{-n}$, we see that the same results holds for any $m \in M$. Hence (e_i) is an orthonormal basis for M.

The other direction is easy and left to the reader. □

Theorem 3.1.17 *Any Banach $\mathbb{Q}_p$-module is potentially orthonormalizable.*

Proof Let M be a Banach $\mathbb{Q}_p$-module with norm $|\ |$. Set $|m|' = \inf_{r \in p^{\mathbb{Z}}, r \geq |m|} r$. Then $|\ |'$ is a norm on M which is equivalent to $|\ |$ and satisfies Hypothesis 3.1.15. Lemma 3.1.16 show that M is potentially orthonormalizable if and only if $\tilde{M}$ has a basis over $\mathbb{F}_p$. The result follows. □

3.1.5 Fredholm's Determinant of a Compact Endomorphism

Let M be a Banach R-module with orthonormal basis $(e_i)_{i \in I}$. Let $\phi \in \mathrm{End}_R(M)$ be a compact operator, of matrix $(a_{ij})_{i,j \in I}$. For any finite subset S of I and any permutation σ of S, we set

$$a_{S,\sigma} = \prod_{i \in S} a_{i\sigma(i)}, \quad c_S = \sum_{\sigma \text{ permutation of } S} \varepsilon(\sigma) a_{S,\sigma},$$

where $\varepsilon(\sigma)$ is the signature of σ.

Since ϕ is compact, if $\epsilon > 0$, there exists by Proposition 3.1.13 a finite subset I_0 of I such that $|a_{i,j}| \leq \epsilon$ for every $i \in I$ and every $j \notin I_0$. For S a finite subset of I, we have the estimate

$$|c_S| \leq |\phi|^{|S \cap I_0|} \epsilon^{|S - (S \cap I_0)|} \tag{3.1.1}$$

since each term $a_{S,\sigma}$ of the sum defining c_S is a product of $|S|$ factors $a_{i\sigma(i)}$, all of them of norm $\leq |\phi|$, and those with $\sigma(i) \notin I_0$ of norm $\leq \epsilon$.

Let $n \geq 1$ be an integer and consider subsets S of I of cardinality n. Let $\epsilon > 0$, and assume also $\epsilon < 1$. By Eq. (3.1.1) we have $|c_S| < \epsilon \max(1, |\phi|)^{n-1}$ for every S not contained in I_0, hence for every S but a finite number. In other words, the sequence $(|c_S|)_{S \subset I, |S|=n}$ converges to 0, and therefore the series $\sum_{S \subset I, |S|=n} c_S$ converges. One sets

$$c_n = \sum_{S \subset I, |S|=n} c_S \text{ and } c_0 = 1.$$

Lemma 3.1.18 *For every real $C > 0$, the sequence $(|c_n|C^n)_{n \in \mathbb{N}}$ converges to* 0.

Proof Let $\epsilon = \min(1/(2C), 1)$ and I_0 finite such that $|a_{i,j}| \leq \epsilon$ if $j \notin I_0$. Let $n > |I_0|$ be an integer. By (3.1.1) we have $|c_n|C^n < C^n \max(1, |\phi|)^{|I_0|} \epsilon^{n-|I_0|} \leq D/2^n$ where $D = \max(1, |\phi|)^{|I_0|} \epsilon^{-|I_0|}$ is independent of n. So $|c_n|C^n < \epsilon$ for n large enough. □

Lemma 3.1.19 *Let ϕ' be another compact operator of M, and let c'_n be defined for ϕ' as we defined c_n for ϕ. Then for $n \geq 1$, $|c_n - c'_n| < \max(|\phi|, |\phi'|)^{n-1} |\phi - \phi'|$.*

Proof Let $(a'_{i,j})$ be the matrix of ϕ'. One has the easy estimate $|a_{S,\sigma} - a'_{S,\sigma}| \leq \max(|\phi|, |\phi'|)^{|S|-1} |\phi - \phi'|$, hence $|c_S - c_{S'}| \leq \max(|\phi|, |\phi'|)^{|S|-1} |\phi - \phi'|$ and $|c_n - c'_n| \leq \max(|\phi|, |\phi'|)^{n-1} |\phi - \phi'|$. The results follows. □

Definition 3.1.20 We denote by $R\{\{T\}\}$ the ring of power series $\sum_{n=0}^{\infty} a_n T^n$ with $a_n \in R$ that *converge everywhere*, in the sense that $|a_n|C^n \to 0$ for every positive real C.

Lemma 3.1.18 says that the series $\sum_{n=0}^{\infty} c_n T^n$ belongs to $R\{\{T\}\}$. We call it the *Fredholm's determinant* of ϕ and denote it by $\det(1 - T\phi)$. Lemma 3.1.19 says that the assignment $\phi \mapsto \det(1 - T\phi)$ is continuous if we provide the set of compact operators with the operator norm and $R\{\{T\}\}$ with the norm of uniform convergence of coefficients (that is the norm $|\sum a_n T^n| = \sup_n |a_n|$).

Note however that at this point, the Fredholm's determinant of ϕ seems to depend on the choice of the orthonormal basis (e_i). We shall soon show (Corollary 3.1.22) that it does not, and which is more that it depends only on the equivalence class of the norm $|\ |$ on M.

If $\phi \in \mathrm{End}_R(M)$ is such that $\phi(M)$ is contained in a finite free sub-module P of M, then we can give another, purely algebraic definition of the power series $\det(1 - T\phi)$, which in this case is a polynomial that we shall denote by $\det(1 - T\phi)_P$: we consider the restriction of ϕ to P, which is an endomorphism of the finite free module P, and define $\det(1 - T\phi)_P$ as the determinant of $(1 - T\phi_{|P})$, which is a polynomial of degree the rank of P. It is not clear at this point that $\det(1 - T\phi)_P$ is independent of P, or coincide with the power series $\det(1 - T\phi)$ constructed above using the basis (e_i).

However, when S is a finite subset of I, and P_S the finite free submodule of I generated by the e_s, $s \in S$, and ϕ has image in P_S, then

$$\det(1 - T\phi)_{P_S} = \det(1 - T\phi) \tag{3.1.2}$$

as is seen immediately by extending the determinant of $\det(1 - T\phi_{|P_S})$ and comparing with the definition of $\det(1 - T\phi)$.

Let now N be an orthonormalizable R-module with basis $(f_j)_{j\in J}$. If $\phi \in \mathrm{Hom}_R(M, N)$ and $\phi' \in \mathrm{Hom}_R(N, M)$ are two operators, $S \subset J$ and $S' \subset I$ two finite sets such that $\phi(M) \subset P_S$ and $\phi'(N) \subset P_{S'}$, then

$$\det(1 - T\phi\phi') = \det(1 - T\phi'\phi) \tag{3.1.3}$$

because the equation reduces by (3.1.2) to $\det(1 - T\phi\phi')_{P_{S'}} = \det(1 - T\phi'\phi)_{P_S}$ and is thus a well-known property of determinants of endomorphisms of finite free modules.

Proposition 3.1.21 *Let M, N be two orthonormalizable Banach R-modules with bases $(e_i)_{i\in I}$ and $(f_j)_{j\in J}$ respectively. Let $\phi \in \mathrm{Hom}_R(M, N)$ be a compact operator and let $u \in \mathrm{Hom}_R(N, M)$. Then $\det(1 - T\phi u) = \det(1 - Tu\phi)$. (Here $\det(1 - T\phi u)$ is computed in the basis (f_j) and $\det(1 - Tu\phi)$ in the basis (e_i).)*

Proof Note that both members make sense, since both $u\phi \in \mathrm{End}_R(M)$ and $\phi u \in \mathrm{End}_R(N)$ are compact by Lemma 3.1.4.

Since $(\pi_S\phi)_{S\subset J,\ S\text{ finite}}$ converges in norm to ϕ (Scholium 3.1.14), and therefore also $(u\pi_S\phi)_S$ to $u\phi$ and $(\pi_S\phi u)_S$ to ϕu, it is enough by Lemma 3.1.19 to prove the result for ϕ replaced by $\pi_S\phi$ where $S \subset J$ is a finite set. In other words, we may and will assume that ϕ has image contained in P_S.

Now let S' be a finite subset of I. The operator $\pi_{S'}u$ has image contained in $P_{S'}$, hence one has by (3.1.3)

$$\det(1 - T\pi_{S'}u\phi) = \det(1 - T\phi\pi_{S'}u).$$

Since $u\phi$ is compact, the sequence $(\pi_{S'}u\phi)_{S'}$ converges to $u\phi$ (Scholium 3.1.14), hence the left hand side $\det(1-T\pi_{S'}u\phi)$ converges to $\det(1-Tu\phi)$ (Lemma 3.1.19). As for the right hand side, it is equal by (3.1.2) to the algebraic determinant $\det(1 - T\phi\pi_{S'}u_{|P_S})$ since ϕ, and therefore $\phi\pi_{S'}u$, have image contained in P_S. But $\pi_{S'}u_{|P_S}$ converge in norm to $u_{|P_S}$, hence $\det(1-T\phi\pi_{S'}u_{|P_S})$ converges to $\det(1-T\phi u_{|P_S}) = \det(1 - T\phi u)$

The proposition follows. □

Corollary 3.1.22 *If $\phi \in \mathrm{End}_R(M)$ is compact, the Fredholm's determinant $\det(1 - T\phi)$ of ϕ depends only on ϕ and on the topological module M, and not on the orthonormal basis (e_i) nor on the norm defining the topology on M.*

Proof We need to prove that if (f_i) is an orthonormal basis for some norm $|\ |'$ equivalent to $|\ |$, and $u \in \mathrm{Hom}_R(M, M)$ is the continuous invertible map sending

(e_i) to (f_i), then $\det(1-T\phi) = \det(1-Tu\phi u^{-1})$. But this follows from Proposition 3.1.21 applied to the continuous operator u and the compact operator ϕu^{-1}. □

3.1.6 Property (Pr)

A Banach module P over R is said to have *property* (Pr) if there exists a Banach module Q such that $P \oplus Q$ is potentially orthonormalizable.

Exercise 3.1.23 Show that a Banach module P has property (Pr) if and only if for every continuous surjection $f : M \to N$, and every continuous map $\alpha : P \to N$ there exists a continuous map $\beta : P \to M$ such that $f\beta = \alpha$.

Proposition 3.1.24 *If a finite R-module has property (Pr), then it is projective.*

Proof If P is a finite R-module, choose a surjective continuous map $f : A^r \to P$ and apply Exercise 3.1.23 to $\alpha = \mathrm{Id}_P : P \to P$. We get $\beta : P \to A^r$ such that $f\beta = \mathrm{Id}_P$, hence $P \simeq \beta(P)$ is a direct summand of A^r and is therefore projective. □

If P have property (Pr), and $\phi \in \mathrm{End}_R(P)$ is compact, one can define $\det(1-T\phi)$ as follows: one chooses Q such that $P \oplus Q$ is potentially orthonormalizable, defines $\tilde{\phi} \in \mathrm{End}_R(P \oplus Q)$ by extending ϕ by zero, and sets $\det(1-T\phi) = \det(1-T\tilde{\phi})$. We need to check that if we have chosen two modules Q and Q' such that $P \oplus Q$ and $P \oplus Q'$ are potentially orthonormalizable, the resulting Fredholm's determinant is the same, but this is true because both are equal to the determinant of the extension by zero of ϕ on $P \oplus Q \oplus Q'$, by Proposition 3.1.21.

It is easy to check that Proposition 3.1.21 is still true for compact morphisms between modules having the property (Pr).

Proposition 3.1.25 *Assume that R is noetherian. Let P be an R-module with property (Pr) and ϕ a compact operator on P such that $1 - \phi$ is nilpotent. Then P is finite and projective.*

Proof If $(1 - \phi)^n = 0$ on P, then by expanding we see that the identity is compact on P. Let Q be such that $M := P \oplus Q$ is potentially orthonormalizable, and let p denote the projector of M onto P alongside Q. Let $u \in \mathrm{End}_R(M)$ be the extension by 0 of the identity of P. Since u is compact, by Scholium 3.1.14 there exists a projector q in $\mathrm{End}_R(M)$ whose image is finite free and such that $|u - qu| < 1/|p|$. Then $pu_{|P} = \mathrm{Id}_P$ and $|pu - pqu| < 1$ so $|\mathrm{Id}_p - pqu_{|P}| < 1$, which shows that the morphism $pqu_{|P} = pq : P \to P$ is invertible. Hence the image $pq(P)$ of that morphism is P. Therefore P is a submodule of $p(q(M))$ which is finite since $q(M)$ is, and P itself is finite. By Proposition 3.1.24, it is also projective. □

3.1.7 Extension of Scalars

Let us begin by recalling the definition of the *completed tensor product* of two Banach R-modules M and N. On $M \otimes_R N$ we define a norm $|z| = \inf \max_{i=1,\dots,r}(|m_i||n_i|)$, where the inf is being taken over all writings $z = \sum_{i=1}^r m_i \otimes n_i$ with $m_i \in M$, $n_i \in N$. The completed tensor product, denoted by $M \hat{\otimes}_R N$ is the completion of $M \otimes_R N$ for this norm. Let us call j the continuous bilinear map $M \times N \to M \hat{\otimes}_R N$ sending (m, n) to $m \otimes n$. Given a Banach R-module P together with a continuous R-bilinear map $f : M \times N \to P$, one can define a continuous map $\tilde{f} : M \hat{\otimes}_R N \to P$ by extending by continuity the map $M \otimes_R N \to P$, $\sum_{i=1}^r m_i \otimes n_i \mapsto \sum_{i=1}^r f(m_i, n_i)$. It is clear that $\tilde{f}$ satisfies $\tilde{f} \circ j = f$ and is the unique continuous map satisfying this; in other words $(M \hat{\otimes}_R N, j)$ is the initial object of the category whose objects are pairs (P, f) where P is a Banach R-module and $f : M \times N \to P$ a continuous R-bilinear map, and morphisms are obviously defined. Moreover one sees on the definition that $|\tilde{f}| \le |f|$.

Consider now a morphism of Banach algebras $R \to R'$. Then for M a Banach module over R, $M \hat{\otimes}_R R'$ is a Banach module over R'.

Lemma 3.1.26 *If $(e_i)_{i \in I}$ is an ON-basis for the Banach R-module M, then $(e_i \otimes 1)_{i \in I}$ is an orthonormal basis of $M \hat{\otimes}_R R'$ up to replacing the norm of that module by an equivalent one.*

Proof Consider the R'-Banach module $c_I(R')$ of Example 3.1.7 and let $j : M \to c_I(R')$ be the continuous R-linear map sending $\sum_{i \in I} a_i e_i$ to $(a_i)_{i \in I}$. Let P be a Banach R-module and $f : M \times R' \to P$ be a continuous bilinear map. Define a map $\tilde{f} : c_I(R') \to P$ by setting $\tilde{f}((r_i')) = \sum f(e_i, r_i')$ (since $|e_i| = 1$, $|r_i'|$ goes to 0, and f is bounded, the sum converges). Then $\tilde{f}$ is a continuous map satisfying $\tilde{f} \circ j' = f$, and is clearly the unique such map. By the universal property, we see that $\tilde{f}$ is an isomorphism $c_I(R') \to M \hat{\otimes} R'$; since $\tilde{f}$ sends the canonical basis of $c_I(R')$ to the family $(e_i \otimes 1)_{i \in I}$ the results follows. □

This results implies that if the R-module M is potentially orthonormalizable, then so is $M \hat{\otimes}_R R'$. The same assertion with potentially orthonormalizable replaced by (Pr) immediately follows.

If $\phi : M \to N$ is a continuous map of Banach R-modules, then by the universal property one can extend uniquely to a continuous morphism of R'-modules $\phi \otimes 1 : M \hat{\otimes}_R R' \to N \hat{\otimes}_R R'$.

Lemma 3.1.27 *Assume that M and N have property (Pr). If ϕ is compact, then so is ϕ', and if $M = N$,* $\det(1 - T\phi')$ *is the image in* $R'\{\{T\}\}$ *of* $\det(1 - T\phi)$.

Proof Let (e_i) and (f_j) be orthonormal bases of M and N. Up to changing the norms without changing the topology (with does not affect the property of being compact nor, by Corollary 3.1.22, the Fredholm's determinant) in $M \otimes_R R'$ and $N \otimes_R R'$, $(e_j \otimes 1)$ and $(f_j \otimes 1)$ are orthonormal bases of those modules. The matrix of $\phi \otimes 1$ in these bases is obviously the same as the matrix of ϕ in the bases (e_i) and (f_j). So $\phi \otimes 1$ is compact by Proposition 3.1.13, and, when $M = N$, $\det(1 - T\phi) = \det(1 - T\phi \otimes 1)$ since those Fredholm determinants are defined using the matrices of ϕ and $\phi \otimes 1$, which are the same. □

Finally, let us prove a result of Serre (cf. [108, Corollary, page 74]) that we shall need in Chap. 6.

Lemma 3.1.28 *Let M and N be two Banach space over $\mathbb{Q}_p$. Then the image of $\mathrm{Hom}_{\mathbb{Q}_p}(M, \mathbb{Q}_p)\hat{\otimes}_{\mathbb{Q}_p} N$ in $\mathrm{Hom}_{\mathbb{Q}_p}(M, N)$ is the subspace of compact operators from M to N.*

Proof The modules M and N are potentially orthonormalizable by Theorem 3.1.17, and changing the norms we can assume that they are orthonormalizable, of orthonormal bases $(e_i)_{i\in I}$ and $(f_j)_{j\in J}$. An element ϕ of $\mathrm{Hom}_{\mathbb{Q}_p}(M, \mathbb{Q}_p)\otimes_{\mathbb{Q}_p} N$ has the form $\phi = \sum_{k=1}^{1} l_k \otimes n_k$, which, considered as an element of $\mathrm{Hom}_{\mathbb{Q}_p}(M, N)$, sends e_i to $\phi(e_i) = \sum_k l_k(e_i) n_k$, whose coordinates in the basis (f_j) are $(\sum_k l_k(e_i) n_{k,j})_{j\in J}$ where $n_k = \sum_{j\in J} n_{k,j} f_j$. One has for every $i \in I$:

$$|\sum_k l_k(e_i) n_{k,j}| < \max_{k\in\{1,\dots,r\}} (|l_k|) \max_{k\in\{1,\dots,r\}} |n_{k,j}|,$$

and the right hand side is independent of i and tends to 0. Hence the sequences $(\sum_k l_k(e_i) n_{k,j})_{j\in J}$ tends to 0 uniformly in i, which means that ϕ is compact by Proposition 3.1.13. Since the subspace of compact operators is closed, it follows that $\mathrm{Hom}_{\mathbb{Q}_p}(M, \mathbb{Q}_p)\hat{\otimes}_{\mathbb{Q}_p} N$ consists of compact operators in $\mathrm{Hom}_R(M, N)$. The converse, that a compact operator belongs to $\mathrm{Hom}_{\mathbb{Q}_p}(M, \mathbb{Q}_p)\hat{\otimes}_{\mathbb{Q}_p} N$ is similar and left to the reader. □

3.2 Everywhere Convergent Formal Series and Riesz's Theory

We keep the running hypothesis of the preceding section: R is a $\mathbb{Q}_p$-Banach algebra with a norm $|\ |$ extending the absolute value of $\mathbb{Q}_p$. However, instead of the multiplicative norm on R, it will be more convenient from now on to work with its additive counterpart, the valuation v_p on R defined as $v_p(f) = -\log|f|/\log p$ (in particular $v_p(p) = 1$). One has $v_p(fg) \geq v_p(f) + v_p(g)$ but this inequality may be strict; in particular, if $f \in R^*$, $v_p(f) + v_p(f^{-1})$ may be non-zero. When it is 0, f is *multiplicative* and $v_p(fg) = v_p(f) + v_p(g)$ for every $g \in R$, cf. Exercise 3.1.1.

We recall the ring $R\{\{T\}\}$ of everywhere convergent formal power series, introduced in the preceding section, cf. Definition 3.1.20. In the language of the valuation v_p, this is the ring of power series $\sum_{n=0}^{\infty} a_n T^n$, with $a_n \in R$, such that $v_p(a_n) - n\nu \to \infty$ for every $\nu \in \mathbb{R}$.

3.2.1 The ν-Valuation and ν-Dominant Polynomials

Let $\nu \in \mathbb{R}$. For $F = \sum_{n=0}^{\infty} a_n T^n \in R[[T]]$ we set $v(F, \nu) = \inf_{n\in\mathbb{N}} v_p(a_n) - n\nu$. It is clear that $F \in R\{\{T\}\}$ is equivalent to having $v(F, \nu) > -\infty$ for all $\nu \in \mathbb{R}$; it is even sufficient to check it for a family of ν's that tends to $+\infty$. For $F, G \in R\{\{T\}\}$, one has $v(F + G, \nu) \geq \min(v(F, \nu), v(G, \nu))$ and $v(FG, \nu) \geq v(F, \nu) + v(G, \nu)$. In other words, $v(\cdot, \nu)$ defines a valuation on $R\{\{T\}\}$ which in turn defines a uniform topology on $R\{\{T\}\}$.

Exercise 3.2.1 Show that the topologies defined by $v(\cdot, \nu)$ and $v(\cdot, \nu')$ are different if $\nu \neq \nu'$.

Exercise 3.2.2 Let F_n be a sequence of power series in $R\{\{T\}\}$. Assume that for every $\nu \in \mathbb{R}$, F_n is Cauchy for the topology defined by $v(\cdot, \nu)$. Then show that there exists an F in $R\{\{T\}\}$ such that for every $\nu \in \mathbb{R}$, (F_n) converges to F for the topology defined by $v(\cdot, \nu)$.

Definition 3.2.3 Let $0 \neq F(T) = \sum_{n=0}^{\infty} a_n T^n \in R\{\{T\}\}$, and $\nu \in \mathbb{R}$. We write $N(F, \nu)$ for the largest integer N such that $v_p(a_N) - N\nu = \inf_{n\in\mathbb{N}}(v_p(a_n) - n\nu)$.

In other words, $N(F, \nu)$ is the unique integer N which has the following property: $v_p(a_n) - n\nu \geq v_p(a_N) - N\nu$ *for all n, with strict inequality whenever* $n > N$. Note that by definition $a_{N(F,\nu)} \neq 0$.

Definition 3.2.4 A polynomial $Q(T) \in R[T]$ is called ν-*dominant* if

(i) Q has degree $N(Q, \nu)$
(ii) The dominant term of Q is invertible.

Lemma 3.2.5 *If $Q(T) \in R[T]$ is a polynomial whose dominant term is invertible, there exists $\nu_0 \in \mathbb{R}$ such that for all $\nu \geq \nu_0$, Q is ν-dominant.*

Proof Write $Q(X) = \sum_{n=0}^{d} a_d T^d$ with a_d invertible. Let $\nu_0 = \max(0, v_p(a_d) - v_p(a_{d-1}), \ldots, v_p(a_d) - v_p(a_0))$. Then if $\nu \geq \nu_0$ and $1 \leq i \leq d$, one has $v_p(a_d) - d\nu = v_p(a_d) - (d-i)\nu - i\nu \leq v_p(a_{d-i}) + \nu_0 - (d-i)\nu - i\nu \leq v_p(a_{d-i}) - (d-i)\nu$ since $\nu_0 - i\nu \leq \nu_0 - \nu \leq 0$. This shows that Q is ν-dominant. □

To get more familiar with the notion of ν-dominance and the invariant $N(F, \nu)$, we study it briefly in the case where $R = L$ is a finite extension of $\mathbb{Q}_p$ provided with its unique valuation extending that of $\mathbb{Q}_p$.

Exercise 3.2.6 Prove Gauss's Lemma: If $F, G \in L\{\{T\}\}$, $N(FG, \nu) = N(F, \nu) + N(G, \nu)$.

Lemma 3.2.7 *Let $F(T) = \sum_n a_n T^n \in L\{\{T\}\}$.*

(i) *If $N(F, \nu) = 0$, then $F(T)$ has no zero $z \in \bar{L}$ with $v_p(z) \geq -\nu$.*
(ii) *If F is a polynomial, F is ν-dominant if and only if all zeros z of F in $\bar{L}$ satisfy $v_p(z) \geq -\nu$.*

Proof For (i), if $v_p(z) \geq -\nu$ and $F(z) = 0$ one has $a_0 = -\sum_{n\geq 1} a_n z^n$, and each term in the RHS is of valuation $v_p(a_n z^n) \geq v_p(a_n) - n\nu > v_p(a_0)$ since $N(F, \nu) = 0$; this is a contradiction.

For (ii), assume F is ν-dominant. If $F(z) = 0$, then the p-valuation of the dominant term of $F(z)$ must be at least as large as the p-valuation of some other term, that is $v_p(a_N z^N) \geq v_p(a_n z^n)$, for some $n < N$, or

$$v_p(a_N) + N v_p(z) \geq v_p(a_n) + n v_p(z).$$

Since $N = N(F, \nu)$, one has

$$v_p(a_n) - n\nu \geq v_p(a_N) - N\nu.$$

Adding these two inequalities and simplifying gives $v_p(z) \geq -\nu$. The converse is easy. □

3.2.2 *Euclidean Division*

Proposition 3.2.8 *Let $F \in R\{\{T\}\}$ and let $B \in R[T]$ a non-zero polynomial whose dominant term is invertible in R. Then there exists a unique $Q \in R\{\{T\}\}$ and $S \in R[T]$ with $\deg S < \deg B$ such that $F = BQ + S$. Furthermore we have, for any $\nu \in \mathbb{R}$, $v(S, \nu) \geq v(F, \nu)$, and if the dominant term of B is multiplicative, $v(Q, \nu) \geq v(F, \nu) - v(B, \nu)$.*

Proof We first observe that it is enough to prove the proposition when B is monic. Indeed, assume the theorem is known for B monic, and let $b \in R^*$ be the dominant term of B, so that $B = bB_0$ with B_0 monic. Then we have a decomposition $F = B_0 Q_0 + S$ with $Q_0 \in R\{\{T\}\}$ and $S \in R[T]$ with $\deg S < \deg B_0$ and $v(S, \nu) \geq v(F, \nu)$, $v(Q_0, \nu) \geq v(F, \nu) - v(B_0, \nu)$. Then setting $Q = b^{-1} Q_0$ we have $F = BQ + S$ and this decompositions satisfies all the stated conditions, namely $\deg S < \deg Q$, $v(S, \nu) \geq v(F, \nu)$ and if b is multiplicative, $v(Q, \nu) \geq v(F, \nu) - v(B, \nu)$. Moreover the uniqueness of such a decomposition follows immediately from the uniqueness in the case where B is monic.

So let us assume henceforth that B is monic.

For uniqueness we need to show that if $BQ + S = 0$ with $\deg S < \deg B$, then $Q = 0$. By Lemma 3.2.5, B is ν-dominant for some $\nu \in \mathbb{R}$, $\nu \geq 0$, so we can write $B = T^d + b_{d-1}T^{d-1} + \cdots + b_0$ with $v_p(b_i) - i\nu \geq -d\nu$ for $i \leq d$. If $Q = \sum q_n T^n$ is non-zero, let $N = N(Q, \nu)$ so that $v_p(q_n) - n\nu > v_p(q_N) - N\nu$ if $n > N$. Then the coefficient of T^{N+d} in BQ is $q_N + \sum_{i=1}^d q_{N+i} b_{d-i}$, and for $1 \leq i \leq d$, $v_p(q_{N+i} b_{d-i}) \geq v_p(q_{N+i}) + v_p(b_{d-i}) > v_p(q_N) + i\nu + (-i)\nu = v_p(q_N)$, hence the coefficient of T^{N+d} in BQ is non-zero, and $BQ + S \neq 0$, a contradiction.

We now make an observation crucial to the proof of existence. Recall that when F is a polynomial, then the existence of polynomials Q and S such that $F = BQ + S$,

$\deg S < \deg B$ is well-known. We claim that in this case, if B is monic, then for any $\nu \in \mathbb{R}$,

$$v(Q, \nu) \geq v(F, \nu) - v(B, \nu), \;\; v(S, \nu) \geq v(F, \nu). \tag{3.2.1}$$

We prove this by induction on $n = \deg F$. For $n < \deg B$, one has $Q = 0$, $S = F$ and the result is clear. So let $n \geq \deg B$ and assume the result is true for all polynomials of degree $< n$. Let $a_n T^n$ be the dominant term of F. Then by induction we can write $F - a_n T^n = BQ_0 + S_0$ with $\deg S_0 < \deg B$ and Q_0, S_0 satisfying (3.2.1), that is $v(Q_0, \nu) \geq v(F - a_n T^n, \nu) - v(B, \nu)$, $v(S_0, \nu) \geq v(F - a_n T^n, \nu)$. Since $v(F - a_n T^n, \nu) \geq v(F, \nu)$, one has:

$$v(Q_0, \nu) \geq v(F, \nu) - v(B, \nu), \;\; v(S_0, \nu) \geq v(F, \nu). \tag{3.2.2}$$

Let T^d be the dominant term of B. We can write $F = B(Q_0 + a_n T^{n-d}) + S_0 - (B - T^d)a_n T^{n-d}$. The polynomial $-(B - T^d)a_n T^{n-d}$ has degree $< n$, so can also be written $BQ_1 + S_1$, with $\deg S_1 < \deg B$ and Q_1, S_1 satisfying (3.2.1), that is $v(Q_1, \nu) \geq v((B - T^d)a_n T^{n-d}, \nu) - v(B, \nu)$, $v(S_1, \nu) \geq v((B - T^d)a_n T^{n-d}, \nu)$. But

$$v(a_n T^{n-d}, \nu) = v_p(a_n) - (n - d)\nu \geq v(F, \nu) - v(B, \nu), \tag{3.2.3}$$

and therefore

$$v((B - T^d)a_n T^{n-d}, \nu) \geq v(B - T^d, \nu) + v(a_n T^{n-d}, \nu) \geq v(B_\nu) + v(F, \nu) - v(B, \nu) = v(F, \nu).$$

Thus

$$v(Q_1, \nu) \geq v(F, \nu) - v(B, \nu), \;\; v(S_1, \nu) \geq v(F, \nu). \tag{3.2.4}$$

Hence $F = B(Q_0 + Q_1 + a_n T^{n-d}) + S_0 + S_1$, and by uniqueness $Q = (Q_0 + Q_1 + a_n T^{n-d})$ and $S = S_0 + S_1$. Now S satisfies (3.2.1) because S_0 and S_1 satisfy (3.2.2) and (3.2.4) respectively, and Q satisfies (3.2.1) by (3.2.2), (3.2.4) and (3.2.3). This completes the induction step and the proof of (3.2.1).

Now we prove the existence. Let $F = \sum_{n=0}^{\infty} a_n T^n \in R\{\{T\}\}$. Write $a_n T^n = BQ_n + S_n$ with $Q_n, S_n \in R[T]$, $\deg S_n < \deg B$. Let $\nu \in \mathbb{R}$. Since $v(a_n T^n, \nu)$ goes to 0, we see by (3.2.1) and Exercise 3.2.2 that the series $\sum_{n=0}^{\infty} Q_n$ and $\sum_{n=0}^{\infty} S_n$ converge respectively to $Q \in R\{\{T\}\}$ and $S \in R\{\{T\}\}$ for the topology defined by $v(\cdot, \nu)$ for any $\nu \in \mathbb{R}$. Actually S is a polynomial of degree $< \deg B$ since the S_n are. Since $F = BQ + S$, the proposition is proved. □

Corollary 3.2.9 *Let $B \in R[T]$ be a non-zero polynomial whose dominant term is invertible in R. The natural morphism of R-algebras $R[T]/(B(T)) \to R\{\{T\}\}/(B(T))$ is an isomorphism.*

Concerning uniqueness our proof gives a little more that what is stated:

Scholium 3.2.10 *Let $F \in R\{\{T\}\}$ and $B \in R[T]$ a polynomial which is ν-dominant for some ν. Then there exists a unique $Q \in R[[T]]$ and $S \in R[T]$ such that $F = BQ + S$ and $v(Q, \nu) > -\infty$.*

3.2.3 Good Zeros

Definition 3.2.11 Let $F \in R\{\{T\}\}$ such that $F(0) = 1$. Let $a \in R$ and $s \in \mathbb{N}$. We say that a is a *good zero of order* s of F if $F(a) = F'(a) = F^{(2)}(a) = \cdots = F^{(s-1)}(a) = 0$ and $F^{(s)}(a)$ is a unit in R.

Exercise 3.2.12 If $F(0) = 1$ and $F(a) = 0$, then a is a unit in R.

If a is a good zero of F of order $s \geq 1$, we can write $F(T) = (1 - a^{-1}T)G_1(T)$ with $G_1(T) \in R\{\{T\}\}$ by Proposition 3.2.8 and we see that a is a good zero of G_1 of order $s-1$. By induction, we can write $F = (1-a^{-1}T)^s G(T)$ with $G(T) \in R\{\{T\}\}$ and $G(a)$ a unit in R.

3.2.4 A Piece of Resultant Theory

Until further notice, R may be any commutative ring.

Following Coleman, we define, if $B \in R[T]$ and $P(T) = 1 - a_1 T + a_2 T^2 + \cdots + (-1)^n a_n T^n \in R[T]$ is of degree n, a polynomial $D(B, P)$ in $R[T]$ as follows: let $t_1, \ldots, t_n$ be n new variables over $R[T]$ and let $s_1, \ldots, s_n$ the symmetric functions in those variables (i.e. $s_1 = t_1 + \cdots + t_n$, $s_n = t_1 \ldots t_n$, etc.) as elements of $R[T][t_1, \ldots, t_n]$. The polynomial $\prod_{i=1}^n (1 - TB(t_i))$ is symmetric in the t_i, hence lies in $R[T][s_n, \ldots, s_n]$; we denote by $D(B, P)$ the image of that product in $R[T]$ by the morphism of $R[T]$-algebras $R[T][s_1, \ldots, s_n] \to R[T]$ sending s_i to a_i. With a slight abuse of notation, we can write like Coleman:

$$D(B, P)(T) = \prod_{i=1}^{n} (1 - TB(t_i)), \text{ where } s_i(t_1, \ldots, t_n) = a_i,\ i = 1, \ldots, n.$$

If R is embedded in an algebraically closed field K, and $t_1, \ldots, t_n$ are the roots of the reciprocal polynomial $P^*(T)$ in K (with multiplicity), then the formula $D(B, P)(T) = \prod_{i=1}^n (1 - TB(t_i))$ is literaly true in K.

The value $D(B, P)(1)$ is essentially a resultant (see [81, §IV.8] for the definition and basic properties of the resultant.):

Lemma 3.2.13 *Let Q be a polynomial such that $Q(0) = 0$, and P a monic polynomial. Then $D(1 - P^*, Q)(1) = Res(Q, P)$.*

Proof Let q_j be the roots of Q^*. Note that Q^* has dominant term 1 hence $Q^*(T) = \prod_j (T - q_j)$. Since Q is monic, one has $\prod_j q_j = (-1)^m Q(0)^{-1}$ and since P^* is monic, one has $\prod_i t_i = P^*(0)$. Then $D(1 - Q^*, P)(1) = \prod_i (1 - (1 - Q^*(t_i)) = Q(0)^n \prod_i Q^*(t_i) = \prod_{i,j}(q_j - t_i) = Q(0)^n (\prod_i t_i)^m (\prod_j q_j)^n \prod_{i,j}(q_i^{-1} - t_j^{-1}) = (-1)^{nm} P^*(0)^m \prod_{i,j}(q_i^{-1} - t_i^{-1})$. Since the q_j^{-1}'s are the roots of Q and the p_i^{-1} are the roots of P, one has

$$D(1 - Q^*, P)(1) = (-1)^{mn}\mathrm{Res}(Q, P) = \mathrm{Res}(P, Q).$$

(See [81, Prop. IV.8.3]) □

Lemma 3.2.14 *Let $B, C, P, Q \in R[T]$ with $P(0) = Q(0) = 1$ and $C(0) = 0$. We have:*

(i) $D(B, PQ) = D(B, P)D(B, Q)$.
(ii) $D(1 - P^*, P)(T) = (1 - T)^n$.

Proof Note that if $f : R' \to R$ is a morphism of rings, then $D(f(B), f(P)) = f(D(B, P))$. Clearly it is enough to prove the formula for a ring of polynomials $R' = \mathbb{Z}[x_1, \ldots, x_m]$ with sufficiently many variables, because then one can define a map f' to f such that B and P in $R[T]$ are in the image by f of polynomials B' and P' of the same degree in $B'[T]$. This ring R' is a domain and it is enough to prove equations (i) and (ii) or an algebraically closed field K containing R'. Thus we can assume that R is an algebraically closed field K.

In this case, formula (i) is obvious, because the set of roots of $P^*Q^* = (PQ)^*$ (with multiplicity) is the union of the sets of roots of P^* and of Q^*. Formula (ii) is clear too since $D(1 - P^*, P)(T) = \prod_{i=1}^n (1 - T(1 - P^*(t_i))) = (1 - T)^n$. □

We now are back to our running hypothesis that R is a Banach algebra over $\mathbb{Q}_p$.

For B fixed, the coefficients of $D(B, P)$ are polynomials in the coefficients a_i of P. They are therefore continuous in P for any of the valuations $v(P, \nu)$. This allows us, if $P = \sum_{i=0}^{\infty} a_i T^i \in R\{\{T\}\}$ is such that $P(0) = 1$, to define $D(B, P) = \lim_{n\to\infty} D(B, P_n)$ with P_n the polynomial $\sum_{i=0}^{n} a_i T^i$, and to check that the formal series $D(B, P)$ belongs to $R\{\{T\}\}$—see Exercise 3.2.2.

The motivation to introduce $D(B, P)$ is given by the

Proposition 3.2.15 *Let M be an R-module satisfying property (Pr) and ϕ be a compact operator of M. Let $Q(T)$ be a polynomial with $Q(0) = 0$. Then $D(Q, \det(1 - T\phi)) = \det(1 - TQ(\phi))$.*

Proof Since $Q(0) = 0$, $Q(\phi)$ is compact and the right hand side of the equality makes sense. To prove it, we are reduced as usual to the case where M is orthonormalizable, and then to the case where ϕ is of finite rank. In this case, the formula is purely algebraic and we shall prove it for any commutative ring R. Actually, standard argument shows that is its enough to prove it when R is an algebraic closed field K. In this case, we can write $\det(1 - T\phi) = (1 - a_1 T)\ldots(1 - a_n T)$ with $a_1, \ldots, a_n \in K^*$ being the eigenvalues of ϕ (with multiplicity counted according to the characteristic polynomial). By definition of D, one has $D(Q, \det(1 - T\phi)) = (1 - Q(a_1)) \dots (1 - Q(a_n))$ which is just $\det(1 - TQ(\phi))$. □

Lemma 3.2.16 *Let $B, Q \in R[T]$, $S \in R\{\{T\}\}$ with $S(0) = Q(0) = 1$.*

(i) $D(B, QS) = D(B, Q)D(B, S)$.
(ii) *If Q and S generate the unit ideal in $R\{\{T\}\}$, then $D(1 - Q^*, S)(1) \in R^*$.*

Proof Assertion (i) follows directly from Lemma 3.2.14(i) by going to the limit. For assertion (ii), let $A, C \in R\{\{T\}\}$ such that $CQ + AS = 1$. Then by passing in the limit in Lemma 3.2.13, we have $D(1 - Q^*, S)(1) = \mathrm{Res}(S, Q)$ where Res has also been extend to convergent power series. Hence we have $1 = D(1 - Q^*, 1) = D(1 - Q^*, CQ + AS) = \mathrm{Res}(CQ + AS, Q) = \mathrm{Res}(AS, Q)$, the latest formule $\mathrm{Res}(CQ+AS, Q) = \mathrm{Res}(AS, Q)$ being well-known in the case of polynomial (see the wikipedia page on *resultant*, or prove it using the definition as a determinant), and easily extended for power series. Thus $1 = \mathrm{Res}(A, Q)\mathrm{Res}(S, Q)$ and (ii) follows. □

Proposition 3.2.17 *Let $P \in R\{\{T\}\}$ with $P(0) = 1$, and suppose that one can write $P = QS$ with $Q \in R[T]$ such that $Q(0) = 1$, and $S \in R\{\{T\}\}$. Assume that the ideal generated by Q and S is the unit ideal. Let n be the degree of Q. Then $D(1 - Q^*, P)(T)$ has a good zero of order n at 1.*

Proof One has $D(1 - Q^*, P)(T) = D(1 - Q^*, Q)(T)D(1 - Q^*, S)(T) = (1 - T)^n D(1 - Q^*, S)(T)$ and $D(1 - Q^*, S)(1)$ is a unit. The proposition follows. □

3.2.5 Riesz's Theory

In this subsection we keep assuming that R is a Banach algebra over $\mathbb{Q}_p$ and we assume in addition that R is noetherian.

We now consider a Banach module M over R which has property (Pr). Let $\phi \in \mathrm{End}_R(M)$ be a compact operator. We denote by $P_\phi(T) \in R\{\{T\}\}$ the Fredholm's determinant of ϕ, $\det(1 - T\phi)$.

Proposition 3.2.18 *Assume that $a \in R$ is a good zero of order s of P_ϕ. Then there exists a unique decomposition $M = N \oplus F$ where N and F are closed ϕ-stable sub-modules of M, N is finite projective and F has property (Pr), and $1 - a\phi$ is nilpotent on N and invertible on F. Moreover, N and F are stable by every operator $u \in \mathrm{End}_R(M)$ that commutes with ϕ.*

Proof The result is trivial when $s = 0$, so let us assume that $s \geq 1$. Let R' be the closure of the subring of $\mathrm{End}_R(M)$ generated by ϕ, so R' is a commutative Banach algebra. Define a power series $Q(T)$ in $R'\{\{T\}\}$ by

$$(1 - T\phi)Q(T) = P_\phi(T). \tag{3.2.5}$$

Applying (3.2.5) to a gives $(1 - a\phi)Q(a) = P_\phi(a) = 0$. Deriving (3.2.5) i times for $i = 1, \ldots, s$ and evaluating at a gives

$$(1 - a\phi)Q^i(a) - i\phi Q^{i-1}(a) = P_\phi^{(i)}(a). \tag{3.2.6}$$

Since $P_\phi^{(i)}(a) = 0$ for $i < s$, we see by induction on i that $(1 - a\phi)^{i+1}Q^{(i)}(a) = 0$ for $i < s$. Since a is a good zero of order s of P_ϕ, $c := P_\phi^{(s)}(a)$ is invertible in R. Equation (3.2.6) for $i = s$ gives $(1 - a\phi)Q^s(a) - i\phi Q^{s-1}(a) = c$ which, if we set $e := c^{-1}(1-a\phi)Q^{(s)}(a)$ and $f := -i\phi Q^{s-1}(a)$ can be written as $e + f = 1$. Since $(1-t\phi)^s Q^{(s-1)}(a) = 0$, we have $e^s f = 0$. We define $p = e^s$, $q = (e+f)^s - e^s$, so we have again $p + q = 1$ and $pq = 0$ since q is divisible by f in $R' \subset \mathrm{End}_R(M)$. It follows that p and q are orthogonal projectors in R'. Let F be the image of p and N the image of q; they are closed submodules of M (as they are also respectively kernels of q and p), they are ϕ-stable (as p and q are in R' and thus commute with ϕ) and $M = N \oplus F$. Since $(1 - a\phi)^s q = 0$, $1 - a\phi$ is nilpotent on N. It follows (Proposition 3.1.25) that N is projective of finite type; the submodule F has (Pr) since it is a direct summand of M. Finally, by elevating the definition of e to the power s we get $(1 - a\phi)^s Q^{(s)}(a)^s = c^s p$ which shows that $1 - a\phi$ is invertible on $\mathrm{Im}\, p = F$. This proves the existence of the decomposition.

For the uniqueness, let $M = N' \oplus F'$ an other decomposition satisfying the required properties. Consider the closed ϕ-stable submodule $N'' = p(N + N')$ of M. On N'', ϕ is nilpotent since it is nilpotent on $N + N'$, and invertible since it is invertible on F. Thus $N'' = 0$ and $N = N'$. That $F = F'$ is proven the same way.

The last sentence is clear since we defined M and F has image of projectors that are power series in ϕ. □

Theorem 3.2.19 *Assume that there exists $Q \in R[T]$ a polynomial with invertible dominant coefficient, and $S \in R\{\{T\}\}$, with $P_\phi = QS$ and $Q(0) = S(0) = 1$. Assume furthermore that Q and S generate the unit ideal of $R\{\{T\}\}$. Then there exists a unique decomposition $M = N \oplus F$ where N and F are closed ϕ-stable sub-modules of M, N is finite projective and F has property (Pr), and $Q^*(\phi)$ is nilpotent on N and invertible on F. Moreover, N and F are stable by every operator $u \in \mathrm{End}_R(M)$ that commutes with ϕ.*

This theorem reduces to the preceding proposition in the case where Q has the form $(1 - aT)^s$, and we will reduce its proof to that case using resultant theory.

Proof Let $\phi' = 1 - Q^*(\phi)/Q^*(0)$. This is a compact operator whose Fredholm's determinant $P_{\phi'}$ satisfies $P_{\phi'} = D(1 - Q^*, P_\phi)$ by Proposition 3.2.15. By Proposition 3.2.17, $P_{\phi'}$ has a good zero of order $\deg Q$ at $T = 1$. The theorem then follows from Proposition 3.2.18 applied to ϕ' with $a = 1$. □

3.3 Adapted Pairs

In this section, we restrict somewhat the setting of the preceding section. We will assume that the Banach algebra R over $\mathbb{Q}_p$ is actually a reduced affinoid algebra, provided with its natural supremum norm extending the absolute value over $\mathbb{Q}_p$. We denote by $\operatorname{Sp} R$ the maximal spectrum of R, and if $x \in \operatorname{Sp} R$ we denote by L_x the residue field at x, which is a finite extension of $\mathbb{Q}_p$. If $f \in R$, we write $f(x) \in L_x$ for the evaluation of f at $x \in \operatorname{Sp} R$. By definition of the supremum norm, $v_p(f) = \inf_{x \in \operatorname{Sp} R} v_p(f(x))$ and by the maximum modulus principle, this infimum is also a minimum. For $f \in R^*$, f is multiplicative (cf. Exercise 3.1.1) if and only if $v_p(f(x))$ is constant on $\operatorname{Sp} R$.

An element $F \in R\{\{T\}\}$ defines a rigid analytic function on the rigid analytic variety $\operatorname{Sp} R \times \mathbf{A}^1_{\mathrm{rig}}$ where $\mathbf{A}^1_{\mathrm{rig}}$ is the rigid analytic affine line over $\mathbb{Q}_p$.

3.3.1 *Strongly ν-Dominant Polynomials*

Let $F \in R\{\{T\}\}$. If $x \in \operatorname{Sp} R$ is a closed point of field of definition $L(x)$, we write $F_x(T)$ for the series $\sum_{n=0}^{\infty} a_n(x) T^n$ which belongs to $L(x)\{\{T\}\}$.

Lemma 3.3.1 *If $N(F_x, \nu)$ is a constant N independent of $x \in \operatorname{Sp} R$, then $N(F, \nu) = N$. Moreover, for $b \in R^*$ and for $x \in \operatorname{Sp} R$, $N(b(x)F_x, \nu) = N(bF, \nu) = N$.*

Proof The hypothesis means that $v_p(a_n(x)) - n\nu \geq v_p(a_N(x)) - N\nu$ for all n and all x, the inequality being strict if $n > N$. Taking the inf on $x \in \operatorname{Sp} R$ gives $v_p(a_n) - n\nu \geq v_p(a_N) - N\nu$ for all n. For a fixed $n > N$, there is an x such that $v_p(a_n(x)) = v_p(a_n)$ by the maximum modulus principle, hence $v_p(a_n) - n\nu = v_p(a_n(x) - n\nu) > v_p(a_N(x)) - N\nu \geq v_p(a_N) - N\nu$. This means that $N(F, \nu) = N$.

To prove the second sentence, note that it is clear that $N(b(x)F_x, \nu) = N(F_x, \nu) = N$ since $b(x) \neq 0$. It follows that $N(bF, \nu) = N$ by the first part. □

Definition 3.3.2 A polynomial $Q(T) \in R[T]$ is called *strongly ν-dominant* if

(i) Q has degree $N(Q, \nu)$
(ii) For all $x \in \operatorname{Sp} R$, $N(Q, \nu) = N(Q_x, \nu)$

Note that if a polynomial $Q(T) \in R[T]$ is *strongly ν-dominant*, then its dominant term $a_{N(Q,\nu)}$ is invertible in R, for if it was not, $a_{N(Q,\nu)}(x)$ would be zero for some x, and we would have $N(Q_x, \nu) \leq \deg Q_x < \deg Q = N(Q, \nu)$. This shows that a strongly ν-dominant polynomial is ν-dominant, and more precisely, that a polynomial Q is strongly ν-dominant if and only if it is ν-dominant and all the Q_x, for $x \in \operatorname{Sp} R$, are ν-dominant.

Exercise 3.3.3 If $Q(T) \in R[T]$ is a polynomial whose dominant term is invertible, there exists $\nu_0 \in \mathbb{R}$ such that for all $\nu \geq \nu_0$, Q is strongly ν-dominant.

Exercise 3.3.4 Let Q be a polynomial of degree N. Assume that $N = N(Q, \nu)$ and that a_N is multiplicative. Show that Q is strongly ν-dominant. Show that the converse is false: a polynomial of degree N can be ν-dominant without satisfying $v_p(a_N) + v_p(a_N^{-1}) = 0$.

Exercise 3.3.5 Let Q be a polynomial of degree N and $a \in R^*$.

1. Show that if a is multiplicative and Q is ν-dominant, then aQ is ν-dominant.
2. Show that if Q is strongly ν-dominant, then aQ is strongly ν-dominant.
3. But show by an example that if a is not multiplicative, aQ may not be ν-dominant even if Q is.

3.3.2 A Canonical Factorization of Everywhere Convergent Power Series

Theorem 3.3.6 *For $F(T) \in R\{\{T\}\}$ with $F(0)$ invertible in R, and $\nu \in \mathbb{R}$, the following are equivalent:*

(i) *One has $N(F_x, \nu) = N(F, \nu)$ for all $x \in Sp\,R$.*
(ii) *There exists a decomposition $F = PG$ in $R\{\{T\}\}$, with $P(0) = 1$, where P is a strongly ν-dominant polynomial of degree $N(F, \nu)$ and $N(G_x, \nu) = 0$ for all $x \in W$.*

Moreover, if those properties hold, the decomposition $F = PG$ of (ii) is unique.

Proof Assume (i). Let $N = N(F, \nu)$. Write $F = \sum_{i=0}^{\infty} a_i T^i$. For $x \in \mathrm{Sp}\,R$, $v_p(a_N(x)) - N\nu \le v_p(a_0) = 0$, hence $v_p(a_N(x)) < +\infty$ and $a_N(x)$ is not zero. Therefore a_N is invertible. If we replace F by $a_N^{-1}F$, then hypothesis (i) is still satisfied by Lemma 3.3.1, and this change doesn't affect the truth of (ii) (if $a_N^{-1}F = PG$ with P and G as in (ii), then $F = P(a_N G)$ and P and $a_N G$ still satisfy the required properties). Hence we may and do assume that $a_N = 1$ below.

We shall construct by induction a sequence of polynomials $P_n \in R[T]$, for $n \ge 1$, satisfying (3.3.1)–(3.3.6) below:

$$P_n \text{ is monic of degree } N \tag{3.3.1}$$

This allows to use Proposition 3.2.8 to find unique $G_n \in R\{\{T\}\}$, $S_n \in R[T]$ such that

$$F = P_n G_n + S_n, \text{ with } \deg S_n < N \tag{3.3.2}$$

We require furthermore for $n \ge 1$ and $\nu \in \mathbb{R}$,

$$v(F - P_n, \nu) > v(F, \nu), \tag{3.3.3}$$

and for $n \geq 2$,

$$P_n = P_{n-1} + S_{n-1}, \tag{3.3.4}$$

$$v(P_n - P_{n-1}, \nu) \geq (n-1)v(F - P_1, \nu) - (n-2)v(F, \nu), \tag{3.3.5}$$

$$v(G_n - G_{n-1}, \nu) \geq n(v(F - P_1, \nu) - v(F, \nu)) \tag{3.3.6}$$

We start the construction by induction by setting $P_1 = \sum_{i=0}^{N} a_i T^i$. The coefficient of T^N in P_1 is $a_N = 1$ so (3.3.1) is satisfied. Since $F - P_1 = \sum_{i>N} a_i T^i, v(F-P_1, \nu) = \inf_{i>N} v_p(a_i) - i\nu > v_p(a_N) - N\nu$ since $N = N(F, \nu)$, and (3.3.3) is proved.

Now assume that for some $n \geq 1$, P_n (and also P_m for $m < n$) is constructed satisfying the conditions above. With G_n and S_n defined by (3.3.2) we set $P_{n+1} = P_n + S_n$, so that (3.3.4) and, since S_n has degree $< N$, (3.3.1) are satisfied for P_{n+1}.

We define G_{n+1} and S_{n+1} by (3.3.2). We write $F - P_{n+1} = P_{n+1}(G_{n+1} - 1) + S_{n+1}$, which is an euclidean division of $F - P_{n+1}$ by P_{n+1} as in Proposition 3.2.8. This proposition gives

$$v(G_{n+1} - 1, \nu) \geq v(F - P_{n+1}, \nu) - v(P_{n+1}, \nu), \tag{3.3.7}$$

$$v(S_{n+1}, \nu) \geq v(F - P_{n+1}, \nu). \tag{3.3.8}$$

Since $P_{n+1} = P_1 + S_1 + S_2 + \ldots S_n$, we see by induction using (3.3.8) that

$$v(F - P_{n+1}, \nu) \geq v(F - P_1, \nu) > v(F, \nu), \tag{3.3.9}$$

so that (3.3.3) is satisfied for P_{n+1} and hence,

$$v(F, \nu) = v(P_{n+1}, \nu). \tag{3.3.10}$$

Using this and (3.3.7):

$$v(G_{n+1} - 1, \nu) \geq v(F - P_1, \nu) - v(F, \nu). \tag{3.3.11}$$

Let us write another euclidean division, namely, $S_n(G_{n+1} - 1) = P_n(G_n - G_{n+1}) - S_{n+1}$ which by Proposition 3.2.8 gives

$$v(S_{n+1}, \nu) \geq v(S_n(G_{n+1} - 1), \nu) \geq v(S_n, \nu) + v(G_{n+1} - 1, \nu) \tag{3.3.12}$$

$$v(G_n - G_{n+1}, \nu) \geq v(S_n, \nu) + v(G_{n+1} - 1, \nu) - v(F, \nu), \tag{3.3.13}$$

from which (3.3.6) and (3.3.5) follow by induction. This completes the construction by induction.

Now let us deduce the existence assertion in (ii). For every $\nu \in \mathbb{R}$, condition (3.3.5) shows, with the help of (3.3.3) with $n = 1$, that the sequence (P_n) is Cauchy for $v(\cdot, \nu)$, hence converges for $v(\cdot, \nu)$ to a limit $P_\infty \in R[T]$ (independent of ν, see Exercise 3.2.2). Similarly condition (3.3.6) shows that G_n also converges to a limit $G_\infty \in R[[T]]$ for $v(\cdot, \nu)$, and that $v(G_\infty, \nu) > -\infty$, while (3.3.6) shows that S_n tends to 0. Going to the limit in (3.3.2), we obtain $F = P_\infty G_\infty$. Let $F = P_\infty G'_\infty + S'_\infty$ be the euclidean division of F by P_∞ given by Proposition 3.2.8. This is also an euclidean division in the sense of Scholium 3.2.10, as is $F = P_\infty G_\infty$, hence by uniqueness $G_\infty = G'_\infty$ and G_∞ is in $R\{\{T\}\}$.

By going to the limit in (3.3.10), $v(P_\infty, \nu) = v(F, \nu)$ and since $N(F, \nu) = N$, and the coefficient of T^N in P_∞ is 1, P_∞ is strongly ν-dominant (Exercise 3.3.4).

Since $F(0) = P_\infty(0)G_\infty(0)$, $P_\infty(0)$ is invertible in R, and we can set $P = P_\infty(0)^{-1}P_\infty$, $G = P_\infty(0)G_\infty$. Then $F = PG$ with $P(0) = 1$. Moreover P is strongly ν-dominant by Exercise 3.3.5. Also for all $x \in \operatorname{Sp} R$, $F_x = P_x G_x$ and $N(F_x, \nu) = N = N(P_x, \nu)$ so $N(G_x, \nu) = 0$. This completes the proof of the existence assertion in (ii).

Let us prove uniqueness. If $F = PG = P'G'$, then for any $x \in \operatorname{Sp} R$ one has $F_x = P_x G_x = P'_x G'_x$. If we prove $P_x = P'_x$, then since R is reduced $P = P'$. So it suffices to prove the uniqueness when R is a finite extension of $\mathbb{Q}_p$, which we assume.

Assume $P \neq P'$. Then $P - P' \neq 0$, hence $(P - P')G \neq 0$ and $G' - G \neq 0$. One has $P'(G' - G) = (P - P')G$, and since $P(0) = P'(0)$, $G(0) = G'(0)$ both $P - P'$ and $G' - G$ are divisible by T, and we can write $P'\frac{G'-G}{T} = \frac{P-P'}{T}G$, hence $N(P', \nu) + N(\frac{G'-G}{T}, \nu) = N(\frac{P-P'}{T}, \nu) + N(G, \nu)$ by Exercise 3.2.6, and $N(P', \nu) \geq N(\frac{P-P'}{T}, \nu)$ since $N(G, \nu) = 0$ by assumption. Also $N(P', \nu) = N(F, \nu)$ by assumption. On the other hand P and P' have degree $N(F, \nu)$ so $\frac{P-P'}{T}$ has degree $\leq N(F, \nu) - 1$, and $N(\frac{P-P'}{T}, \nu) \leq N(F, \nu)$, hence $N(F, \nu) \leq N(F, \nu) - 1$, a clear contradiction, which proves the uniqueness and completes the proof of (ii).

Now we prove (ii) implies (i). From $F = PG$ we get, for every $x \in W$, $F_x = P_x G_x$ hence $N(F_x, \nu) = N(P_x, \nu)N(G_x, \nu)$. By assumption $N(G_x, \nu) = 0$ and $N(P_x, \nu) = N(P, \nu) = \deg P = N(F, \nu)$ since P is strongly ν-dominant of degree $N(F, \nu)$, hence $N(F_x, \nu) = N(F, \nu)$. □

Corollary 3.3.7 *Assume that L is a finite extension of $\mathbb{Q}_p$. Let $F(T) \in L\{\{T\}\}$ with $F(0) = 1$, and $\nu \in \mathbb{R}$. Then there exists a unique decomposition $F = PG$, with $P \in L[T]$, $G \in L\{\{T\}\}$, $P(0) = G(0) = 1$, P ν-dominant of degree $N(F, \nu)$, $N(G, \nu) = 0$. Moreover the zeros z in $\bar{L}$ of $F(T)$ with $v_p(z) \geq -\nu$, are the same, with multiplicity, than the zeros of P, and their number is $N(F, \nu)$.*

Proof The first assertion follows from (ii) of the above theorem since (i) is obviously satisfied. The second sentence then follows from Lemma 3.2.7 . □

3.3.3 Adapted Pairs

We begin by giving a rigid-analytic geometric interpretation of Theorem 3.3.6. As above R is a reduced affinoid; we set $W = \operatorname{Sp} R$.

Fix an $F \in R\{\{T\}\}$ and assume $F(0) = 1$. Let Z be the analytic subvariety of $W \times \mathbf{A}^1_{\mathrm{rig}}$ cut out by $F(T)$. Let f be the projection map $Z \to W$. For $\nu \in \mathbb{R}$, let Z_ν be the affinoid of Z of points z whose component in $\mathbf{A}^1_{\mathrm{rig}}$ have $v_p(z) \geq -\nu$, and let f_ν be the restriction of f to Z_ν. We record some elementary properties of the map $f_\nu : Z_\nu \to W$.

Lemma 3.3.8 *The fiber at $x \in W$ of f_ν is a finite scheme of degree $N(F_x, \nu)$, whose points are the z such that $F_x(z) = 0$ and $v_p(z) \geq -\nu$. In particular, f_ν is quasi-finite. Moreover f_ν is quasi-compact, separated and flat.*

Proof Using a decomposition $F_x = P_x G_x$ as in Corollary 3.3.7, we see that the fiber at x is defined by the equation $P_x(z) = 0$, which is equivalent to $F_x(z) = 0$ and $v_p(z) \geq -\nu$. The degree of this finite scheme is the number of zeros of P_x in an algebraic closure of $L(x)$ (counted with multiplicity), that is $\deg P_x = N(P_x, \nu) = N(F_x, \nu)$. Since $N(F_x, \nu)$ is always finite, it follows that f_ν is quasi-finite. That f_ν is separated and quasi-compact is clear.

It remains to prove that f_ν is flat. We may assume that $\nu = 0$ by rescaling, hence that $F \in R\langle T\rangle$, and we just need to prove that $R\langle T\rangle/(F)$ is flat over R. Since $R\langle T\rangle$ is flat, this can be done by applying [89, Theorem 22.6], which reduces to proving that for every maximal ideal $\mathfrak{m}$ of $R\langle T\rangle$, the image of F in $R\langle T\rangle/(\mathfrak{m} \cap R)R\langle T\rangle = R/(\mathfrak{m} \cap R)\langle T\rangle$ is not a zero-divisor. But the last ring is a domain, and the image of F is non-zero (since $F(0) = 1$), so we are done. □

Proposition 3.3.9 *For $F(T) \in R\{\{T\}\}$ with $F(0) = 1$, and $\nu \in \mathbb{R}$, the following are equivalent:*

(i) *The map $f_\nu : Z_\nu \to W$ is finite.*
(ii) *There exists a decomposition $F = PG$ in $R\{\{T\}\}$, with $P(0) = G(0) = 1$, where P is a strongly ν-dominant polynomial of degree $N(F, \nu)$ and $N(G_x, \nu) = 0$ for all $x \in W$.*
(iii) *One has $N(F_x, \nu) = N(F, \nu)$ for all $x \in W$,*

If those properties hold, the decomposition $F = PG$ as in (ii) is unique, the ideal (P, G) is the unit ideal of $R\{\{T\}\}$, Z_ν is disconnected from its complement in Z (that is, there is an idempotent $e \in \mathcal{O}(Z)$ such that Z_ν is defined by $e = 1$) and $f_\nu : Z_\nu \to W$ is finite flat surjective of degree $\deg Q = N(F, \nu)$.

Proof If (i) holds, since f_ν is finite flat, its fibers all have the same rank, in other words the power series F_x all have the same number (with multiplicity) of zeros z satisfying $v_p(z) \geq -\nu$, which means that $N(F_x, \nu)$ is constant on W by Corollary 3.3.7, hence that $N(F_x, \nu) = N(F, \nu)$ by Lemma 3.3.1, and (iii) holds.

Also, if (iii) holds, then the fibers of $f : Z_\nu \to W$ all have the same degrees $N(F, \nu)$. Since f_ν is flat, separated and quasi-compact, f_ν is finite by a result of Conrad [47, Theorem A.1.2] and (i) holds.

We have already seen the equivalence of (ii) and (iii) in Theorem 3.3.6. Hence the equivalence between (i), (ii) and (iii). The uniqueness of the decomposition in (ii) has also been proved in Theorem 3.3.6.

When they hold, the natural injective map $j : Z_\nu \to Z$ is finite, hence Z_ν is the support of the coherent sheaf $j_* \mathcal{O}_{Z_\nu}$, and such a support is always Zariski closed. Since Z_ν is also an open admissible subdomain of Z, it is disconnected from its complement in Z. If we define a function a on Z equal to 0 on Z_ν and $1/P$ on its complement (since P has no zero on that complement), and a function b equal to $1/G$ on Z_ν (since G has no zero on Z_ν) and 0 on its complement, then $aP+bG = 1$, hence the ideal (P, G) is the unit ideal. The other assertions are clear. □

Definition 3.3.10 A power series $F(T) \in R\{\{T\}\}$ with $F(0) = 1$ being fixed, we say that $\nu \in \mathbb{R}$ is *adapted* to F if the conditions of the above proposition are satisfied.

Exercise 3.3.11 If $F(T)$ and $G(T)$ are in $R\{\{T\}\}$ with $F(0) = G(0) = 1$, and ν is adapted for FG, then ν is adated for F and for G.

When F is fixed, for X an affinoid open of W, of affinoid ring R_X, we call $F_X(T)$ the image of $F(T)$ in $R_X\{\{T\}\}$ and define $Z_X = f^{-1}(X)$. Clearly Z_X is also the analytic subvariety in $X \times \mathbf{A}^1_{\text{rig}}$ cut by F_X. We shall say that ν is *adapted* to X (or that X and ν are *adapted*, or that (X, ν) is an adapted pair) if ν is adapted to F_X. In this case $Z_{X,\nu} := Z_X \cap Z_\nu$ is finite flat surjective over X, and we have a decomposition $F_X(T) = Q_X(T) G_X(T)$ in $R_X\{\{T\}\}$ as in Proposition 3.3.9.

Of course, this notion is useful only if we can prove that there are enough adapted pairs (X, ν).

Proposition 3.3.12 *For any real ν, and any $x \in W$, there exists an admissible affinoid neighborhood X of x such that (X, ν) is adapted.*

Proof Let $N = N(F_x, \nu)$. As $v_p(a_n) - n\nu$ goes to infinity, there exists an $n_0 > N$ such that for all $n > n_0$, $v_p(a_n) - n\nu > v_p(a_N(x)) - N\nu$. Let X be the subset of W consisting of the points $y \in W$ such that for all n, $0 \le n \le n_0$ and $n \ne N$,

$$v_p(a_n(y)) \ge v_p(a_N(x)) - N\nu + n\nu + 1$$

and $v_p(a_N(y)) = v_p(a_N(x))$. Then X is clearly an affinoid admissible subdomain of W, and it is plain that for all $y \in X$, $N(F_y, \nu) = N$. Thus $N(F_X, \nu) = N$ by Lemma 3.3.1 and (X, ν) is adapted. □

Corollary 3.3.13 *The family of affinoid subdomains $Z_{X,\nu}$ of Z for X an affinoid of Z and ν adapted to X is a covering of Z.*

A stronger version of this corollary is true. It is not needed for the construction and for establishing the main properties of eigenvarieties below, but we shall use it to prove the crucial fact that an eigenvariety is separated.

Theorem 3.3.14 *The covering of the preceding corollary is an admissible covering.*

For this we need a lemma.

Lemma 3.3.15 *Let X be an open affinoid of W, $\nu \in \mathbb{R}$ adapted to X. There exists an open affinoid X' containing X and a real number $\nu' > \nu$ adapted to X', such* *that $N(F_X, \nu) = N(F_{X'}, \nu')$, and such that $W - X'$ can be covered by finitely many open affinoids that do not meet X.*

Proof Write $N = N(F, \nu)$. Since $N(F_x, \nu) = N$ for all $x \in X$, a_N does not vanish on x, and by the maximum modulus principle applied to a_N^{-1}, $M := \sup_{x \in X} v_p(a_N(x)) - N\nu$ is finite.

Since $v(F, \nu + 1)$ is finite, there is an $n_0 > N$ such that

$$v_p(a_n) - n(\nu + 1) > M + 1 \text{ for } n \geq n_0. \tag{3.3.14}$$

For $x \in X$, one has, since $N(F_x, \nu) = N$,

$$v_p(a_n(x)) - n\nu \geq v_p(a_N(x)) - N\nu \text{ if } n < N, \tag{3.3.15}$$

$$v_p(a_n(x)) - n\nu > v_p(a_N(x)) - N\nu \text{ if } N < n < n_0. \tag{3.3.16}$$

By choosing a real number ν' between ν and $\nu+1$ but close to ν, one can ensure that inequalities (3.3.15) and (3.3.16) stay true with ν replaced by ν', and that (3.3.15) becomes strict. That is, for every $x \in X$,

$$v_p(a_n(x)) - n\nu' > v_p(a_N(x)) - N\nu' \text{ if } n < n_0, n \neq N. \tag{3.3.17}$$

Also, since $\nu' < \nu + 1$, (3.3.14) gives

$$v_p(a_n) - n\nu' > M + 1 \text{ for } n \geq n_0. \tag{3.3.18}$$

By the maximum modulus principle, there exists $\epsilon > 0$ such that for every $x \in X$,

$$v_p(a_n(x)) - n\nu' \geq v_p(a_N(x)) - N\nu' + \epsilon \text{ if } n < n_0, n \neq N \tag{3.3.19}$$

Let X' be the subset of W consisting of the points x such that

$$v_p(a_n(x)) - n\nu' \geq v_p(a_N(x)) - N\nu' + \epsilon/2 \text{ if } n < n_0, n \neq N, \tag{3.3.20}$$

and

$$v_p(a_N(x)) - N\nu' \leq M + 1. \tag{3.3.21}$$

A point $x \in X$ satisfies (3.3.20) because of (3.3.19) and it satisfies (3.3.21) by definition of M. Thus $X \subset X'$. Moreover X' is clearly an affinoid subdomain of W, and (X', ν') is adapted by (3.3.20) and (3.3.18) together with (3.3.21).

Now for n any integer between 1 and n_0 and not equal to N, let U_n be the affinoid defined by the equation

$$v_p(a_n(x)) - n\nu' \leq v_p(a_N(x)) - N\nu' + \epsilon/2 \tag{3.3.22}$$

and let U_N be defined by $M + 1/2 \leq v_p(a_N(x)) - Nv'$. Then clearly the U_n's do not meet X, but the affinoids U_n and X' together cover W. □

We now prove Theorem 3.3.14. Since Z is admissibly covered by the Z_ν, it suffices to prove that for each ν, Z_ν is covered by a finite collection of affinoid of the form $Z_{X,\nu'}$ with (X, ν') adapted.

We shall prove that *for every affinoid* $W = \operatorname{Sp} R$, *integer* $n_0 \geq 0$, *real* $\nu > 0$, *power series* $F \in R\{\{T\}\}$ *with* $F(0) = 1$ *such that* $N(F_x, \nu) \leq n_0$ *for every* $x \in W$, Z_ν *is covered by a finite collection of affinoid subdomains of the form* $Z_{X,\nu'}$ *with* (X, ν') *adapted*. The method of proof is induction on n_0.

When $n_0 = 0$, the result is trivial. Assume it is known for $n_0 - 1$. Note that for any n, the set X_n defined by the condition $N(F_x, \nu) \geq n$ is an affinoid subset of W. The affinoid X_{n_0} is the set of x such that $N(F_x, \nu) = n_0$, hence (X_{n_0}, ν) is adapted. Applying Lemma 3.3.15, let (X', ν') be an adapted pair, with X' an affinoid containing X and such that $W - X'$ can be covered by finitely many open affinoids $U_1, \ldots, U_k$ that do not meet X, and $\nu' > \nu$. On each of the affinoid U_i, we have $N(F_x, \nu') \leq n_0 - 1$ since U_i does not meet X_{n_0}. Hence by the induction hypothesis $Z_{U_i,\nu'}$ can be covered by finitely many affinoids $Z_{X''_{i,j},\nu''_{i,j}}$ with $X''_{i,j}$ open affinoid of U_i, $\nu''_{i,j} > \nu'$, $(X''_{i,j}, \nu''_{i,j})$ adapted. Then the union of all those affinoid $Z_{X''_{i,j},\nu''_{i,j}}$ together with $Z_{X',\nu'}$ contains Z_ν. This completes the proof of Theorem 3.3.14.

3.4 Submodules of Bounded Slope

In this section we keep the assumption of the preceding subsection: R is a reduced affinoid ring whose normalized valuation is denoted by v_p; if $x \in \operatorname{Sp} R = X$, the residue field at x is called $L(x)$.

Definition 3.4.1 Let M be a Banach R-module satisfying property (Pr) and U_p a compact R-linear operator of M. Let $F(T) = \det(1 - TU_p) \in R\{\{T\}\}$ be the Fredholm's determinant of U_p. We assume that ν is adapted for F (Definition 3.3.10). Then by Prop 3.3.9, there is a unique decomposition $F = PG$ in $R\{\{T\}\}$, with P a strongly ν-dominant polynomial of degree $N(F, \nu)$, G such that $N(G_x, \nu) = 0$ for all $x \in \operatorname{Sp} R$, and $P(0) = G(0) = 1$. Moreover P and G generate the unit ideal. According to Theorem 3.2.19, there exists a unique ϕ-stable decomposition $M = N \oplus F$ with N, F ϕ-stable closed submodules of M such that $P^*(U_p)$ is nilpotent on N and invertible on F. We call the module N the *submodule of slope less than* ν of M, and denote it by $M^{\leq \nu}$.

Proposition 3.4.2 *Under the same hypotheses as in the above definition, the module* $M^{\leq \nu}$ *is a finite projective module over* R. *It is stable by every endomorphism of* M *that commutes with* ϕ. *The formation of* $M^{\leq \nu}$ *commute with base change, in the sense that if* $R \to R'$ *is a morphism of affinoid rings over* L, *then* ν *is adapted for the image of* $F(T)$ *in* $R'\{\{T\}\}$, *and* $(M \hat{\otimes}_R R')^{\leq \nu} = M^{\leq \nu} \otimes_R R'$ *as submodules of* $M \hat{\otimes}_R R'$.

Proof That $M^{\leq \nu}$ is finite projective, and stable by every endomorphism that commutes with ϕ directly results from its definition and Theorem 3.2.19. For the commutation with base change, note that $U'_p := U_p \hat{\otimes} 1$ is a compact operator of $M \hat{\otimes}_R R'$ whose power series is the image $F'(T)$ of $F(T)$ in $R'\{\{T\}\}$ by Lemma 3.1.27. The decomposition $F = QG$ in $R\{\{T\}\}$ defines a decomposition $F' = Q'G'$ which obviously still satisfies (ii) of Prop 3.3.9. Hence ν is still adapted for F', and $P^*(U'_p) = P^*(U_p)\hat{\otimes}1$. It follows then from the uniqueness of the decomposition in Theorem 3.2.19 that $(M\hat{\otimes}_R R')^{\leq \nu} = M^{\leq \nu} \otimes_R R'$. □

Lemma 3.4.3 *If* $0 \to M_1 \xrightarrow{f} M_2 \xrightarrow{g} M_3 \to 0$ *is an exact sequence of Banach spaces provided with compact operators* U_p, *and* f *and* g *are continuous linear maps commuting with* U_p, *then* $0 \to M_1^{\leq \nu} \xrightarrow{f} M_2^{\leq \nu} \xrightarrow{g} M_3^{\leq \nu} \to 0$ *is exact.*

Proof This is clear since f and g preserve the canonical decompositions $M_i = M_i^{\leq \nu} \oplus F_i$ for $i = 1, 2, 3$. □

The phrase "submodule of slope less than ν" is in part explained by the following easy result:

Lemma 3.4.4 *Assume that* $R = L$ *is a finite extension of* $\mathbb{Q}_p$, *normed by the normalized valuation* v_p. *Then in the situation of the definition above,* $M^{\leq \nu}$ *is the largest* U_p*-stable subspace of* M *on which the roots (in an algebraic closure of* L*) of the characteristic polynomial of* U_p *all have valuation* $\leq \nu$.

Proof By definition $P^*(U_p)$ is nilpotent on $M^{\leq \nu}$, so the characteristic polynomial χ_{U_p} of U_p on $M^{\leq \nu}$ divides a power of $P^*(U_p)$. By Lemma 3.2.7, all the roots of $P(U_p)$ have valuation $\geq -\nu$, hence all the roots of $P^*(U_p)$ and of χ_{U_p} have valuation $\leq \nu$. To show that $M^{\leq \nu}$ is the largest finite-dimensional ϕ-stable subspace with this property, assume there is an U_p-stable subspace $N \subset M$ containing M^{ν}. Remember the decomposition $M = M^{\leq \nu} \oplus F$ of Definition 3.4.1. The projection of N onto F alongside $M^{\leq \nu}$ is a non-zero U_p-stable sub-module of N on which $Q^*(U_p)$ is nilpotent. Since $Q^*(U_p)$ is invertible on F, the projection of N on F must be zero, that is $N = M^{\leq \nu}$. □

We shall now give a simple characterization of $M^{\leq \nu}$, in the spirit of Lemma 3.4.4. Since we will not use it in this book, the reader may safely skip to the next section.

Lemma 3.4.5 *Let* F *be a Banach* R*-module having property (Pr),* $N \subset F$ *a finite-type sub-module. If for every* $x \in \mathrm{Sp}\, R$ *the natural map* $N_x \to F_x$ *is zero, then* $N = 0$. *(Here* N_x *is the fiber of* N *at* x, *that is* $N \otimes_R L(x)$, *and similarly for* F_x*)*

Proof We first assume that F is finite free, say of rank n. Then the hypothesis implies that $(F/N)_x = F_x$ for all $x \in \mathrm{Sp}\, R$. Hence the finite module F/N has rank n at every closed point of Spec R, hence has rank n at every point of Spec R since closed points are dense (affinoid rings are Jacobson) and the rank of $(F/N)_x$ is semi-continuous as a function of $x \in \mathrm{Spec}\, R$ and bounded above by the rank of F_x which is n. By Nakayama this means that F/N is projective, and thus N is direct summand in F, which implies that the maps $N_x \to F_x$ are injective. By hypothesis, it follows that $N_x = 0$ for all x, and since N is of finite type we conclude that $N = 0$.

We now assume only that F is orthonormalizable, with orthonormal basis $(e_i)_{i \in I}$. For S a finite subset of I, we denote by F_S the submodule of F generated by $(e_i)_{i \in S}$ and by π_S the projection of F onto F_S alongside other vectors of the basis. Then the hypothesis implies that $\pi_S(N)_x \to (F_S)_x$ is the zero map for all $x \in \operatorname{Sp} R$, so by the above result applied to F_S, $\pi_S(N) = 0$, and since it is true for all finite S, $N = 0$. The case where F is potentially orthonormalizable trivially follows.

Finally, if F has (Pr), then $F \oplus F'$ is potentially orthonormalizable for some Banach module F', and if we see N as a sub-module of $F \subset F \oplus F'$ then $N_x \to F_x \oplus F'_x$ is zero, so $N = 0$ by the preceding case. □

Proposition 3.4.6 *Same hypothesis as in Definition 3.4.1. The submodule $M^{\leq \nu}$ of M is the largest finite type sub-module M' of M which is U_p-stable, and such that for every $x \in \operatorname{Sp} R$, all the roots (in an algebraic closure of $L(x)$) of the characteristic polynomial of U_p on the image of M'_x in M_x have valuation $\leq \nu$.*

Proof The image of $M_x^{\leq \nu}$ in M_x is isomorphic to $(M_x)^{\leq \nu}$ by Proposition 3.4.2. Hence the roots of the characteristic polynomial of U_p acting on that image have valuation $\leq \nu$ by Lemma 3.4.4.

Let M' be a finite type sub-module of M which is U_p-stable, and such that for every $x \in \operatorname{Sp} R$, all the roots (on an algebraic closure of $L(x)$) of the characteristic polynomial of U_p on the image of N_x in M_x have valuation $\leq \nu$. Using the decomposition $M = M^{\leq \nu} \oplus F$ and calling π the projection onto F along $M^{\leq \nu}$, let us set $N = \pi(M') \subset F$. Then N is a finite type sub-module of F, and for every $x \in \operatorname{Sp} R$, the action of $P^*(U_p)$ on the image of N_x in F_x is nilpotent, and since this action is invertible on F_x, this image is 0. By the lemma above, we conclude that $N = 0$, hence $M' \subset M^{\leq \nu}$. □

3.5 Links

Definition 3.5.1 If R is an affinoid ring, M and M' be two Banach R-modules satisfying property (Pr) with morphisms $\psi : \mathcal{H} \to \operatorname{End}_R(M)$ and $\psi' : \mathcal{H} \to \operatorname{End}_R(M')$ such that $\psi(U_p)$ and $\psi'(U_p)$ are compact, we say that M and M' are *linked* if $\psi(U_p)$ and $\psi'(U_p)$ have the same characteristic power series $F(T)$, and for every monic polynomial $Q \in R[T]$ with invertible constant term, $\ker Q(\psi(U_p))$ and $\ker Q(\psi'(U_p))$ are isomorphic as R-modules with action of $\mathcal{H}$.

It is clear that to be linked is an equivalence relation. Observe that if M and M' are linked, then ν is adapted for M if and only if it is adapted for M', and then the two modules $M^{\leq \nu}$ and $M'^{\leq \nu}$ are isomorphic as R-modules with action of $\mathcal{H}$.

Lemma 3.5.2 *Let M and M' be as in the definition above. Assume that there are two continuous homomorphism of R-modules and $\mathcal{H}$-modules $f : M \to M'$ and*

$g : M' \to M$ such that $g \circ f = \psi(U_p)$ and $f \circ g = \psi(U'_p)$. If either f or g is compact, then M and M' are linked.

Proof That the characteristic power series of $\psi(U_p)$ and $\psi'(U_p)$ are the same results from the equality in $R\{\{T\}\}$: $\det(1 - Tfg) = \det(1 - Tgf)$ proved in Lemma 3.1.21. Moreover f maps $\ker Q(\psi(U_p))$ into $\ker Q(\psi'(U_p))$ since it is an $\mathcal{H}$-homomorphism and g maps $\ker Q(\psi'(U_p))$ into $\ker Q(\psi(U_p))$. The composition $g \circ f = \psi(U_p)$ on $\ker Q(\psi(U_p))$ is invertible since Q has invertible constant term, and so is $f \circ g$. Thus $\ker Q(\psi(U_p))$ and $\ker Q(\psi'(U_p))$ are isomorphic. □

An interesting special case is the following:

Lemma 3.5.3 *Let M, M' be two R-modules with $\mathcal{H}$-actions as above. Assume that there are two homomorphisms of R-modules and $\mathcal{H}$-modules $f : M \to M'$ and $g : M' \to M$ such that f is injective and continuous, g is compact, and $f \circ g = \psi(U'_p)$. Then M and M' are linked. Moreover, if f factors as a composition $M \hookrightarrow M'' \hookrightarrow M'$, where M'' is as in the definition above and the two morphisms in this factorization are injective continuous morphism of R-modules and $\mathcal{H}$-modules, then M and M' are also linked to M''.*

Proof For the first assertion, we observe that $f \circ g \circ f = \psi(U'_p) \circ f = f \circ \psi(U_p)$ hence by injectivity of f, $g \circ f = \psi(U_p)$ and we can apply Lemma 3.5.2. The second assertion follows form the first by replacing M by M'', f by the injection $M'' \hookrightarrow M'$, and g by the composition of $g : M' \to M$ with the injection $M \to M''$. □

Lemma 3.5.4 *If M and M' are linked, and $\operatorname{Sp} R' \to \operatorname{Sp} R$ is a morphism of affinoid, them $M \hat{\otimes}_R R'$ and $M' \hat{\otimes}_R R'$ are also linked.*

Proof This is clear from Lemma 3.1.27. □

3.6 The Eigenvariety Machine

3.6.1 Eigenvariety Data

An *eigenvariety data* over a finite extension L of $\mathbb{Q}_p$ is the data of

(ED1) A commutative ring $\mathcal{H}$, with a distinguished element U_p.
(ED2) A reduced rigid analytic variety $\mathcal{W}$ over L, and an admissible covering $\mathfrak{C}$ by admissible affinoid open subsets of $\mathcal{W}$.
(ED3) For every admissible affinoid $W = \operatorname{Sp} R$ of $\mathcal{W}$, $W \in \mathfrak{C}$, a Banach module M_W satisfying property (Pr), and a ring homomorphism $\psi_W : \mathcal{H} \to \operatorname{End}_R(M_W)$.

satisfying the following conditions:

(EC1) For every $W \in \mathfrak{C}$, $\psi_W(U_p)$ is compact on M_W.

(EC2) For every pair of affinoid subdomains $W' = \mathrm{Sp}\, R' \subset W = \mathrm{Sp}\, R \subset \mathcal{W}$, $W, W' \in \mathfrak{C}$, the R'-Banach modules $M_W \hat{\otimes}_R R'$ and $M_{W'}$ are *linked* (cf. Definition 3.5.1)

In (ED3) it would suffice to give the modules M_W up to a link, as our constructions shall depend only on the link-equivalence classes of the modules M_W.

Given an eigenvariety data, and a closed point $w \in \mathcal{W}$ of field of definition $L(w)$, we can define the *fiber* M_w as follows: choose an affinoid admissible subdomain in $\mathfrak{C}$, $W = \mathrm{Sp}\, R$, containing w, and set $M_w = M_W \otimes_R L(w)$, where the implicit morphism $R \to L(w)$ is the evaluation at w. Then the space M_w inherits a linear action of $\mathcal{H}$, denoted by $\psi_w : \mathcal{H} \to \mathrm{End}_{L(w)}(M_w)$, and it is together with that action, well defined (that is, independent of the choice of W) up to a link (by (ED2) and Lemma 3.5.4). In particular, the *subspace of finite slope* $M_w^{\#}$ of M_w defined as

$$M_w^{\#} = \bigcup_{\nu<\infty} M_w^{\leq \nu},$$

is absolutely well-defined.

With no loss of generality, we can and, when convenient, will assume that if $X \subset W$ is an inclusion of affinoid subdomains of $\mathcal{W}$, with $W \in \mathfrak{C}$, then $X \in \mathfrak{C}$ as well. It suffices to define M_X as $M_W \otimes_R R'$ if $W = \mathrm{Sp}\, R$ and $X = \mathrm{Sp}\, R'$, which makes M_X well-defined up to a link.

3.6.2 *Construction of the Eigenvariety*

For any admissible affinoid subdomain $W = \mathrm{Sp}\, R$ in $\mathfrak{C}$, let $F_W \in R\{\{T\}\}$ be the characteristic power series of U_p acting on M_W (that is $F_W(T) = \det(1 - T\psi_W(U_p))$.) To any such W, and $\nu \in \mathbb{R}$ adapted to W (that is, adapted to F_W), we attach a *local piece of the eigenvariety* which consists of the following:

(a) An L-affinoid variety $\mathcal{E}_{W,\nu}$, called *the local piece.*
(b) A finite morphism $\kappa : \mathcal{E}_{W,\nu} \to W$, called *the weight map.*
(c) A morphism of rings $\psi : \mathcal{H} \to \mathcal{O}(\mathcal{E}_{W,\nu})$.

The construction is as follows: since ν is adapted, we can define the finite flat R-submodule $M_W^{\leq \nu}$. Since $\mathcal{H}$ is commutative, it stabilizes $M_W^{\leq \nu}$ by Proposition 3.4.2. We define $\mathcal{T}_{W,\nu}$ to be the eigenalgebra (cf. Sect. 2.2) of $\mathcal{H}$ acting on $M_W^{\leq \nu}$. In other words, $\mathcal{T}_{W,\nu}$ is defined as the R-subalgebra of $\mathrm{End}_R(M_W)$ generated by $\psi_W(\mathcal{H})$. Then $\mathcal{T}_{W,\nu}$ is a finite R-module, hence an affinoid algebra since R is, and we define $\mathcal{E}_{W,\nu} = \mathrm{Sp}\, \mathcal{T}_{W,\nu}$. The structural map defines a finite morphism of rigid affinoid variety $\kappa : \mathcal{E}_{W,\nu} = \mathrm{Sp}\, \mathcal{T}_{W,\nu} \to W = \mathrm{Sp}\, R$. And we have by construction a natural ring homomorphism $\psi : \mathcal{H} \to \mathcal{T}_{W,\nu} = \mathcal{O}(\mathcal{E}_{W,\nu})$.

Lemma 3.6.1 *Let $W' \subset W$ be an inclusion of admissible open affinoids of $\mathcal{W}$, with $W, W' \in \mathfrak{C}$. Let $\nu' \leq \nu$, and assume that (W, ν) and (W', ν') are adapted. Then* *there exists a unique open immersion $\mathcal{E}_{W',\nu'} \hookrightarrow \mathcal{E}_{W,\nu}$ compatible with the weight maps κ and the maps ψ from $\mathcal{H}$. Its image is the affinoid open subset of $\mathcal{E}_{W,\nu}$ defined by the conditions $v_p(U_p(z)) \leq \nu'$, $\kappa(z) \in W'$.*

Proof Note that since ν is adapted to W, it is also adapted to W'. By considering the intermediate (W', ν), we can reduce to the two cases where $W = W'$ and where $\nu = \nu'$.

In the case $W = W' = \mathrm{Sp}\, R$, for simplicity of notation write $\mathcal{T} = \mathcal{T}_{W,\nu}$ and $\mathcal{T}' = \mathcal{T}_{W,\nu'}$ for the Hecke algebras attached to $M^{\leq\nu}$ and $M^{\leq\nu'}$ respectively. Write $F = PG$, resp. $F = P'G'$ the decomposition given by Proposition 3.3.9 for ν and ν' respectively. Since $N(G_x, \nu) = 0$ for all $x \in W$, $N(G_x, \nu') = 0$ as well hence $N(P_x, \nu') = N(P'_x, \nu')$ and by Proposition 3.3.9 attached to P and ν', we have a decomposition $P = P'P''$ with $(P', P'') = 1$. Let M'' be $\ker Q''^*(U_p)$, so that by Propositions 3.3.9 and 3.4.2, $M^{\leq\nu} = M'_{\leq\nu'} \oplus M''$ and the two projectors e and f corresponding to this decomposition belongs to $\psi(\mathcal{H})$. If $\mathcal{T}''$ is the eigenalgebra attached to $\mathcal{T}'$, the natural surjective morphisms of R-algebras (compatible with the morphisms from $\mathcal{H}$) $\mathcal{T} \to \mathcal{T}'$ and $\mathcal{T} \to \mathcal{T}''$ (cf. Exercise 2.3.6) define a map $\mathcal{T} \to \mathcal{T}' \times \mathcal{T}''$ which is clearly injective (cf. Exercise 2.3.6), but in this case also surjective since if $t' \in \mathcal{T}'$ is the image of $t_1 \in \mathcal{T}$, and $t'' \in \mathcal{T}''$ is the image of $t_2 \in \mathcal{T}''$, them (t', t'') is the image of $t_1 e + t_2 f \in \mathcal{T}$. Thus $\mathcal{T} \to \mathcal{T}' \times \mathcal{T}''$ is an isomorphism, and the closed immersion $\mathrm{Sp}\,\mathcal{T}' = \mathcal{E}_{W,\nu'} \hookrightarrow \mathrm{Sp}\,\mathcal{T} = \mathcal{E}_{W,\nu}$ is also an open immersion, clearly compatible with the morphism to W and from $\mathcal{H}$.

We now deal with the case where $W' = \mathrm{Sp}\, R' \subset W = \mathrm{Sp}\, R$ but $\nu = \nu'$. We have $(M_W \hat{\otimes}_R R')^{\leq\nu} \simeq M_{W'}^{\leq\nu}$ as $\mathcal{H}$-modules, since $M_W \hat{\otimes}_R R'$ and $M_{W'}$ are linked by hypothesis. Therefore their eigenalgebras are isomorphic. On the other hand we also have $(M_W \hat{\otimes}_R R')^{\leq\nu} = M_W^{\leq\nu} \otimes_R R'$ by Proposition 3.4.2. Then, since R' is R-flat (cf. [48, page 21]), by Proposition 2.4.1 we see that the Hecke-algebra of $M_W^{\leq\nu} \otimes_R R'$ is $\mathcal{T}_{W,\nu} \hat{\otimes}_R R'$. Combining the above, we have constructed an isomorphism (obviously R-linear and compatible with the morphisms from $\mathcal{H}$):

$$\mathcal{T}_{W,\nu} \hat{\otimes}_R R' = \mathcal{T}_{W',\nu}$$

which shows that $\mathcal{E}_{W',\nu}$ is the open affinoid of $\mathcal{E}_{W,\nu}$ defined by the condition $\kappa(z) \in W'$.

Finally, in each case the isomorphism we have constructed is unique since the eigenalgebras are generated by $\psi(\mathcal{H})$ over R. □

The eigenvariety is then constructed by gluing the local pieces:

Definition 3.6.2 An *eigenvariety* for the above eigenvariety data, consists of

(a) A rigid analytic variety $\mathcal{E}/L$ (the *eigenvariety* proper).
(b) A locally finite map $\kappa : \mathcal{E} \to \mathcal{W}$ (the *weight map*).
(c) A morphism of rings $\psi : \mathcal{H} \to \mathcal{O}(\mathcal{E})$, that sends U_p to an invertible function.

such that for any affinoid subdomain W of $\mathcal{W}$ in $\mathfrak{C}$, and real ν adapted to W, the open subvariety $\mathcal{E}(W, \nu)$ of $\kappa^{-1}(W) \subset \mathcal{E}$ defined by $v_p(\psi(U_p)) \leq \nu$, is isomorphic (as analytic variety over W with a map ψ form $\mathcal{H}$ to their ring of functions) to the local piece $\mathcal{E}_{W,\nu}$ constructed above, and such that the $\mathcal{E}(W,\nu)$ form an admissible covering of $\mathcal{E}$.

Note that the isomorphisms $\mathcal{E}(W,\nu) \simeq \mathcal{E}_{W,\nu}$ whose existences are required by the definition are necessarily unique.

Theorem 3.6.3 *For any given eigenvariety data, an eigenvariety exists and is unique up to unique isomorphism.*

Proof We first prove the existence, by gluing, in two steps. First let W be an open admissible affinoid of $\mathcal{W}$, in $\mathfrak{C}$. We will construct an 'eigenvariety' $\mathcal{E}_W$ with the same data and properties as $\mathcal{E}$ but with $\mathcal{W}$ always replaced by W.

We consider the family of affinoid varieties $\mathcal{E}_{X,\nu}$ for $X \subset W$ an open affinoid subdomain, and ν a real number adapted to X. We define a gluing data on this family as follows: given two $\mathcal{E}_{X,\nu}$ and $\mathcal{E}_{X',\nu'}$ then by Lemma 3.6.1 $(X \cap X', \min(\nu, \nu'))$ is adapted, and $\mathcal{E}_{X\cap X',\min(\nu,\nu')}$ can be seen as an open affinoid subset of both $\mathcal{E}_{X,\nu}$ and $\mathcal{E}_{X',\nu'}$, in a unique way respecting the maps form $\mathcal{H}$, and the weight maps. Hence we have defined two open affinoids, one of $\mathcal{E}_{X,\nu}$ and one of $\mathcal{E}_{X',\nu'}$, and an isomorphism between them. It is clear by uniqueness that those isomorphisms satisfy the cocycle condition. Thus, by Bosch et al. [27, Prop. 9.3.2/1] we can glue the $\mathcal{E}_{X,\nu}$'s to form a rigid analytic variety $\mathcal{E}_W$, of which the $\mathcal{E}_{X,\nu}$'s are an admissible covering by admissible open affinoid subdomains. Then the maps $\kappa : \mathcal{E}_{X,\nu} \to X$ glue into a single map $\kappa : \mathcal{E}_W \to W$ by Bosch et al. [27, Prop. 9.3.1/1]. Also the maps $\psi : \mathcal{H} \to \mathcal{O}(\mathcal{E}_{X,\nu})$ glue into a map $\psi : \mathcal{H} \to \mathcal{O}(\mathcal{E}_W)$ because the $\mathcal{E}_{X,\nu}$'s form an admissible covering. The uniqueness of $\mathcal{E}_W$ up to unique isomorphism (compatible with κ and ψ) is obvious.

Finally, to construct $\mathcal{E}$ we have to glue the $\mathcal{E}_W$'s. The only minor difficulty is that because $\mathcal{W}$ is not itself an affinoid, the intersection of two admissible affinoid subdomains W and W' of $\mathcal{W}$ may not be affinoid. But $W \cap W'$ is an admissible open subdomain and can be admissibly covered by affinoids W_i. Therefore, we can define an isomorphism between the open admissible subdomains $\kappa^{-1}(W \cap W')$ in $\mathcal{E}_W$ and $\mathcal{E}'_W$ by gluing (using again [27, Prop. 9.3.1/1]) the isomorphism between the $\kappa^{-1}(W_i)$ in $\mathcal{E}_W$ and $\mathcal{E}_{W'}$ obtained by identifying the two $\kappa^{-1}(W_i)$'s considered as a subdomain of $\mathcal{E}_W$ or $\mathcal{E}_{W'}$ obtained from identifying them to the eigenvariety $\mathcal{E}_{W_i}$. Since by uniqueness those isomorphisms obviously satisfy the gluing condition, [27, Prop 9.2.1/1] and [27, Prop 9.3.1/1] gives us finally the eigenvariety $\mathcal{E}$ by gluing the $\mathcal{E}_W$, with a map $\kappa : \mathcal{E} \to \mathcal{W}$ gluing the maps $\kappa : \mathcal{E} \to W$ and a morphism of rings $\psi : \mathcal{H} \to \mathcal{O}(\mathcal{E})$ gluing the morphisms $\psi : \mathcal{H} \to \mathcal{O}(\mathcal{E}_W)$. The eigenvariety $\mathcal{E}$ with those maps obviously is unique up to unique isomorphism and satisfies all the properties stated in the theorem. □

Exercise 3.6.4 Prove the following easy functoriality results for eigenvarieties.

1. Show that the eigenvariety (with its maps κ and ψ) depends only of the link classes of the modules M_W up to unique isomorphism.

2. Show that the eigenvariety (with its maps κ and ψ) is not changed (up to unique isomorphism) if we replace the covering $\mathfrak{C}$ by a finer covering.
3. Consider two eigenvariety data with same (ED1) and (ED2) but such that for all $W \in \mathfrak{C}$, the modules M_W (for the first data) and M'_W (for the second data) are such that we have an embedding $M_W \hookrightarrow M'_W$ of Banach R-modules $\mathcal{H}$-modules with closed image.
Let $\mathcal{E}$ and $\mathcal{E}'$ be the eigenvarieties constructed with those data. Show that there is a unique closed immersion $\mathcal{E} \to \mathcal{E}'$ compatible with κ and ψ.
4. Same as 3. but we assume that for all $W \in \mathfrak{C}$, we have embeddings $M_W \hookrightarrow M'_W \hookrightarrow M^2_W$ of Banach R-modules and $\mathcal{H}$-modules with closed images. Then there is a unique isomorphism compatible with κ and ψ between $\mathcal{E}$ and $\mathcal{E}'$.

3.7 Properties of Eigenvarieties

The most important property, which is the reason eigenvarieties deserve their name, is the following:

Theorem 3.7.1 *Let w be a point of $\mathcal{W}$, of field of definition $L(w)$. For any finite extension L' of $L(w)$, the set of L'-points z of $\mathcal{E}$ such that $\kappa(z) = w$ is in natural bijection with the systems of eigenvalues of $\mathcal{H}$ appearing in $M_w \otimes_{L(w)} L'$ such that the eigenvalue of U_p is non-zero. The bijection attaches to z the system of eigenvalues $\psi_z : \mathcal{H} \to L'$, $T \mapsto \psi(T)(z)$ for $T \in \mathcal{H}$.*

Note that since M_w is well-defined up to a link, the set of systems of eigenvalues of $\mathcal{H}$ appearing in $M_w \otimes_{L(w)} L'$ such that the eigenvalues of U_p is non-zero is well-defined.

Proof Obviously it is enough to prove that for all $\nu \in \mathbb{R}$, $z \mapsto \psi_z$ is a bijection between the set of L'-points of $\mathcal{E}$ with $\kappa(z) = w$ and $v_p(U_p(z)) \leq \nu$ and systems of eigenvalues of $\mathcal{H}$ appearing in $M_w \otimes_{L(w)} L'$ such that the eigenvalues of U_p have valuation $\leq \nu$. Choose an affinoid admissible neighborhood $W = \operatorname{Sp} R$ of $w \in \mathcal{W}$ such that ν is adapted to W (Proposition 3.3.12). It is enough to prove the last assertion with $\mathcal{E}$ replaced by its local piece $\mathcal{E}_{W,\nu}$, since on the one hand, every L'-point of $\mathcal{W}$ satisfying the desired condition is in $\mathcal{E}_{W,\nu}$, and on the other hand, $(M_w \hat{\otimes}_{L(w)} L')^{\leq \nu} = (M_W^{\leq \nu}) \otimes_R L'$ (the implied map being $R \to L(w) \hookrightarrow L$) by Proposition 3.4.2. Now by Lemma 2.4.1, since L' is a field, the L'-points of $\mathcal{E}_{W,\nu}$ above w are the same as the L'-points of the eigenalgebra of $\mathcal{H}$ acting on M_w. The result thus follows from Corollary 2.5.10. □

Exercise 3.7.2 Consider two eigenvariety data with same (ED1) and (ED2) but with different (ED3): for $W \in \mathfrak{C}$ we call M_W and M'_W the $\mathcal{H}$-modules of the data. Show that $W \mapsto M''_W = M_W \oplus M'_W$ defines also an eigenvariety data. Let us call $\mathcal{E}$, $\mathcal{E}'$ and $\mathcal{E}''$ the eigenvarieties defined by those data. There are unique injective map $\mathcal{E} \hookrightarrow \mathcal{E}''$ and $\mathcal{E}' \hookrightarrow \mathcal{E}''$ compatible with κ and ψ by Exercise 3.6.4. Show that $\mathcal{E}''$ is the union of the image of $\mathcal{E}$ and $\mathcal{E}'$.

We say that a compact operator U_p on a (Pr) Banach space M over a finite extension L of $\mathbb{Q}_p$ is *quasi-nilpotent* if for every a in a finite extension L' of L, $1 - au$ is invertible on $M \otimes_L L'$. By Corollary 3.3.7 and Proposition 3.2.18, this is equivalent to the condition $\det(1 - TU_p) = 1$.

Corollary 3.7.3 *The image of $\kappa : \mathcal{E} \to \mathcal{W}$ consists of all the points w such that the action of U_p on M_w is not quasi-nilpotent. It is a Zariski-open subset of $\mathcal{W}$.*

Proof A point w is in the image of κ if and only if there is a system of $\mathcal{H}$-eigenvalues (over an algebraic closure $\bar{L}$ of $L(w)$) of finite slope in M_w. If there is such a system, there is a non-zero eigenvector for U_p in $M_w \otimes \bar{L}$ with a non-zero eigenvalue, and U_p is not quasi-nilpotent. Conversely, if U_p is not quasi-nilpotent, it has by Riesz's theory a non-zero eigenvalue in $\bar{L}$, say of valuation ν, and there is an $\mathcal{H}$-eigenvector in $M_{\bar{w}}^{\leq \nu} \otimes \bar{L}$.

To prove the second assertion, for all $W = \operatorname{Sp} R \in \mathfrak{C}$, write $1 + \sum_{i=1}^{\infty} a_{i,W} T^i$ the Fredholm's power series of U_p on M_W. By (EC2), the $a_{i,W}$ glues to define functions $a_i \in \mathcal{O}(W)$. For $w \in \mathcal{W}$, the Fredholm's series of U_p on M_w is $1 + \sum_{i=1}^{\infty} a_i(w) T^i$. Hence by the above, w is in the image of κ if and only if at least one of the $a_i(w)$ is non-zero. The image of κ is thus Zariski-open, as the complement of the analytic subspace defined by the a_i's □

We shall give below (Proposition 3.7.7), under some mild hypothesis on $\mathcal{W}$, a result precising the second assertion of that corollary.

To state the next property, we shall need to recall some facts about the delicate notion of *irreducible component* of a rigid analytic space X. When $X = \operatorname{Sp} R$ is affinoid, one defines naturally its rigid analytic *irreducible components* as the closed subspaces $X_i = \operatorname{Sp} R/\mathfrak{p}_i$ of X, where the $\mathfrak{p}_i$ are the (finitely many, since R is noetherian) minimal prime ideals of R. In general, irreducible components have been defined by Conrad (cf. [46]), in a rather indirect way : for X a rigid analytic spaces one defines an *irreducible* component of X as the image in X of a **connected** component of the **normalization** of X. One says that X is *irreducible* if it has only one irreducible component.

Lemma 3.7.4

(i) *The space X is the union of its irreducible components X_i, which may be infinitely many.*
(ii) *A subset Z of X is Zariski-dense if and only if $Z \cap X_i$ is Zariski-dense in X_i for every i.*
(iii) *Every irreducible component X_i of X is irreducible.*
(iv) *If X is irreducible, it has a dimension, which is the common dimension of all rings of functions R of its admissible open affinoids $U = \operatorname{Sp} R$.*
(v) *The intersection $X_i \cap U$ of an irreducible component X_i with an open admissible affinoid U of X is either empty or a union of irreducible components of U.*
(vi) *If X is irreducible, any non-empty admissible open affinoid U is Zariski-dense in X.*

(vii) *If $f : X \to X'$ is a finite map, the image of an irreducible component of X by f is an irreducible closed analytic subspace of X'.*

Proof For (v), see [46, 2.2.9]. For (vii) see [46, 2.2.3]. The other properties follow easily from the definition. □

One says that a rigid analytic space X is *equidimensional of dimension n* if all its irreducible components have dimension n. By (iv) and (v), this is equivalent to saying that for all admissible open affinoids $U = \operatorname{Sp} R$, R is equidimensional of dimension n in the algebraic sense, that is to say all the domains $R/\mathfrak{p}_i$ for $\mathfrak{p}_i$ minimal prime ideals have Krull dimension n.

Proposition 3.7.5 *Assume that $\mathcal{W}$ is equidimensional of dimension n. Then so is $\mathcal{E}$.*

Proof Since being equidimensional of dimension n is a local property as we just saw, this results from Proposition 2.3.3. □

Proposition 3.7.6 *If $\mathcal{W}$ is reduced and irreducible, the local rings of $\mathcal{E}$ have no associated primes.*

Proof If $z \in \mathcal{E}$, $x = \kappa(z) \in \mathcal{W}$ and $W = \operatorname{Sp} R \in \mathcal{C}$ an affinoid containing x, then the local ring of $\mathcal{E}$ at z is by construction a localization of the eigenalgebra associated of the R-module M_W. Since R is a domain by assumption, the result then follows from Proposition 2.3.4. □

The next proposition proves the few properties of eigenvarieties for which we need the full force of Theorem 3.3.14.

Proposition 3.7.7

(i) *The map $\kappa \times \psi(U_p)^{-1} : \mathcal{E} \to \mathcal{W} \times \mathbf{A}^1_{rig}$ is finite.*
(ii) *If $\mathcal{W}$ is separated, the eigenvariety $\mathcal{E}$ is separated.*
(iii) *Assume that $\mathcal{W}$ is equidimensional, and that a subset of $\mathcal{W}$ is Zariski-open in $\mathcal{W}$ if and only if its intersection with every $W \in \mathfrak{C}$ is Zariski-open in W. If D is an irreducible component of $\mathcal{E}$, then $\kappa(D)$ is Zariski-open in $\mathcal{W}$.*

Proof For (i) it is enough to prove that for all affinoid subdomains $W = \operatorname{Sp} R \subset \mathcal{W}$ the map $\kappa \times \psi(U_p)^{-1} : \mathcal{E}_W \to W \times \mathbf{A}^1_{\mathrm{rig}}$ is finite. This map factors through a map $\mathcal{E}_W \to Z_W$, where Z_W is the closed subvariety of $W \times \mathbf{A}^1_{\mathrm{rig}}$ cut out by the Fredholm's power series $F_W(T)$ of U_p acting on M_W and it is enough to show that $\mathcal{E}_W \to Z_W$ is a finite map. Since Z_W is admissibly covered by the local pieces $Z_{X,\nu}$ for adapted pairs $X \subset W$, $\nu \in \mathbb{R}$ (Theorem 3.3.14), and the inverse image of $Z_{X,\nu}$ by $\kappa \times \psi(U_p)^{-1}$ is $\mathcal{E}_{X,\nu}$, it suffices to show that the natural map $\mathcal{E}_{X,\nu} \to Z_{X,\nu}$ is finite. But the composition $\mathcal{E}_{X,\nu} \to Z_{X,\nu} \to X$ is finite by construction of $\mathcal{E}_{X,\nu}$ and (i) follows.

Assertion (ii) follows from (i) since a finite map is separated and $\mathcal{W} \times \mathbf{A}^1_{\mathrm{rig}}$ is separated.

Let us prove (iii). By the hypothesis made on $\mathcal{W}$, it suffices to show that $\kappa(D \cap \mathcal{E}_W)$ is Zariski-open in W for $W \in \mathfrak{C}$. By Lemma 3.7.4, $D \cap \mathcal{E}_W$ is a union of irreducible components of $\mathcal{E}_W$ (or is empty, in which case there is nothing to prove), so we may assume that $D \cap \mathcal{E}_W$ is an irreducible component D_W of $\mathcal{E}_W$. Since $\mathcal{E}_W \to Z_W$ is a finite map, it sends D_W to a closed analytic subspace of Z_W of same dimension $\dim \mathcal{E}_W = \dim W$, irreducible by Lemma 3.7.4. Since Z_W is equidimensional of dimension $\dim \mathcal{W} = \dim W$ (same proof as Proposition 3.7.5), the image of D_W in Z_W is an irreducible component Z'_W of Z_W, hence is the locus cut out in $W \times \mathbf{A}^1_{\mathrm{rig}}$ by a power series $F'(T) = 1 + \sum_{i=1}^{\infty} a_i T^i \in R\{\{T\}\}$. The image by the first projection map $Z_W \times W$ is the complementary in W of the closed analytic subspaces defined by the equations $a_i = 0$, hence is Zariski-open. But this is also trivially $\kappa(D_W)$. □

Remark 3.7.8 The condition in (c) of the above proposition is satisfied for instance when $\mathcal{W}$ is affinoid, or when $\mathcal{W}$ is an open ball of radius 1 in $\mathbf{A}^1_{\mathrm{rig}}$, for in that case a proper subspace is Zariski-closed in $\mathcal{W}$ if and only if its intersection with each of the closed ball of radius $r < 1$ is finite, that is, Zariski-closed in that ball.

For the definition of *nested*, and the easy proof of the next proposition, we refer the reader to [13, Definition 7.2.10 and Lemma 7.2.11]:

Proposition 3.7.9

(i) *The eigenvariety $\mathcal{E}$ is nested.*
(ii) *If $\mathcal{O}(\mathcal{E})$ is provided with the coarsest locally convex topology such that the restriction maps $\mathcal{O}(\mathcal{E}) \to \mathcal{O}(U)$ for every affinoid subdomain $U \subset \mathcal{E}$, and if $\mathcal{E}$ is reduced, then the subring $\mathcal{O}(\mathcal{E})^0$ of power-bounded functions in $\mathcal{O}(\mathcal{E})$ is compact.*

A criterion for $\mathcal{E}$ to be reduced is given in the next section.

3.8 A Comparison Theorem for Eigenvarieties

In this section we present two very useful theorems of Chenevier: one gives a sufficient condition for an eigenvariety to be reduced, and the other allows to compare two eigenvarieties. Both rely on the notion of *classical structure*.

3.8.1 Classical Structures

Definition 3.8.1 Let $\mathcal{W}$ be a reduced rigid analytic space over $\mathbb{Q}_p$, and $X \subset \mathcal{W}$. We say that X is *very Zariski-dense* in $\mathcal{W}$ if for every $x \in X$ there is a basis of open affinoid neighborhoods V of x in X such that $X \cap V$ is Zariski-dense in V.

Exercise 3.8.2 Let $\mathcal{W} = \mathbf{A}^1_{\mathrm{rig}}$ be the rigid affine line. Show that any set of the form $a + b\mathbb{N}$ with $a, b \in \mathbb{Z}$, $b \neq 0$ is very Zariski-dense.

Exercise 3.8.3 Give an example of a Zariski-dense subset X of $\mathbf{A}^1_{\mathrm{rig}}$ which is not very Zariski-dense.

Exercise 3.8.4 Let $\mathcal{W}$ be the rigid affine plane $(\mathbf{A}^1_{\mathrm{rig}})^2$. Find a set $X \subset \mathcal{W}$ which is Zariski-dense but such that for all affinoid open subsets V of $\mathcal{W}$, $X \cap V$ is not Zariski-dense in V.

Now let us fix an eigenvariety data as in Sect. 3.6.1, that is (ED1) a ring $\mathcal{H}$ with a distinguished $U_p \in \mathcal{H}$, (ED2) a reduced rigid space $\mathcal{W}$ with an admissible covering $\mathfrak{C}$, and (ED3) Banach modules M_W with an action of $\mathcal{H}$ for every $W \in \mathfrak{C}$, satisfying conditions (EC1) and (EC2).

Definition 3.8.5 A *classical structure* on an eigenvariety data is the data of

(CSD1) a very Zariski-dense subset $X \subset \mathcal{W}$;
(CSD2) for every $x \in X$, an $\mathcal{H}$-module and finite-dimensional $L(x)$-vector space M_x^{cl}.

such that

(CSC) For every real ν, let X_ν be the set of points $x \in X$ such that there exists an embedding of $\mathcal{H}$-module $M_x^{\leq \nu} \hookrightarrow M_x^{\mathrm{cl}}$. Then X_ν is very Zariski-dense in $\mathcal{W}$.

Given an eigenvariety data and a classical structure as above, we say that a point z in $\mathcal{E}$ is *classical* if $\kappa(z)$ belongs to X and the system of eigenvalues ψ_z appears in $M^{\mathrm{cl}}_{\kappa(z)} \otimes_{L(\kappa(z))} L(z)$.

Proposition 3.8.6 *Classical points are very Zariski dense in $\mathcal{E}$.*

Proof First we prove that this set is Zariski-dense. It is enough to show (by Lemma 3.7.4) that classical points are Zariski-dense in any irreducible component D of $\mathcal{E}$. By Proposition 3.7.7(iii), $\kappa(D)$ is Zariski open in $\mathcal{W}$, hence there is an $x \in X$ that belongs to $\kappa(D)$. Let us fix $z \in D$ such that $\kappa(z) = x$ and $\nu \in \mathbb{R}$ such that $\nu > v_p(U_p(z))$. By (CSC) and Proposition 3.3.12 there is an affinoid $W = \operatorname{Sp} R$ in $\mathfrak{C}$ containing x such that (W, ν) is adapted, and $X_\nu \cap W$ is Zariski dense in W. One has $z \in \mathcal{E}_{W,\nu}$ and by definition of Z_ν every point in $\mathcal{E}_{W,\nu}$ that lies above a point of $X_\nu \cap W$ is classical. It follows, by the following lemma applied to the map $\kappa : \operatorname{Spec} \mathcal{T}_{W,\nu} \to \operatorname{Spec} R$ (where $W = \operatorname{Sp} R$ and $\mathcal{E}_{W,\nu} = \operatorname{Sp} \mathcal{T}_{W,\nu}$), that classical points are dense in $\mathcal{E}_{W,\leq\nu}$, hence in $\mathcal{E}_{W,\nu} \cap D$ because it is a union of irreducible components of $\mathcal{E}_{W,\nu}$ (Lemma 3.7.4). Since $\mathcal{E}_{W,\nu} \cap D$ is a non-empty affinoid (it contains z) of the irreducible space D, it is Zariski-dense in D (Lemma 3.7.4), and this completes the proof that Z is Zariski-dense.

Actually this also shows that classical points are very Zariski-dense, because for z a classical point of weight x, the above argument provides an affinoid neighborhood of z on which very classical points are Zariski-dense. □

Lemma 3.8.7 *Let $f : X \to Y$ be a finite flat morphism of noetherian schemes, and $Z \subset Y$ a Zariski-dense subset. Then $f^{-1}(Z)$ is Zariski-dense in X.*

Proof Since f is finite flat, and the schemes noetherian, f maps irreducible components X_i of X surjectively to irreducible components $Y_{j(i)}$ of Y, and the generic point of $X_{(i)}$ is the only point of $X_{(i)}$ that maps to the generic point of $Y_{j(i)}$. In particular, one may assume that X and Y are irreducible. Let T be the Zariski-closure of $f^{-1}(Z)$. Since f is closed, $f(T)$ is closed in Y, but it also contains Z which is dense, so $f(T) = Y$. Hence there is a point in T that maps to the generic point of Y, and this point necessarily is the generic point of X. Since T is closed, $T = X$. □

3.8.2 A Reducedness Criterion

Theorem 3.8.8 *Consider an eigenvariety data with a classical structure as above. Assume that the local rings of $\mathcal{W}$ have no embedded primes. If for every $x \in X$, M_x^{cl} is semi-simple as an $\mathcal{H}$-module, then the eigenvariety $\mathcal{E}$ is reduced.*

We show that if z is a classical point of $\mathcal{E}$, there is an open affinoid neighborhood of z in $\mathcal{E}$ which is reduced. Set $x = \kappa(z) \in X$ and choose a real number $\nu > v_p(U_p(z))$. By (CSC) and Proposition 3.3.12 there is an affinoid $W = \mathrm{Sp}\, R$ in $\mathfrak{C}$ containing x such that (W, ν) is adapted, and $X_\nu \cap W$ is Zariski dense in W. For $x \in X_\nu \cap W$, by definition of X_ν, $M_x^{\leq \nu} = M_x^{\mathrm{cl}, \leq \nu}$, hence is a semi-simple $\mathcal{H}$-module. Since $\mathcal{T}_{W,\nu}$ is the eigenalgebra of $\mathcal{H}$ acting on the module M_W, Proposition 2.9.1 shows that $\mathcal{T}_{W,\nu}$ is a reduced algebra, hence that $\mathcal{E}_{W,\nu}$ is reduced.

To deduce that $\mathcal{E}$ is reduced, it thus suffices, in view of Proposition 3.7.6, to apply the following technical lemma of Chenevier:

Lemma 3.8.9 *If Z is a rigid analytic space over $\mathbb{Q}_p$ whose local rings at every point z have no embedded primes, and are reduced for z in a Zariski dense subset of Z. Then Z is reduced.*

For the proof, see [35, Lemma 3.11].

3.8.3 A Comparison Theorem

Theorem 3.8.10 *Assume that the space $\mathcal{W}$ has an admissible cover, refining $\mathfrak{C}$, by open affinoid subdomains Sp A with A relatively factorial.*[2] *Suppose that we have two eigenvariety datas with the same data (ED1) and (ED2) (that is the same $\mathcal{H}$, same U_p, same $\mathcal{W}$ and same $\mathfrak{C}$), but different Banach modules (ED3), that we shall denote by M_W and M'_W (for $W \in \mathfrak{C}$), and different maps $\phi_W : \mathcal{H} \to \mathrm{End}(M_W)$ and $\phi'_W : \mathcal{H} \to \mathrm{End}(M'_W)$.*

[2] An affinoid ring A is *relatively factorial* if A is a domain and if $f \in A\langle T_1, \dots, T_n\rangle$ is such that $f(0) = 1$, then (f) is a product of principal prime ideals.

Let us call $\mathcal{E}$ and $\mathcal{E}'$ the two eigenvarieties attached to those data. Assume that those two eigenvarieties are each provided with a classical structure with the same set X (CSD1), but with different $\mathcal{H}$-modules M_x^{cl} and M'^{cl}_x.

Suppose that for every x in X, there exists an $\mathcal{H}$-equivariant injective map:

$$M_x^{cl,ss} \hookrightarrow M'^{cl,ss}_x. \tag{3.8.1}$$

Then there exists a unique closed embedding $\mathcal{E}^{red} \hookrightarrow \mathcal{E}'^{red}$ compatible with the weight maps to $\mathcal{W}$ and with the maps $\mathcal{H} \to \mathcal{O}(\mathcal{E}^{red})$ and $\mathcal{H} \to \mathcal{O}(\mathcal{E}'^{red})$. The image of $\mathcal{E}^{red}$ is a union of irreducible components of $\mathcal{E}'^{red}$. Moreover, for every $w \in \mathcal{W}$, there exists an embedding of $\mathcal{H}$-modules $M_w^{\#,ss} \hookrightarrow M'^{\#,ss}_w$ (see Sect. 3.6.1 for the notation).

We are only going to give a complete proof of this very important theorem when the data also satisfy the hypothesis of Theorem 3.8.8 (and in particular, the eigenvarieties $\mathcal{E}$ and $\mathcal{E}'$ are reduced). These hypotheses will always be satisfied in the applications in this book. For the complete proof, see [35, Section 3].

Let us denote by $\mathcal{O}(\mathcal{W})$ the ring of global functions over $\mathcal{W}$. If h is any element of $\mathcal{H}$, then the operators $\psi_W(hU_p)$ on M_W and $\psi'_W(hU_p)$ on M'_W (for $W = \mathrm{Sp}\, R \in \mathfrak{C}$) are compact, and therefore they have a Fredholm determinant $\det(1 - T\psi_W(hU_p))$ and $\det(1 - T\psi_W(hU_p))$ in $R\{\{T\}\}$. When W runs in $\mathfrak{C}$, those determinants glue together and define elements P_{hU_p} and P'_{hU_p} in $\mathcal{O}(\mathcal{W})\{\{T\}\}$.

The **first step** consists in proving that $P_{hU_p} \mid P'_{hU_p}$, for every $h \in \mathcal{H}$, using (3.8.1). This is done in details [35, Prop. 3.2]. We just give here the main ideas.

If $Z_{hU_p} \subset \mathcal{W} \times \mathcal{A}^1$ and $Z'_{hU_p} \subset \mathcal{W} \times \mathcal{A}^1$ are the (possibly non-reduced) analytic varieties defined by P_{hU_p} and P'_{hU_p}, we can reformulate the aim of the first step as showing that Z_{hU_p} is a closed sub-variety of Z'_{hU_p}.

It is easy to prove that $Z^{\mathrm{red}}_{hU_p} \subset Z'^{\mathrm{red}}_{hU_p}$. Indeed, let us say that a point z in Z_{hU_p} is *$\psi(hU_p)$-classical* if $\kappa(z) \in X$ and $\psi(hU_p)(z)$ is an eigenvalue of hU_p in $M_{\kappa(z)}$. The $\psi(hU_p)$-classical points are Zariski dense in Z_{hU_p} by the same proof than that of Proposition 3.8.6 (or also by applying this proposition with $\mathcal{H}$ changed to $Z[hU_p]$ and U_p changed to hU_p). But condition (3.8.1) ensures that those classical points also belong to Z'_{hU_p}.

The inclusion $Z^{\mathrm{red}}_{hU_p} \subset Z'^{\mathrm{red}}_{hU_p}$ follows. Under the hypothesis of Theorem 3.8.8, this implies that $Z_{hU_p} \subset Z'_{hU_p}$, and completes the proof of the first step. Proving the same inclusion in general requires more work, involving the unique factorization into a product of irreducible elements of the elements of $\mathcal{O}(\mathcal{W})\{\{T\}\}$, proved by Conrad [48] under the hypothesis that $\mathcal{W}$ is covered by relatively factorial affinoids.

The **second step** consists in proving that for every $w \in \mathcal{W}$, and $\nu \in \mathbb{R}$, there is an embedding of $\mathcal{H}$-modules $M_w^{\leq\nu,\mathrm{ss}} \hookrightarrow M'^{\leq\nu,\mathrm{ss}}_w$. Note that this already implies the last assertion of the theorem. The second step is proved as follows.

By classical representation theory over finite dimensional vector spaces, it suffices for this to prove that

$$\forall h \in \mathcal{H},\ \det(1 - T\psi_w(h)_{|M_w^{\leq \nu}}) \mid \det(1 - T\psi'_w(h)_{|M'^{\leq \nu}_w}) \text{ in } L_w\{\{T\}\}. \tag{3.8.2}$$

We claim that actually it is enough to prove (3.8.2) when

$$\text{The characteristic polynomial } \psi_w(h) \text{ (resp. } \psi'_w(h)) \text{ acting on } M_w^{\leq \nu} \text{ (resp. } M'^{\leq \nu}_w) \text{ has all its roots of valuation 0.} \tag{3.8.3}$$

Indeed, for $h \in \mathcal{H}$, any element of the form $1 + \lambda h$ with $\lambda \in L_w$ sufficiently small satisfies property (3.8.3) and (3.8.2) for such elements h implies that $\det(1 - T\psi_w(1 + \lambda h)_{|M_w^{\leq \nu}}) \mid \det(1 - T\psi'_w(1+\lambda h)_{|M'^{\leq \nu}_w}))$ in $L[T, \lambda]$; looking at the dominant term gives (3.8.2).

For $h \in \mathcal{H}$, by Step 1, we know that $P_{hU_p,w} \mid P'_{hU_p,w}$ in $L(w)\{\{T\}\}$. Consider the canonical decompositions $P_{hU_p,w} = QG$ and $P'_{hU_p,w} = Q'G'$ of Corollary 3.3.7, where Q is strongly ν-dominant of degree $N(P_{hU_p,w}, \nu)$, $N(G, \nu) = 0$ and $Q(0) = G(0) = 1$, and similarly for Q' and G'. Then from $P_{hU_p,w} \mid P'_{hU_p.w}$ we deduce that $Q \mid Q'$. Now if h satisfies (3.8.3), then $\det(1 - T\psi_w(hU_p)_{|M_w^{\leq \nu}}) = Q$ and $\det(1 - T\psi_w(h'U_p)_{|M'^{\leq \nu}_w}) = Q'$. Hence we get for such elements h:

$$\det(1 - T\psi_w(hU_p)_{|M_w^{\leq \nu}}) \mid \det(1 - T\psi'_w(hU_p)_{|M'^{\leq \nu}_w}) \tag{3.8.4}$$

If $\lambda \in \mathbb{Z}$ small enough p-adically, then $h + \lambda$ still satisfies (3.8.3), so by (3.8.4)

$$\det(1 - T(\psi_w(hU_p) + \lambda\psi_w(U_p))_{|M_w^{\leq \nu}}) \mid \det(1 - T(\psi'_w(hU_p) + \lambda\psi'_w(U_p))_{|M'^{\leq \nu}_w}) \tag{3.8.5}$$

Lemma 3.8.11 *Let L be a field, A and B (resp. A' and B') be two commuting square matrices over L, with B invertible. Assume that*

$$\det(1 - T(AB + \lambda B)) \mid \det(1 - T(A'B' + \lambda B')) \text{ in } L[T]$$

for infinitely many λ in L. Then

$$\det(1 - TA) \mid \det(1 - TA') \text{ in } L[T].$$

Proof Let $a_i, b_i, i = 1, \ldots, r$ (resp. $a'_i, b'_i, i = 1, \ldots, r'$) be the eigenvalues of A and B (resp. A' and B') in the algebraic closure of L, repeated according to their multiplicity. Since A and B commute, and A' and B' as well, the hypothesis implies for infinitely many λ

$$\prod_{i=1}^{r}(1 - Ta_ib_i - T\lambda b_i) \mid \prod_{i=1}^{r'}(1 - Ta'_ib'_i - T\lambda b'_i) \text{ in } L[T],$$

which means that for every $i \in \{1, \dots, r\}$ and for infinitely many λ, there is a $j \in \{1, \dots, r'\}$ such that $a_i b_i + \lambda b_i = a'_j b'_j + \lambda b'_j$. Since there are only finitely many possible j, it is still true that for infinitely many λ, $a_i b_i + \lambda b_i = a'_j b'_j + \lambda b'_j$ with the same j. But this implies $b_i = b'_j$, and $a_i b_i = a'_j b'_j$ hence (since $b_i \neq 0$), $a_i = a'_j$. Thus the multiset of eigenvalues of A is included into the one of A', and the lemma follows. □

The lemma and (3.8.5) imply (3.8.2) and completes the second step of the proof.

The **third step** begins by constructing an admissible covering of $\mathcal{E}$ by admissible open affinoids of the form $\mathcal{E}_{W,\nu}$ where the (W, ν) are adapted both for P_{U_p} and P'_{U_p}. For such a pair (W, ν) with $W = \operatorname{Sp} R$, there exists a unique closed immersion $\mathcal{E}^{\text{red}}_{W,\nu} = \operatorname{Sp} \mathcal{T}^{\text{red}}_{W,\nu} \to \mathcal{E}'^{\text{red}}_{W,\nu} = \operatorname{Sp} \mathcal{T}'^{\text{red}}_{W,\nu}$ compatible with the weight maps and the morphisms from $\mathcal{H}$: the uniqueness is obvious, and for the existence, remember that $\mathcal{T}_{W,\nu}$ (resp. $\mathcal{T}'_{W,\nu}$) is the eigenalgebra constructed on $M_W^{\leq \nu}$ (resp. $M'^{\leq \nu}_W$), so Theorem 2.9.2, whose hypothesis is satisfied by step 2, gives a surjective morphism of $\mathcal{H}$-algebra and R-modules $\mathcal{T}'^{\text{red}}_{W,\nu} \to \mathcal{T}^{\text{red}}_{W,\nu}$. By uniqueness we can glue these morphisms to get a closed immersion $\mathcal{E} \hookrightarrow \mathcal{E}'$ compatible with the weight maps and the morphisms from $\mathcal{H}$.

This completes the proof of the first sentence of the Theorem.

The **last step** that remains is to prove that $\mathcal{E}^{\text{red}}$ is a union of irreducible components of $\mathcal{E}'$.

Lemma 3.8.12 *Let R be a noetherian domain, T an R-algebra that is finite torsion-free as an R-module, and $(\mathfrak{p}_s)_{s \in S}$ the complete finite list of minimal prime ideals of T. Let I an ideal of T. Then T/I is reduced and torsion-free as an R-module, if and only if, $I = \cap_{s \in S'} \mathfrak{p}_s$ for some subset S' of S.*

In particular, T^{red} is also finite torsion-free over R.

Proof Note that for every s, $\mathfrak{p}_s \cap R = (0)$ by Cohen's first theorem.

Assume that $I = \cap_{s \in S'} \mathfrak{p}_s$. Let $t \in T$, $x \in R - \{0\}$, such that $xt \in I$. For every $s \in S'$, $xt \in \mathfrak{p}_s$, but $x \notin \mathfrak{p}_s$, so $t \in \mathfrak{p}_s$, and thus $t \in I$. It follows that T/I is torsion-free as an R-module. To prove that T/I is reduced, just observe that if $t^n \in I$ for some $n \geq 1$, then $t^n \in \mathfrak{p}_s$ for every $s \in S'$, hence $t \in \mathfrak{p}_s$ for every $s \in S'$, hence $t \in I$.

The "in particular" follows, by applying the above to the radical of T, which is $\cap_{s \in S} \mathfrak{p}_s$. We may therefore, for the converse, assume that T is reduced.

Let I be an ideal of T such that T/I is reduced, and torsion-free as an R-module. Let P be an associated prime of T/I as a T-module. Then $P \cap R$ is an associated prime of T/I as an R-module (by Eisenbud [59, Cor. 3.2]), hence is 0. Therefore P is a minimal prime ideal of T. Thus the set of associated primes of T/I is a subset $(\mathfrak{p}_s)_{s \in S'}$ of the set of minimal prime ideals of T. Consider a primary decomposition $I = \cap_{s \in S'} Q_s$, where each ideal Q_s is $\mathfrak{p}_s$-primary. Since R/I is reduced, $I = \sqrt{I} = \sqrt{\cap_{s \in S'} Q_s} = \cap_{s \in S'} \sqrt{Q_s} = \cap_{s \in S'} \mathfrak{p}_S$. □

It remains to prove that the image of $\mathcal{E}^{\text{red}}$ is a union of irreducible components of $\mathcal{E}'^{\text{red}}$. Let $\mathcal{I}$ denote this image. By construction, there is an admissible covering of $\mathcal{E}'$ by admissible open affinoids of the form $\mathcal{E}'_{W,\nu} = \operatorname{Spec} \mathcal{T}'_{W,\nu}$, and of $\mathcal{E}$ by admissible open affinoids of the form $\mathcal{E}_{W,\nu} = \operatorname{Spec} \mathcal{T}_{W,\nu}$, with $W = \operatorname{Sp} R$ the maximal spectrum of locally factorial (in particular, a domain) affinoid ring R, and $\mathcal{T}_{W,\nu}$, $\mathcal{T}'_{W,\nu}$ finite and torsion-free over R such that $\mathcal{I} \cap \mathcal{E}'^{\text{red}}_{W,\nu}$ is the image of $\mathcal{E}^{\text{red}}_{W,\nu}$ by a closed embedding into $(\mathcal{E}'_{W,\nu})^{\text{red}}$. Since the ring $\mathcal{T}^{\text{red}}_{W,\nu}$ and $\mathcal{T}'^{\text{red}}_{W,\nu}$ are also finite torsion-free over R, the preceding lemma tells us that $\mathcal{I} \cap (\mathcal{E}'_{W,\nu})^{\text{red}}$ is a finite union of irreducible components of $(\mathcal{E}_{W,\nu})^{\text{red}}$.

Now let $x \in \mathcal{I}$, and U an admissible affinoid domain containing x such that $\mathcal{I} \cap U$ is a finite union of irreducible components of $\mathcal{E}' \cap U$. Let J be one of these finite irreducible component, containing x, and $\mathcal{J}$ the irreducible component of $\mathcal{E}'$ containing it. Then $\mathcal{J} \cap U$ is a finite union of irreducible components (by Lemma 3.7.4(v)) and J is one of then. It follows that for any point y in J sufficiently close to x but not equal to x, y belongs to only one component of $\mathcal{J} \cap U$, namely J. On some sufficiently small admissible affinoid neighborhood U' of y, $\mathcal{I} \cap U' = J \cap U' = \mathcal{J} \cap U'$. But this subset of $\mathcal{J}$ is Zariski-dense in $\mathcal{J}$ by Lemma 3.7.4(vi), and it follow that the closed set $\mathcal{I}$ contains the whole component $\mathcal{J}$ of $\mathcal{E}'$.

We have therefore proven that for any $x \in \mathcal{I}$, there is an irreducible component of $\mathcal{E}$ (namely $\mathcal{J}$) containing x and contained in $\mathcal{I}$. This exactly means that $\mathcal{I}$ is a union of irreducible components.

This completes our sketch of the proof of Theorem 3.8.10.

We note a simple corollary:

Corollary 3.8.13 *Same hypothesis but we assume that for every x in X, there exists an isomorphism of $\mathcal{H}$-modules*

$$M_x^{cl,ss} \simeq M'^{cl,ss}_x.$$

Then there exists a unique isomorphism $\mathcal{E} \simeq \mathcal{E}'$ compatible with the weight maps to $\mathcal{W}$ and with the maps $\mathcal{H} \to \mathcal{O}(\mathcal{E})$ and $\mathcal{H} \to \mathcal{O}(\mathcal{E}')$.

3.9 A Simple Generalization: The Eigenvariety Machine for Complexes

In this section we present a simple generalization of the eigenvariety machine, which takes as input not a single $\mathcal{H}$-module, but a complex of $\mathcal{H}$-modules, and returns a rigid analytic space that is, morally, the eigenvariety associated to the cohomology of that complex, or *cohomological eigenvariety*. This generalization is often necessary to construct eigenvarieties of higher dimension. It may even be useful in the theory of the eigencurve. In this book, we will use it in the final chapter.

3.9.1 Data for a Cohomological Eigenvariety

To fix ideas, in this section we call *complex* (in an abelian category $\mathcal{A}$) a cohomological complex (M^n, d_n), with the M^n objects of $\mathcal{A}$ and the $d_n : M^n \to M^{n+1}$ morphisms in $\mathcal{A}$ such that $d_{n+1}d_n = 0$ for all n, and with the property that $M^n = 0$ for all but finitely many n. The complex is usually denoted by $(M^\bullet)$, and the cohomology $H^n(M^\bullet)$ is defined as $\ker d_{n+1}/\operatorname{Im} d_n$, with the total cohomology $H^\bullet(M^\bullet)$ defined as the direct sum

$$H^\bullet(M^\bullet) = \oplus_{n\in\mathbb{Z}} H^n(M^\bullet).$$

A *cohomoligical eigenvariety data* over a finite extension L of $\mathbb{Q}_p$ is the data of

(ECD1) A commutative ring $\mathcal{H}$, with a distinguished element U_p.
(ECD2) A reduced rigid analytic variety $\mathcal{W}$ over L, and an admissible covering $\mathfrak{C}$ by admissible affinoid open subsets of $\mathcal{W}$.
(ECD3) For every admissible affinoid $W = \operatorname{Sp} R$ of $\mathcal{W}$, $W \in \mathfrak{C}$, a cohomological complex $M^\bullet_W$ of Banach modules M^n_W satisfying property (Pr), and a ring homomorphism $\psi_W : \mathcal{H} \to \operatorname{End}_R(M^\bullet_W)$.

satisfying the following conditions:

(ECC1) For every $W \in \mathfrak{C}$, $\psi_W(U_p)$ is compact on each of the M^n_W.
(ECC2) For every pair of affinoid subdomains $W' = \operatorname{Sp} R' \subset W = \operatorname{Sp} R \subset \mathcal{W}$, $W, W' \in \mathfrak{C}$, and every integer n the R'-Banach modules $M^n_W \hat{\otimes}_R R'$ and $M^n_{W'}$ are *linked* (cf. Definition 3.5.1)

To this data we attach a *cohomological eigenvariety*, as explained in the following subsection.

3.9.2 Construction of the Cohomological Eigenvariety

For any admissible affinoid subdomain $W = \operatorname{Sp} R$ in $\mathfrak{C}$, let $F^n_W \in R\{\{T\}\}$ be the Fredholm determinant of U_p acting on M^n_W (that is $F_W(T) = \det(1 - T\psi^n_W(U_p))$, where ψ^n_W is the map $\mathcal{H} \to \operatorname{End}_R(M^n_W)$ deduced by restriction from ψ_W.) To any such W, and $\nu \in \mathbb{R}$ adapted to F^n_W for every n, we attach a *local piece of the eigenvariety* which consists of the following:

(a) An L-affinoid variety $\mathcal{E}_{W,\nu}$, called *the local piece*.
(b) A finite morphism $\kappa : \mathcal{E}_{W,\nu} \to W$, called *the weight map*.
(c) A morphism of rings $\psi : \mathcal{H} \to \mathcal{O}(\mathcal{E}_{W,\nu})$.

The construction is as follows: since ν is adapted to every F^n_W, we can define the finite flat R-submodule $(M^n_W)^{\leq\nu}$. Since $\mathcal{H}$ is commutative, it stabilizes $(M^n_W)^{\leq\nu}$ by Proposition 3.4.2, and since d_n commutes with the action of U_p, $(M^\bullet_W)^{\leq\nu}$ is a complex of finite projective R-modules.

We define $\mathcal{T}_{W,\nu}$ to be the eigenalgebra of $\mathcal{H}$ acting on $H^\bullet((M_W^\bullet)^{\leq\nu})$, the total cohomology of that complex. That is, $\mathcal{T}$ is the sub-algebra of $\mathrm{End}_R(H^\bullet((M_W^\bullet)^{\leq\nu})$ generated by $\psi(\mathcal{H})$. (Note that we are outside of the context of Sect. 2.2 since the R-module $H^\bullet((M_W^\bullet)^{\leq\nu})$ is certainly finite, but may not be projective.)

Since $H^\bullet((M_W^\bullet)^{\leq\nu})$ is a finite R-module, $\mathcal{T}_{W,\nu}$ is finite over R, hence an affinoid ring. We thus can define $\mathcal{E}_{W,\nu} = \mathrm{Sp}\,\mathcal{T}_{W,\nu}$. The structural map defines a finite morphism of rigid affinoid variety $\kappa : \mathcal{E}_{W,\nu} = \mathrm{Sp}\,\mathcal{T}_{W,\nu} \to W = \mathrm{Sp}\,R$. And we have by construction a natural ring homomorphism $\psi_{W,\nu} : \mathcal{H} \to \mathcal{T}_{W,\nu} = \mathcal{O}(\mathcal{E}_{W,\nu})$.

Remark 3.9.1 Observe that we first took the $\leq \nu$-slope part of the modules M_W^n, getting a finite complex of finite flat R-modules, and then took the cohomology of that complex. We didn't go the other way around, by first taking the cohomology of the complex $M_W^\bullet$, and then the $\leq \nu$ slope-part. Indeed, the cohomology space $H^n(M_W^\bullet)$, being a sub-quotient of M_W^n, has no reason to be a Banach space, let alone to have property (Pr); and we have only defined the notion of ν being adapted, and the notion of slope-$\leq \nu$ part, for a Banach module satisfying (Pr).

We observe that Lemma 3.6.1 remains true in this context. Indeed, in its proof, the only case that requires modification is the construction of a closed immersion $\mathcal{E}_{W,\nu} \to \mathcal{E}_{W,\nu'}$, compatible with the maps to $W = \mathrm{Sp}\,R$ and the maps from $\mathcal{H}$, where $\nu < \nu'$ are both adapted to W. But in this case we just have to observe that the complex of finite flat R-modules with $\mathcal{H}$-action $(M_W^\bullet)^{\leq\nu}$ is just the image of the complex $(M_W^\bullet)^{\leq\nu'}$ by the functor $N \to N^{\leq\nu}$, which is exact (Lemma 3.4.3). Hence a natural $\mathcal{H}$-compatible injection $H^\bullet((M_W^\bullet)^{\leq\nu}) \hookrightarrow H^\bullet((M_W^\bullet)^{\leq\nu'})$, which defines a surjective map $\mathcal{T}_{W,\nu'} \to \mathcal{T}_{W,\nu}$, hence the desired closed immersion $\mathcal{E}_{W,\nu} \to \mathcal{E}_{W,\nu'}$.

We then glue those local pieces $\mathcal{E}_{W,\nu}$ to construct a full eigenvariety $\mathcal{E}$, with a weight map $\kappa : \mathcal{E} \to \mathcal{W}$ glueing the structural maps $\mathcal{E}_{W,\nu} \to W$, and a morphism of algebras $\psi : \mathcal{H} \to \mathcal{O}(\mathcal{E})$, exactly as in Sect. 3.9.2.

3.9.3 Properties of the Cohomological Eigenvariety

By assumption, for n fixed, the series $F_W^n(T) \in R\{\{T\}\}$ agree for the various $W = \mathrm{Spec}\,R \in \mathfrak{C}$, and glue to define a powerseries $F^n(T) \in \mathcal{O}(\mathcal{W})\{\{T\}\}$. Calling Z_W the zero locus of the power series $F(T) = \prod_n F^n(T)$, Theorem 3.3.14 tells us that Z is admissibly covered by subset $Z_{W,\nu}$ with (W, ν) admissible for F, hence for each of the F_n (Exercise 3.3.11). Hence we get a finite map $\mathcal{E} \to Z$, which implies that $\mathcal{E}$ is a separated rigid analytic variety.

However, the reader should not expect $\mathcal{E}$ to be equidimensional when W is. Indeed, the modules $H^\bullet((M_W^\bullet)^{\leq\nu})$ are not in general projective over R. Actually the following result gives useful information about these modules:

Proposition 3.9.2 *Let $W = \mathrm{Spec}\,R$ in $\mathfrak{C}$, S any R-algebra, and $\nu \in \mathbb{R}$ admissible for all the M_W^n, there exists an Hecke-equivariant convergent spectral sequence*

$$E_{i,j}^2 = \mathrm{Ext}_R^i(H^j((M_W^\bullet)^{\leq\nu}), S)) \Rightarrow H^{i+j}((M_W^\bullet)^{\leq\nu} \otimes_R S).$$

Proof This is just the spectral sequence of hypercohomology of the functor $- \otimes_R S$ on the complex $(M_W^\bullet)^{\leq \nu}$, taking into account that the functor $N \mapsto N^{\leq \nu}$ is exact. □

Let w be a point of $\mathcal{W}$, of field of definition L, and ν a real such that there exists an admissible neighborhood W of x with (W, ν) adapted for F_n for every n. Write $(M_w^n)^{\leq \nu}$ for the L-vector space (with action of $\mathcal{H}$) $(M_W^n)^{\leq \nu} \otimes_R L$ which is independent of W by assumption (ECC2). Let $\mathcal{T}_{w,\nu}$ be the sub-algebra acting of $\mathrm{End}(H^\bullet((M_w^\bullet)^{\leq \nu}))$ generated by $\mathcal{H}$. Then

Corollary 3.9.3 *There is a unique isomorphism $\mathcal{T}_{w,\nu}^{red} \simeq (\mathcal{T}_{W,\nu} \otimes_R L)^{red}$, compatible with the morphisms from $\mathcal{H}$ to both sides.*

Proof Up to replacing L by a finite extension, we only need to show that the systems of eigenvalues of $\mathcal{H}$ appearing in the finite-dimensional semi-simple L-algebra $\mathcal{T}_{w,\nu}^{\mathrm{red}}$ are the same (without multiplicity) as the one appearing in the finite-dimensional semi-simple L-algebra $(\mathcal{T}_{W,\nu} \otimes_R L)^{\mathrm{red}}$. Using Corollary 2.5.10 and Prop 2.4.1, it suffices to show that the system of $\mathcal{H}$-eigenvalues appearing in $\mathrm{End}(H^\bullet((M_w^\bullet)^{\leq \nu}))$ are the same (without taking care of multiplicity) that the ones appearing in $\mathrm{End}(H^\bullet((M_W^\bullet)^{\leq \nu})) \otimes_R L$.

Now choose a character $\lambda : \mathcal{H} \to L^*$ and localize the entire spectral sequence of the preceding proposition (for $S = L$), which is $\mathcal{H}$-equivariant, at λ. This localized spectral sequence abuts at $E_{i,j}^\infty$ such that $\oplus_{i,j} E_{i,j}^\infty = (H^\bullet((M_w^\bullet)^{\leq \nu}))_\lambda$. Thus saying that λ appears in $H^\bullet((M_w^\bullet)^{\leq \nu})$ is the same as saying that some $E_{i,j}^\infty$ is not zero.

The first page is by definition $E_{i,j}^2 = \mathrm{Ext}_R^i(H^j((M_W^\bullet)_\lambda^{\leq \nu}, L))$, so if λ appears in $H^\bullet((M_w^\bullet)^{\leq \nu})$, this E^2-page is non-zero, so for some (i, j), $\mathrm{Ext}_R^i(H^j((M_W^\bullet)^{\leq \nu})_\lambda, L) \neq 0$ hence $H^j((M_W^\bullet)^{\leq \nu})_\lambda \neq 0$, which means that λ appears in $H^j((M_W^\bullet)^{\leq \nu}) \otimes L$. Conversely, if λ appears in $H^\bullet((M_W^\bullet)^{\leq \nu}) \otimes L$, there is a **maximal** integer j such that λ appears in $H^j((M_W^\bullet)^{\leq \nu}) \otimes_R L$, that is such that $H^j((M_W^\bullet)^{\leq \nu})_\lambda \neq 0$. The term $E_{0,j}^2$ is thus non-zero, but the terms $E_{2_{i,j}}$ are zero for $i > 0$. Thus $E_{0,j}^\infty = E_2^{0,j}$ is non-zero, and λ appears in $H^\bullet((M_w^\bullet)^{\leq \nu})$ as wanted. □

Theorem 3.9.4 *Let w be a point of $\mathcal{W}$, of field of definition $L(w)$. For any finite extension L' of $L(w)$, the set of L'-points z of $\mathcal{E}$ such that $\kappa(z) = w$ is in natural bijection with the systems of eigenvalues of $\mathcal{H}$ appearing in $H^\bullet(M_w^\bullet) \otimes_{L(w)} L'$ such that the eigenvalue of U_p is non-zero. The bijection attaches to z the system of eigenvalues $\psi_z : \mathcal{H} \to L'$, $T \mapsto \psi(T)(z)$ for $T \in \mathcal{H}$.*

This is easily proven like Theorem 3.7.1, using the preceding corollary.

3.9.4 Classical Structures and an Application to Reducedness

Definition 3.9.5 A *classical structure* on a cohomological eigenvariety data is the data of

(CSD1) a very Zariski-dense subset $X \subset \mathcal{W}$;

(CSD2) for every $x \in X$, an $\mathcal{H}$-module and finite-dimensional $L(x)$-vector space M_x^{cl}.

such that

(CSC) For every real ν, let X_ν be the set of points $x \in X$ such that there exists an embedding of $\mathcal{H}$-module $H^\bullet(M_x^\bullet)^{\leq \nu} \hookrightarrow M_x^{\mathrm{cl}}$. Then X_ν is very Zariski-dense in $\mathcal{W}$.

Given a cohomological eigenvariety data and a classical structure as above, we say that a point z in $\mathcal{E}$ is *classical* if $\kappa(z)$ belongs to X and the system of eigenvalues ψ_z appears in $M_{\kappa(z)}^{\mathrm{cl}} \otimes_{L(\kappa(z))} L(z)$.

Proposition 3.9.6 *Classical points are very Zariski dense in $\mathcal{E}$.*

The proof of 3.9.6 adapts without difficulty.

Theorem 3.9.7 *Consider a cohomological eigenvariety data with a classical structure as above. Assume that the local rings of $\mathcal{W}$ have no embedded primes. If for every $x \in X$, M_x^{cl} is semi-simple as an $\mathcal{H}$-module, then the eigenvariety $\mathcal{E}$ is reduced.*

Again, the proof of Theorem 3.8.8 adapts immediately.

Remark 3.9.8 Their seem to be serious difficulties, however, in adapting Chenevier's camparison theorem in the cohomological context. They lie in the fact that the global Fredholm determinant $F_{U_p} \in \mathcal{O}(W)\{\{T\}\}$, and more generally F_{hU_p} for $h \in \mathcal{H}$, are constructed from the complexes $M_W^\bullet$, not from their cohomology, and therefore may have zeros that are eigenvalues on one of the M^n but not on the cohomoogy of the complex. On the other hand trying to define an F_{hU_p} attached to the cohomology of the complex seems difficult. Without these global objects, the proof of Chenevier's theorem breaks down.

3.10 Notes and References

Most of the results of this chapter are due to Serre, Lazard, Coleman, Mazur, Buzzard and Chenevier. Only the presentation and some generalizations or variants are due to the author of this book.

The largest part of the material presented here is due to R. Coleman. He was the first, in [38] to construct families of finite slope modular forms and from this families of eigenforms or local pieces of the eigencurve, providing a generalization of the 10-year older work of Hida on ordinary family of modular forms that had been conjectured in the meantime by Mazur and Gouvêa. To this aim, Coleman had to develop a Riesz's theory for compact operators over certain type of affinoid rings, which he did, extending much earlier works of Lazard and Serre over a base field. Results of Coleman were soon after globalized by himself and Mazur, who defined and constructed the Eigencurve in [40]. The method to construct the eigencurve from Coleman's modules of families of finite slope modular forms was later generalized and axiomatized by Buzzard in [29]. At about the same time, Chenevier used it in [34] to construct eigenvarieties for inner forms of GL_n that are compact at infinity.

More precisely, the material in Sect. 3.1 mostly follows [29, §2] which itself follows closely [38], but we take a somewhat shorter route, which allows us to slightly weaken noetherian hypothesis. The material of Sects. 3.2.1, 3.2.2, 3.3.1, and 3.3.2 is a generalization to the case where the base is a Banach algebra of results of Lazard [83] in the case of a base field. The proofs here in general follow Lazard's proofs, with the necessary adaptations. Section 3.2.4 follows very closely Coleman [38] and Sect. 3.2.5 is a simple generalization of [108]. Section 3.3.3 which defines the notion of adapted pair prepares for a variant of Buzzard's construction of the eigenvariety, where the local pieces are "cylinders" $\mathcal{E}_{W,\nu}$ easier to deal with in the applications than the more general local pieces used by Buzzard. The proof of Theorem 3.3.14 is inspired by that of [29, Theorem 4.6]. Sections 3.4, 3.5, and 3.6 presents that construction which parallels that of Buzzard. The material of Sect. 3.7 is due to Chenevier [34] who generalizes in part Coleman–Mazur [40] and that of Sect. 3.8 is entirely due to Chenevier [35].

The generalization presented in Sect. 3.9 is based on constructions and results of [5, 123] and [65].

Part II
Modular Symbols and L-Functions

Chapter 4
Abstract Modular Symbols

In this chapter, we introduce the fundamental notion of *modular symbols* with values in a *system of coefficients* V, an abelian group with an action of a submonoid of $GL_2(\mathbb{Q})$. Modular symbols will be used throughout the rest of this book, with increasingly sophisticated systems of coefficients V—first in the next chapter, Chap. 5, with V being a symmetric power of the standard representation of GL_2, and then in the subsequent chapters, with V being various spaces of modules of distributions. Modular symbols with this kind of coefficients have strong connections respectively with classical modular forms, and with families of p-adic modular forms and their p-adic L-functions.

This chapter will discuss the general properties enjoyed by modular symbols with all kind of systems of coefficients. We study their definition, their behavior under base change and the action of Hecke operators they enjoy. We also study their connection with algebraic topology, namely the identification of groups of modular symbols with systems of coefficients V with cohomology with compact support of the open modular curve with value in local coefficients $\tilde{V}$ attached to V. Fundamental for the sequel, we define and study various natural pairings between spaces of modular symbols.

4.1 The Notion of Modular Symbols

Let Δ be the abelian group of divisors on $\mathbb{P}^1(\mathbb{Q})$, that is the group of formal sums $\sum_{x \in \mathbb{P}^1(\mathbb{Q})} n_x\{x\}$, with $n_x \in \mathbb{Z}$ and $n_x = 0$ for all x but finitely many. Let Δ_0 be the subgroup of divisors of degree 0, that is such that $\sum_x n_x = 0$. The natural action of $GL_2(\mathbb{Q})$ on $\mathbb{P}^1(\mathbb{Q})$, $\gamma \cdot x = \frac{ax+b}{cx+d}$ if $\gamma = \begin{pmatrix} a & b \\ c & d \end{pmatrix}$, induces an action of $GL_2(\mathbb{Q})$ on Δ and on Δ_0, respecting the group structures.

J. Bellaïche, *The Eigenbook*, Pathways in Mathematics,
https://doi.org/10.1007/978-3-030-77263-5_4

Exercise 4.1.1

1. Prove Manin's lemma: Δ_0 is generated, as a left $\mathbb{Z}[\mathrm{SL}_2(\mathbb{Z})]$-module, by the divisor $\{\infty\} - \{0\}$. You can proceed as follows:
   (a) Prove Pick's lemma: a non-flat triangle with vertices in $\mathbb{Z}^2$ has area $\geq 1/2$, with equality if and only if the three vertices are the only points of $\mathbb{Z}^2$ in the triangle.
   (b) Let a/c and b/d be two irreducible fractions. Show that there exists $\gamma \in \mathrm{SL}_2(\mathbb{Z})$ such that $\gamma(\{\infty\} - \{0\}) = \{a/c\} - \{b/d\}$ if and only if the triangle of vertices $(0, 0)$, (a, c), (b, d) has area $1/2$.
   (c) Conclude.
2. Let $\tau = \begin{pmatrix} 0 & -1 \\ 1 & -1 \end{pmatrix}$ and $\sigma = \begin{pmatrix} 0 & -1 \\ 1 & 0 \end{pmatrix}$. By Manin's lemma, the morphism of left $\mathbb{Z}[\mathrm{SL}_2(\mathbb{Z})]$-modules $M : \mathbb{Z}[\mathrm{SL}_2(\mathbb{Z})] \to \Delta_0$, $x \mapsto x \cdot (\{\infty\} - \{0\})$ is surjective. Show that the elements $1 + \tau + \tau^2$ and $1 + \sigma$ of $\mathbb{Z}[\mathrm{SL}_2(\mathbb{Z})]$ belong to the kernel of M.
3. The kernel of M is actually generated, as a left ideal of $\mathbb{Z}[\mathrm{SL}_2(\mathbb{Z})]$, by $1 + \tau + \tau^2$ and $1 + \sigma$. Just read the original proof of this fact: [87, Theorem 1.9].

Exercise 4.1.2 Using Exercise 4.1.1, prove that if Γ is a finite index subgroup of $\mathrm{SL}_2(\mathbb{Z})$, then Δ_0 is finitely presented as a left $\mathbb{Z}[\Gamma]$-module.

Exercise 4.1.3 Using question 1 of Exercise 4.1.1, prove Pick's theorem: *if P is a convex polygon whose all vertices are in $\mathbb{Z}^2$, then the area of P is $I + \frac{F}{2} - 1$ where I is the number of points of $\mathbb{Z}^2$ in the interior of P, and F the number of points of $\mathbb{Z}^2$ on its boundary.* Is the result still true for non-convex polygons?

Let Σ be a submonoid of $\mathrm{GL}_2(\mathbb{Q})$ and let V a right Σ-module, i.e. an abelian group with a Σ-action on the right (denoted by $v \mapsto v_{|\sigma}$). There is a right action of Σ on $\mathrm{Hom}(\Delta_0, V)$ given by

$$(\phi_{|\sigma})(D) = \phi(\sigma \cdot D)_{|\sigma} \tag{4.1.1}$$

for σ in Σ. We also define a right action of Σ on $\mathrm{Hom}(\Delta, V)$ by the same formula.

Definition 4.1.4 If Γ is a subgroup of Σ, we define the group of *V-valued modular symbols for* Γ as the subgroup of $\mathrm{Hom}(\Delta_0, V)$ fixed by the action of Γ:

$$\mathrm{Symb}_\Gamma(V) = \mathrm{Hom}(\Delta_0, V)^\Gamma.$$

Let R be a commutative ring. Observe that if V is an R-module with a right Σ-action that preserves its R-module structure, then $\mathrm{Symb}_\Gamma(V)$ has an obvious structure of R-module. The construction $V \mapsto \mathrm{Symb}_\Gamma(V)$ is functorial, from the categories of R-modules with right Σ-actions to the category of R-modules.

Lemma 4.1.5 *The functor $V \mapsto \mathrm{Symb}_\Gamma(V)$ is left-exact and commutes with flat base change $R \to R'$ (that is, if V is an R-module with a right Σ-action, there is a natural isomorphism $\mathrm{Symb}_\Gamma(V) \otimes_R R' \simeq \mathrm{Symb}_\Gamma(V \otimes_R R')$ when R' is R-flat).*

Proof The left-exactness is clear since both $V \mapsto \mathrm{Hom}(\Delta_0, V)$ and the functor of Γ-invariants are left exact. For the second part, first note that $\mathrm{Symb}_\Gamma(V) = \mathrm{Hom}_{\mathbb{Z}[\Gamma]}(\Delta_0, V)$ if we give V its left Γ-structure $\gamma \cdot v = v_{|\gamma^{-1}}$. Now consider a presentation $\mathbb{Z}[\Gamma]^{(J)} \overset{\mu}{\to} \mathbb{Z}[\Gamma]^{(I)} \to \Delta_0 \to 0$ of Δ_0 as $\mathbb{Z}[\Gamma]$-module (Exercise 4.1.2 tells us that if Γ is a finite index subgroup of $\mathrm{SL}_2(\mathbb{Z})$, there is even a finite presentation, but we do not need this fact). Therefore, $\mathrm{Symb}_\Gamma(V) = \ker(V^{(I)} \overset{\mu^*}{\to} V^{(J)})$, where μ^* is the obvious map defined from μ. Since the formation of μ^* obviously commutes with arbitrary base change $R \to R'$, the formation of its kernel commutes with any flat base change. □

Remark 4.1.6 Note that there is no hypotheses of finiteness on V nor on the map $R \to R'$ in the above lemma. We shall have to apply this lemma to huge V's and all kind of flat base changes.

Exercise 4.1.7 Assume that R is noetherian and that V is finite type as an R-module. Show that $\mathrm{Symb}_\Gamma(V)$ is also a finite type R-module.

Definition 4.1.8 We shall denote by $\mathrm{BSymb}_\Gamma(V)$ the image of the restriction map (called the *boundary map*) $\mathrm{Hom}(\Delta, V)^\Gamma \to \mathrm{Hom}(\Delta_0, V)^\Gamma = \mathrm{Symb}_\Gamma(V)$. An element of $\mathrm{BSymb}_\Gamma(V)$ is called a *boundary modular symbol.*

Exercise 4.1.9 Show that the kernel of the boundary map is V^Γ. Show that the formation of $\mathrm{BSymb}_\Gamma(V)$ commute with flat base change.

4.2 Action of the Hecke Operators on Modular Symbols

We suppose given a submonoid Σ of $\mathrm{GL}_2(\mathbb{Q})$, a subgroup Γ of Σ, and a right Σ-module V. We assume that Γ and Σ satisfy hypothesis (2.6.3), namely

$$\forall s \in \Sigma,\ \Gamma s \Gamma \text{ is the union of } \textit{finitely many} \text{ left } \Gamma\text{-cosets } \Gamma s_i.$$

This condition holds, for any Σ, when Γ is a finite index subgroup of $\mathrm{SL}_2(\mathbb{Z})$ or any conjugate of such a subgroup: cf. [113, §3].

Since $\mathrm{Symb}_\Gamma(V)$ is by definition the group of Γ-invariants in the module $\mathrm{Hom}(\Delta_0, V)$ on which Σ acts, there is an action (on the right) of the Hecke ring $\mathcal{H}(\Sigma, \Gamma)$, see Sect. 2.6.2. The construction $V \mapsto \mathrm{Symb}_\Gamma(V)$ is a functor form the category of right Σ-modules to the category of right-$\mathcal{H}(\Sigma, \Gamma)$-modules. When V is an R-module, and the action of Σ is by R-linear operators, then obviously the Hecke operators are R-linear on $\mathrm{Symb}_\Gamma(V)$.

When Γ is a congruence subgroup such that $\Gamma_1(N) \subset \Gamma \subset \Gamma_0(N)$ for some N, we use the classical notation T_l for $[\Gamma \begin{pmatrix} 1 & 0 \\ 0 & 1 \end{pmatrix} \Gamma]$, replaced by U_l when l divides N.

Exercise 4.2.1 Let Γ be a group such that $\Gamma_1(N) \subset \Gamma \subset \Gamma_0(N)$. If V is $\mathbb{Q}$-vector space, with a right action of Σ, and $n \in \mathbb{Z}$, we let $V(n)$ be the group V with action of $s \in \Sigma$ multiplied by $(\det s)^n$. That is, if $v \in V$, $v_{|V(n)}s = (\det s)^n v_{|s}$. Note that the Γ-action on V and $V(n)$ are the same, so $\mathrm{Symb}_\Gamma(V) = \mathrm{Symb}_\Gamma(V(n))$. However, the action of the Hecke operators are different. Show that the action of T_l for $l \nmid N$ (resp. U_l for $l \mid N$) on $\mathrm{Symb}_\Gamma(V(n))$ is l^n (resp. p^n) times its action on $\mathrm{Symb}_\Gamma(V)$.

4.3 Reminder on Cohomology with Local Coefficients

This section is a reminder on the homology and cohomology of a topological space *with coefficients in a local system*, or for short, *with local coefficients*. We suppose the reader familiar with singular homology and cohomology with constant coefficients, as exposed in classical textbooks such as [66, Chapters 2 and 3]. Homology and cohomology with local coefficients were invented by Steenrod in [118]. For a more modern treatment, see [117]. For the convenience of the reader, our treatment is self-contained, containing all needed definitions and proofs, except for the proof of Poincaré's duality.

4.3.1 *Local Systems*

We fix a commutative ring R. Let M be an R-module. If X is a topological space, the *constant sheaf* M_X attached to M is defined as the sheaf in R-modules attached to the presheaf which sends any open subset $U \subset X$ to M. A *local system* (or *locally constant sheaf*) of R-modules on a space X is a sheaf in R-modules G on X such that there exists an R-module M and a covering of X by open sets U such that $G_{|U} \simeq M_U$ as sheaves of R-modules over U. The R-module M is then isomorphic to the stalk at any point of the local system G, hence is uniquely determined by G up to isomorphism. We call M the *fiber* of G.

If $f : Y \to X$ is a continuous map, and G a local system on X, then the pull-back f^*G is a local system on Y, with the same fiber. We denote by $G(f)$ the module of global sections on Y of f^*G. In other words, $G(f) = \varinjlim_{U \text{ open},\ f(Y) \subset U \subset X} G(U)$. The restriction maps $G(X) \to G(U)$ for U an open set of X containing $f(Y)$ induces a map $f^* : G(X) \to G(f)$, called the *pullback by* f. More generally, if $h : Z \to Y$ is a continuous map, the map $h^* : G(f) = (f^*G)(Y) \to G(h \circ f) = (f^*G)(h)$ is called *pullback by* h.

Exercise 4.3.1 Let G be a local system of fiber M on a topological space X, $f : Y \to X$ a continuous map. Show that if Y is connected, $G(f)$ is isomorphic to a sub-module of M.

4.3.2 Local Systems and Representations of the Fundamental Group

We now assume that X is connected, locally path-connected, and semi-locally simply connected, which are the standard conditions under which X is known to have a universal cover $\pi : \tilde{X} \to X$. We choose a point $x \in X$.

If M is an R-module with a right action of $\pi_1(X, x)$, we can attach to M a local system $\tilde{M}$ of X, of fiber M, as follows. We consider the product space $\tilde{X} \times M$ where M is given the discrete topology, and let $\pi_1(X, x)$ acts on it by $g(x, m) = (gx, gm)$. The quotient space $X_M := (\tilde{X} \times M)/\pi_1(X, x)$ inherits from π a continuous map $\pi_M : X_M \to X$ and we define a sheaf $\tilde{M}$ over X as the sheaf of continuous sections of π_M, that is $\tilde{M}(U) = \{s : U \to X_M,\ s \text{ continuous},\ \pi_M \circ s = \mathrm{Id}_U\}$.

It is clear that the construction $M \to \tilde{M}$ defines a functor from the category of R-modules with R-linear action of $\pi_1(X, x)$ to the category of locally constant sheaves of R-modules on X.

In the two exercises below you are asked to construct a quasi-inverse functor to $M \to \tilde{M}$, showing that this functor is an equivalence of category.

Exercise 4.3.2 Show that every local system on $X := [0, 1]$ of fiber M is a constant sheaf M_X.

Exercise 4.3.3 Assume that X is a path-connected topological space. Let G be a locally free sheaf on X.

1. If γ is a path in X from x to y, explain how to define using γ an isomorphism $i_\gamma : G_x \to G_y$, and show that i_γ depends only of the homotopy class in γ.
2. If γ' is a path from y to z, show that $i_{\gamma'} \circ i_\gamma = i_{\gamma \circ \gamma'}$ where $\gamma \circ \gamma'$ is the concatenation of γ and γ'. In particular, if $x \in X$, we have a natural action of $\pi_1(X, x)$ on G_x called the *monodromy* action.
3. Show that if $x \in X$ is fixed, you have defined a functor $G \mapsto G_x$ from the category of locally constant sheaves to the category of abelian groups with an action of $\pi^1(X, x)$. Explain how this functor depends on the choice of the base point x.
4. Show that if X is connected, locally path-connected and semi-locally simply connected, the above functor is an equivalence of categories, with the functor $M \to \tilde{M}$ constructed above as quasi-inverse.

If G is a locally constant sheaf of R-modules on X, we write $G^\vee$ for the sheaf $\underline{\mathrm{Hom}}(G, R_X)$. In other words, $G^\vee$ is the sheaf on X such that $G^\vee(U) = \mathrm{Hom}(G_{|U}, R_U)$, the Hom here being the R-module of morphisms of sheaves of R-modules over U. It is clear that $G^\vee$ is again a locally free sheaf of R-modules, and that its fiber is $M^\vee := \mathrm{Hom}_R(M, R)$. When X is connected, and M is a finite projective R-module, the natural map $m : G(X) \otimes G^\vee(X) \to R$ defines a perfect bilinear pairing between $G(X) = M$ and $G^\vee(X) = M^\vee$.

4.3.3 Singular Simplices

We denote as usual by Δ^q for $q \geq 0$ the standard q-simplex, with $q + 1$ vertices numbered $0, 1, \ldots, q$.

Definition 4.3.4 If X is a topological space, a *singular q-simplex* of X is a continuous map $\sigma : \Delta^q \to X$.

For S a subset of $\{0, \ldots, q\}$ with $s \leq q$ elements, let $\phi_{q,S}$ be the affine map sending the s-simplex Δ^s to Δ^q, which sends the vertices $0, 1, \ldots, s$ to the vertices of Δ^q numbered in S, in increasing order. If $S = \{0, \ldots, q\} - \{i\}$, we write ϕ_q^i instead of $\phi_{q,S}$.

If σ is a singular q-simplex of X, then $\sigma \circ \phi_{q,S}$ is a singular s-simplex of X, which we shall denote for shortness as σ_S, and σ^i for $\sigma \circ \phi_q^i$.

Lemma 4.3.5 *Let X be a topological space and σ a singular q-simplex. Let G be a locally constant sheaf on X. Let S be a subset of $\{0, \ldots, q\}$ with s elements. Then the morphism of R-modules $\phi_{q,S}^* : G(\sigma) \to G(\sigma_S)$ is an isomorphism.*

Proof Since Δ^q satisfies the hypothesis of Exercise 4.3.3 and is simply connected, the locally constant sheaf $\sigma^* G$ on Δ^q is constant, and so is $\sigma_S^* G$ on Δ^s. Since both Δ^q and Δ^s are contractible, choosing a point x of Δ^s, $G(\sigma)$ and $G(\sigma_S)$ are both naturally isomorphic to the fiber of $\sigma^* G$ at $\phi_{q,S}(x)$ and of $\sigma_S^* G$ at x respectively, and since $\phi_{q,S}^*$ induces the identity map on these fibers, it is an isomorphism $G(\sigma) \to G(\sigma_S)$. □

In the situation of the above lemma, if $g \in G(\sigma_S)$ we call *extension* of g to σ (or to Δ^q) the preimage of g in $G(\sigma)$ by the isomorphism $\phi_{q,S}^*$, and we denote it by $\bar{g}$.

4.3.4 Homology with Local Coefficients

For $q \geq 0$, we define $C_q(X, G)$ as the set of formal sums $\sum g_\sigma \sigma$ where σ runs among singular q-simplices $\Delta_q \to X$ and g_σ is an element of the R-module $G(\sigma)$ which is 0 for almost all σ. The addition $\sum g_\sigma \sigma + \sum g'_\sigma \sigma := \sum (g_\sigma + g'_\sigma)\sigma$ and the multiplication by $r \in R$: $r(\sum g_\sigma \sigma) := \sum (r g_\sigma)\sigma$ obviously give $C_q(X, G)$

a structure of R-module. For $q \geq 1$, we define a morphism of R-modules $\partial : C_q(X, G) \to C_{q-1}(X, G)$ by setting

$$\partial(g_\sigma \sigma) = \sum_{i=0}^{q} (-1)^i (\phi_q^i)^* (g_\sigma) \sigma^i.$$

The verification that $\partial\partial = 0$ is easy, and we denote by $H_\bullet(X, G)$ the homology of the complex $(C_\bullet(X, G), \partial)$. The groups $H_q(X, G)$ are called the *homology groups* of X with values in G.

If $s : G \to G'$ is a morphisms of local systems over X, and σ is a q-simplex on X, then s maps $G(\sigma)$ to $G'(\sigma)$ and defines a morphism $s_* : C_q(X, G) \to C_q(X, G')$ sending $\sum g_\sigma \sigma$ to $\sum s(g_\sigma)\sigma$. This morphism is compatible with ∂, hence induces a morphism $s_* : H_q(X, G) \to H_q(X, G')$. This construction clearly makes $G \mapsto H^q(X, G)$ a covariant functor in G.

Exercise 4.3.6 Assume that X is connected, locally path-connected and semi-locally simply connected. Let $x \in X$, M be an R-module with a right action of $\pi_1(X, x)$ and $\tilde{M}$ the local system attached to it. Show that $H_0(X, \tilde{M}) = M_\Gamma$, where $M_\Gamma := M/\langle m_{|\gamma} - m, m \in M, \gamma \in \Gamma\rangle$ is the space of *co-invariants*.

4.3.5 Cohomology with Local Coefficients

For $q \geq 0$, we define $C^q(X, G)$ as the set of functions assigning to every singular q-simplex $\sigma : \Delta^q \to X$ an element $u(\sigma) \in G(\sigma)$. Addition and scalar multiplication of functions give $C^q(X, G)$ a structure of R-module. For $q \geq 0$ we define a map $d : C^q(X, G) \to C^{q+1}(X, G)$ by setting, for every singular $(q+1)$-simplex σ of X,

$$du(\sigma) = \sum_{i=0}^{q+1} (-1)^i \overline{u(\sigma^i)},$$

where $\overline{u(\sigma^i)}$ is the extension of $u(\sigma^i)$ to Δ^{q+1} (see Sect. 4.3.3). Then again one checks easily that $dd = 0$, and we denote by $H^\bullet(X, G)$ the cohomology of the complex $(C^\bullet(X, G), d)$. The groups $H^q(X, G)$ are called the *cohomology groups* of X with values in G.

As for homology, a map $u : G \to G'$ of local systems over X induces a map $u_* : H^q(X, G) \to H^q(X, G')$.

Let $f : X' \to X$ be a continuous map. For σ a q-cochain of X', $f \circ \sigma$ is a q-cochain of X, and we have a pullback map $f^* : G(f \circ \sigma) \to (f^*G)(\sigma)$. We define a map $f^* : C^q(X, G) \to C^q(X', f^*G)$ by $f^*(u)(\sigma) = f^*(u(f \circ \sigma))$ for every q-cochain σ of X'. The maps f^* are compatible with d and induce maps $f^* : H^q(X, G) \to H^q(X', f^*G)$.

4.3.6 Relative Homology and Relative Cohomology

We shall also need, occasionally, the *relative cohomology* of a pair (X, Y) where X is a topological space and Y a subspace of X. We simply define $C_q(X, Y, G)$ as the quotient $C_q(X, G)/C_q(Y, G)$, and $C^q(X, Y, G)$ as the kernel of the restriction map $C^q(X, G) \to C^q(Y, G)$. The *relative homology* $H_\bullet(X, Y, G)$ (resp. *relative cohomology* $H^\bullet(X, Y, G)$) is defined as the homology (resp. cohomology) of the complex $C_\bullet(X, Y, G)$ (resp. $C^\bullet(X, Y, G)$). From the short exact sequence of complexes $0 \to C^\bullet(X, Y, G) \to C^\bullet(X, G) \to C^\bullet(Y, G) \to 0$ one gets a long exact sequence

$$\cdots \to H^q(X, Y, G) \to H^q(X, G) \to H^q(Y, G) \to H^{q+1}(X, Y, G) \to \dots \tag{4.3.1}$$

Let us also note the easy excision property: if U is a subspace of X whose closure lies in the interior of Y,

$$H^q(X, Y, G) = H^q(X - U, Y - U, G). \tag{4.3.2}$$

4.3.7 Formal Duality Between Homology and Cohomology

Assume that G is a locally constant sheaf on a topological space X whose fiber is a finite projective R-module.

If $u \in C^q(X, G^\vee)$, one can attach to u an R-linear form, denoted $\tilde{u}$, from $C_q(X, G)$ to R, by setting

$$\tilde{u}(\sum_\sigma g_\sigma \sigma) = \sum_\sigma m(g_\sigma \otimes u(\sigma))$$

where $m : G(\sigma) \otimes G^\vee(\sigma) \to R$ is the natural map. Note that the sum is finite since only finitely many of the sections g_σ are non-zero. We have thus defined a map $u \mapsto \tilde{u}$, $C^q(X, G^\vee) \to \mathrm{Hom}_R(C_q(X, G), R)$ which is easily seen to be an isomorphism, since the maps $m : G(\sigma) \otimes G^\vee(\sigma) \to R$ are perfect pairings. Moreover it is clear that the transposes of the boundary maps δ becomes, via these isomorphism, the coboundary maps d.

The maps $C^q(X, G^\vee) \to \mathrm{Hom}_R(C_q(X, G), R)$ induce natural morphisms of R-modules:

$$H^q(X, G^\vee) \to \mathrm{Hom}(H_q(X, G), R) \tag{4.3.3}$$

Indeed, an element of $H^q(X, G^\vee)$ is the class $[z]$ of an element $z \in C^q(X, G^\vee)$ such that $dz = 0$. Such an element z is a morphism of R-modules $\tilde{z} : C_q(X, G) \to R$, and the condition $dz = 0$ is equivalent to $\tilde{z}\delta = 0$, i.e. $\tilde{z}$ is trivial on the submodules of boundaries $B_q(X, G) \subset C_q(X, G)$. Restricting $\tilde{z}$ to $Z_q(X, G) = \ker \delta \subset C_q(X, G)$ defines thus an element of $\mathrm{Hom}(Z_q(X, G)/B_q(X, G), R) = \mathrm{Hom}(H_q(X, G), R)$. It is easy to check that this element only depends on the class $[z]$ of z in $H^q(X, G^\vee)$, and the map $H^q(X, G^\vee) \to \mathrm{Hom}(H_q(X, G), R)$ is the morphism of R-modules (4.3.3).

Theorem 4.3.7 (Universal Coefficient Theorem) *Assume that R is a PID, that G is a locally constant sheaf of R-modules whose fiber is a finite free R-module, and that the modules $H_q(X, G)$ are finitely generated for $q \geq 0$. Then the morphism (4.3.3) fits into an exact sequence*

$$\mathrm{Ext}^1_R(H_{q-1}(X, G), R) \to H^q(X, G^\vee) \to \mathrm{Hom}_R(H_q(X, G), R) \to 0$$

Proof The proof is standard. The reader may consult [66, pages 191–195] for a very detailed proof in the special case where G is the constant sheaf R. The case of general locally constant G needs only very simple adaptations. For instance, to prove the surjectivity of (4.3.3), one observes that the module $C_{q-1}(X, G)$ is a direct sum of modules of the form $G(\sigma)$ (for σ a singular simplex), which by Exercise 4.3.1 are submodules of the fiber of G, hence are free R-modules (since R is a PID). Thus $C_{q-1}(X, G)$ is free and, since R is a PID, its submodule $B_{q-1}(X, G)$ is free. Therefore the exact sequence $0 \to Z_q(X, G) \to C_q(X, G) \stackrel{\partial}{\to} B_{q-1}(X, G) \to 0$ is split. This shows that any map in $\mathrm{Hom}(Z_q(X, G), R)$ extends to a map in $\mathrm{Hom}(C_q(X, G), R)$, which is precisely the surjectivity of (4.3.3). We leave the adaptation of the proof of the exactness at $H^q(X, G)$ to the reader. □

Corollary 4.3.8 *Under the hypothesis of Theorem 4.3.7, we have an isomorphism $H^q(X, G^\vee)/torsion \stackrel{\simeq}{\to} \mathrm{Hom}_R(H_q(X, G), R)$. If R is a field, we have an isomorphism $H^q(X, G^\vee) \stackrel{\simeq}{\to} \mathrm{Hom}_R(H_q(X, G), R)$*

Proof Since R is a PID, and $H_{q-1}(X, G)$ is finitely generated, it is a sum of modules isomorphic to R and to R/I with I a non-zero ideal. Thus $\mathrm{Ext}^1_R(H_{q-1}(X, G), R)$ is a sum of modules of the form $\mathrm{Ext}^1_R(R, R) = 0$ and $\mathrm{Ext}^1_R(R/I, R) = R/I$, and threefore is torsion. Hence the kernel of surjective map $H^q(X, G) \to \mathrm{Hom}_R(H_q(X, G), R)$ is torsion, and because $\mathrm{Hom}_R(H_q(X, G), R)$ is torsion-free, this kernel is the entire torsion of $H^q(X, G)$. □

4.3.8 Cohomology with Compact Support and Interior Cohomology

For $u \in C^q(X, G)$, the *support* of the cochain u is defined as the closure of $\bigcup_{\sigma\ q\text{-simplex of } X, u(\sigma) \neq 0} \sigma(\Delta^q)$ in X, and is denoted by $\mathrm{supp}\, u$. Let $C^q_c(X, G)$ be the submodule of $C^q(X, G)$ consisting of cochains u such that $\mathrm{supp}\, u$ is compact. Then

$C_c^\bullet(X, G)$ is a subcomplex of $C^\bullet(X, G)$. We denote its q-th cohomology group by $H_c^q(X, G)$ and call it the q-th *cohomology with compact support* of (X, G).

There are natural maps

$$\beta : H_c^q(X, G) \to H^q(X, G)$$

induced by the inclusion maps $C_c^q(X, G) \hookrightarrow C^q(X, G)$. Their images are denoted by $H_!^q(X, G)$ and are called the *interior cohomology* groups of (X, G).

Let $f : X' \to X$ be a continuous map. Assume moreover that f is proper, i.e. $f^{-1}(K)$ is compact for every compact subset K of X. If $u \in C^q(X, G)$, then the cochain f^*u of $C^q(X', f^*G)$ has support

$$\operatorname{supp} f^*u \subset \overline{\bigcup_{\sigma\ q-\text{simplex of } X',\, f\circ u(\sigma)\neq 0} \sigma(\Delta^q)} \subset f^{-1}(\operatorname{supp} u).$$

Thus if $u \in C_c^q(X, G)$, then

$$f^*u \in C_c^q(X, f^*G).$$

We therefore have a natural map $f^* : H_c^q(X, G) \to H_c^q(X', f^*G)$, which by construction is compatible with $f^* : H^q(X, G) \to H^q(X', f^*G)$. The map f^* thus restricts to a map $H_!^q(X, G) \to H_!^q(X', f^*G)$.

One has $C_c^q(X, G) = \varinjlim_{K \text{ compact}} C^q(X, X - K, G)$ by definition, and thus, since filtered inductive limits are exact in the category of abelian groups,

$$H_c^q(X, G) = \varinjlim_{K \text{ compact}} H^q(X, X - K, G). \tag{4.3.4}$$

Proposition 4.3.9 *Assume that X is compact, and that Y is a closed subset of X. Then we have a natural isomorphism $H_c^q(X - Y, G) = H^q(X, Y, G)$.*

Proof By (4.3.4), $H_c^q(X - Y, G) = \varinjlim_{K\subset X-Y,\ K\text{compact}} H^q(X - Y, (X - Y) - K, G)$. By excision (4.3.2), $H^q(X - Y, (X - Y) - K, G) = H^q(X, X - K, G)$. Since X is compact, when K runs among compact subsets of $X - Y$, $X - K$ runs among open neighborhoods of Y. Hence

$$H_c^q(X - Y, G) = \varinjlim_{Y\subset U\subset X,\ U\text{open}} H^q(X, U, G) = H^q(X, Y, G).$$

(For the last equality, see [116, Theorem 1].) □

4.3.9 Cup-Products and Cap-Products

Let G and G' be two locally constant sheaves over a topological space X, and $G \otimes G'$ their tensor product (as sheaves of R-modules), which is also a locally constant sheaf on X. Let a, b, q be non-negative integers such that $a + b = q$. Let A be the subset $\{0, 1, \ldots, a\}$ and B be the subset $\{a, a+1, \ldots, q\}$ of $\{0, \ldots, q\}$.

If $u \in C^a(X, G)$ and $v \in C^b(X, G')$, we define $u \cup v \in C^q(X, G \otimes G')$ for $q = a + b$ by setting, for every q-singular simplex $\sigma : \Delta^q \to X$, $(u \cup v)(\sigma) = u(\sigma_A) \otimes v(\sigma_B)$. We observe that if either u or v has compact support, so does $u \cup v$.

Since $d(u \cup v) = d(u) \cup v + (-1)^a u \cup dv$, we obtain *cup-products* maps $\cup : H^a(X, G) \otimes H^b(X, G') \to H^{a+b}(X, G \otimes G')$ and similar (and compatible) maps $\cup : H^a_c(X, G) \otimes H^b(X, G') \to H^q_c(X, G \otimes G')$, $u \otimes v \mapsto u \cup v$ and $\cup : H^a(X, G) \otimes H^b_c(X, G') \to H^q_c(X, G \otimes G')$.

If $f : X' \to X$ is a continuous map, then $f^*(G \otimes G') = f^*(G) \otimes f^*(G')$ and it follows from the definitions that the diagram

$$\begin{array}{ccc} H^a(X, G) \otimes H^b(X, G') & \xrightarrow{\ \cup\ } & H^q(X, G \otimes G') \\ \Big\downarrow{\scriptstyle f^* \otimes f^*} & & \Big\downarrow{\scriptstyle f^*} \\ H^a(X', f^*G) \otimes H^b(X', f^*G') & \xrightarrow{\ \cup\ } & H^q(X, f^*(G \otimes G')) \end{array} \tag{4.3.5}$$

is commutative. When f is proper, the same diagrams where the groups H^a and H^q are replaced by H^a_c and H^q_c, or by $H^a_!$ and $H^q_!$ are also commutative.

If $u \in H^a(X, G)$ and $v \in H^b(X, G')$, then $u \cup v \in H^q(X, G \otimes G')$ and $v \cup u \in H^q(X, G' \otimes G)$. Using the natural isomorphism $G \otimes G' \xrightarrow{\simeq} G' \otimes G$, the last two cohomology groups can be naturally identified, and after this identification one has

$$u \cup v = (-1)^{ab} v \cup u. \tag{4.3.6}$$

The same holds for $u \in H^a_c$, etc. The proof is exactly the same has in the case of singular cohomology with constant coefficients (see [66, Theorem 3.14]), so we omit it.

Still assuming $q = a + b$, we shall now define a bilinear map $\cap : C_q(X, G) \otimes_R C^a(X, G') \to C_b(X, G \otimes G')$ as follows. It suffices by additivity to define $(c_\sigma \sigma) \cap u$ with σ a q-singular simplex $\Delta^a \to X$, $c_\sigma \in G(\sigma)$, and $u \in C^a(X, G')$. We set

$$(c_\sigma \sigma) \cap u = \phi^*_{q,B}(c_\sigma \otimes \overline{u(\sigma_A)})\sigma_B.$$

In the above definition, $u(\sigma_A)$ is an element of $G'(\sigma_A)$, its extension $\overline{u(\sigma_A)}$ is an element of $G'(\sigma)$ and the tensor product $(c_\sigma \otimes \overline{u(\sigma_A)})$ is an element of $(G \otimes G')(\sigma)$, that $\phi^*_{q,B}$ (see Sect. 4.3.3) restricts to an element of $(G \otimes G')(\sigma_B)$.

Having defined the *cap-product* $\cap : C_q(X, G) \otimes_R C^a(X, G') \to C_b(X, G \otimes G')$ we check easily that $\partial(c \cap u) = (-1)^b(\partial c \cap u - c \cap \partial u)$, which shows that the cap-prodcut induces a map, also called cap-product:

$$\cap : H_q(X, G) \otimes_R H^a(X, G') \to H_b(X, G \otimes G').$$

One has the following relation between cup and cap products, as is easily checked on the definitions:

Lemma 4.3.10 *For all* $u \in H^a(X, G^\vee \otimes (G')^\vee)$, $v \in H^b(X, G)$, $c \in H_{a+b}(X, G')$,

$$(u \cup v)(c) = v(u \cap c).$$

4.3.10 Poincaré Duality

Let X be a topological real manifold of dimension n. We recall that the orientation sheaf $\mathbb{Z}_w$ on X is the sheaf $U \mapsto H_n(X, X - U, \mathbb{Z})$ (relative homology). This sheaf is locally constant of fiber $\mathbb{Z}$ and attached, when X is connected, to a morphism $\pi_1(X, x) \to \mathrm{Aut}(\mathbb{Z}) = \{\pm 1\}$. An *orientation* of X, if it exists, is a global section of $\mathbb{Z}_X$ on X, or equivalently, an isomorphism of sheaves $\mathbb{Z}_w \to \mathbb{Z}_X$ over X. If R is a commutative ring, we call R_w the sheaf $\mathbb{Z}_w \otimes_{\mathbb{Z}} R$ which is a locally free sheaf of R-modules.

The most general form of Poincaré duality for homology and cohomology of local systems is the following theorem. For a proof, see [117, Theorem 10.3].

Theorem 4.3.11 *Let X be a connected manifold of dimension n.*

(i) *The homology group $H_n(X, R_w)$ is canonically isomorphic to R.*

We denote by $[X]$ the element corresponding to 1 in $H_n(X, \mathbb{Z}_w)$, and by $[X]_R$ its image in $H_n(X, R_w)$ for any ring R.

(ii) *For any integer q such that $0 \le q \le n$ and any locally constant sheaf of R-modules G on X, the cap-product with $[X]_R$ defines an isomorphism:*

$$H^q_c(X, G) \xrightarrow{\simeq} H_{n-q}(X, G \otimes R_w).$$

We deduce

Corollary 4.3.12 *Let X be an orientable connected manifold of dimension n. Let R be a PID, and let G be a locally free sheaf of R-modules whose fiber is a finite projective R-module. We have $H^n_c(X, R_X) \simeq R$, the isomorphism depending of the choice of an orientation of X. The cup-product map*

$$H^q_c(X, G) \otimes H^{n-q}(X, G^\vee) \to H^n_c(X, R_X) \simeq R$$

induces a perfect pairing after moding out $H_c^q(X, G)$ and $H^{n-q}(X, G^\vee)$ by their torsion.

Proof Let us fix an orientation on X, hence an isomorphism $H_n(X, R_X) \simeq H_n(X, R_w) \simeq R$ which allows us to see $[X]$ as an element of $H_n(X, R_X)$. By Lemma 4.3.10, the maps

$$H_c^q(X, G) \otimes H^{n-q}(X, G^\vee) \stackrel{(-\cap[X])\otimes \mathrm{Id}}{\longrightarrow} H_{n-q}(X, G) \otimes H^{n-q}(X, G^\vee \otimes R_w) \to R$$

and

$$H_c^q(X, G) \otimes H^{n-q}(X, G^\vee) \stackrel{\cup}{\longrightarrow} H_c^n(X, R_X) \stackrel{-\cap[X]}{\longrightarrow} H_0(X, R_X) = R_X$$

are equal. The first map is a perfect pairing after moding out by torsion by Theorem 4.3.11 and Corollary 4.3.8. Hence so is the second map. □

4.3.11 Singular Cohomology and Sheaf Cohomology

This subsection supposes that the reader has some familiarity with the cohomology of sheaves as defined for example in [64]. Let us just recall that if X is a topological space, and Γ_X is the left-exact functor *global sections* from the category of sheaves of abelian groups on X to the category of abelian groups, the *sheaf cohomology groups* of a sheaf $\mathcal{F}$ are defined as $R^n\Gamma_X(\mathcal{F})$ where $R^n\Gamma_X$, $n \geq 0$ are the derived functors of Γ_X.

Theorem 4.3.13 *Let X be a locally contractible and hereditary paracompact*[1] *topological space, G a local system on X. Then the singular cohomology groups $H^n(X, G)$ are canonically isomorphic to the sheaf cohomology groups $R^n\Gamma_X(\mathcal{F})$.*

While this theorem was certainly known since the beginning of sheaf cohomology theory, I don't know a complete reference for this: [101] contains a proof in the case where G is a constant sheaf. We recall the proof (as presented by Sella [107]) to show it works for general local systems G.

We call $\tilde{C}^n(-, G)$ the sheafication of the presheaf $U \mapsto C^n(U, G)$. It is easy to see, since X is locally contractible, that the complex of sheaves $\tilde{C}^\bullet(-, G)$ is a resolution of G.

Lemma 4.3.14 *The sheaves $\tilde{C}^n(-, G)$ are flabby, and the morphism of complex of abelian groups $C^\bullet(X, G) \to \tilde{C}^\bullet(X, G)$ is a quasi-isomorphism.*

[1] This means that every open subset of X is paracompact.

It is clear that this lemma implies the theorem, for it implies that $R^\bullet\Gamma_X(\mathcal{F})$ is the cohomology of the complex $\tilde{C}^\bullet(X, G)$, hence also of the complex $C^\bullet(X, G)$, that is $H^\bullet(X, G)$.

To prove the lemma, we introduce, for every open subset U of X, and every covering $\mathcal{U}$ of U by open subsets of U, the group $C^n_{\mathcal{U}}(U, G)$ as the group of functions u assigning to every singular n-simplex $\sigma : \Delta^n \to U$ of U *whose image falls in one of the open sets of the covering* $\mathcal{U}$ an element $u(\sigma) \in G(\sigma)$. It is obvious that the restriction map $C^n(U, G) \to C^n_{\mathcal{U}}(U, G)$ is surjective and it follows from a standard barycentric subdivision argument that the morphism of complexes $C^\bullet(U, G) \to C^\bullet_{\mathcal{U}}(U, G)$ is a quasi-isomorphism. Moreover, since U is paracompact, $\tilde{C}^n(U, G) = \varinjlim_{\mathcal{U}} C^n_{\mathcal{U}}(U, G)$ (cf. [107, Prop 0.2]). It follows that the morphism $C^\bullet(U, G) \to \tilde{C}^\bullet(U, G)$ is a quasi-isomorphism, and that the maps $C^n(U, G) \to \tilde{C}^n(U, G)$ are surjective. Since this also applies to $U = X$, and since the restriction maps $C^n(X, G) \to C^n(U, G)$ are clearly surjective, it follows that the maps $\tilde{C}^n(X, G) \to \tilde{C}^n(U, G)$ are surjective; in other words, the sheaves $\tilde{C}^n(-, G)$ are flabby. This completes the proof of the lemma, and of the theorem.

We recall (cf. [62, Théorème 2.9.1, page 138]) that for $\mathcal{F}$ any sheaf of abelian groups on X, and $Y \subset X$ any locally closed subset with $i : Y \hookrightarrow X$ the inclusion, there exists a unique sheaf $\mathcal{F}_Y$ such that $(\mathcal{F}_Y)_{|Y} = i^{-1}\mathcal{F}$ and $(\mathcal{F}_Y)_{|X-Y} = 0$. The functor $\mathcal{F} \mapsto \mathcal{F}_Y$ is exact. One has $(\mathcal{F}_Y) = i_! i^{-1}\mathcal{F}$ and when Y is closed in X, this becomes $\mathcal{F}_Y = i_* i^{-1}\mathcal{F}$ and one has an exact sequence $0 \to \mathcal{F}_Y \to \mathcal{F} \to \mathcal{F}_{X-Y} \to 0$ (cf. [62, Théorème 2.9.3, page 140])

Theorem 4.3.15 *Let X be a locally contractible and hereditary paracompact topological space, G a local system on X. Let $Z \subset X$ be a closed subset, $i : Z \hookrightarrow X$ the inclusion. Then the relative singular cohomology groups $H^n(X, Z, G)$ are canonically isomorphic to the sheaf cohomology groups $R^n\Gamma_X(G_{X-Z})$.*

Proof Consider the exact sequence $0 \to \tilde{C}^n(-, G)_{X-Z} \to \tilde{C}^n(-, G) \to \tilde{C}^n(-, G)_Z \to 0$. We note that $C^n(-, G)_Z$ is flabby by Godement [62, Théorème 3.1.2 and Corollaire 1 of Théorème 3.3.1, page 151]. Thus, we have an exact sequence of complexes $0 \to \tilde{C}^\bullet(-, G)_{X-Z} \to \tilde{C}^\bullet(-, G) \to \tilde{C}^\bullet(-, G)_Z \to 0$, whose middle term and right term are flasque resolution of G and G_Z, while left term is a resolution of G_{X-Z}. It follows that the groups $R^n\Gamma_X(G_{X-Z})$ can be computed as the cohomology of the bicomplex whose first row is $\tilde{C}^\bullet(X, G)$ and second row $\tilde{C}^\bullet(Z.G)$ (all other rows being 0), and this cohomology is easily seen to be the one of $C^n(X, Z, G)$. □

4.4 Modular Symbols and Cohomology

In this section, we fix a submonoid Σ of $\mathrm{GL}_2(\mathbb{Q})$, a right Σ-module V, and a subgroup Γ of Σ, which satisfies (2.6.3).

4.4.1 Right Action on the Cohomology

We begin by extending the classical action of $\mathrm{GL}_2^+(\mathbb{Q})$ on the Poincaré upper half-plane $\mathcal{H}$ (cf.(2.6.1)) to $\mathrm{GL}_2(\mathbb{Q})$ as follows. If $\gamma = \begin{pmatrix} a & b \\ c & d \end{pmatrix} \in \mathrm{GL}_2(\mathbb{Q})$, and $z \in \mathcal{H}$, then

$$\gamma \cdot z = \begin{cases} (az+b)/(cz+d) & \text{if } \det \gamma > 0 \\ (a\bar{z}+b)/(c\bar{z}+d) & \text{if } \det \gamma < 0 \end{cases}$$

We set $\delta(\bar{\mathcal{H}}) = \mathbb{P}^1(\mathbb{Q})$ (the set of *cusps*) and we endow $\bar{\mathcal{H}} := \mathcal{H} \cup \delta(\bar{\mathcal{H}})$ with the topology which induces on $\mathcal{H}$ the usual topology, and such that a cusp $s \in \mathbb{P}^1(\mathbb{Q}) - \{\infty\}$ has for system of neighborhoods $\{s\} \cup$ the interior of circles in $\mathcal{H}$ tangent to the line $\mathrm{Im} z = 0$ at s, and the cusp ∞ has for system of neighborhoods $\{\infty\} \cup \{z \in \mathcal{H},\ \mathrm{Re}\, z > A\}$ for $A \in \mathbb{R}_+$. Then it is easy to check that the continuous action of $\mathrm{GL}_2(\mathbb{Q})$ on $\mathcal{H}$ extends to a continuous action on $\bar{\mathcal{H}}$. Note that it is clear from the definition that $\delta(\bar{\mathcal{H}})$ is closed in $\bar{\mathcal{H}}$, and discrete with its induced topology. In particular, $\bar{\mathcal{H}}$ is not compact (since $\delta(\bar{\mathcal{H}}) = \mathbb{P}^1(\mathbb{Q})$ would be discrete and compact, hence finite.)

Let $V_{\mathcal{H}}$ and $V_{\bar{\mathcal{H}}}$ be the constant sheaves on $\mathcal{H}$ and $\bar{\mathcal{H}}$ attaches to the abelian group V. We define a right action of Σ on the cohomology groups $H^i(\mathcal{H}, V_{\mathcal{H}})$, $H^i(\bar{\mathcal{H}}, V_{\bar{\mathcal{H}}})$, $H^i(\bar{\mathcal{H}}, \delta(\bar{\mathcal{H}}), V_{\bar{\mathcal{H}}})$ in the natural way. For instance, on the latter space, it is defined precisely this way: if σ in Σ, then there is a map $\sigma_* : H^q(\bar{\mathcal{H}}, \delta(\bar{\mathcal{H}}), V_{\bar{\mathcal{H}}}) \to H^q(\bar{\mathcal{H}}, \delta(\bar{\mathcal{H}}), V_{\bar{\mathcal{H}}})$ induced by the map $v \mapsto v_{|\sigma}$ on $V_{\bar{\mathcal{H}}}$, and a map $\sigma^* : H^q(\bar{\mathcal{H}}, \delta(\bar{\mathcal{H}}), V_{\bar{\mathcal{H}}}) \to H^q(\bar{\mathcal{H}}, \delta(\bar{\mathcal{H}}), V_{\bar{\mathcal{H}}})$ since the functors H^q are contravariant for continuous maps. The action of σ on $c \in H^q(\bar{\mathcal{H}}, \delta(\bar{\mathcal{H}}), V_{\bar{\mathcal{H}}})$ is $c_{|\sigma} = \sigma^* \sigma_* c$.

4.4.2 Modular Symbols and Relative Cohomology

Theorem 4.4.1 (Ash-Stevens) *There is a canonical isomorphism of right Σ-modules, functorial in V,*

$$\mathrm{Hom}(\Delta_0, V) = H^1(\bar{\mathcal{H}}, \delta(\bar{\mathcal{H}}), V_{\bar{\mathcal{H}}}).$$

This isomorphism induces an isomorphism of $\mathcal{H}(\Sigma, \Gamma)$-modules

$$Symb_\Gamma(V) \simeq H^1(\bar{\mathcal{H}}, \delta(\bar{\mathcal{H}}), V_{\bar{\mathcal{H}}})^\Gamma.$$

Proof The long exact sequence of cohomology attached to a pair gives us

$$0 \to H^0(\bar{\mathcal{H}}, V_{\bar{\mathcal{H}}}) \to H^0(\delta(\bar{\mathcal{H}}), V_{\bar{\mathcal{H}}}) \to H^1(\bar{\mathcal{H}}, \delta(\bar{\mathcal{H}}), V_{\bar{\mathcal{H}}}) \to 0 \quad (4.4.1)$$

Indeed, the preceding term is $H^0(\bar{\mathcal{H}}, \delta(\bar{\mathcal{H}}), V_{\mathcal{H}})$ which is 0 since $\bar{\mathcal{H}}$ is connected and $\delta(\bar{\mathcal{H}})$ not empty, and the next term is $H^1(\bar{\mathcal{H}}, V_{\mathcal{H}})$ which is 0 since $\bar{\mathcal{H}}$ is contractible. Note that the morphisms of this sequence are compatible with the natural right Σ-actions on the terms.

On the other hand, the exact sequence of abelian groups $0 \to \Delta_0 \to \Delta \to \mathbb{Z} \to 0$ is split, so it gives rise to a short exact sequence, also compatible with the right Σ-actions

$$0 \to V \to \mathrm{Hom}(\Delta, V) \to \mathrm{Hom}(\Delta_0, V) \to 0. \quad (4.4.2)$$

We claim that the two exact sequences (4.4.1) and (4.4.2) are isomorphic as exact sequences of left Σ-modules. To see this, it suffices to construct compatible Σ-equivariant isomorphisms between the first terms and between the second terms of the two exact sequences; but $H^0(\bar{\mathcal{H}}, V_{\bar{\mathcal{H}}}) = V$ since $\bar{\mathcal{H}}$ is connected, and $H^0(\delta(\bar{\mathcal{H}}), V_{\bar{\mathcal{H}}}) = H^0(\mathbb{P}^1(\mathbb{Q}), V_{\bar{\mathcal{H}}}) = \mathrm{Hom}(\Delta, V)$ since Δ is the free group generated by $\mathbb{P}^1(\mathbb{Q})$.

Therefore, the third terms of the two exact sequences are isomorphic (as Σ-modules):

$$\mathrm{Hom}(\Delta_0, V) = H^1(\bar{\mathcal{H}}, \delta(\bar{\mathcal{H}}), V_{\mathcal{H}}).$$

Taking the Γ-invariants, we get

$$\mathrm{Symb}_\Gamma(V) = H^1(\bar{\mathcal{H}}, \delta(\bar{\mathcal{H}}), V_{\bar{\mathcal{H}}})^\Gamma.$$

□

4.4.3 *Modular Symbols and Cohomology with Compact Support*

In this subsection, we keep the notation of the preceding subsection, but we assume in addition that the subgroup Γ and the module V satisfy those three conditions:

$$\Gamma \text{ is discrete, and for every } z \in \mathcal{H}, \text{ the closure of the orbit } \Gamma z \text{ in } \bar{\mathcal{H}} \text{ contains } \delta(\bar{\mathcal{H}}). \quad (4.4.3)$$

$$\text{The only torsion elements in } \Gamma \text{ are Id and possibly } -\mathrm{Id}. \quad (4.4.4)$$

$$\text{If } -\mathrm{Id} \in \Gamma, \text{ then } -\mathrm{Id} \text{ acts trivially on } V. \quad (4.4.5)$$

We define $\bar{\Gamma}$ as $\Gamma/\Gamma \cap \{\mathrm{Id}, \pm\mathrm{Id}\}$. Condition (4.4.3) means that $\bar{\Gamma}$ is a fuchsian group of the first kind, and condition (4.4.4) means that $\bar{\Gamma}$ has no elliptic point, hence acts freely on $\mathcal{H}$. Thus the quotient $\mathcal{H}/\bar{\Gamma}$ has a natural structure of Riemann surface, which we call Y_Γ. As is well known [113], Y_Γ is the complement of a finite set of points, $\delta(\bar{\mathcal{H}})/\bar{\Gamma}$ (the *set of cusps of* Y_Γ) in a unique compact Riemann surface X_Γ which is homeomorphic to $\bar{\mathcal{H}}/\bar{\Gamma}$.

By condition 4.4.5, the action of Γ on V factors through $\bar{\Gamma}$, and one can attach to the $\bar{\Gamma}$-module V a local system $\tilde{V}$ on Y_Γ (cf. Sect. 4.3.1).

Theorem 4.4.2 *There is a natural isomorphism of $\mathcal{H}(\Sigma, \Gamma)$-modules*

$$H^1(\bar{\mathcal{H}}, \delta(\bar{\mathcal{H}}), V_{\mathcal{H}})^\Gamma = H^1_c(Y_\Gamma, \tilde{V}).$$

Hence

$$Symb_\Gamma(V) = H^1_c(Y_\Gamma, \tilde{V}).$$

Proof Let us denote by i both the inclusions $\mathcal{H} \hookrightarrow \bar{\mathcal{H}}$ and $Y_\Gamma \hookrightarrow X_\Gamma$, and by π the natural quotient map $\bar{\mathcal{H}} \to X_\Gamma$ and its restriction $\mathcal{H} \to Y_\Gamma$. Thus $i \circ \pi = \pi \circ i$ as a map $\mathcal{H} \to X_\Gamma$. Recall that is $\mathcal{F}$ is a $\bar{\Gamma}$-sheaf on $\bar{\mathcal{H}}$ (resp. $\mathcal{H}$), $\pi_*^{\bar{\Gamma}} \mathcal{F}$ is the sheaf on X_Γ (resp. Y_Γ) whose sections on an open U are the $\bar{\Gamma}$-invariant sections of $\mathcal{F}$ on $\pi^{-1}(U)$. It is easy to see that $\pi_*^{\bar{\Gamma}} V_{\mathcal{H}}$ is the local system $\tilde{V}$ on Y_Γ.

By Theorem 4.3.15, we have a natural isomorphism of Σ-modules

$$H^1(\bar{\mathcal{H}}, \delta(\bar{\mathcal{H}}), V_{\bar{\mathcal{H}}}) = H^1(\bar{\mathcal{H}}, i_! V_{\mathcal{H}}). \tag{4.4.6}$$

Note that the sheaf $i_! V_{\mathcal{H}}$ on $\bar{\mathcal{H}}$ is a $\bar{\Gamma}$-sheaf that lies over the sheaf $\pi_* i_! V_{\mathcal{H}} = i_! \pi_* V_{\mathcal{H}} = i_! \tilde{V}$ on X_Γ. By Grothendieck [64, Theorem, 5.2.1], there are two spectral sequences converging to the same limit, whose second pages are respectively

$$\begin{aligned} I^2_{p,q} &= R^p \Gamma_{X_\Gamma}(R^q \pi_*^{\bar{\Gamma}}(i_! V_{\mathcal{H}})) \\ II^2_{p,q} &= H^p(\Gamma, H^q(\bar{\mathcal{H}}, i_! V_{\mathcal{H}})) \end{aligned} \tag{4.4.7}$$

Since π has discrete fibers, $R^q \pi_*^{\bar{\Gamma}}(i_! V_{\mathcal{H}})) = H^q(\Gamma, \pi_*(i_! V_{\mathcal{H}})) = H^q(\Gamma, i_! \tilde{V}) = i_! H^q(\Gamma, \tilde{V})$ and thus this sheaf has no non-zero global sections, and one sees that $I^2_{0,q} = 0$ for every q. Also $I^2_{1,0} = H^1(X_\Gamma, i_! \tilde{V})$.

Similarly $i_! V_{\mathcal{H}}$ has no non-zero global section, hence $II^2_{p,0} = 0$ for every p. We have $II^2_{0,1} = H^1(\bar{\mathcal{H}}, i_! V_{\mathcal{H}})^\Gamma$.

Comparing the H^1-term of the common limit of the two spectral sequences (4.4.7), we thus get

$$H^1(\bar{\mathcal{H}}, i_! V_{\mathcal{H}})^\Gamma = H^1(X_\Gamma, i_! \tilde{V}),$$

that is, using (4.4.6),

$$H^1(\bar{\mathcal{H}}, \delta(\bar{\mathcal{H}}), V_{\bar{\mathcal{H}}})^\Gamma = H^1(X_\Gamma, i_! \tilde{V}).$$

Using Theorem 4.3.15 for the right hand side and then Proposition 4.3.9, we get a natural isomorphisms of $\mathcal{H}(\Sigma, \Gamma)$-modules

$$H^1(\bar{\mathcal{H}}, \delta(\bar{\mathcal{H}}), V_{\bar{\mathcal{H}}})^\Gamma = H^1_c(Y_\Gamma, \tilde{V}).$$

This completes the proof.[2] □

Corollary 4.4.3 *We have a long exact sequence*

$$0 \to V^\Gamma \to \mathrm{Hom}(\Delta, V)^\Gamma \to \mathit{Symb}_\Gamma(V) = H^1_c(Y_\Gamma, \tilde{V}) \xrightarrow{\beta} H^1(Y_\Gamma, \tilde{V}),$$

where the last map is the natural map from cohomology with compact support to cohomology.

Proof This exact sequence is just a part long exact sequence attached to the short exact sequence of Γ-modules (4.4.2). □

Corollary 4.4.4 *If* $0 \to V_1 \to V_2 \to V_3 \to 0$ *is an exact sequence of* Γ*-modules, then we have a long exact sequence*

$$\begin{aligned} 0 &\to \mathit{Symb}_\Gamma(V_1) \to \mathit{Symb}_\Gamma(V_2) \to \mathit{Symb}_\Gamma(V_3) \\ &\to H^2_c(Y_\Gamma, \tilde{V}_1) \to H^2_c(Y_\Gamma, \tilde{V}_2) \to H^2_c(Y_\Gamma, V_3) \to 0. \end{aligned}$$

Proof Follows from the long exact sequence for H^i_c, using that $H^3_c(Y_\Gamma, \tilde{V}_1) = 0$ since Y_Γ is a real manifold of dimension 2. □

Corollary 4.4.5 *If* $0 \to V_1 \to V_2 \to V_3 \to 0$ *is an exact sequence of* Γ*-modules, then we have a long exact sequence*

$$0 \to \mathit{Symb}_\Gamma(V_1) \to \mathit{Symb}_\Gamma(V_2) \to \mathit{Symb}_\Gamma(V_3) \to (V_1)_\Gamma \to (V_2)_\Gamma \to (V_3)_\Gamma \to 0,$$

where $(V_i)_\Gamma$ *are the co-invariants of* V_i *for* Γ*, see Exercise 4.3.6.*

Proof This is Corollary 4.4.4 combined with Poincaré's duality (Theorem 4.3.11) and Exercise 4.3.6. □

Corollary 4.4.6 *Let* R *be a ring, and assume that* V *is a torsion-free* R*-module, and that the action of* Σ *on* V *is* R*-linear. Then* $\mathit{Symb}_\Gamma(V) = H^1_c(Y_\Gamma, \tilde{V})$ *is a torsion-free* R*-module. If* $x \in R$ *is not a zero divisor, and if* $V_\Gamma[x] = 0$ *(where* V_Γ *are the co-invariants of* V *for* Γ*, see Exercise 4.3.6), then we have* $\mathit{Symb}_\Gamma(V/xV) = \mathit{Symb}_\Gamma(V) \otimes_R R/(x)$.

---

[2] I am thankful to Johannes Huisman for explaining this proof to me, on mathoverflow, question 317640.

Proof Let $x \in R$ be an element which is not a divisor of 0. The sequence $0 \to V \overset{\times x}{\to} V \to V/xV \to 0$ is exact since V is torsion free. The long exact sequence for H^i_c (Corollary 4.4.4) gives an exact sequence

$$H^i_c(Y_\Gamma, \tilde{V}) \otimes_R R/xR \to H^i_c(Y_\Gamma, \tilde{V}/x\tilde{V}) \to H^{i+1}_c(Y_\Gamma, \tilde{V})[x] \to 0. \quad (4.4.8)$$

For $i = 0$, noting that $H^0_c(Y_\Gamma, \tilde{V}/x\tilde{V}) = 0$ since Y_Γ is connected, we get that $H^1_c(Y_\Gamma, \tilde{V})[x] = 0$, hence the first result.

By Poincaré duality (Theorem 4.3.11) and Exercise 4.3.6, $H^2_c(Y_\Gamma, \tilde{V})[x] = H_0(Y_\Gamma, V)[x] = V_\Gamma[x]$, so the second result follows from the same exact sequence for $i = 1$.

□

We have a similar result for cohomology.

Proposition 4.4.7 *Let R be a ring, and assume that V is a torsion-free R-module, and that the action of Σ on V is R-linear. If $x \in R$ is not a zero divisor, and if $(V/xV)^\Gamma = 0$ (or if the action of Γ on V is trivial), then we have $H^1(Y_\Gamma, \tilde{V}/x\tilde{V}) = H^1(Y_\Gamma, \tilde{V}) \otimes_R R/(x)$ and $H^1(Y_\Gamma, \tilde{V})[x] = 0$.*

Proof We have an exact sequence

$$H^i(Y_\Gamma, \tilde{V}) \otimes_R R/xR \to H^i(Y_\Gamma, \tilde{V}/x\tilde{V}) \to H^{i+1}(Y_\Gamma, \tilde{V})[x] \to 0. \quad (4.4.9)$$

For $i = 0$, in the case where V has trivial Γ-action, the first arrow is an isomorphism, hence $H^1(Y_\Gamma, \tilde{V})[x] = 0$. Otherwise, since $H^0(Y_\Gamma, \tilde{V}/x\tilde{V}) = (V/xV)^\Gamma$ is 0 by assumption, we again have $H^1(Y_\Gamma, \tilde{V})[x] = 0$.

Since $H^2(Y_\Gamma, \tilde{V}) = 0$ because Y_Γ is a non-compact manifold of dimension 2, the same exact sequence for $i = 2$ gives that $H^1(Y_\Gamma, \tilde{V}/x\tilde{V}) = H^1(Y_\Gamma, \tilde{V}) \otimes_R R/(x)$.

□

4.4.4 Pairings on Modular Symbols

We keep assuming in thus subsection, the conditions (4.4.3)–(4.4.5) on Γ and V. In addition, let R be a ring, and we assume that V is an R-module, and that the action of Γ on V is R-linear. The module $V^\vee = \mathrm{Hom}_R(V, R)$ also has an R-linear Γ-action, defined as usual by $l_{|\gamma}(v) = l(v_{|\gamma^{-1}})$.

We define an R-linear pairing, denoted Π_1 (Π for Poincaré),

$$\Pi_1 : \mathrm{Symb}_\Gamma(V) \times H^1(Y_\Gamma, \tilde{V}^\vee) \to R. \quad (4.4.10)$$

by defining $\Pi_1(\phi, y)$ as the image of $\phi \cup y \in H^2_c(Y_\Gamma, V \otimes V^\vee)$ (identifying ϕ with an element of $H^1_c(Y_\Gamma, \tilde{V})$ by Theorem 4.4.2) in $H^2_c(Y_\Gamma, R) = R$ (this isomorphism given by Theorem 4.3.11 using the isomorphism $R_w \simeq R_{Y_\Gamma}$ given by the orientation on Y_Γ coming from its complex structure).

In view of (4.3.6), the pairing Π is antisymmetric: for $\phi \in \mathrm{Symb}_\Gamma(V), \phi' \in \mathrm{Symb}_\Gamma(V')$,

$$\Pi(\phi', \phi) = -\Pi(\phi, \phi') \tag{4.4.11}$$

Proposition 4.4.8 *The pairing Π_1 is non-degenerate in each of the following cases:*

(i) *R is a field and V is a finite-dimensional vector space.*
(ii) *R is a PID, V is a finite projective R-module, and either the action of Γ on V is trivial or for every prime element π of R, $(V^\vee/\pi V^\vee)^\Gamma = 0$.*

Proof The assertion in case (i) is just Poincaré duality (see Corollary 4.3.12). The same result says that when R is a PID and V finite projective as in case (ii), Π is non-degenerate after moding out $\mathrm{Symb}_\Gamma(V)$ and $H^1(Y_\Gamma, \tilde{V})$ by their torsion. But under the hypotheses in (ii), those R-modules are already torsion-free by Corollary 4.4.6 and Proposition 4.4.7. □

The map $\beta : \mathrm{Symb}_\Gamma(V^\vee) = H^1_c(Y_\Gamma, \tilde{V}^\vee) \to H^1(Y_\Gamma, \tilde{V})$, allows us to deduce a pairing

$$\Pi : \mathrm{Symb}_\Gamma(V) \times \mathrm{Symb}_\Gamma(V^\vee) \to R, \tag{4.4.12}$$

sending (ϕ, ϕ') to $\Pi(\phi, \phi') := \Pi_1(\phi, \beta(\phi'))$. This pairing Π is compatible with the pairing Π_1, that is there is a commutative diagram

$$\begin{array}{ccc} \mathrm{Symb}_\Gamma(V) \times H^1(Y_\Gamma, \tilde{V}^\vee) & \longrightarrow & R \\ {\scriptstyle \mathrm{Id}\times\beta}\uparrow & & \uparrow{\scriptstyle =} \\ \mathrm{Symb}_\Gamma(V) \times \mathrm{Symb}_\Gamma(V^\vee) & \longrightarrow & R \end{array} \tag{4.4.13}$$

Remember (Sect. 4.3.8) that the *interior cohomology* $H^1_!(Y_\Gamma, V)$ is defined as the image of β (in the context of modular symbols, it is also called *parabolic cohomology* and sometimes denoted by H^i_p). For $\phi \in \mathrm{Symb}_\Gamma(V), \phi' \in \mathrm{Symb}_\Gamma(V')$, $\Pi(\phi, \phi')$ depends only on ϕ and $\beta(\phi')$ by definition. But in view of the antisymmetry of Π, it actually depends on ϕ only via $\beta(\phi)$.

We can thus define a pairing

$$\Pi_! : H^1_!(Y_\Gamma, \tilde{V}) \times H^1_!(Y_\Gamma, \tilde{V}^\vee) \to R$$

by setting $\Pi_!(x, y) = \Pi(\phi, \phi')$ for any $\phi \in \mathrm{Symb}_\Gamma(V)$, $\phi' \in \mathrm{Symb}_\Gamma(V)$ such that $\beta(\phi) = x, \beta(\phi') = y$. One also has $\Pi_!(x, y) = \Pi_1(\phi, y)$.

The question of the non-degeneracy of the pairing $\Pi_!$ is somewhat more subtle than for the pairing Π_1. We write Γ_∞ for the intersection $\Gamma \cap \begin{pmatrix} 1 & \mathbb{Z} \\ 0 & 1 \end{pmatrix}$.

Theorem 4.4.9 *Assume that either*

(i) *R is a field and V is a finite-dimensional vector space.*
(ii) *R is a PID, V is a finite projective R-module, and either the action of Γ on V is trivial or for every prime element π of R, $(V^\vee/\pi V^\vee)^\Gamma = 0$, $V_\Gamma[\pi] = 0$, $V^{\Gamma_\infty}/\pi V^{\Gamma_\infty} = (V/\pi V)^{\Gamma_\infty}$ and $V^\Gamma/\pi V^\Gamma = (V/\pi V)^\Gamma$*

Then $H^1_!(Y_\Gamma, V)$ is a direct summand in the finite-type projective module $H^1(Y_\Gamma, V)$ and the pairing $\Pi_!$ is non-degenerate

Proof The first assertion is trivial if R is a field. For the second, if $y \in H^1_!(Y_\Gamma, \tilde{V})$, then there is an $x \in H^1_c(Y_\Gamma, \tilde{V})$ such that $\Pi_1(x, y) \neq 0$, and thus $\Pi_!(\beta(x), y) \neq 0$. Since R is a field, this is enough to prove the non-degeneracy of $\Pi_!$.

Let us now assume that we are under hypothesis (ii). Consider the exact sequence from Cor 4.4.3:

$$0 \to V^\Gamma \to \operatorname{Hom}(\Delta, V)^\Gamma \to H^1_c(Y_\Gamma, \tilde{V}) \xrightarrow{\beta} H^1(Y_\Gamma, \tilde{V}).$$

It is an exact sequence of modules of finite type over V (the finiteness of the H^1 results from standard theorems on cohomology), and under our hypotheses, all those modules are projective. But the formation of this exact sequence commutes with the base change $R \to R/\pi$: in the case where the action of Γ is trivial, for the H^1 term it follows from Proposition 4.4.7 and our hypothesis $(V/\pi V)^\Gamma = 0$; for the H^1_c term, it follows from Corollary 4.4.6 and our hypotheses $V_\Gamma[\pi] = 0$; for the $\operatorname{Hom}(\Delta, V)^\Gamma$ term, it follows from our hypothesis on V^{Γ_∞} since this term is clearly a sum of finitely many copies of V^{Γ_∞}, and for V^Γ it is directly one of our hypothesis. In the case where Γ acts trivially on V, this is even simpler. Thus, this exact sequence stays exact after tensoring by R/π for all π. This implies easily that $H^1_!(Y_\Gamma, \tilde{V}) = \operatorname{Im}\beta$ is a direct summand in H^1.

Now consider the linear map $H^1_!(Y_\Gamma, \tilde{V}) \to H^1_!(Y_\Gamma, \tilde{V})^\vee$ induced by Π. The same argument as in the case of a field shows that this map is injective. Let $l \in H^1_!(Y_\Gamma, \tilde{V})^\vee$. Since $H^1_!(Y_\Gamma, \tilde{V})$ is direct summand in $H^1(Y_\Gamma, \tilde{V})$, l extends to a linear form $\tilde{l} \in H^1(Y_\Gamma, \tilde{V})^\vee$. Since Π_1 is non-degenerate under our hypotheses, there is an $x \in H^1_c(Y_\Gamma, \tilde{V})$ such that $\tilde{l}(y) = \Pi_1(x, y)$. In particular, $l(y) = \Pi(\beta(x), y)$ for $y \in H^1_!(Y_\Gamma, \tilde{V})$, which shows the surjectivity of the map. □

Remark 4.4.10 All the results of Sects. 4.4.3 and of 4.4.4 apply when Γ is a congruence subgroup of $\mathrm{SL}_2(\mathbb{Z})$ that does not contain $-\mathrm{Id}$, for any Γ-module V, since such a subgroup satisfies conditions (4.4.3), (4.4.4), as well as (4.4.5) for any V. They also apply when Γ is *any* congruence subgroup (even if it contains $-\mathrm{Id}$) provided that the multiplication by 6 is invertible in V. Indeed, we can introduce the normal subgroup $\Gamma' = \Gamma \cap \Gamma(3)$ of Γ, whose index is divisible only by the prime numbers 2 and 3. Since $-\mathrm{Id} \notin \Gamma(3)$, the results proved here apply to Γ' and easy and standard arguments allow to deduce that they also hod for Γ provided that in the modules V considered, the multiplication by 6 is invertible.

4.5 Notes and References

The results in this chapter are due to Eichler, Shimura [111–113], Ash-Stevens [4] and Hida [68]. We have tried to provide a unified presentation, using only basic algebraic topology, and avoid detours sometimes taken in the literature (for example, using the fact that the topological spaces Y_Γ have a structure of algebraic varieties over $\mathbb{C}$, the comparison theorem between usual and étale cohomologies, and the deep theorems of SGA 4 to prove things about modular symbols.)

Chapter 5
Classical Modular Symbols, Modular Forms, L-functions

Concerning spaces of modular forms and Hecke operators acting on them, we use the same notations and conventions recalled in Sect. 2.6.

5.1 On a Certain Monoid and Some of Its Modules

In this section, we introduce a certain submonoid S of $\mathrm{GL}_2(\mathbb{Q})$ containing $\mathrm{SL}_2(\mathbb{Z})$, and some S-modules $\mathcal{V}_k$ and $\mathcal{P}_k$, which will be needed to define the modules of *classical modular symbols* and the action of the Hecke operators upon them.

5.1.1 *The Monoid S*

We consider the following monoid:

$$S = \mathrm{GL}_2(\mathbb{Q}) \cap M_2(\mathbb{Z}). \tag{5.1.1}$$

On this monoid we define an anti-involution

$$\gamma \mapsto \gamma' = \det(\gamma)\gamma^{-1}. \tag{5.1.2}$$

Explicitly, $\gamma' = \begin{pmatrix} d & -b \\ -c & a \end{pmatrix}$ if $\gamma = \begin{pmatrix} a & b \\ c & d \end{pmatrix}$. The *anti* in anti-involution means that $(\gamma_1\gamma_2)' = \gamma_2'\gamma_1'$.

This anti-involution allows us to see any left-S-module as a right-S-module, and conversely : if S acts on V of the left, we define an action on the right by letting γ acts on V on the right through the left-action of γ' on V, and conversely: for

J. Bellaïche, *The Eigenbook*, Pathways in Mathematics,
https://doi.org/10.1007/978-3-030-77263-5_5

$v \in V, \gamma \in S$

$$\gamma \cdot v = v_{|\gamma'}. \tag{5.1.3}$$

Using this involution, we can also define the *contragredient* of a right (to fix ideas) S-module V (say, over a ring R): it is the right R-module $V^\vee = \mathrm{Hom}_R(V, R)$ provided with the following action of S: for every $l \in V^\vee$, and every $\gamma \in S$, $v \in V$,

$$l_{|\gamma}(v) = l(v_{|\gamma'}).$$

In terms of the right action on V attached to the left-action, this formula is simply

$$l_{|\gamma}(v) = l(\gamma \cdot v).$$

Obviously the natural map $V \to (V^\vee)^\vee$ is a morphism of S-modules. We note that if we restrict the S-action to the subgroup $\mathrm{SL}_2(\mathbb{Z})$ of S, then $V^\vee$ is the usual contragredient of the representation V.

We also set

$$S^+ = \mathrm{GL}_2^+(\mathbb{Q}) \cap M_2(\mathbb{Z}). \tag{5.1.4}$$

This submonoid of S is stable by the involution $s \mapsto s'$.

5.1.2 The S-modules $\mathcal{P}_k$ and $\mathcal{V}_k$

Let $k \geq 0$ be an integer. For any commutative ring R, let $\mathcal{P}_k(R)$ be the R-module of polynomials in one variable z with degree at most k. So $\mathcal{P}_k(R)$ is free of rank $k + 1$.

There is a well-known right-action of S on $\mathcal{P}_k(R)$, given by a formula similar to the one used in the definition of modular forms (but with k replaced by $-k$):

$$P_{|\gamma}(z) = (cz + d)^k P\left(\frac{az + b}{cz + d}\right) \text{ for } \gamma = \begin{pmatrix} a & b \\ c & d \end{pmatrix}. \tag{5.1.5}$$

The associated left action has a slightly less familiar form:

$$\gamma \cdot P = P_{|\gamma'}(z) = (a - cz)^k P\left(\frac{dz - b}{a - cz}\right). \tag{5.1.6}$$

Note that $\mathcal{P}_k(R') = \mathcal{P}_k(R) \otimes_R R'$ as (left or right) S-modules.

Exercise 5.1.1 Let $\mathcal{P}'_k(R)$ be the R-module of homogeneous polynomials of degree k in two variables X, Y provided with the right S-action $Q_{|\gamma}(X, Y) = Q(aX + bY, cX + dY)$.

1. Show that $Q(X, Y) \mapsto Q(z, 1)$ defines an isomorphism of right S-modules $\mathcal{P}'_k(R) \to \mathcal{P}_k(R)$.
2. Show that the map $t_k : \mathrm{Sym}^k \mathcal{P}_1(R) \to \mathcal{P}^k(R)$ that sends $P_1 \odot \cdots \odot P_k$ to $P_1 \dots P_k$, where the P_i are in $\mathcal{P}_1(R)$, is an isomorphism of right S-modules. (If V is any R-module, we use the notation $x_1 \odot \cdots \odot x_n$ for the image of $x_1 \otimes \cdots \otimes x_n$ of $V^{\otimes n}$ in $\mathrm{Sym}^n V$.)

We define $\mathcal{V}_k(R) = \mathcal{P}_k(R)^\vee = \mathrm{Hom}_R(\mathcal{P}_k(R), R)$ as the contragredient of $\mathcal{P}_k(R)$. Hence the action of $\gamma \in S$ on $l \in \mathcal{V}_k(R)$ is given by

$$l_{|\gamma}(P) = l(\gamma \cdot P) = l(P_{|\gamma'}), \ \ \forall P \in \mathcal{P}_k(R).$$

Hence

$$l_{|\gamma}(P_{|\gamma}) = \det(\gamma)^k l(P) \tag{5.1.7}$$

Again, $\mathcal{V}_k(R') = \mathcal{V}_k(R) \otimes_R R'$ as right S-modules. The linear forms $l_i \in \mathcal{V}_k(R)$ defined for $i = 0, \dots, k$ by $l_i(z^j) = \delta_{i,j}$ for $j = 0, \dots, k$ form a basis of $\mathcal{V}_k(R)$.

Exercise 5.1.2

1. Show that the map $s_k : \mathrm{Sym}^k \mathcal{V}_1(R) \to (\mathrm{Sym}^k \mathcal{P}_1(R))^\vee$ that sends $\lambda_1 \odot \cdots \odot \lambda_k$ to the linear form $P_1 \odot \cdots \odot P_k \mapsto \sum_{\sigma \in S_k} \prod_{i=1}^k \lambda_i(P_{\sigma(i)})$ is a well-defined morphism of S-modules.
2. Consider the map $t_k^\vee \circ s_k : \mathrm{Sym}^k \mathcal{V}_1(R) \to (\mathrm{Sym}^k \mathcal{P}_1(R))^\vee \xrightarrow{t_k^\vee} \mathcal{P}_k(R)^\vee = \mathcal{V}_k(R)$ where t_k is as defined in question 2 of Exercise 5.1.1. Show that this map is an isomorphism of S-module if $k!$ is invertible in R.

Lemma 5.1.3 *The morphism of groups $\Theta_k : \mathcal{V}_k(R) \to \mathcal{V}_k^\vee(R) = \mathcal{P}_k(R)$ defined by*

$$\Theta_k(l_i) = (-1)^i \binom{k}{i} z^{k-i} \ \textit{for}\ i = 1, \dots, k,$$

is a morphism of right S-modules, which satisfies

$${}^t\Theta_k = (-1)^k \Theta_k. \tag{5.1.8}$$

It is an isomorphism if $k!$ is invertible in R.

Proof When $k = 1$, $\Theta_1(l_0) = z$ and $\Theta_1(l_1) = -1$. The form $(l_0)_{|\gamma}$ sends 1 to $l_0(a - cz) = a$ and z to $l_0(dz - b) = -b$, so $(l_0)_{|\gamma} = al_0 - bl_1$ and $\Theta_1((l_0)_{|\gamma}) = az + b = z_{|\gamma}$. Similarly, $(l_1)_{|\gamma} = -cl_0 + dl_1$ and $\Theta_1((l_1)_{|\gamma}) = cz + d = 1_{|\gamma}$. Hence Θ_1 is S-equivariant. Also the matrix of Θ_1 in the bases (l_0, l_1) of $\mathcal{V}_k(\mathbb{Z})$ and its dual basis $(1, z)$ of $\mathcal{P}_k(\mathbb{Z})$ is $\begin{pmatrix} 0 & -1 \\ 1 & 0 \end{pmatrix}$, so ${}^t\Theta_1 = -\Theta_1$.

Now for general k, using the notations of Exercises 5.1.1 and 5.1.2, it is clear that $t_k \circ (\mathrm{Sym}^k \Theta_1) = \Theta_k \circ s_k$. To prove that Θ_k is S-equivariant, it is enough to do it when $R = \mathbb{Z}$ since the formation of Θ_k commutes with base change, and it is actually enough to do it for $R = \mathbb{Z}[1/k!]$ since $\mathbb{Z} \subset \mathbb{Z}[1/k!]$. In this case, both s_k and t_k are S-equivariant isomorphisms, and so is $\mathrm{Sym}^k\Theta_1$, so Θ_k is also an S-equivariant isomorphism. It is easy to see that ${}^t\Theta_k = (-1)^k\Theta_k$, and this completes the proof of the lemma. □

Exercise 5.1.4 Let $N \geq 1$ be an integer, Γ be a congruence subgroup such that $\Gamma_1(N) \subset \Gamma$. Then $\Gamma_\infty := \begin{pmatrix} 1 & \mathbb{Z} \\ 0 & 1 \end{pmatrix} \subset \Gamma$.

1. If p is a prime $> k$, then $\mathcal{P}_k(\mathbb{F}_p)^{\Gamma_\infty} = \mathbb{F}_p$
2. Let p be a prime number not dividing $Nk!$. Show that the reduction mod p map $\Gamma \to \mathrm{SL}_2(\mathbb{Z}/p\mathbb{Z})$ is surjective. Deduce that $\mathcal{P}_k(\mathbb{F}_p)^{\Gamma} = 0$ if $k > 0$. Also show that $\mathcal{P}_k(\mathbb{Q})^{\Gamma} = 0$ if $k > 0$.
3. If $Nk!$ is invertible in a ring R, and $k > 0$, then the submodule of co-invariants $\mathcal{P}_k(R)_\Gamma$ is 0.

5.2 Classical Modular Symbols

In this section, R will be a commutative ring in which 6 is invertible, and Γ will be a congruence subgroup of $\mathrm{SL}_2(\mathbb{Z})$. By Remark 4.4.10, all the results of last chapter apply to Γ and any R-module V with left Γ-action.

As already observed Γ and $\mathrm{GL}_2(\mathbb{Q})$ satisfies condition (2.6.3), so the Hecke algebra $\mathcal{H}(S, \Gamma)$ (and its subalgebra $\mathcal{H}(S^+, V)$) are defined, and acts on V^Γ for any right S-module V.

5.2.1 *Definition*

Definition 5.2.1 The module of *classical modular symbols* (for Γ, of weight k, over R) is $\mathrm{Symb}_\Gamma(\mathcal{V}_k(R))$.

Thus $\mathrm{Symb}_\Gamma(\mathcal{V}_k(R))$ is an R-module endowed with an action of the Hecke-algebra $\mathcal{H}(S, \Gamma)$.

5.2.2 *The Standard Pairing on Classical Modular Symbols*

Same assumptions as in the preceding subsections. Using the pairings Π, Π_1 and $\Pi_!$ we define pairings

$$\begin{aligned} \mathrm{Symb}_\Gamma(\mathcal{V}_k(R)) \times \mathrm{Symb}_\Gamma(\mathcal{V}_k(R)) &\to R \\ (\phi, \phi') &\mapsto (\phi, \phi') = \Pi(\phi, (\Theta_k)_*(\phi')) \end{aligned} \tag{5.2.1}$$

$$\begin{aligned} \operatorname{Symb}_\Gamma(\mathcal{V}_k(R)) \times H^1(Y_\Gamma, \tilde{\mathcal{V}}_k(R)) &\to R \\ (\phi, \phi') &\mapsto (\phi, \phi')_1 = \Pi_1(\phi, (\Theta_k)_*(\phi')) \end{aligned} \tag{5.2.2}$$

$$\begin{aligned} H^1_!(Y_\Gamma, \tilde{\mathcal{V}}_k(R)) \times H^1_!(Y_\Gamma, \tilde{\mathcal{V}}_k(R)) &\to R \\ (x, y) &\mapsto (x, y)_! = \Pi_!(x, (\Theta_k)_*(y)) \end{aligned} \tag{5.2.3}$$

which we call the *standard pairings on classical modular symbols*. Here $(\Theta_k)_*$ is the map $\operatorname{Symb}_\Gamma(\mathcal{V}_k(R)) \to \operatorname{Symb}_\Gamma(\mathcal{V}_k(R)^\vee)$ induced by the map Θ_k on coefficients. Those pairings are related by the formulas

$$(\phi, \phi') = (\phi, \beta(\phi'))_1 = (\beta(\phi), \beta(\phi'))_!,$$

for all $\phi, \phi' \in \operatorname{Symb}_\Gamma(\mathcal{V}_k(R))$. By (5.1.8), and (4.4.11) one has

$$(\phi, \phi') = (-1)^{k+1}(\phi', \phi) \tag{5.2.4}$$

and for $x, y \in H^1_!(Y_\Gamma, \tilde{\mathcal{V}}_k(R))$:

$$(x, y)_! = (-1)^{k+1}(y, x)_! \tag{5.2.5}$$

Theorem 5.2.2 *Assume that $\Gamma_1(N) \subset \Gamma$ for some $N \geq 1$, and that either R is an $\mathbb{F}_p$-algebra with $p > k$, or that R is a flat extension of $\mathbb{Z}[1/Nk!]$. Then $Symb_\Gamma(V_k(R))$, $H^1_!(Y_\Gamma, \tilde{V}_k(R))$ and $H^1(Y_\Gamma, \tilde{V}_k(R))$ are finite projective module over R and the pairing $(-,-)_1$ and $(-,-)_!$ are non-degenerate.*

Proof When $R = \mathbb{F}_p$, we know by Proposition 4.4.8 and Theorem 4.4.9 that $\Pi_!$ and Π is non-degenerate, and since $p > k$, we know that Θ_k is an isomorphism. The results follows in this case.

When $R = \mathbb{Z}[1/Nk!]$, the conclusion of those results and of Proposition 4.4.7 and Corollary 4.4.6 will give the result, provided we check that hypotheses (ii) of those statement hold. If $k = 0$, then $\mathcal{V}_k(R) = R$ is the trivial representation and the hypothesis (ii) is satisfied. Let us assume that $k > 0$. In this case, $\mathcal{V}_k(R) \simeq \mathcal{P}_k(R)$ and we need to show that for p a prime not dividing $Nk!$, $\mathcal{P}_k(\mathbb{F}_p)^\Gamma = 0$, which follows from question 2 of Exercise 5.1.4; that $\mathcal{P}_k(R)_\Gamma[p] = 0$, which follows from question 3 of the same; that $\mathcal{P}_k(R)^{\Gamma_\infty}/p\mathcal{P}_k(R)^{\Gamma_\infty} = \mathcal{P}_k(\mathbb{F}_p)^{\Gamma_\infty}$ which follows by question 1; and that $\mathcal{P}_k(R)^\Gamma/p\mathcal{P}_k(R)^\Gamma = \mathcal{P}_k(\mathbb{F}_p)^\Gamma$ which follows again from question 2. □

5.2.3 Adjoint of Hecke Operators for the Standard Pairing

The following important result describes the adjoint of Hecke operators for the standard pairing.

Theorem 5.2.3 *Let $x, y \in Symb_\Gamma(\mathcal{V}_k(R))$ and $s \in S$, then*

$$(x_{|[\Gamma s\Gamma]}, y) = (-1)^{sgn(\det s)}(x, y_{|[\Gamma s'\Gamma]}),$$

where (we recall) $s' = s^{-1}\det s$.

Since $\mathrm{Symb}_\Gamma(\mathcal{V}_k(R)) \subset \mathrm{Symb}_\Gamma(\mathcal{V}_k(R) \otimes_{\mathbb{Z}} \mathbb{Q}) = \mathrm{Symb}_\Gamma(\mathcal{V}_k(R)) \otimes_{\mathbb{Z}} \mathbb{Q}$, it suffices to prove the theorem when R is a $\mathbb{Q}$-algebra. In this case, the action of S on $\mathcal{V}_k(R)$ and $\mathcal{P}_k(R)$ extends to an action of $\mathrm{GL}_2(\mathbb{Q})$, defined with the same formulas, and the proof of Lemma 5.1.3 shows that Θ_k is $\mathrm{GL}_2(\mathbb{Q})$-equivariant.

Let $s \in S$ and $\Gamma_s = s^{-1}\Gamma s$. Since $\Gamma_s \subset \mathrm{GL}_2(\mathbb{Q})$ acts on the right on $\mathcal{V}_k(R)$, $\mathrm{Symb}_{\Gamma_s}(\mathcal{V}_k(R))$ is defined and the map $x \mapsto x_{|s}$ is an isomorphism of $\mathrm{Symb}_\Gamma(\mathcal{V}_k(R))$ onto $\mathrm{Symb}_{\Gamma_s}(\mathcal{V}_k(R))$. Since Γ_s, being conjugate of Γ, satisfies conditions (4.4.3) and (4.4.4), the constructions and results of Sect. 4.4.3 and of Sect. 4.4.4 apply to Γ_s and in particular we have pairings $(-,-)$,$\Pi(-,-)$ attached to Γ_s that we will not distinguish notationally from the pairings for Γ, the context making it clear which pairing is used.

$$\begin{aligned}
(x_{|s}, y_{|s}) &= \Pi(x_{|s}, (\Theta_k)_* y_{|s}) \\
&= \Pi(s^* s_* x, s^*(\Theta_k)_* s_* y) \text{ (cf. Sect. 4.4.1)} \\
&= s^*\Pi(s_* x, (\Theta_k)_* s_* y) \in H^2_c(Y_{\Gamma_s}, R) \text{ (cf. (4.3.5))} \\
&= (-1)^{\mathrm{sgn}(\det s)}\Pi(s_* x, s^*(\Theta_k)_* s_*(y)) \in R \text{ (see below)} \\
&= (-1)^{\mathrm{sgn}(\det s)}\Pi(s_* x, s_*(\Theta_k)_* y) \text{ (since } \Theta_k \text{ is } S\text{-equivariant)} \\
&= (-1)^{\mathrm{sgn}(\det s)}(\det s)^k \Pi(x, \Theta_k(y)) \text{ (by (5.1.7))} \\
&= (-1)^{\mathrm{sgn}(\det s)}(\det s)^k (x, y).
\end{aligned}$$

The justification for the line missing one is that the isomorphism $H^2(Y_{\Gamma_s}, R) \simeq R$ is changed to its opposite when the orientation of Y_{Γ_s} is reversed, and the action of s on $\mathcal{H}$, hence on Y_{Γ_s}, preserves the orientation if $\det s > 0$ and reverses it if $\det s < 0$.

Now let $z \in \mathrm{Symb}_{\Gamma_s}(\mathcal{V}_k(R))$. One has

$$(x_{|s}, z) = (-1)^{\mathrm{sgn}(\det s)}(x, z_{|s'}). \tag{5.2.6}$$

Indeed, note that $(z_{|s'})_{|s} = z_{|s's} = z_{|\det s} = (\det s)^k z$. Therefore

$$\begin{aligned}
(x_{|s}, z) &= (\det s)^{-k}(x_{|s}, (z_{|s'})_{|s}) \\
&= (\det s)^{-k}(\det s)^k(-1)^{\mathrm{sgn}(\det s)}(x, z_{|s'})
\end{aligned}$$

To conclude the proof we choose finitely many $s_i \in \Gamma s\Gamma$ such that $\Gamma s\Gamma = \coprod_s \Gamma s_i = \coprod_i s_i\Gamma$. This is possible by Diamond and Shurman [56, Lemma 5.5.1]. Since $\Gamma \subset \mathrm{SL}_2(\mathbb{Z})$, $\det s_i = \det s$ for all i, and also $\Gamma' = \Gamma$ so that we have $\Gamma s'\Gamma = (\Gamma s\Gamma)' = \coprod \Gamma s_i'$. Therefore

$$(x_{|[\Gamma s\Gamma]}, y) = \sum_i (x_{|s_i}, y) = (-1)^{\det s}(x, y_{|s_i'}) = (-1)^{\det s}(x, y_{|[\Gamma s'\Gamma]})$$

5.3 Classical Modular Symbols and Modular Forms

In this section, Γ is a congruence subgroup of $\mathrm{SL}_2(\mathbb{Z})$.

5.3.1 *Modular Forms and Real Classical Modular Symbols*

Let f be a cuspidal modular form for Γ of weight $k+2$. We define a homomorphism of groups $\phi_f : \Delta_0 \to \mathcal{V}_k(\mathbb{R})$ by setting

$$\phi_f(D) = \left(P \mapsto \mathrm{Re}\left(\int_D f(z)P(z)dz\right)\right).$$

Some explanation may be necessary. if $D = \{b_1\} + \cdots + \{b_n\} - \{a_1\} - \cdots - \{a_n\}$, then $\int_D f(z)P(z)dz := \int_{a_1}^{b_1} f(z)P(z)dz + \cdots + \int_{a_n}^{b_n} f(z)P(z)dz$. Here $\int_a^b$ means the integral along the geodesic from a to b. That the integrals converge follows from the cuspidality of f, and that their sum is independent of the writing of D as above follows from Cauchy's homotopy independence of path integrals. Since $\mathrm{Re}\left(\int_D f(z)P(z)dz\right)$ is a real number depending linearly of $P \in \mathcal{P}_k(\mathbb{R})$, the map $P \mapsto \mathrm{Re}\left(\int_D f(z)P(z)dz\right)$ is an element of $\mathcal{V}_k(\mathbb{R})$.

Lemma 5.3.1 *The morphism ϕ_f belongs to $Symb_\Gamma(\mathcal{V}_k(\mathbb{R}))$. The map $f \mapsto \phi_f : S_{k+2}(\Gamma, \mathbb{C}) \to Symb_\Gamma(\mathcal{V}_k(\mathbb{R}))$ is an $\mathbb{R}$-linear map commuting with the actions of $\mathcal{H}(S^+, \Gamma)$.*

Remember that while $\mathcal{H}(S, \Gamma)$ acts on modular symbols, only its sub-algebra $\mathcal{H}(S^+, \Gamma)$ acts on the spaces of modular forms (see Sect. 2.6.5).

Proof We need to show that for all $D \in \Delta_0$, $\gamma \in \Gamma$, we have $\phi_f(\gamma \cdot D)_{|\gamma} = \phi_f(D)$, We compute, for $P \in \mathcal{P}_k(\mathbb{R})$:

$$\begin{aligned}\phi_f(\gamma \cdot D)_{|\gamma}(P) &= \phi_f(\gamma \cdot D)(\gamma \cdot P)\\ &= \mathrm{Re}\left(\int_{\gamma \cdot D} f(z)(\gamma \cdot P)(z)dz\right)\end{aligned}$$

$$= \mathrm{Re}\left(\int_{\gamma\cdot D} f(z)(a-cz)^k P(\gamma^{-1}\cdot z)dz\right)$$

$$= \mathrm{Re}\left(\int_{\gamma\cdot D} (a-cz)^{-k-2} f(\gamma^{-1}\cdot z)(a-cz)^k P(\gamma^{-1}\cdot z)(a-cz)^2 d(\gamma^{-1}\cdot z)\right)$$

$$= \mathrm{Re}\left(\int_{\gamma\cdot D} f(\gamma^{-1}\cdot z)P(\gamma^{-1}\cdot z)d(\gamma^{-1}\cdot z)\right)$$

$$= \mathrm{Re}\left(\int_{D} f(z)P(z)dz\right) = \phi_f(D)(P).$$

The fact that the map $f \mapsto \phi_f$ is $\mathbb{R}$-linear is obvious.

To prove that the map commutes with the action of the Hecke operators, we only need to check that for $\gamma \in S^+ := S \cap \mathrm{GL}_2(\mathbb{Q})$, $(\phi_f)_{|\gamma} = \phi_{f_{|_{k+2}\gamma}}$. This is a slightly more general computation that the one we just did:

$$\begin{aligned}
\phi_f(\gamma\cdot D)_{|\gamma}(P) &= \phi_f(\gamma\cdot D)(\gamma\cdot P)\\
&= \mathrm{Re}\left(\int_{\gamma\cdot D} f(z)(\gamma\cdot P)(z)dz\right)\\
&= \mathrm{Re}\left(\int_{\gamma\cdot D} f(z)(a-cz)^k P(\gamma'\cdot z)dz\right)\\
&= \mathrm{Re}\left(\int_{\gamma\cdot D} (f_{|\gamma})_{|\gamma^{-1}}(z)(a-cz)^k P(\gamma'\cdot z)(a-cz)^2(\det\gamma)^{-1}d(\gamma'\cdot z)\right)\\
&= \mathrm{Re}\left(\int_{\gamma\cdot D} ((\det\gamma)^{-1}(a-cz))^{-k-2}(\det\gamma)^{-k-1} f_{|\gamma}(\gamma' z)\right.\\
&\qquad\qquad \left.(a-cz)^k P(\gamma'\cdot z)(a-cz)^2(\det\gamma)^{-1}d(\gamma'\cdot z)\right)\\
&= \mathrm{Re}\left(\int_{\gamma\cdot D} f_{|\gamma}(\gamma' z)P(\gamma'\cdot z)d(\gamma'\cdot z)\right)\\
&= \mathrm{Re}\left(\int_{D} f_{|\gamma}(z)P(z)dz\right) = \phi_{f_{|\gamma}}(D)(P).
\end{aligned}$$

(In the above computation, we have used that $\gamma^{-1}\cdot z = \gamma'\cdot z$.) □

Recall that the Peterson Hermitian product is defined on $S_{k+2}(\Gamma, \mathbb{C})$ by

$$(f, g)_\Gamma = \int_{\mathcal{H}/\Gamma} f(z)\overline{g(z)}y^k\,dx\,dy, \tag{5.3.1}$$

where $z = x + iy$. One has $(f, g)_\Gamma = \overline{(g, f)_\Gamma}$.

Proposition 5.3.2 *Let f, g in $S_{k+2}(\Gamma, \mathbb{C})$. We have*

$$(\phi_f, \phi_g) = (2i)^{k-1}[(f, g)_\Gamma + (-1)^{k+1}(g, f)_\Gamma].$$

In other words,

$$(\phi_f, \phi_g) = \begin{cases} (2i)^k \mathrm{Im}(f, g)_\Gamma & \text{if } k \text{ is even} \\ 2(2i)^{k-1}\mathrm{Re}(f, g)_\Gamma & \text{if } k \text{ is odd} \end{cases}$$

Proof We may assume that Γ acts freely on $\mathcal{H}$.

For the sake of pedagogy, let us first give the proof when $k = 0$. In this case the formula to prove is that $(\phi_f, \phi_g) = \mathrm{Im}(f, g)_\gamma$. Now if we follow ϕ_f through the map $\mathrm{Symb}_\Gamma(\mathbb{R}) = H^1_c(Y(\Gamma), \mathbb{R}) \subset H^1(Y(\Gamma), \mathbb{R}) = H^1_{\mathrm{dR}}(Y(\Gamma), \mathbb{R})$ we see that ϕ_f is the cohomology class that sends a closed path c in $Y(\Gamma)$ to $\mathrm{Re} \int_c f(z)\, dz$, where $\int_c f(z) dz := \int_{\tilde{c}} f(z) dz$, where $\tilde{c}$ is any path in $\mathcal{H}$ lifting c (that $\int_{\tilde{c}} f(z) dz$ is independent of the lift results from the computation of Lemma 5.3.1).

This cohomology class is represented by the differential form $(\mathrm{Re} f)\, dx - (\mathrm{Im} f)\, dy$. This follows from the computation, if c is the path $x(t) + iy(t)$ for $t \in [0, 1]$:

$$\mathrm{Re} \int_c f(z)\, dz = \mathrm{Re} \int_0^1 f(x(t), y(t))(x'(t) + iy'(t))dt = \int_0^1 ((\mathrm{Re} f)x'(t) - (\mathrm{Im} f)y'(t))dt.$$

Therefore, to compute the pairing (ϕ_f, ϕ_g), we only have to integrate over $\mathcal{H}/\Gamma$ the 2-form $((\mathrm{Re} f)\, dx - (\mathrm{Im} f)\, dy) \wedge ((\mathrm{Re} g)\, dx - (\mathrm{Im} g)\, dy) = (\mathrm{Re} g \mathrm{Im} f - \mathrm{Re} f \mathrm{Im} g)(dx \wedge dy)$, thus we get $(\phi_f, \phi_g) = \int_{\mathcal{H}/\Gamma} (\mathrm{Re} g \mathrm{Im} f - \mathrm{Re} f \mathrm{Im} g) dx\, dy$, but this is exactly $\mathrm{Im}(f, g)_\Gamma$.

Now we treat the case of general k. If we follow ϕ_f through the map

$$\mathrm{Symb}_\Gamma(\mathcal{V}_K(R)) = H^1_c(Y(\Gamma), \tilde{\mathcal{V}}_k(R)) \subset H^1(Y(\Gamma), \tilde{\mathcal{V}}_K(R)) = H^1_{\mathrm{dR}}(Y(\Gamma), \tilde{\mathcal{V}}_k(R))$$

we see that ϕ_f is the cohomology class that sends a closed path c in $Y(\Gamma)$ to the linear form $P \mapsto \mathrm{Re} \int_c f(z)P(z)\, dz$ on $\mathcal{P}_k(\mathbb{R})$, where $\int_c f(z)P(z)\, dz$, and the same computation as above shows that this cohomology class is represented by the $\mathcal{V}_k(\mathbb{C})$-valued Γ-invariant differential form on $\mathcal{H}$:

$$\omega_f = \sum_{i=0}^k \mathrm{Re}(f(z)z^i) l_i dx - \sum_{i=0}^k \mathrm{Im}(f(z)z^i)\, l_i\, dy.$$

Thus

$$\omega_f \wedge (\Theta_k)_*(\omega_g) = \sum_{i,j} [(\mathrm{Im}(f(z)z^i)\mathrm{Re}(g(z)z^j) - \mathrm{Re}(f(z)z^i)\mathrm{Im}(g(z)z^j))]\, l_i \otimes (\Theta_k)_*(l_j)\, dxdy$$
$$= \sum_{i,j} \mathrm{Im}(f(z)\bar{g}(z)z^i\bar{z}^j)\, l_i \otimes (\Theta_k)_*(l_j)\, dxdy.$$

The image of this Γ-invariant $\mathcal{V}_k(\mathbb{R}) \otimes \mathcal{P}_k(\mathbb{R})$-valued 2-form on $\mathcal{H}$ by the tautological map $\mathcal{V}_k(\mathbb{R}) \otimes \mathcal{P}_k(\mathbb{R}) \to \mathbb{R}$ is the Γ-invariant real-valued 2-form

$$\sum_{i=0}^{k} \mathrm{Im}\left(f(z)\bar{g}(z)z^i\bar{z}^{k-i}(-1)^{k-i}\binom{k}{i}\right) = \mathrm{Im}(f(z)\bar{g}(z)(z-\bar{z})^k)\, dxdy.$$

Finally,

$$(\phi_f, \phi_g) = \int_{\mathcal{H}/\Gamma} \mathrm{Im}(f(z)\bar{g}(z)(z-\bar{z})^k)\, dxdy$$
$$= \begin{cases} 2^k i^k \mathrm{Im} \int_{\mathcal{H}/\Gamma} f(z)\bar{g}(z)y^k dxdy & \text{if } k \text{ is even} \\ 2^k i^{k-1} \mathrm{Re} \int_{\mathcal{H}/\Gamma} f(z)\bar{g}(z)y^k dxdy & \text{if } k \text{ is odd} \end{cases}$$

□

Exercise 5.3.3 Prove the formula

$$(\phi_f, \phi_{i^{k-1}g}) = 2^k \mathrm{Re}(f, g)_\Gamma.$$

Corollary 5.3.4 *The map $S_{k+2}(\Gamma, \mathbb{C}) \to \mathrm{Symb}_\Gamma(\mathcal{V}_k(\mathbb{R}))$ is injective. Actually, even the composition $S_{k+2}(\Gamma, \mathbb{C}) \to H^1_!(\Gamma, \mathcal{V}_k(\mathbb{R}))$ of this map with β is injective.*

Proof Let $f \in S_{k+2}(\Gamma, \mathbb{C})$. Assume $\beta(\phi_f) = 0$. Then for any $g \in S_{k+2}(\Gamma, \mathbb{C})$, we have $\mathrm{Re}(f, g)_\Gamma = 2^{-k}(\phi_f, \phi_{i^{k-1}g}) = 2^{-k}(\beta(\phi_f), \beta(\phi_{i^{k-1}g})) = 0$. Applying this to ig we get that $\mathrm{Im}(f, g)_\Gamma = 0$ so $(f, g)_\Gamma = 0$ for all g. Since the Peterson's product is perfect, $f = 0$. □

Actually we have more.

Theorem 5.3.5 *The map $f \mapsto \beta(\phi_f)$, $S_{k+2}(\Gamma, \mathbb{C}) \to H^1_!(\Gamma, \mathcal{V}_k(\mathbb{R}))$ is an $\mathbb{R}$-linear isomorphism commuting with the action of $\mathcal{H}(S^+, \Gamma)$.*

Since we already know that this map is injective, we only need to prove that

$$\dim_{\mathbb{R}} H^1_!(\Gamma, \mathcal{V}_k(\mathbb{R})) \le \dim S_{k+2}(\Gamma, \mathbb{C}).$$

When $k = 0$, $\mathcal{V}_0(\mathbb{R}) = \mathbb{R}$. In this case, the elements of $S_2(\Gamma, \mathbb{C})$ are the holomorphic differentials 1-forms on $X(\Gamma)$ (the compactification of $Y(\Gamma)$ as a smooth Riemann surface). The complex dimension of this space is the genus g of $X(\Gamma)$, so $2g$ for the real dimension. On the other hand, the morphism β is actually composition $H^1_c(Y(\Gamma), \mathbb{R}) \to H^1(X(\Gamma), \mathbb{R}) \to H^1(Y(\Gamma), \mathbb{R})$ where the first morphism is given by covariant functoriality of $H^i_c(-, \mathbb{R})$, and the second by the contravariant functoriality of $H^i(-, \mathbb{R})$. So $\dim_{\mathbb{R}} H^1_!(\Gamma, \mathbb{R}) \leq \dim_{\mathbb{R}} H^1(X(\Gamma), \mathbb{R}) = 2g$ and we are done.

The general case is proved by a similar but more complicated computation, where we have to take the number of cups into account: [113, (8.2.23)].

Remark 5.3.6 Note that we have proved $H^1_!(\Gamma, \mathbb{R}) \simeq H^1(X(\Gamma), \mathbb{R})$. The same results holds for $\mathbb{R}$ replaced by $\mathcal{V}_k(\mathbb{R})$.

5.3.2 Modular Forms and Complex Classical Modular Symbols

Let us first recall the notion of the conjugate $\bar{W}$ of a $\mathbb{C}$-vector space W. As an abelian group, and even as a real vector space $\bar{W}$ is equal to W, except that we denote by $\bar{w} \in \bar{W}$ the element $w \in W$ when we want to make clear that we see it as an element of $\bar{W}$. We define a $\mathbb{C}$-scalar multiplication on $\bar{W}$ by the rule $\lambda\bar{w} = \overline{\bar{\lambda} w}$. This provides $\bar{W}$ with a structure of $\mathbb{C}$-vector space. Note that the construction $W \mapsto \bar{W}$ is functorial, as a $\mathbb{C}$-linear map $f : W \to W'$ is also a $\mathbb{C}$-linear map $f : \bar{W} \to \bar{W}'$.

Exercise 5.3.7 Suppose given a commutative ring $\mathcal{H}$ that acts $\mathbb{C}$-linearly on W. Then $\mathcal{H}$ acts on $\bar{W}$ by functoriality. Let χ be a character $\mathcal{H} \to \mathbb{C}$. Do we have $\dim_{\mathbb{C}} W[\chi] = \dim_{\mathbb{C}} \bar{W}[\chi]$?

Are the $\mathbb{C}$-vector spaces W and $\bar{W}$ isomorphic? Obviously yes, since they have the same real dimension, hence the same complex dimension. Are they canonically isomorphic? No, they are not. However, if we are given an $\mathbb{R}$-structure $W_{\mathbb{R}}$ on W, that is a sub-$\mathbb{R}$-vector space such that the natural map $W_{\mathbb{R}} \otimes_R \mathbb{C} \to W$ is an isomorphism, then this $\mathbb{R}$-structure defines a canonical $\mathbb{C}$-linear isomorphism $c : W \to \bar{W}$ as follows: any $w \in W$ can be written uniquely $w_1 + i w_2$ with $w_1, w_2 \in W_{\mathbb{R}}$ (we write $w_1 = \mathrm{Re} w$ and $w_2 = \mathrm{Im} w$ in this case) and we send w to

$$c(w) = \overline{w_1 - i w_2} = \overline{w_1} + i \overline{w_2} \in \bar{W}. \tag{5.3.2}$$

Exercise 5.3.8 Check that c is really a $\mathbb{C}$-linear isomorphism $W \to \bar{W}$.

Lemma 5.3.9

(i) *Let W be a $\mathbb{C}$-vector space. Then there is a canonical and functorial $\mathbb{C}$-linear isomorphism*

$$W \otimes_{\mathbb{R}} \mathbb{C} \simeq W \oplus \bar{W}, \quad w \otimes \lambda \to (\lambda w, \lambda \bar{w}).$$

(Here the complex structure on $W \otimes_{\mathbb{R}} \mathbb{C}$ is the one coming from the second factor, $\mathbb{C}$.)

(ii) *Let X be another $\mathbb{C}$-vector space, with a real structure $X_{\mathbb{R}}$. Let h be a $\mathbb{C}$-linear map $W \to X$, so that* Re h *is an $\mathbb{R}$-linear map $W \to X_{\mathbb{R}}$. Then*

$$\operatorname{Re} h \otimes 1 : W \otimes_{\mathbb{R}} \mathbb{C} \to X_{\mathbb{R}} \otimes_{\mathbb{R}} \mathbb{C}$$

is, modulo the obvious identifications, the map

$$W \oplus \bar{W} \to X,\ (w, \overline{w'}) \mapsto \frac{1}{2}(h(w) + c^{-1}(h(w'))).$$

(Let us insist on the meaning of $c^{-1}(h(w'))$: by functoriality, h induces (in fact, is) a $\mathbb{C}$-linear map $\bar{W} \to \bar{X}$, and c^{-1} is a $\mathbb{C}$-linear isomorphism $\bar{X} \to X$.)

Proof Let's prove (i). The map is obviously $\mathbb{C}$-linear. Let us check that it is injective. Any element of $W \otimes_{\mathbb{R}} \mathbb{C}$ can be written $w \otimes 1 + w' \otimes i$. Such an element is sent to $(w + iw', \bar{w} + i\bar{w}') = (w + iw', \overline{w - iw'})$. If that element is 0, then w and w' are 0, which shows the injectivity of our map. Since the source and target have clearly the same dimension, this proves (i).

Let us prove (ii): we have $(\operatorname{Re} h \otimes 1)(w \otimes 1) = \operatorname{Re} h(w)$, and $(\operatorname{Re} h \otimes 1)((iw) \otimes i) = i\operatorname{Re} h(iw) = -i\operatorname{Im} h(w)$. Since the isomorphism $W \otimes_{\mathbb{R}} \mathbb{C} = W \oplus W$ identifies $w \otimes 1$ to $(w, \bar{w})$ and $iw \otimes i$ to $(-w, \bar{w})$, the map $\operatorname{Re} h \otimes 1$ sends $(w, \bar{w})$ to $\operatorname{Re} h(w)$ and $(-w, \bar{w})$ to $-i\operatorname{Im} h(w)$. Therefore it sends $(2w, 0)$ to $\operatorname{Re} h(w) + i\operatorname{Im} h(w) = h(w)$ and $(0, 2w')$ to $\operatorname{Re} h(w') - i\operatorname{Im} h(w') = c^{-1}(h(w'))$. □

We shall apply this lemma to $W = S_{k+2}(\Gamma)$ and $X = \operatorname{Symb}_{\Gamma}(\mathcal{V}_k(\mathbb{C}))$. The space X has a real structure $X_{\mathbb{R}} = \operatorname{Symb}_{\Gamma}(\mathcal{V}_k(\mathbb{R}))$, and the two maps Re and Im, $X \to X_{\mathbb{R}} \subset X$ are easily described: If $\Phi \in \operatorname{Symb}_{\Gamma}(\mathcal{V}_k(\mathbb{C}))$, then $(\operatorname{Re}\Phi)(D)$ is the $\mathbb{C}$-linear map on $\mathcal{P}_k(\mathbb{C})$ that sends a **real** polynomial $P \in \mathcal{P}_k(\mathbb{R})$ to $\operatorname{Re}(\Phi(D)(P))$, and similarly for Im.

Now we have a map $h : f \mapsto \phi_f^1 : W \to X$ where ϕ_f^1 is defined, for every $D \in \Delta_0$ and every $P \in \mathcal{P}_k(\mathbb{C})$, by

$$\phi_f^1(D)(P) = \int_D f(z)P(z)dz \tag{5.3.3}$$

Obviously $\operatorname{Re} h$ is the map $f \mapsto \phi_f$ considered in the above section. Also note that by definition $c^{-1}h : \overline{S_{k+2}(\Gamma, \mathbb{C})} \to \operatorname{Symb}_{\Gamma}(\mathcal{V}_k(\mathbb{C}))$ sends $\bar{f}$ to the symbol $\phi_{\bar{f}}^2$ defined by

$$\phi_{\bar{f}}^2(D)(P) = \overline{\int_D f(z)P(z)dz} \text{ for all } P \in \mathcal{P}_k(\mathbb{R}), \text{ hence} \tag{5.3.4}$$

$$\phi_{\bar{f}}^2(D)(P) = \int_D \overline{f(z)}P(\bar{z})d\bar{z} \text{ for all } P \in \mathcal{P}_k(\mathbb{C}) \tag{5.3.5}$$

Theorem 5.3.10 *The $\mathbb{C}$-linear map*

$$S_{k+2}(\Gamma, \mathbb{C}) \oplus \overline{S_{k+2}(\Gamma, \mathbb{C})} \to Symb_\Gamma(\mathcal{V}_k(\mathbb{C})) \tag{5.3.6}$$

$$(f, \bar{g}) \mapsto \phi_f^1 + \phi_{\bar{g}}^2 \tag{5.3.7}$$

is an injective $\mathbb{C}$-linear map commuting with the actions of $\mathcal{H}(S^+, \Gamma)$.

The composition of this application with $\beta : Symb_\Gamma(\mathcal{V}_k(\mathbb{C})) \to H^1_!(\Gamma, \mathcal{V}_k(\mathbb{C}))$ is a $\mathbb{C}$-linear isomorphism, commuting with the actions of $\mathcal{H}(S^+, \Gamma)$,

$$S_{k+2}(\Gamma, \mathbb{C}) \oplus \overline{S_{k+2}(\Gamma, \mathbb{C})} \simeq H^1_!(\Gamma, \mathcal{V}_k(\mathbb{C})). \tag{5.3.8}$$

The injective map (5.3.6) can be extended to a $\mathbb{C}$-linear isomorphism, commuting with the actions of $\mathcal{H}(S^+, \Gamma)$,

$$S_{k+2}(\Gamma, \mathbb{C}) \oplus \overline{S_{k+2}(\Gamma, \mathbb{C})} \oplus BSymb_\Gamma(\mathcal{V}_k(\mathbb{C})) \simeq Symb_\Gamma(\mathcal{V}_k(\mathbb{C})). \tag{5.3.9}$$

Proof By the lemma, the application (5.3.6) (resp. (5.3.8)) is the same as $2(\mathrm{Re}\, h) \otimes 1$ (resp. $2\beta(Re\, h) \otimes 1$), hence the injectivity (resp. bijectivity) and Hecke-compatibility of this application results from Corollary 5.3.4 (resp. from Theorem 5.3.5). Then (5.3.9) follows from Corollary 4.4.3 and Definition 4.1.8. □

We also need to keep track of the pairing in our complexification process. First an abstract lemma in the spirit of Lemma 5.3.9.

Lemma 5.3.11 *Let W be a complex vector space with an Hermitian pairing $(\, , \,)_\Gamma$ (conjugate-linear in the second variable).*

The $\mathbb{R}$-linear symmetric pairing $\mathrm{Re}(\, , \,)_\Gamma$ *on W, extended by linearity into a $\mathbb{C}$-linear symmetric pairing on $W \otimes_{\mathbb{R}} C = W \oplus W$ is given by the formula*

$$\mathrm{Re}((w_1, \bar{w}_2), (w_1', \bar{w}_2'))_\Gamma = \frac{1}{2}((w_1, w_2')_\Gamma + (w_1', w_2)_\Gamma).$$

The $\mathbb{R}$-linear antisymmetric pairing $Im(\, , \,)_\Gamma$ on W, extended by linearity into a $\mathbb{C}$-linear antisymmetric pairing on $W \otimes_{\mathbb{R}} C = W \oplus W$ is given by the formula

$$\mathrm{Im}((w_1, \bar{w}_2), (w_1', \bar{w}_2'))_\Gamma = \frac{-i}{2}((w_1, w_2')_\Gamma - (w_1', w_2)_\Gamma).$$

Proof Since $w \otimes 1$ in $W \otimes \mathbb{C}$ is identified with $(w, \bar{w})$ in $W \oplus \bar{W}$, we have

$$\mathrm{Re}((w, \bar{w}), (w', \overline{w'}))_\Gamma = \mathrm{Re}(w, w')_\Gamma.$$

The same result applied to iw instead of w gives

$$i\mathrm{Re}((w, -\bar{w}), (w', \overline{w'}))_\Gamma = \mathrm{Re}(iw, w')_\Gamma.$$

Adding the first equality and $-i$ times the second gives

$$2\mathrm{Re}((w,0),(w',\overline{w'}))_\Gamma = \mathrm{Re}(w,w')_\Gamma - i\mathrm{Re}i(w,w')_\Gamma = (w,w')_\Gamma.$$

The same equality with w' replaced by iw' gives

$$2\mathrm{Re}((w,0),(w',-\overline{w'}))_\Gamma = -(w,w')_\Gamma.$$

Summing the two equalities above we get

$$\mathrm{Re}((w,0),(w',0)])_\Gamma = 0$$

and taking instead the difference gives

$$2\mathrm{Re}((w,0),(0,\bar{w}'))_\Gamma = (w,w')_\Gamma.$$

The formula for the symmetric bilinear map $\mathrm{Re}(\ ,\ )_\Gamma$ follows. The formula for $\mathrm{Im}(\ ,\ )_\Gamma$ is proved similarly. □

Proposition 5.3.12 *Let $f, g \in S_{k+2}(\Gamma, \mathbb{C})$. Then*

$$(\phi^1_f, \phi^1_g) = 0 \tag{5.3.10}$$

$$(\phi^2_{\bar{f}}, \phi^2_{\bar{g}}) = 0 \tag{5.3.11}$$

$$(\phi^1_f, \phi^2_{\bar{g}}) = (2i)^{k-1}(f, g)_\Gamma. \tag{5.3.12}$$

Proof Let assume that k is odd, the proof with k even being similar. The map $f \mapsto \phi_f$, $S_{k+2}(\Gamma, \mathbb{R}) \to \mathrm{Symb}_\Gamma(\mathcal{V}_k(\mathbb{C}))$ is compatible with the symmetric bilinear form $2(2i)^{k-1}\mathrm{Re}(\ ,\ )_\Gamma$ on the source and $(\ ,\ )$ on the target. Hence its complexification,

$$f \mapsto (\phi^1_f, \phi^2_{\bar{f}}),\ S_{k+2}(\Gamma, \mathbb{C}) \to \mathrm{Symb}_\Gamma(\mathcal{V}_k(\mathbb{C})) \oplus \overline{\mathrm{Symb}_\Gamma(\mathcal{V}_k(\mathbb{C}))}$$

is compatible with the complexification of those symmetric bilinear maps, which means that

$$(\phi^1_f + \phi^2_{\bar{f}}, \phi^1_g + \phi^2_{\bar{g}}) = 2(2i)^{k-1}\mathrm{Re}((f, \bar{f}), (g, \bar{g}))_\Gamma$$

Expanding the left hand side and rewriting the right hand side using the above lemma gives

$$(\phi^1_f, \phi^1_g) + (\phi^1_f, \phi^2_{\bar{g}}) + (\phi^2_{\bar{f}}, \phi^1_g) + (\phi^2_{\bar{f}}, \phi^2_{\bar{g}}) = (2i)^{k-1}((f,g)_\Gamma + (g,f)_\Gamma).$$

Identifying the linear in f, anti-linear in g parts of the two sides of the equation gives (5.3.12), and identifying the linear in f and g parts of the two sides, or their anti-linear parts give the two other equations. □

5.3.3 The Involution ι, and How to Get Rid of the Complex Conjugation

In this paragraph, R is as before a ring where 6 is invertible, but from now on we shall fix an integer $N \geq 1$ and assume that Γ is a congruence subgroup such that

$$\Gamma_1(N) \subset \Gamma \subset \Gamma_0(N). \tag{5.3.13}$$

Remember that the map $\begin{pmatrix} a & b \\ c & d \end{pmatrix} \mapsto d$ identifies $\Gamma_0(N)/\Gamma_1(N)$ with $(\mathbb{Z}/N\mathbb{Z})^*$ so for a given $\mathbb{Z}$, the Γ satisfying (5.3.13) are in bijection with the subgroups of $(\mathbb{Z}/\mathbb{N}\mathbb{Z})^*$.

As in Sect. 2.6.4, we shall denote by $\mathcal{H}$ the sub-algebra of $\mathcal{H}(S^+, \Gamma)$ generated by the double classes $[\Gamma \begin{pmatrix} 1 & 0 \\ 0 & l \end{pmatrix} \Gamma]$ for l any prime (defining the Hecke operators T_l or U_l according to whether $l \nmid N$ or $l \mid N$), and for $a \in (\mathbb{Z}/N\mathbb{Z})^*$, and γ_a any matrix in $\Gamma_0(N)$ whose upper-left entry is $a \pmod N$, the double classes $[\Gamma\gamma_a\Gamma]$ (defining the diamond operators $\langle a \rangle$).

The matrix $\iota = \begin{pmatrix} 1 & 0 \\ 0 & -1 \end{pmatrix} \in S$ defines a double class $[\Gamma\iota\Gamma]$ which acts on $\mathrm{Symb}_\Gamma(V)$, where V is any R-module with a right action of S.

Lemma 5.3.13 *For any $\phi \in \mathrm{Symb}_\Gamma(V)$, one has $\phi_{|\Gamma\iota\Gamma} = \phi_{|\iota}$. The maps $\phi \mapsto \phi_\iota$ is an involution of $\mathrm{Symb}(\Gamma(V))$ and commutes with the action of $\mathcal{H}$.*

Proof Conjugating a square matrix by ι does not change its diagonal terms and multiply its anti-diagonal terms by -1. In particular ι normalizes $\Gamma_1(N)$ and $\Gamma_0(N)$, and also Γ since Γ is characterized among groups satisfying (5.3.13) by the subgroup of $(\mathbb{Z}/N\mathbb{Z})^*$ forms by the upper-left entry of its element. Therefore $\Gamma\iota\Gamma = \Gamma\iota$ and $\phi_{|\Gamma\iota\Gamma} = \phi_{|\iota}$. Since $\iota^2 = 1$, the map $\phi \mapsto \phi_\iota$ is an involution. Since $\iota\begin{pmatrix} 1 & 0 \\ 0 & l \end{pmatrix}\iota = \begin{pmatrix} 1 & 0 \\ 0 & l \end{pmatrix}$, ι commutes with the T_l and U_l, and since $\iota\gamma_a\iota$ is another matrix in $\Gamma_0(N)$ with upper-left coefficient equal to $a \pmod N$, ι commutes with $\langle a \rangle$. Hence ι commutes with $\mathcal{H}$. □

We define $\mathrm{Symb}^{\pm}_\Gamma(V)$ as the subgroup of $\mathrm{Symb}_\Gamma(V)$ on which ι acts by ± 1 and we have $\mathrm{Symb}_\Gamma(V) = \mathrm{Symb}^+_\Gamma(V) \oplus \mathrm{Symb}^-_\Gamma(V)$.

The space $S_{k+2}(\Gamma, \mathbb{C})$ has a canonical real structure $S_{k+2}(\Gamma, \mathbb{R})$, the subspace of forms whose Fourier coefficients are all in $\mathbb{R}$ (cf. [113, Theorems 3.52] or [93, Theorem 4.5.19]) Using this real structure, we can identify canonically the space $S_{k+2}(\Gamma, \mathbb{C})$ with its conjugate $\overline{S_{k+2}(\Gamma, \mathbb{C})}$. Explicitly, this identification is as follows:

Definition 5.3.14 For $f \in S_{k+2}(\Gamma, \mathbb{C})$ we define an holomorphic function $f_\rho : \mathcal{H} \to \mathbb{C}$ by $f_\rho(z) = \overline{f(-\bar{z})}$.

Lemma 5.3.15 *For any $\gamma \in S^+$, we have $(f_\rho)_\gamma = (f_{\iota\gamma\iota})_\rho$. In particular, $f_\rho \in S_{k+2}(\Gamma, \mathbb{C})$.*

If $f(z) = \sum_{n=1}^{\infty} a_n q^n$, then $f_\rho(z) = \sum_{n=1}^{\infty} \overline{a_n} q^n$.

The linear isomorphism $c : S_{k+2}(\Gamma, \mathbb{C}) \mapsto \overline{S_{k+2}(\Gamma, \mathbb{C})}$ defined by this real structure (see (5.3.2)) is the map $f \mapsto \overline{f_\rho}$ (where $\overline{f_\rho}$ is, according to our conventions, f_ρ seen as an element of $\overline{S_{k+2}(\Gamma, \mathbb{C})}$), and this map is compatible with the action of $\mathcal{H}$.

Proof For any $\gamma = \begin{pmatrix} a & b \\ c & d \end{pmatrix} \in S^+$, we have (remembering that a, b, c, d are real):

$$\begin{aligned}(f_\rho)_{|\gamma}(z) &= (cz+d)^{-k-2} f_\rho(\frac{az+b}{cz+d}) \\ &= (cz+d)^{-k-2} \overline{f(-\frac{a\bar{z}+b}{c\bar{z}+d})}\end{aligned}$$

and, since $\iota\gamma\iota = \begin{pmatrix} a & -b \\ -c & d \end{pmatrix}$,

$$\begin{aligned}(f_{|\iota\gamma\iota})_\rho(z) &= [(-cz+d)^{-k-2} f(\frac{az-b}{-cz+d}))]_\rho \\ &= \overline{(c\bar{z}+d)^{-k-2} f(\frac{-a\bar{z}-b}{c\bar{z}+d})} \\ &= (cz+d)^{-k-2} \overline{f(-\frac{a\bar{z}+b}{c\bar{z}+d})}\end{aligned}$$

Hence $(f_\rho)_\gamma = (f_{\iota\gamma\iota})_\rho$. Since ι normalizes Γ, it follows that $f_\rho \in S_{k+2}(\Gamma, \mathbb{C})$. If $f(z) = \sum a_n e^{2i\pi nz}$, then $f(-\bar{z}) = \sum a_n e^{-2i\pi n\bar{z}} = \sum_n a_n \overline{e^{2i\pi nz}}$ and therefore $f_\rho(z) = \sum \overline{a_n} e^{2i\pi nz}$. In particular, $f \mapsto f_\rho$ fixes the real structure $S_{k+2}(\Gamma, \mathbb{R})$ and the other assertions follow easily. □

Using this identification, our somewhat twisted linear map

$$\overline{S_{k+2}(\Gamma, \mathbb{C})} \to \mathrm{Symb}_\Gamma(\mathcal{V}_k(\mathbb{C})), \ \bar{g} \mapsto \phi_g^2$$

defines a linear map

$$\begin{aligned} S_{k+2}(\Gamma, \mathbb{C}) &\to \mathrm{Symb}_\Gamma(\mathcal{V}_k(\mathbb{C})) \\ g &\mapsto \phi^3_g := -\phi^2_{\overline{g_\rho}} \end{aligned}$$

Lemma 5.3.16

$$\phi^3_g(D)(P) = \int_{\ominus D} g(z)P(-z)dz \textit{ for all } g \in S_{k+2}(\Gamma, \mathbb{C}),\ D \in \Delta_0,\ P \in \mathcal{P}_k(\mathbb{C}).$$

Here, $D \mapsto \ominus D$ is the linear map on Δ_0 sending $\{\infty\} - \{a\}$ to $\{\infty\} - \{-a\}$ for $a \in \mathbb{Q}$.

Proof Since $g \mapsto \phi^3_g$ is $\mathbb{C}$-linear, and the Cauchy line integral is also $\mathbb{C}$-linear in its integrand, we only have to prove the formula for $g \in S_{k+2}(\Gamma, \mathbb{R})$ and $P \in \mathcal{P}_k(\mathbb{R})$. By additivity we may assume that $D = \{\infty\} - \{a\}$

$$\begin{aligned} \phi^3_g(D)(P) &= -\phi^2_{\bar{g}}(D)(P) \text{ (by definition)} \\ &= -\overline{\int_D g_\rho(z)P(z)dz} \text{ (by (5.3.4))} \\ &= -\overline{\int_0^\infty g_\rho(a+iy)P(a+iy)idy} \\ &= \int_0^\infty g(-a+iy)P(a-iy)idy \text{ using } \overline{g_\rho(z)} = g(-\bar{z}) \text{ and } \overline{P(z)} = P(\bar{z})) \\ &= \int_{\ominus D} g(z)P(-z)dz \end{aligned}$$

□

Proposition 5.3.17 *For any $f \in S_{k+2}(\Gamma, \mathbb{C})$, we have $(\phi^1_f)_{|\iota} = \phi^3_f$. The map $f \mapsto \phi^1_f$ commutes with the action of $\mathcal{H}(S^+, \Gamma)$ and the map $f \mapsto \phi^3_f$ commutes with $\mathcal{H}$. Therefore, we have an injective linear application compatible with $\mathcal{H}$:*

$$S_{k+2}(\Gamma, \mathbb{C}) \oplus S_{k+2}(\Gamma, \mathbb{C}) \to \mathit{Symb}_\Gamma(\mathcal{V}_k(\mathbb{C})),\ (f, g) \mapsto \phi^1_f + \phi^3_g.$$

Its image admits $\mathit{BSymb}_\Gamma(\mathcal{V}_k(\mathbb{C}))$ as an $\mathcal{H}(S^+, \Gamma)$-stable supplementary.

The nice thing with that application then there is no complex conjugation around anymore.

Proof We have $\phi_{|\iota}(D)(P) = \phi(\ominus D)(P(-z))$ by a trivial computation, hence the first assertion follows from the above Lemma. That $f \mapsto \phi^1_f$ commutes with $\mathcal{H}(S^+, \Gamma)$ can be proved by the exact same proof as Lemma 5.3.1 (or alternatively, deduced from it). If $T \in \mathcal{H}(S^+, \Gamma)$ it follows from the above that

$$(\phi^3_f)_T = (\phi^1_f)_{|\iota T} = (\phi^1_f)_{|\iota T\iota} = (\phi^1_{f_{|\iota T\iota}})_{|\iota} = \phi^3_{f_{|\iota T\iota}} \tag{5.3.14}$$

and if $T \in \mathcal{H}$, $\iota T\iota = T$ by Lemma 5.3.13. The injectivity of the displayed application follows from Theorem 5.3.10 and its compatibility with $\mathcal{H}$ from what we have just proven. □

Proposition 5.3.18 *Let $f, g \in S_{k+2}(\Gamma, \mathbb{C})$. Then*

$$(\phi^1_f, \phi^1_g) = 0 \tag{5.3.15}$$

$$(\phi^3_f, \phi^3_g) = 0 \tag{5.3.16}$$

$$(\phi^1_f, \phi^3_g) = -(2i)^{k-1}(f, g_\rho)_\Gamma. \tag{5.3.17}$$

Proof This follows immediately from Proposition 5.3.18 and the definition of ϕ_3. □

We now consider a basis of the space generated by ϕ^1_f and ϕ^3_f on which ι acts diagonally: we let

$$\phi^\pm_f = \frac{1}{2}(\phi^1_f \pm \phi^3_f) = \frac{1}{2}\left(\int_D f(z)P(z)\,dz \pm \int_{\ominus D} f(z)P(-z)\,dz\right).$$

We can split Theorem 5.3.10 or Proposition 5.3.17 in half:

Theorem 5.3.19 *The map $f \mapsto \phi_{f^\pm}$ is a $\mathbb{C}$-linear, compatible with $\mathcal{H}$, and injective linear map $S_{k+2}(\Gamma, \mathbb{C}) \to \mathrm{Symb}^\pm_\Gamma(\mathcal{V}_k(\mathbb{C}))$. Its image admits $\mathrm{BSymb}^\pm_\Gamma(\mathcal{V}_k(\mathbb{C}))$ as an $\mathcal{H}$-stable supplementary.*

5.3.4 The Endomorphism W_N and the Corrected Scalar Product

We keep assuming that $\Gamma_1(N) \subset \Gamma \subset \Gamma_0(N)$ and that R is a commutative ring where 6 is invertible. The subalgebra $\mathcal{H}$ of $\mathcal{H}(S^+, \Gamma)$ is the one generated by the operators T_l, U_l for l prime and the diamond operators, as above.

Let us recall the classical theory of the operator W_N, which plays a fundamental role in the functional equation of the complex L-function of a modular form (see e.g. [113, chapter 3] or below, Sect. 5.4.1).

Let $W_N = \begin{pmatrix} 0 & 1 \\ -N & 0 \end{pmatrix} \in S^+ = \mathrm{GL}_2^+(\mathbb{Q}) \cap M_2(\mathbb{Z})$. In particular, we can define an Hecke operator $[\Gamma W_N \Gamma]$ on the spaces of modular forms $M_{k+2}(\Gamma, R)$ and on the spaces of modular symbols $\mathrm{Symb}_\Gamma(V)$ (for V any R-module with a right action of S^+).

Observe that

$$W_N \begin{pmatrix} a & b \\ c & d \end{pmatrix} W_N^{-1} = \begin{pmatrix} d & -c/N \\ -Nb & a \end{pmatrix}. \tag{5.3.18}$$

Lemma 5.3.20 *The matrix W_N normalizes Γ, and the operator $[\Gamma W_N \Gamma]$ on $M_{k+2}(\Gamma, L)$ or $\mathrm{Symb}_\Gamma(\mathcal{V}_k(L))$ is just the action of the matrix W_N. Its square is the multiplication by $(-N)^k$. In particular, this Hecke operator acts invertibly if N is invertible in R.*

Proof It is clear on (5.3.18) that W_N normalizes $\Gamma_1(N)$ and $\Gamma_0(N)$.The image of Γ in $\Gamma_0(N)/\Gamma_1(N) = (\mathbb{Z}/N\mathbb{Z})^*$ is the subgroup formed by the a mod N when $\begin{pmatrix} a & b \\ c & d \end{pmatrix}$ runs in Γ, while by (5.3.18) the image of $W_N \Gamma W_N^{-1}$ is the subgroup formed by the d mod N when $\begin{pmatrix} a & b \\ c & d \end{pmatrix}$ runs in Γ. Since $ad \equiv 1 \pmod N$, these two subgroups are equal, hence $\Gamma = W_N \Gamma W_N^{-1}$.

The other assertions follow trivially. □

Note in particular that if γ is diagonal, $W_N \gamma W_N^{-1} = \gamma'$ (where $\gamma' = (\det \gamma)\gamma^{-1}$ as always). Hence

$$\text{for } s \in S \text{ diagonal, } W_N[\Gamma s \Gamma]W_N^{-1} = [\Gamma s' \Gamma]. \tag{5.3.19}$$

Definition 5.3.21 We define the *corrected scalar product* $[\ ,\ ]_\Gamma$ on $S_{k+2}(\Gamma, \mathbb{C})$ and $[\ ,\ ]$ on $\mathrm{Symb}_\Gamma(\mathcal{V}_k(R))$

$$[f, g]_\Gamma = (f, g_{|W_N})_\Gamma, \quad [\phi_1, \phi_2] = (\phi_1, (\phi_2)_{|W_N})$$

Lemma 5.3.22 *The corrected scalar product $[\ ,\ ]$ on $\mathrm{Symb}_\Gamma(\mathcal{V}_k(R))$ is anti-symmetric.*

Proof Since $W_N^2 = -N$, $W_N' = \det(W_N)W_N^{-1} = NW_N^{-1} = -W_N$, and

$$\begin{aligned}
[\phi_1, \phi_2] &= (\phi_1, (\phi_2)_{|W_N}) \\
&= ((\phi_1)_{|W_N'}, \phi_2) \\
&= ((\phi_1)_{|-W_N}, \phi_2) \\
&= (-1)^k((\phi_1)_{|W_N}, \phi_2) \\
&= (-1)^k(-1)^{k+1}(\phi_2, (\phi_1)_{|W_N}) \\
&= -[\phi_2, \phi_1]
\end{aligned}$$

□

Lemma 5.3.23 *The operators $T \in \mathcal{H}$ are self-adjoint for $[\ ,\ ]_\Gamma$ and $[\ ,\ ]$. In other words, for $T \in \mathcal{H}$, one has $[f_{|T}, g]_\Gamma = [f, g_{|T}]_\Gamma$ for all $f, g \in M_{k+2}(\Gamma, L)$ and similarly for modular symbols.*

Proof We give the proof only for the pairing $[\ ,\ ]_\Gamma$ on $M_{k+2}(\Gamma, L)$ since the other case is entirely similar.

Write $T = [\Gamma s \Gamma]$ with $s \in S^+$. Then

$$\begin{aligned}[f_{|T}, g]_\Gamma &= (f_{[\Gamma s\Gamma]}, g_{|W_N})_\Gamma \\ &= (f, g_{|W_N[\Gamma s'\Gamma]})_\Gamma \\ &= [f, g_{|W_N[\Gamma s'\Gamma]W_N^{-1}}]_\Gamma.\end{aligned}$$

Now if $T = T_l$ or U_l we can take s diagonal, so $W_N[\Gamma s'\Gamma]W_N^{-1} = [\Gamma s\Gamma] = T$. If $T = \langle a \rangle$ is a diamond operator, $a \in (\mathbb{Z}/N\mathbb{Z})^*$, then $T = [\Gamma s \Gamma]$ where s is any matrix in $\Gamma_0(N)$ with lower-right term congruent to $a \pmod N$, and $W_N s' W_N^{-1}$ is a matrix in $\Gamma_0(Np)$ has the same diagonal terms than s, hence $W_N[\Gamma s'\Gamma]W_N^{-1} = \langle a \rangle$ and the result follows. □

Proposition 5.3.24 *If f is a newform in $M_{k+2}(\Gamma, \mathbb{C})$ then*

$$f_{|W_N} = W(f) f_\rho$$

where $W(f)$ is a non-zero scalar in the number field K_f (the subfield of $\mathbb{C}$ generated by the coefficients of f) satisfying $|W(f)| = N^k$.

If f has moreover trivial nebentypus (that is, is a modular form for $\Gamma_0(N)$), then $f_{|W_N} = W(f) f$ and $W(f) = \pm N^{k/2}$.

Proof Write $f = q + \sum_{n\geq 2} a_n q^n$. Remember that $T_l = [\Gamma s \Gamma]$ with $s = \begin{pmatrix} 1 & 0 \\ 0 & l \end{pmatrix}$ and set $T_l' = [\Gamma s' \Gamma]$ with $s' = a\begin{pmatrix} l & 0 \\ 0 & 1 \end{pmatrix}$.

Since f is a newform and the operators T_l commute with the operators $T_{l'}'$, f is an eigenform for the T_l', say $T_l' f = \lambda_l f$. From $(f_{|T_l}, f)_\Gamma = (f, f_{|[\Gamma s \Gamma]})_\Gamma$ we see that $\lambda_l = \bar{a}_l$. By (5.3.19), we have $T_\ell(f_{|W_N}) = \lambda_l f_{|W_n} = \bar{a}_l f_{|W_n}$.

On the other hands, f_ρ is the newform of eigenvalues $\bar{a}_\ell$. By the strong multiplicity one theorem, $f_{|W_n} = W(f) f_\rho$ for some $W(f) \in \mathbb{C}$. We have $(f, f)_\Gamma = (f_\rho, f_\rho)_\Gamma = |W(f)|^{-2}(f_{|W_N}, f_{|W_N})_\Gamma = |W(f)|^{-2}(f_{|W_N W_N'}, f)_\Gamma = |W(f)|^{-2} N^k (f, f)_\Gamma$ using that $W_N W_N' = N\mathrm{Id}$, hence $|W(f)| = N^{k/2}$.

Finally, when f has a trivial nebentypus, then N is even and the a_l are real, so $f = f_\rho$, and $f_{|W_N} = W(f) f$. Since $W_N^2 = -N\mathrm{Id}$ acts like the multiplication by $(-N)^k = N^k$, one has $W(f) = \pm N^{k/2}$. □

Remark 5.3.25 When f has trivial nebentypus, the sign $W(f)/N^{k/2}$ is called the *root number of* f and is the sign of the functional equation satisfied by $L(f, s)$. See below, Theorem 5.4.3.

Lemma 5.3.26 *If $f \in S_{k+2}(\Gamma, \mathbb{C})$, we have*

$$[\phi_f^+, \phi_f^-] = (-i)^{k-1} 2^{k-2} (f, (f_{|W_N})_\rho)_\Gamma.$$

When f is a newform,

$$[\phi_f^+, \phi_f^-] = (-i)^{k-1} 2^{k-2} \mathrm{Re}(W(f))(f, f)_\Gamma.$$

Proof We have $\phi^1_{f_{|W_N}} = (\phi^1_f)_{|W_N}$ by Proposition 5.3.12 since $W_N \in S^+$, and thus $\phi^3_{f_{|W_N}} = \phi^1_{f_{|\iota W_n \iota}} = (\phi^1_f)_{|(-W_N)} = (-1)^k (\phi^3_f)_{|W_N}$

Since by definition $\phi_f^\pm = \frac{1}{2}(\phi^1_f \pm \phi^3_f)$ we compute

$$\begin{aligned}
[\phi_f^+, \phi_f^-] &= \frac{1}{4}[\phi^1_f + \phi^3_f, \phi^1_f - \phi^3_f] \\
&= -\frac{1}{2}[\phi^1_f, \phi^3_f] \\
&= -\frac{1}{2}(\phi^1_f, (\phi^3_f)_{|W_N}) \\
&= (-1)^{k+1}\frac{1}{2}(\phi^1_f, \phi^3_{f_{|W_N}}) \\
&= (-1)^{k-1}(2i)^{k-1}\frac{1}{2}(f, (f_{|W_N})_\rho)_\Gamma \\
&= (-i)^{k-1}2^{k-2}(f, (f_{|W_N})_\rho)_\Gamma
\end{aligned}$$

using Proposition 5.3.18. In the case where f is a newform, one applies Proposition 5.3.24. □

5.3.5 *Boundary Modular Symbols and Eisenstein Series*

In this section, we still assume that $\Gamma_1(N) \subset \Gamma \subset \Gamma_0(N)$.

Theorem 5.3.27 *We have a natural perfect pairing (,)*

$$BSymb_\Gamma(\mathcal{V}_k(\mathbb{C})) \times \mathcal{E}_{k+2}(\Gamma, \mathbb{C}) \to \mathbb{C}$$

where $\mathcal{E}_{k+2}(\Gamma, \mathbb{C}) := M_{k+2}(\Gamma, \mathbb{C})/S_{k+2}(\Gamma, \mathbb{C})$ is the space of Eisenstein series of weight $k+2$ and level Γ which satisfies, for all $s \in S^+$,

$$(x_{|[\Gamma s \Gamma]}, y) = (x, y_{|[\Gamma s' \Gamma]})$$

Proof We begin by constructing a map

$$f \mapsto u_f$$
$$M_{k+2}(\Gamma, \mathbb{C}) \to H^1(\Gamma, \mathcal{V}_k(\mathbb{C})),$$

where $H^1(\Gamma, \mathcal{V}_k(\mathbb{C}))$ is the first group cohomology space of Γ with coefficients $\mathcal{V}_k(\mathbb{C})$, as follows:

Choose a point $x \in \mathcal{H}$. Define u_f as the map $\Gamma \to \mathcal{V}_k(\mathbb{C})$ such that for $\gamma \in \Gamma$,

$$u_f(\gamma)(P) = \int_x^{\gamma x} f(z)P(z)dz.$$

One has

$$\begin{aligned} u_f(\gamma\gamma')(P) &= \int_x^{\gamma\gamma' x} f(z)P(z)dz \\ &= \int_x^{\gamma x} f(z)P(z)dz + \int_{\gamma x}^{\gamma\gamma' x} f(z)P(z)dz \\ &= u_f(\gamma)(P) + \int_x^{\gamma' x} f(\gamma^{-1}z)P(\gamma^{-1}z)d(\gamma^{-1}(z)) \\ &= u_f(\gamma)(P) + \int_x^{\gamma' x} f(z)P_{|\gamma^{-1}(z)}dz \\ &= u_f(\gamma)(P) + u_f(\gamma')_{|\gamma}(P) \end{aligned}$$

Thus u_f is a cocycle. Moreover changing the point $x \in \mathcal{H}$ into a point $x' \in \mathcal{H}$ adds to u_f the coboundary attached to the element $P \mapsto \int_x^{x'} f(z)P(z)dz$ of $\mathcal{V}_k(\mathbb{C})$. Thus u_f, as an element of $H^1(\Gamma, \mathcal{V}_k(\mathbb{C}))$ is independent of x.

Using the identification $H^1(\Gamma, \mathcal{V}_k(\mathbb{C})) = H^1(Y_\Gamma, \tilde{\mathcal{V}}_k(C))$, we see u_f as a morphism $M_{k+2}(\Gamma, \mathbb{C}) \to H^1(Y_\Gamma, \tilde{\mathcal{V}}_k(C))$. It is easy to see that this morphism is injective and $\mathcal{H}(S^+, \Gamma)$-equivariant.

Now consider the space $\mathcal{E}_{k+2}(\Gamma, \mathbb{C})$ as a subspace of $M_{k+2}(\Gamma, \mathbb{C})$ (the kernel of the Peterson product). The restriction of $f \mapsto u_f$ embeds $\mathcal{E}_{k+2}(\Gamma, \mathbb{C})$ into $H^1(Y_\Gamma, \tilde{\mathcal{V}}_k(\mathbb{C}))$. We also have an embedding $H^1_!(Y_\Gamma, \tilde{\mathcal{V}}_k(\mathbb{C})) \subset H^1(Y_\Gamma, \tilde{\mathcal{V}}_k(\mathbb{C}))$. The sum map

$$H^1_!(Y_\Gamma, \tilde{\mathcal{V}}_k(\mathbb{C})) \oplus \mathcal{E}_{k+2}(\Gamma, \mathbb{C}) \to H^1(Y_\Gamma, \tilde{\mathcal{V}}_k(\mathbb{C}))$$

is still injective, as no system of Hecke-eigenvalues of an Eisenstein series may appear in $H^1_!(Y_\Gamma, \tilde{\mathcal{V}}_k(\mathbb{C}))$ by Theorem 5.3.10 and the Hecke estimates of eigenvalues of cuspidal modular forms. By equality of dimensions between the target of the source and this map (for example when $k = 0$, both sides have dimension $2g + c - 1$

where g is genus of the Riemann surface X_Γ, and c the number of cusps for Γ), it is an isomorphism.

Combining this and Corollary 4.4.3, we have an exact sequence of Hecke-modules:

$$0 \to \mathrm{BSymb}_\Gamma(\mathcal{V}_k(\mathbb{C})) \to \mathrm{Symb}_\Gamma(\mathcal{V}_k(\mathbb{C})) \to H^1(Y_\Gamma, \tilde{\mathcal{V}}_k(\mathbb{C})) \to \mathcal{E}_{k+2}(\Gamma, \mathbb{C}) \to 0$$

The pairing $(\ ,\ )$ between $\mathrm{Symb}_\Gamma(\mathcal{V}_k(\mathbb{C}))$ and $H^1(Y_\Gamma, \tilde{\mathcal{V}}_k(\mathbb{C}))$ induces a pairing between $\mathrm{BSymb}_\Gamma(\mathcal{V}_k(\mathbb{C}))$ and $\mathcal{E}_{k+2}(\Gamma, \mathbb{C})$, as follows: for $x \in \mathrm{BSymb}_\Gamma(\mathcal{V}_k(\mathbb{C}))$ that we see as an element of $\mathrm{Symb}_\Gamma(\mathcal{V}_k(\mathbb{C}))$ and $y \in \mathcal{E}_{k+2}(\Gamma, \mathbb{C})$, the number (x, y') is independent of the lift y' of y in $H^1(Y_\Gamma, \tilde{\mathcal{V}}_k(\mathbb{C}))$. To see this, let y'' be another lift, so that $y' - y''$ is the image of an element z in $\mathrm{Symb}_\Gamma(\mathcal{V}_k(\mathbb{C}))$. But $(x, y' - y'') = (x, z) = \pm(z, x)$ and the latter is 0 since the image of x in $H^1(Y_\Gamma, \tilde{\mathcal{V}}_k(\mathbb{C}))$ is 0. Obviously then, the pairing between $\mathrm{BSymb}_\Gamma(\mathcal{V}_k(\mathbb{C}))$ and $\mathcal{E}_{k+2}(\Gamma, \mathbb{C})$ is perfect. The proves Theorem 5.3.27, since its last assertion follows from Proposition 5.2.3. □

Exercise 5.3.28 Describe a basis of eigenforms for all Hecke operators for the space $\mathrm{BSymb}_\Gamma(\mathcal{V}_k(\mathbb{C}))$ when $k = 0$ and $\Gamma = \Gamma_0(N)$ where N is a product of r distinct primes $l_1, \ldots, l_r$.

Corollary 5.3.29 *As $\mathcal{H}$-modules, $BSymb_\Gamma(\mathcal{V}_k(\mathbb{C}))$, $\mathcal{E}_{k+2}(\Gamma, \mathbb{C})$ and $\mathcal{E}_{k+2}(\Gamma, \mathbb{C})^\vee$ are isomorphic.*

Proof For l not dividing N (resp. dividing N), let T_l' (resp. U_l') be $[\Gamma\begin{pmatrix} l & 0 \\ 0 & 1 \end{pmatrix}\Gamma]$. Let $\mathcal{H}'$ be the commutative subalgebra of $\mathcal{H}(S^+, \Gamma)$ generated by the T_l''s, the U_l''s and the Diamond operators. Fix an isomorphism $\psi : \mathcal{H} \to \mathcal{H}'$ by sending T_l on T_l', U_l and U_l' and $\langle a \rangle$ on $\langle a^{-1} \rangle$. Note that in each case ψ sends $[\Gamma s \Gamma]$ to $[\Gamma s' \Gamma]$. Thus the theorem gives us an isomorphism of vector spaces $\mathcal{E}_{k+2}(\Gamma, \mathbb{C}) \to \mathrm{BSymb}_\Gamma(\mathcal{V}_k(\mathbb{C}))^\vee$ which is compatible with the $\mathcal{H}'$-structure on the source and the $\mathcal{H}$-structure on the target (where $\mathcal{H}'$ is identified with $\mathcal{H}$ using ψ).

Now we claim that $\mathcal{E}_{k+2}(\Gamma, \mathbb{C})$ as an $\mathcal{H}$-module is isomorphic to itself as an $\mathcal{H}'$-module, when we identify $\mathcal{H}$ and $\mathcal{H}'$ using ψ. Let $W_N = \begin{pmatrix} 0 & 1 \\ -N & 0 \end{pmatrix}$ as in the preceding section. Then a straightforward computation shows that $f \mapsto f_{|W_N}$ is an isomorphism of vector spaces of $\mathcal{E}_{k+2}(\Gamma, \mathbb{C})$ onto itself, and that $(Tf)_{|W_N} = \psi(T) f_{|W_N}$ for any $T \in \mathcal{H}$ (this is [93, Theorem 4.5.5]), which proves the claim.

Hence we get a genuine $\mathcal{H}$-module isomorphism $\mathcal{E}_{k+2}(\Gamma, \mathbb{C}) \simeq \mathrm{BSymb}_\Gamma(\mathcal{V}_k(\mathbb{C}))^\vee$. Dualyzing, we get $\mathcal{E}_{k+2}(\Gamma, \mathbb{C})^\vee \simeq \mathrm{BSymb}_\Gamma(\mathcal{V}_k(\mathbb{C}))$. To conclude, we just recall that by Corollary 2.6.20, $\mathcal{E}_{k+2}(\Gamma, \mathbb{C})^\vee \simeq \mathcal{E}_{k+2}(\Gamma, \mathbb{C})$ as $\mathcal{H}$-modules. □

Exercise 5.3.30 Check directly that $\mathrm{Symb}_{\Gamma_0(N)}(\mathcal{V}_k(\mathcal{V})) \simeq \mathcal{E}_{k+2}(\Gamma_0(N), \mathbb{C})$ as $\mathcal{H}$-modules when $k = 0$, N a product of distinct primes, using your description of Exercise 5.3.28.

Remark 5.3.31 The reader should be warned that while we have constructed for all N a canonical isomorphism

$$\mathcal{E}_{k+2}(\Gamma_1(N), \mathbb{C}) \to \mathrm{BSymb}_{\Gamma_1(N)}(\mathcal{V}_k(\mathbb{C}))$$

these isomorphisms are **not compatible** for different N in the following natural sense: for l a prime non-dividing N let V_l be the Hecke operator $l^{-1-k}[\Gamma_1(N)\begin{pmatrix} l & 0 \\ 0 & 1 \end{pmatrix}\Gamma_1(Nt)]$ which maps $\mathcal{E}_{k+2}(\Gamma_1(N), \mathbb{C})$ to $\mathcal{E}_{k+2}(\Gamma_1(Nl), \mathbb{C})$ (sending $f(z)$ to $f(lz)$), and $\mathrm{BSymb}_{\Gamma_1(N)}(\mathcal{V}_k(\mathbb{C}))$ to $\mathrm{BSymb}_{\Gamma_1(Nl)}(\mathcal{V}_k(\mathbb{C}))$. Then the diagram

$$\begin{array}{ccc} \mathcal{E}_{k+2}(\Gamma_1(N), \mathbb{C}) & \xrightarrow{\simeq} & \mathrm{BSymb}_{\Gamma_1(N)}(\mathcal{V}_k(\mathbb{C})) \\ \downarrow{\scriptstyle \mu V_l} & & \downarrow{\scriptstyle V_l} \\ \mathcal{E}_{k+2}(\Gamma_1(Nl, \mathbb{C}) & \xrightarrow{\simeq} & \mathrm{BSymb}_{\Gamma_1(Nl)}(\mathcal{V}_k(\mathbb{C})) \end{array}$$

is not commutative, where $\mu \in \mathbb{C}^*$ is any normalization factor.

Exercise 5.3.32 Prove the remark by showing that in the situation of Exercise 5.3.28, the map $lV_l : \mathrm{BSymb}_{\Gamma_0(N)}(\mathbb{C}) \to \mathrm{BSymb}_{\Gamma_0(Nl)}(\mathbb{C})$ is the natural inclusion for all l.

Definition 5.3.33 If f is a new form in $\mathcal{E}_{k+2}(\Gamma, \mathbb{C})$, we denote by $\epsilon(f)$ the eigenvalue of the involution ι on the boundary modular symbol corresponding to f by the isomorphism 5.3.29. We refer to this number as the *sign* of the new Eisenstein series f.

In other words, $\epsilon(f)$ is the ι-eigenvalues of the unique boundary symbols with the same Hecke eigenvalues than f. The following proposition determines the sign. For the proof, we refer the reader to [14].

Proposition 5.3.34 *The sign of a normal new Eisenstein series (cf. Sect. 2.6.4) $E_{k+2,\tau,\psi}$ is $\tau(-1)$. The sign of an exceptional Eisenstein series $E_{2,\ell}$ is* 1.

5.3.6 Summary

We summarize the main results of this section by the following version of the theorem of Eichler-Shimura:

Theorem 5.3.35 *Let $\Gamma = \Gamma_1(N)$ and $\mathcal{H}$ the polynomial ring generated by the Hecke operators T_l (for $l \nmid N$) U_l (for $l \mid N$), and $\langle a \rangle$ (for $a \in (\mathbb{Z}/N\mathbb{Z})^*$). We* *have a canonical isomorphism of $\mathcal{H}$-modules*

$$Symb_\Gamma(\mathcal{V}_k(\mathbb{C})) = S_{k+2}(\Gamma, \mathbb{C}) \oplus \bar{S}_{k+2}(\Gamma, \mathbb{C}) \oplus \mathcal{E}_{k+2}(\Gamma, \mathbb{C}) = S_{k+2}(\Gamma, \mathbb{C}) \oplus M_{k+2}(\Gamma, \mathbb{C})$$

which transform the ι involution into the involution that acts by $+1$ on the first summand $S_{k+2}(\Gamma, \mathbb{C})$, -1 and the second summand, and by some involution on $\mathcal{E}_{k+2}(\Gamma, \mathbb{C})$. We have canonical $\mathcal{H}$-embedding

$$S_{k+2}(\Gamma, \mathbb{C}) \hookrightarrow Symb^{\pm}_\Gamma(\mathcal{V}_k(\mathbb{C})) \hookrightarrow M_{k+2}(\Gamma, \mathbb{C}).$$

Finally, all those results hold when $\mathbb{C}$ is replaced everywhere by any field of characteristic 0, excepted that the isomorphisms are not anymore canonical.

Proof The isomorphisms over $\mathbb{C}$ are obtained trivially by putting together Theorem 5.3.19 and Corollary 5.3.29. The embeddings follow. Finally, to prove that there are also such isomorphisms or embeddings over any field L of characteristic 0, we only need to consider the case $N = \mathbb{Q}$. In this case, it follows from the following elementary statement: *if M and N are two finite-dimensional $\mathbb{Q}$-vector spaces provided with an action of a ring $\mathcal{H}$, such that $M \otimes \mathbb{C} \simeq N \otimes \mathbb{C}$ as $\mathbb{C}$-vector spaces and $\mathcal{H}$-modules, then $M \simeq N$ as $\mathbb{Q}$-vector spaces and $\mathcal{H}$-modules.* Indeed, choosing a basis of M and N, a $\mathbb{Q}$-linear application $f : M \to N$ is described by its matrix $(x_{i,j})$. The fact that $f : M \to N$ is an isomorphism, as well as the fact that it commutes with the action of $\mathcal{H}$ are expressed by a system of linear equations satisfied by the $x_{i,j}$ with coefficients in $\mathbb{Q}$. By hypothesis, this systems has solution in $\mathbb{C}$. Hence by Gauss method it has a solution in $\mathbb{Q}$. □

Through the Atkin-Lehner's theory does not carry over for modular symbols (cf. Remark 5.3.31), many consequences of it carries trivially using the theorem above:

Corollary 5.3.36 *Let L be any field of characteristic 0. Let $\mathcal{H}_0$ be the polynomial algebra generated by the T_l $(l \nmid N)$ and the diamond operators $\langle a \rangle$ $(a \in (\mathbb{Z}/N\mathbb{Z})^*)$. Let λ be a system of eigenvalues $\mathcal{H}_0 \to L$ appearing in $Symb^{\pm}_{\Gamma_1(N)}(\mathcal{V}_k(L))$. Then the set of divisors of N such that λ appears in $Symb^{\pm}_{\Gamma_1(N)}(\mathcal{V}_k(L))$ has a minimal element N_0, that we call the* minimal level *of λ as in the classical Atkin-Lehner's theory (Sect. 2.6.4). In particular the notions of λ being new and old, l-new and l-old make sense. Also,* $\dim Symb^{\pm}_{\Gamma_1(N)}(\mathcal{V}_k(L))[\lambda] = \sigma(N/N_0)$.

Exercise 5.3.37 Let Γ be a congruence subgroup and k be an even integer, In this exercise, we give a precise description of the boundary eigensymbols of level Γ and even weight k.

Let u, v be two relatively prime integers. Define $\phi_{k,u,v} \in \mathrm{Hom}(\Delta, \mathcal{V}_k)$ to be supported in the Γ-orbit of u/v and such that for $\gamma = \begin{pmatrix} a & b \\ c & d \end{pmatrix} \in \Gamma$,

$$\phi_{k,u,v}\left(\gamma \cdot \frac{u}{v}\right)(P(z)) = P\left(\gamma \cdot \frac{u}{v}\right)(cu+dv)^k. \qquad (5.3.20)$$

1. Show that $\phi_{k,u,v}$ is well-defined and that

$$\phi_{k,u,v} \in \mathrm{Hom}(\Delta, \mathcal{V}_k)^{\Gamma_1(N)}.$$

2. Show that $\phi_{k,u,v}$ depends only of the Γ-equivalence class of u/v in $\mathbb{P}^1(\mathbb{Q})$ (i.e. of the Γ-cusp u/v) and that the $\phi_{k,u.v}$ when u/v runs in $\mathbb{P}^1(\mathbb{Q})/\Gamma$ are a basis of $\mathrm{Hom}(\Delta, \mathcal{V}_k)^{\Gamma_1(N)}$.
3. Now let $N \geq 1$ be an integer and p a prime not dividing N. Consider the case $\Gamma = \Gamma_1(N) \cap \Gamma_0(p)$. Let $\mathrm{Hom}(\Delta, \mathcal{V}_k)^{\Gamma_1(N)}_{\mathrm{ord}}$ be the subspace of $\mathrm{Hom}(\Delta, \mathcal{V}_k)^{\Gamma_1(N)}$ generate by the $\phi_{k,u,v}$ where u/v is in the $\Gamma_0(p)$-equivalence class of 0, that is when u/v is a p-integer. Show that $\mathrm{Hom}(\Delta, \mathcal{V}_k)^{\Gamma_1(N)}_{\mathrm{ord}}$ is stable by the operator U_p, and that all eigenvalues of U_p in $\mathrm{Hom}(\Delta, \mathcal{V}_k)^{\Gamma_1(N)}_{\mathrm{ord}}$ are p-adic integers. Show also that the eigenvalues of U_p on $\mathrm{Hom}(\Delta, \mathcal{V}_k)^{\Gamma_1(N)}/\mathrm{Hom}(\Delta, \mathcal{V}_k)^{\Gamma_1(N)}_{\mathrm{ord}}$ have all p-adic valuation $k+1$. (Compare Exercise 2.6.15.)

5.4 Applications of Classical Modular Symbols to L-functions and Congruences

5.4.1 *Reminder About L-functions*

Let $f = \sum_{n\geq 1} a_n q^n \in S_{k+2}(\Gamma_1(N), \mathbb{C})$. Let us recall that the L-function of f is the analytic function $L(f,s) = \sum_{n\geq 1} a_n/n^s$; the sum converges (absolutely uniformly over all compact) if $\mathrm{Re}\, s > k/2+2$, using the Hecke's easy estimate on the a_n, and actually if $\mathrm{Re}\, s > k/2+3/2$ using the estimate given by Ramanujan's generalized conjecture proved by Deligne.

Proposition 5.4.1 *We have, for* $\mathrm{Re}\, s > k/2+2$,

$$\int_0^\infty f(iy)y^{s-1}dy = \frac{\Gamma(s)}{(2\pi)^s}L(f,s) \qquad (5.4.1)$$

Proof Leaving to the reader the easy arguments justifying the permutation of the sum and the integral on our domain of convergence, we compute

$$\begin{aligned}\int_0^\infty f(iy)y^{s-1}dy &= \sum_{n>0}\int_0^\infty a_n e^{-2\pi ny}y^{s-1}dy\\ &= \sum_{n>0}\int_0^\infty \frac{a_n}{(2\pi n)^s}e^{-t}t^{s-1}\,dt \quad \text{(change of variable } t = 2\pi ny)\\ &= \frac{\Gamma(s)}{(2\pi)^s}\sum_{n>0}\frac{a_n}{n^s}\end{aligned}$$

□

Corollary 5.4.2 *The function $L(f,s)$ has an analytic extension as an entire holomorphic function over $\mathbb{C}$, and the formula (5.4.1) is valid for all $s \in \mathbb{C}$.*

Proof Since $f(iy)$ has exponential decay both when $y \to \infty$ and $y \to 0+$, $\int_0^\infty f(iy)y^{s-1}dy$ converges for all values of s and defines an entire function. Since $\Gamma(s)$ has no zero, the results follows. □

Theorem 5.4.3 (Functional Equation) *If f is a newform in $S_{k+2}(\Gamma,\mathbb{C})$, one has*

$$\frac{\Gamma(s)}{(2\pi)^s}W(f)\overline{L(f,\bar{s})} = \frac{(-1)^k N^s}{i^{k+2}}\frac{\Gamma(k+2-s)}{(2\pi)^{k+2-s}}L(f,k+2-s). \tag{5.4.2}$$

If moreover f has trivial nebentypus, $W(f) = \pm N^{k/2}$ and the functional equation becomes

$$\frac{\Gamma(s)}{(2\pi)^s}L(f,s) = \pm\frac{(-1)^k N^{s-k/2}}{i^{k+2}}\frac{\Gamma(k+2-s)}{(2\pi)^{k+2-s}}L(f,k+2-s). \tag{5.4.3}$$

At the point $s = k/2+1$, the function $L(f,k+2-s)$ has a zero of even order (or no zero) if $\pm 1 = 1$, and a zero of odd order if $\pm 1 = -1$.

Proof For the Hecke operator W_N defined in Sect. 5.3.4, one has

$$f_{|W_N}(z) = \frac{(-1)^k}{(Nz)^{k+2}}f\left(\frac{-1}{Nz}\right).$$

Hence

$$\frac{\Gamma(s)}{(2\pi)^s}L(f_{|W_N},s) = \frac{(-1)^k}{(Ni)^{k+2}}\int_0^\infty \frac{1}{y^{k+2}}f\left(\frac{i}{Ny}\right)y^{s-1}dy.$$

Make the change of variables $y' = 1/(Ny)$. We get

$$\frac{\Gamma(s)}{(2\pi)^s} L(f_{|W_N}, s) = \frac{(-1)^k N^s}{i^{k+2}} \int_0^\infty y^{k+1-s} f(iy) dy$$
$$= \frac{(-1)^k N^s}{i^{k+2}} \frac{\Gamma(k+2-s)}{(2\pi)^{k+2-s}} L(f, k+2-s).$$

Since $f_{|W_N} = W(f) f_\rho$ by Proposition 5.3.24, $L(f_{|W_N}, s) = L(f_\rho, s) = \overline{L(f, \bar{s})}$, one gets (5.4.2), and (5.4.3) follows. The last assertion is clear since $\Gamma(s)$ has no pole nor zero at $s = k/2 + 1$. □

If f is a normalized eigenform for all Hecke operators in $S_{k+2}(\Gamma_1(N), \epsilon)$, its L-function has an Euler product:

$$L(f, s) = \prod_{p\mathbb{N}} (1 - a_p p^{-s} + \epsilon(p) p^{1-2s})^{-1} \prod_{p|N} (1 - a_p p^{-s})^{-1} \qquad (5.4.4)$$

Slightly more generally, let $\chi : \mathbb{Z} \to \mathbb{C}$ be a Dirichlet character (let us recall that this means there is an $m > 0$ such that χ factors as a map $\chi : \mathbb{Z}/m\mathbb{Z} \to \mathbb{C}$, with $\chi(a) = 0$ if $a \notin (\mathbb{Z}/m\mathbb{Z})^*$, and $\chi_{|(\mathbb{Z}/m\mathbb{Z})^*}$ a character. The smallest such m is called the *conductor* of χ).

We define the *twisted L-function* $L(f, \chi, s) = \sum_{n>0} a_n \chi(n)/n^s$.

Actually, this L-function may also be seen as the (untwisted) L-function of a modular form (hence it has an analytic continuation as en entire function as well). To see this, introduce $f_\chi(z) := \sum_{n \geq 1} \chi(n) a_n e^{2\pi i n z}$. Then an easy computation shows that

$$f_\chi(z) = \frac{1}{\tau(\bar{\chi})} \sum_{a (\text{mod } m)} \bar{\chi}(a) f\left(z + \frac{a}{m}\right),$$

where $\tau(\bar{\chi})$ is the Gauss sum $\sum_{a\,(\text{mod } m)} \bar{\chi}(a) e^{2i\pi a/m}$, and that f_χ is a modular form of level $\Gamma_1(Nm)$.

Exercise 5.4.4 Do that computation.

Obviously, $L(f_\chi, s) = L(f, \chi, s)$. Therefore, by the same computation as above:

Lemma 5.4.5

$$\frac{\Gamma(s)}{(2\pi)^s} L(f, \chi, s) = \frac{1}{\tau(\bar{\chi})} \sum_{a (\text{mod } m)} \bar{\chi}(a) \int_0^\infty f(iy + a/m) y^{s-1} dy.$$

As Tate's thesis showed, the right way to think of $L(f, \chi, s)$ is as an extended L-function of f, in the variable (χ, s). Actually, one can think of (χ, s) as a continuous character on the idèles class group $\mathbb{A}_\mathbb{Q}^*/\mathbb{Q}^*$ with values in $\mathbb{C}^*$, as follows. First $\chi : (\mathbb{Z}/m\mathbb{Z})^* \to \mathbb{C}^*$ can be seen as a character of $\mathrm{Gal}(\mathbb{Q}(\zeta_m)/\mathbb{Q})$ of $\mathrm{Gal}(\bar{\mathbb{Q}}/\mathbb{Q})$, that is by Class Field Theory as a character on $\mathbb{A}_{\mathbb{Q}}^*/\mathbb{Q}^*$, trivial moreover on $\mathbb{R}_+^*$. Or more directly, using that the inclusion $\prod_l \mathbb{Z}_l^* \subset \mathbb{A}_{\mathbb{Q}}^*$ realizes an isomorphism $\prod_l \mathbb{Z}_l^* = \mathbb{A}_{\mathbb{Q}}^*/\mathbb{Q}^*\mathbb{R}_*^+$, we see χ as a character of $\prod_l \mathbb{Z}_l^*$, hence of the idèle classes, by the natural surjection $\prod_l \mathbb{Z}_l^* \to (\mathbb{Z}/m\mathbb{Z})^*$ whose kernel is $\prod_{l \nmid m} \mathbb{Z}_l^* \times \prod_{l \nmid m}(1 + l^{v_l(m)}\mathbb{Z}_l)$. In any case, we see χ as a character of $\mathbb{A}_{\mathbb{Q}}^*/\mathbb{Q}^*$ trivial on $\mathbb{R}_+^*$. Now we identifies (χ, s) as the character $\chi|\ |^s$ on $\mathbb{A}_{\mathbb{Q}}^*/\mathbb{Q}^*$. Actually, any continuous character $\mathbb{A}_{\mathbb{Q}}^*/\mathbb{Q}^* \to \mathbb{C}^*$ corresponds this way to one and only one pair (χ, s) where χ is a Dirichlet character and $s \in \mathbb{C}$.

Exercise 5.4.6 Prove the last assertion.

If f is a normalized eigenform for all Hecke operators in $S_{k+2}(\Gamma_1(N), \epsilon)$, its L-function has an Euler product:

$$L(f, \chi, s) = \prod_{p\mathbb{N}}(1 - a_p\chi(p)p^{-s} + \epsilon(p)\chi^2(p)p^{1-2s})^{-1} \prod_{p|N}(1 - a_p\chi(p)p^{-s})^{-1} \tag{5.4.5}$$

5.4.2 *Modular Symbols and L-functions*

We can now use what we have done to prove a deep arithmetic result about the values of the L-function of a modular eigenform.

In this section, let $\Gamma = \Gamma_1(N)$ and let $f \in S_{k+2}(\Gamma)$ a normalized eigenform for all the Hecke operators in $\mathcal{H}$. We call $\lambda : \mathcal{H} \to \mathbb{C}$ be the corresponding character: $Tf = \lambda(T)f$ for all $T \in \mathcal{H}$.

We recall the modular symbols $\phi_f^\pm \in \mathrm{Symb}_\Gamma(\mathcal{V}_k(\mathbb{C}))$. Since $f \mapsto \phi_f^\pm$ is $\mathcal{H}$-equivariant (and $\mathbb{C}$-linear) we have $\phi_f^\pm \in \mathrm{Symb}_\Gamma^\pm(\mathcal{V}_k(\mathbb{C}))[\lambda]$. We have more precisely:

Lemma 5.4.7 *The dimension of* $\mathrm{Symb}_\Gamma^\pm(\mathcal{V}_k(\mathbb{C}))[\lambda]$ *is* 1.

Proof We have a natural embedding $S_{k+2}(\Gamma, \mathbb{C})[\lambda] \hookrightarrow \mathrm{Symb}_\Gamma^\pm(\mathcal{V}_k(\mathbb{C}))[\lambda]$ whose cokernel is a direct summand in $\mathcal{E}_{k+2}(\Gamma, \mathbb{C})^\vee[\lambda]$ by Theorem 5.3.19 and Corollary 5.3.29. But the space $\mathcal{E}_{k+2}(\Gamma, \mathbb{C})[\lambda]$ is 0 (hence also $\mathcal{E}_{k+2}(\Gamma, \mathbb{C})^\vee[\lambda]$) since by the Peterson estimate, there is no Eisenstein series which is an eigenform with the same eigenvalues as a cuspidal form. And $S_{k+2}(\Gamma, \mathbb{C})[\lambda]$ has dimension 1 by multiplicity one (Corollary 2.6.18). □

Let K_f be the number field generated by the $\lambda(T)$, $T \in \mathcal{H}$. Then

Lemma 5.4.8 *The dimension over* K_f *of* $\mathrm{Symb}_\Gamma^\pm(\mathcal{V}_k(K_f))[\lambda]$ *is* 1.

Proof Since λ takes values in K_f, this follows form the preceding lemma and Exercise 2.5.5. □

Definition 5.4.9 We define *the periods* $\Omega_f^{\pm}$ of f as the non-zero complex numbers such that

$$\phi_f^{\pm}/\Omega_f^{\pm} \in \mathrm{Symb}_\Gamma^{\pm}(\mathcal{V}_k(K_f))[\lambda].$$

Obviously, those numbers exist and are well-determined up to multiplication by an element of K_f^* (each of them).

Remark 5.4.10 Sometimes we want to work integrally rather than rationally, and have a smaller indeterminacy on the periods $\Omega_f^{\pm}$. Let $\mathcal{O}_f$ be the ring of integer of f. We know that $\lambda(\mathcal{H}) \subset \mathcal{O}_f$, so $\mathrm{Symb}_\Gamma^{\pm}(\mathcal{V}_k(\mathcal{O}_f))[\lambda]$ is an $\mathcal{O}_f$-lattice in the one-dimensional space $\mathrm{Symb}_\Gamma^{\pm}(\mathcal{V}_k(K_f)][\lambda]$. The minor irritating issue we are facing is that since $\mathcal{O}_f$ is a Dedekind domain, but not necessarily principal, $\mathrm{Symb}_\Gamma^{\pm}(\mathcal{V}_k(\mathcal{O}_f))[\lambda]$ might not be free of rank one. It is at any rate projective of rank one.

When it happens that $\mathrm{Symb}_\Gamma^{\pm}(\mathcal{V}_k(\mathcal{O}_f))[\lambda]$ is free of rank one, we can define $\Omega_f^{\pm}$ by the requirement that $\phi_f^{\pm}/\Omega_f^{\pm}$ is a generator of $\mathrm{Symb}_\Gamma^{\pm}(\mathcal{V}_k(\mathcal{O}_f))[\lambda]$. Thus the periods $\Omega_f^{\pm}$ are well defined up to a unit in $\mathcal{O}_f^*$. For example, if f has coefficient in $\mathbb{Q}$, then $K_f = \mathbb{Q}$, $\mathcal{O}_f = \mathbb{Z}$ which is principal, so $\mathrm{Symb}_\Gamma^{\pm}(\mathcal{V}_k(\mathbb{Z}))[\lambda]$ is free of rank one, and the periods $\Omega_f^{\pm}$ are well defined up to a sign.

When $\mathrm{Symb}_\Gamma^{\pm}(\mathcal{V}_k(\mathcal{O}_f))[\lambda]$ is not free, we have several solutions, depending on what we want to do. For example we can replace K_f by a finite extension K' on which this module becomes free (for example, the Hilbert Class Field of K_f will do), and have $\Omega_f^{\pm}$ well defined up to an element in $\mathcal{O}_{K'}^*$. Or we can work at one prime at a time: for $\mathfrak{p}$ a prime of $\mathcal{O}_f$, the localization $\mathcal{O}_{f,(\mathfrak{p})}$ is principal, so it is possible to require that $\Omega_f^{\pm}$ be chosen such that $\phi_f^{\pm}/\Omega_f^{\pm}$ is in $\mathrm{Symb}_\Gamma^{\pm}(\mathcal{V}_k(\mathcal{O}_f))[\lambda]$, and generates that module after localizing at $\mathfrak{p}$. Such a requirement fixes $\Omega_f^{\pm}$ up to multiplication by an element of $K_\mathfrak{p}^*$ whose $\mathfrak{p}$-valuation is 0, which is often sufficient in the applications. When $\Omega^{\pm}$ is chosen this way, we shall say that it is $\mathfrak{p}$-*normalized* (that's our terminology, not a standard one).

Theorem 5.4.11 (Manin-Shokurov) *For any Dirichlet character χ, any integer j such that $0 \le j \le k$, we have*

$$\frac{L(f, \chi, j+1)}{\Omega_f^{\pm}(i\pi)^{j+1}} \in K_f[\chi],$$

where the sign $\pm$ is chosen such that $\pm 1 = (-1)^j \chi(-1)$, and $K_f[\chi]$ is the extension of K_f generated by the image of χ.

Proof We have for j an integer, $0 \le j \le k$,

$$\int_0^\infty f(iy + a/m)y^i dy = i^{-j-1}\int_{a/m}^\infty f(z)(z-a/m)^j dz$$
$$= i^{-j-1}\phi_f^1(\infty - \{a/m\})((z-a/m)^j).$$

Hence using Lemma 5.4.5,

$$\frac{\Gamma(j+1)i^{j+1}\tau(\bar{\chi})}{(2\pi)^{j+1}}L(f,\chi,j+1) = \sum_{a(\text{mod } m)} \bar{\chi}(a)\phi_f^1(\{\infty\} - \{a/m\})((z-a/m)^j)$$

We note that the right hand side is multiplied by $(-1)^j\chi(-1)$ if the occurrence of ϕ_f^1 in it is replaced by $(\phi_f^1)_{|\iota}$. Hence we can replace ϕ_f^1 by $\phi_f^\pm$ with $\pm 1 = (-1)^j\chi(-1)$ in that formula. We thus have

$$\frac{j!\tau(\bar{\chi})}{(-2i\pi)^{j+1}\Omega_f^\pm}L(f,\chi,j+1) = \sum_{a(\text{mod } m)} \bar{\chi}(a)\frac{\phi_f^\pm(\{\infty\} - \{a/m\})((z-a/m)^j)}{\Omega_f^\pm} \tag{5.4.6}$$

and the results follows from Definition 5.4.9. □

Exercise 5.4.12 Let $\mathfrak{p}$ be a prime of $\mathcal{O}_f$ and assume that $\Omega^\pm$ is $\mathfrak{p}$-normalized (notations of Remark 5.4.10). Show that the numbers $\frac{j!\tau(\bar{\chi})}{(-2i\pi)^{j+1}\Omega_f^\pm}L(f,\chi,s)$ (when $0 \le j \le k$, χ is a Dirichlet character, and $\chi(-1)(-1)^j = \pm 1$) are $\mathfrak{p}$-integers if the conductor m of χ is prime to $\mathfrak{p}$. Show also that when $k = 0$, at least one of those number has $\mathfrak{p}$-valuation 0.

5.4.3 Scalar Product and Congruences

In this paragraph, we present as an application a weak version of a nice theorem of Hida (cf. [68] and [69]). We continue to assume $\Gamma = \Gamma_1(N)$. Let f be a newform in $S_{k+2}(\Gamma, \mathbb{C})$.

Theorem 5.4.13 *Let $\mathfrak{p}$ be a prime ideal of $\mathcal{O}_f$, of residual characteristic p. Assume that p does not divides $6Nk!$. Then the following are equivalent:*

(a) *When $\Omega_f^\pm$ are chosen $\mathfrak{p}$-normalized, $v_\mathfrak{p}(W(f)(f,f)_\Gamma/(i^{k-1}\Omega_f^+\Omega_f^-)) > 0$.*
(b) *There exists a cuspidal normalized eigenform $g \in S_{k+2}(\Gamma, \mathbb{C})$ with coefficients in a finite extension K' of K_f, and a prime $\mathfrak{p}'$ of K' above $\mathfrak{p}$, such that $g \ne f$ and*

$$g \equiv f \pmod{\mathfrak{p}'}.$$

Proof Let $\mathcal{O}_\mathfrak{p}$ be the localized ring at $\mathfrak{p}$ of the ring of integers $\mathcal{O}_f$ of K_f. Choosing $\Omega_f^\pm$ $\mathfrak{p}$-normalized means that $\phi_f^\pm/\Omega_f^\pm$ is a generator of $\mathrm{Symb}_\Gamma^\pm(\mathcal{V}_k(\mathcal{O}_\mathfrak{p}))[\lambda_f]$. Equivalently, since 2 is invertible in $\mathcal{O}_\mathfrak{p}$, ϕ_f^+/Ω_f^+ and ϕ_f^-/Ω_f^- form a basis of $\mathrm{Symb}_\Gamma(\mathcal{V}_k(\mathcal{O}_p))[\lambda_f]$.

Let us write $M = H_!^1(\Gamma, \mathcal{V}_k(\mathcal{O}_\mathfrak{p}))$, so that $M \otimes K_f = H_!^1(\Gamma, \mathcal{V}_k(K_f))$. We provide M with the corrected scalar product $[\ ,\ ]$ (Definition 5.3.21). It is a perfect pairing (by Theorem 5.2.2 and Lemma 5.3.20, using that $6Nk!$ is invertible in $\mathcal{O}_\mathfrak{p}$), for which the Hecke operators of $\mathcal{H}$ are self-adjoint (Lemma 5.3.23).

Let $A = \mathrm{Symb}_\Gamma(\mathcal{V}_k(K_f))[\lambda_f]$ which is two-dimensional subspace of $M \otimes K_f = H_!^1(\Gamma, \mathcal{V}_k(K_f))$ since f is cuspidal. Obviously, $M \cap A = \mathrm{Symb}_\Gamma(\mathcal{V}_k(\mathcal{O}_\mathfrak{p}))[\lambda]$. Let B be the orthogonal of A in $M \otimes K_f$.

Since ϕ_f^+/Ω_f^+ and ϕ_f^-/Ω_f^- form a basis of $M \cap A$, we can very easily compute the discriminant of the restriction of the scalar product to $M \cap A$, using Lemma 5.3.26: it is $2^{-2k}(-1)^{1-k}W(f)(f,f)_\Gamma/(\Omega_f^+\Omega_f^-)^2$. Since this discriminant is non zero, we see that $M \otimes K_f = A \oplus B$. We can thus define the congruence module $C = M/(M \cap A) \oplus (M \cap B)$ and since the scalar product on M is a perfect pairing we know by Proposition 2.7.16 that this module C is non-zero if and only if the discriminant of the scalar product on $M \cap A$ has positive $\mathfrak{p}$ valuation, that is if and only if (a) holds.

But on the other hand, the fact that the congruence module C is non-zero is equivalent (cf. Proposition 2.7.15) to the existence of a modular symbol ϕ' in B such that $\phi' \equiv a\phi_f^+ + b\phi_f^- \pmod{\mathfrak{p}}$, with $a, b \in \mathcal{O}_\mathfrak{p}$, not both in $\mathfrak{p}$. By Deligne-Serre's lemma (up to allowing finite extension K' of K_f and replacement of $\mathfrak{p}$ be a prime $\mathfrak{p}'$ of K' dividing $\mathfrak{p}$) this is equivalent to the same thing with ϕ' assume to be an eigenvector for all Hecke operators. Such an eigenvector ϕ' defines a cuspidal normalized eigenform $g \in S_{k+2}(\Gamma, K')$ with the same Hecke eigenvalues. Hence C is non-zero if and only if there exists a cuspidal normalized eigenform g whose Hecke eigenvalues are congruent to those of f. In our case, this is also equivalent to $g \equiv f \pmod{\mathfrak{p}}$, for the two residual form g mod $\mathfrak{p}$ and f mod p in $S_{k+2}(\Gamma, \mathcal{O}_\mathfrak{p}/\mathfrak{p})$ will have the same eigenvalues, hence the same Fourier coefficients, hence will be equal by the q-development principle.

In other words, $C \neq 0$ is equivalent to (b). This proves the theorem. □

5.5 Notes and References

The most important part of this chapter, namely the identification of spaces of modular forms with cohomology groups of modular curves is due to Eichler and Shimura: see [58, 111, 113]. The reformulation of this identification in terms of modular symbols is due to Ash and Stevens: see [4]. Modular symbols (more precisely what we call *classical* modular symbols in this book) had been invented by Manin in relation to the special values of L-functions of modular forms, and their theory had been extended by Shokurov, Amice-Vélu, Mazur-Tate-Teitelbaum:

see [1, 88, 92, 115]. The results about boundary symbols and Eisenstein series are mostly due to the author and Dasgupta: see [14]. Some of the results about the scalar products are due to Hida [68], as well as the idea of the final theorem: [68, 69].

This material is standard basic knowledge to specialists in the arithmetic theory of modular forms, but its dispersal in many old papers written in German, French, Russian and English often makes it difficult to learn. An important part of it (but nor modular symbols nor Eisenstein series) is covered in the first part of see [68]. We have tried here to give a complete and truly self-contained treatment, which we hope will be useful to the students.

Chapter 6
Rigid Analytic Modular Symbols and p-Adic L-functions

In all this chapter we fix an arbitrary number p.

6.1 Rigid Analytic Functions and Distributions

In all this section, R denotes a commutative $\mathbb{Q}_p$-Banach algebra with a norm denoted $|\ |$ extending the usual absolute value on $\mathbb{Q}_p$, so that $|p| = 1/p$ (as in Sect. 3.1).

6.1.1 Some Modules of Sequences and Their Dual

For r any positive real number, we shall denote by $c_r(R)$ (resp. $b_r(R)$) the Banach R-module of sequences $(a_n)_{n\geq 0}$, with $a_n \in R$ and $|a_n|r^n \to 0$ when $n \to \infty$ (resp. with $|a_n|r^n$ bounded above), with the norm $|(a_n)| = \sup_{n\geq 0} |a_n|r^n$. When $r = 1$, we shall write[1] $c(R)$ for $c_1(R)$ (resp. $b(R)$ for $b_1(R)$).

Note that the Cauchy product makes $c_r(R)$ and $b_r(R)$ Banach algebras over R.

Lemma 6.1.1 *The natural map $c_r(\mathbb{Q}_p)\hat{\otimes}_{\mathbb{Q}_p} R \to c_r(R)$ is an isometry, and is compatible with the R-algebra structure. In particular, for any continuous morphism $R \to R'$ of $\mathbb{Q}_p$-Banach algebras, we have an isomorphism $c_r(R)\hat{\otimes}_R R' = c_r(R')$ of R'-Banach algebras.*

Proof The first statement is proved exactly as Lemma 3.1.26. The second follows from the first and the associativity of completed tensor product. □

[1]The very same module $c(R)$ was denoted by $c_{\mathbb{N}}(R)$ in Example 3.1.7. Here we shall not need to consider modules of sequences indexed by sets larger than $\mathbb{N}$, so we drop the $\mathbb{N}$ from the notation.

J. Bellaïche, *The Eigenbook*, Pathways in Mathematics,
https://doi.org/10.1007/978-3-030-77263-5_6

Lemma 6.1.2 *If there is a multiplicative element*[2] ρ *in* R^* *with* $|\rho| = r$*, then there are isometric isomorphisms of* R*-algebras* $c_r(R) \simeq c(R)$ *and* $b_r(R) \simeq b(R)$.

Proof By Exercise 3.1.1, $|a_n\rho^n| = |a_n||\rho|^n = |a_n|r^n$. Hence we can define a map $c_r(R) \to c(R)$ that sends (a_n) to $(a_n\rho^n)$. It is obviously an isometric isomorphism of algebras. Same proof for $b_r(R)$ and $b(R)$. □

Lemma 6.1.3 *The Banach* R*-module* $c_r(R)$ *is always potentially orthonormalizable.*[3] *If there is a multiplicative element* $\rho \in L^*$ *such that* $|\rho| = r$*, then* $c_r(R)$ *is orthonormalizable.*

Proof The $\mathbb{Q}_p$-Banach space $c_r(\mathbb{Q}_p)$ is potentially orthonormalizable as is any $\mathbb{Q}_p$-Banach space by a result of Serre, cf. Theorem 3.1.17. Then the R-Banach module

$$c_r(R) = c_r(\mathbb{Q}_p)\hat{\otimes}_{\mathbb{Q}_p} R$$

is potentially orthonormalizable by Lemma 3.1.26. The second assertion follows from Lemma 6.1.2 since $c(R)$ is ON-able by definition. □

Exercise 6.1.4 Show that $c_{p^{1/2}}(\mathbb{Q}_p)$ is not orthonormalizable.

Lemma 6.1.5 *There are natural isometric isomorphisms of Banach* R*-modules*

$$b_{\frac{1}{r}}(R) = \mathrm{Hom}_R(c_r(R), R) = \mathrm{Hom}_{\mathbb{Q}_p}(c_r(\mathbb{Q}_p), R) \tag{6.1.1}$$

Proof If $b = (b_n)_{n\in\mathbb{N}} \in b_{\frac{1}{r}}(R)$, one defines an R-linear form l_b on $c_r(R)$ by setting $l_b((a_n)) = \sum_n a_n b_n$. The sum converges since $|a_n|r^n$ goes to 0 and $|b_n|/r^n$ is bounded, so $a_n b_n$ goes to 0. The operator norm of l_b is clearly $|b|$. Conversely, any continuous linear form l on $c(R)$ defines a sequence $b = (b_n)$ by setting $b_n = l(e_n)$ where $e_n \in c_r(R)$ is the sequence $(e_{n,m})_{m\in\mathbb{N}}$ with $e_{n,m} = 1$ if $m = n$ and $e_{n,m} = 0$ if $m \neq n$. That sequence (b_n) satisfies $|b_n| \leq |l||e_n| = |l|r^n$, hence belongs to $b_{\frac{1}{r}}(R)$. One checks immediately that $l = l_b$. This constructs the morphism $b_{\frac{1}{r}}(R) \to \mathrm{Hom}_R(c_r(R), R)$ and proves that it is an isomorphism. The second isomorphism is constructed the same way. □

Let us define a Banach R-module $b'_r(R)$ as $b_r(\mathbb{Q}_p)\hat{\otimes}_{\mathbb{Q}_p} R$. The map $b_r(\mathbb{Q}_p) \subset b(R)$ induces an injective norm-preserving $b'_r(R) \hookrightarrow b_r(R)$, which allows us to see $b'_r(R)$ as a closed submodule of $b_r(R)$ with the induced norm. Using Lemma 6.1.5, one can also see $b'_{\frac{1}{r}}(R)$ as a closed submodule of $\mathrm{Hom}_{\mathbb{Q}_p}(c_r(\mathbb{Q}_p), R)$.

Proposition 6.1.6 *The sub-module* $b'_{\frac{1}{r}}(R)$ *of* $\mathrm{Hom}_{\mathbb{Q}_p}(c_r(\mathbb{Q}_p), R)$ *contains exactly the linear maps from* $c_r(\mathbb{Q}_p)$ *to* R *that are compact (that is, completely continuous).*

---

[2]cf. Exercise 3.1.1.

[3]For the definition of *potentially orthonormalizable*, cf. *supra*, Definition 3.1.6.

Proof This follows from Lemma 3.1.28. □

Corollary 6.1.7 *The inclusion $b'_r(R) \subset b_r(R)$ is proper if and only if R is infinite-dimensional over $\mathbb{Q}_p$.*

Proof If R is finite-dimensional, every linear map in $\mathrm{Hom}_{\mathbb{Q}_p}(c_r(\mathbb{Q}_p), R)$ has finite rank hence is compact. If R is infinite-dimensional, there are linear maps in $\mathrm{Hom}_{\mathbb{Q}_p}(c_r(\mathbb{Q}_p), R)$ that are non-compact. □

If $\tau \in \mathrm{End}_R(c_r(R))$, we denote by ${}^t\tau \in \mathrm{End}_R(\mathrm{Hom}_R(c_r(R), R)) = \mathrm{End}_R(b_{\frac{1}{r}}(R))$ the *transpose* map: ${}^t\tau(l) = l \circ \tau$.

Proposition 6.1.8 *Let $\tau_0 \in \mathrm{End}_{\mathbb{Q}_p}(c_r(\mathbb{Q}_p))$, and let $\tau = \tau_0 \hat{\otimes} \mathrm{Id}_R \in \mathrm{End}_R(c_r(R))$. Then ${}^t\tau$ stabilizes the submodule $b'_{\frac{1}{r}}(R)$ of $b_{\frac{1}{r}}(R)$.*

Proof Under the identification $b_{\frac{1}{r}}(R) = \mathrm{Hom}_{\mathbb{Q}_p}(c_r(\mathbb{Q}_p), R)$ of Lemma 6.1.5, ${}^t\tau$ is identified with the endomorphism of $\mathrm{Hom}_{\mathbb{Q}_p}(c_r(\mathbb{Q}_p), R)$ that sends u to $u \circ \tau_0$. If u is a compact linear map, so is $u \circ \tau_0$ (cf. Lemma 3.1.4), hence the result by Proposition 6.1.6. □

Proposition 6.1.9 *Let $a \in c_r(R)$, and let $\tau_a \in \mathrm{End}_R(c_r(R))$ be the multiplication by a. Then ${}^t\tau_a$ stabilizes the submodule $b'_{\frac{1}{r}}(R)$ of $b_{\frac{1}{r}}(R)$.*

Proof By Lemma 6.1.1, a is the limit in $c_r(R)$ of a sequence of elements $a_k \in c_r(\mathbb{Q}_p) \otimes R$ (algebraic tensor product) when $k \to \infty$. Hence ${}^t\tau_a$ is the limit in $\mathrm{End}_R(b_{\frac{1}{r}}(R))$ of the sequence ${}^t\tau_{a_k}$. Since $b'_{\frac{1}{r}}(R)$ is closed in $b_{\frac{1}{r}}(R)$ it suffices to show that ${}^t\tau_{a_k}$ stabilizes $b'_{\frac{1}{r}}(R)$. In other words, we may assume that $a \in c_r(\mathbb{Q}_p) \otimes R$. We may further assume that a is of the form $a_0 \otimes x$ with $a_0 \in c_r(\mathbb{Q}_p)$ and $x \in R$. Thus $\tau_a = x\tau_{a_0}$ stabilizes $b'_{\frac{1}{r}}(R)$ by Proposition 6.1.8. □

6.1.2 Modules of Functions over $\mathbb{Z}_p$

Definition 6.1.10 For $r \in |\mathbb{C}_p^*| = p^{\mathbb{Q}}$, we define $\mathcal{A}[r](R)$ as the R-module of functions $f : \mathbb{Z}_p \to R$ such that for every $e \in \mathbb{Z}_p$, there exists a power series

$$\sum_{n=0}^{\infty} a_n(e)(z-e)^n,$$

with $a_n(e) \in R$, that converges on the closed ball $\bar{B}(e, r) = \{z \in \mathbb{C}_p,\ |z-e| \le r\}$ in $\mathbb{C}_p$, and that coincides with f on $\bar{B}(e, r) \cap \mathbb{Z}_p$.

Definition 6.1.11 For $f \in \mathcal{A}[r](R)$, we set $\|f\|_r = \sup_{e \in \mathbb{Z}_p} \sup_n |a_n(e)| r^n$.

Proposition 6.1.12 *In the definition of $\mathcal{A}[r](R)$, it suffices to check the condition for a finite set E of $e \in \mathbb{Z}_p$ such that the balls $\bar{B}(e, r)$ cover $\mathbb{Z}_p$. Moreover, for such a set E, one has $\|f\|_r = \sup_{e\in E} \sup_n |a_n(e)|r^n$.*

Proof Saying that the power series $\sum_{n=0}^{\infty} a_n(e)(z - e)^n$ converges on $\bar{B}(e, r)$ is equivalent to saying that $\lim_{n\to+\infty} |a_n(e)|r^n = 0$. If $e' \in \bar{B}(e, r)$, then $\bar{B}(e', r) = \bar{B}(e, r)$. Set

$$a_n(e') = \sum_{m=0}^{\infty} \frac{(m + n)!a_{m+n}(e)(e' - e)^m}{m!n!}. \tag{6.1.2}$$

This series converges since $|a_{m+n}(e)(e'-e)^m| \le |a_{m+n}(e)|r^{m+n}r^{-n}$ and for n fixed, $m \to \infty$, $|a_{m+n}(e)|r^{m+n}$ goes to 0 since f is in $\mathcal{A}[r]$. We have

$$|a_n(e')r^n| \le \sup_{m\ge0} |a_{m+n}(e)|r^m r^n = \sup_{m\ge n} |a_m(e)|r^m,$$

which shows that the series $\sum a_n(e')(z-e')^n$ converges on $\bar{B}(e, r)$ to the same limit as $\sum_n a_n(e)(z - e)^n$, and that $\sup_n |a_n(e)|r^n \ge \sup_n |a_n(e')|r^n$. By symmetry, we have equality. This proves that in the definition of $\mathcal{A}[r]$ and of $\|f\|_r$, it is sufficient to work with one e in each closed ball of radius r. Since $\mathbb{Z}_p$ is compact, it is covered by finitely many such balls. It is clear that $\mathcal{A}[r](R)$ is complete. □

It is easy to see that $\|\ \|_r$ is a norm on $\mathcal{A}[r](R)$ which makes it a Banach algebra over R.

Exercise 6.1.13 Let $x \in R^*$ such that $|x - 1| < 1$.

1. Show that the function $\mathbb{Z} \to R$, $n \mapsto x^n$ extends uniquely to a continuous function $\mathbb{Z}_p \to R$ denoted $z \mapsto x^z$. Show that for $z, z' \in \mathbb{Z}_p$, $x^{z+z'} = x^z x^{z'}$.
2. Show that $(z \mapsto x^z) \in \mathcal{A}[r](R)$ for every $r < 1 - |1 - x|$.

Exercise 6.1.14 Show that the algebra $R[z]$ of polynomials in z is dense in $\mathcal{A}[r](R)$ for all r.

Before going on, let us record an estimate (essentially the Taylor-Laplace estimate) the we shall need later:

Scholium 6.1.15 *For f in $\mathcal{A}[r](R)$, $e, e' \in \mathbb{Z}_p$ such that $|e' - e| \le r$, $n, N \in \mathbb{N}$, $N \ge n$ we have*

$$\left| a_n(e') - \sum_{k=n}^{N} \frac{k!a_k(e)(e' - e)^{k-n}}{(k - n)!n!} \right| \le r^{-N-1}|e' - e|^{N+1-n}\|f\|_r.$$

Proof Equation (6.1.2), after the change of variables $k = m + n$, gives:

$$a_n(e') = \sum_{k=n}^{\infty} \frac{k! a_k(e)(e' - e)^{k-n}}{(k-n)!n!}.$$

For $k > N$, one has

$$\left| \frac{k! a_k(e)(e' - e)^{k-n}}{(k-n)!n!} \right| \leq |a_k(e)||e' - e|^{k-N-1}|e' - e|^{N+1-n}$$
$$\leq r^{k-N-1}|a_k(e)||e' - e|^{N+1-n} \leq r^{-N-1}\|f\|_r|e' - e|^{N+1-n}.$$

□

Corollary 6.1.16 *The choice of a finite subset E of $\mathbb{Z}_p$ as in the above proposition determines a natural isometric isomorphism of normed R-modules*

$$\mathcal{A}[r](R) = c_r(R)^E. \tag{6.1.3}$$

Proof Send f to $(a_n(e))_{n \in \mathbb{N}, e \in E}$. □

Corollary 6.1.17 *The formation of $\mathcal{A}[r](R)$ commutes with base change, in the following sense: $\mathcal{A}[r](R)\hat{\otimes}_R R' = \mathcal{A}[r](R')$ for every R-Banach algebra R'.*

Proof This follows from the above corollary and Lemma 6.1.1. □

Exercise 6.1.18 Define $B[\mathbb{Z}_p, r] = \{z \in \mathbb{C}_p, \exists a \in \mathbb{Z}_p, |z - a| \leq r\}$ (a finite union of closed balls of radius r, and only one ball of center 0 and radius r when $r \geq 1$).

1. Show that an $f \in \mathcal{A}[r](R)$ defines a continuous map $B[\mathbb{Z}_p, r] \to R$.
2. Show that $\|f\|_r = \sup_{z \in B[\mathbb{Z}_p, r]} |f(z)|$.
3. Show that $B[\mathbb{Z}_p, r]$ has a natural structure of affinoid space over $\mathbb{Q}_p$ and that $\mathcal{A}[r](\mathbb{Q}_p)$ is the ring of global analytic functions on this space.

Lemma 6.1.19 *The Banach R-module $\mathcal{A}[r](R)$ is always potentially ON-able. If there is a multiplicative element $\rho \in R^*$ such that $|\rho| = r$, then $\mathcal{A}[r](R)$ is actually ON-able.*

Proof This follows from Corollary 6.1.16 and Lemma 6.1.3. □

There is a natural map $\mathcal{A}[r_1](R) \to \mathcal{A}[r_2](R)$ for $r_1 > r_2$ which sends a function f on $\mathbb{Z}_p$ to itself. This map is clearly continuous, injective, and has dense image (since its image contains all polynomials). We shall call such a map a *restriction* map, since in the interpretation of Exercise 6.1.18 of $\mathcal{A}[r](R)$ as the space of functions on $B[\mathbb{Z}_p, r]$, this map is indeed the restriction of global analytic functions from $B[\mathbb{Z}_p, r_1]$ to $B[\mathbb{Z}_p, r_2]$.

Moreover

Lemma 6.1.20 *The restriction maps* $\mathcal{A}[r_1](R) \to \mathcal{A}[r_2](R)$ *for* $r_1 > r_2$ *are compact.*

Proof We first prove the result when there exist multiplicative elements $\rho_1, \rho_2 \in R^*$ such that $|\rho_1| = r_1$ and $|\rho_2| = r_2$. Choose a set E that satisfies the condition of Proposition 6.1.14 for both r_1 and r_2. Through the identifications $\mathcal{A}[r_1](R) = c_{r_1}(R)^E \simeq c(R)^E$ and $\mathcal{A}[r_2](R) = c_{r_2}(R)^E \simeq c(R)^E$ given by Corollary 6.1.16, the restriction map becomes the diagonal map $c(R)^E \to c(R)^E$ that on each component is the map $c(R) \to c(R)$, $(a_n) \mapsto ((r_2/r_1)^n a_n)$. Since $(r_2/r_1)^n$ goes to 0 when n goes to ∞, this map is compact by Proposition 3.1.13.

We next prove the result for any real numbers $r_1 > r_2 > 0$, but for $R = L$ any finite extension of $\mathbb{Q}_p$ (provided with its absolute value as norm). By Lemma 3.1.27, it suffices to prove the result after replacing L by a finite extension, so we can assume that there exists real numbers r_1' and r_2' with $r_1 > r_1' > r_2' > r_2$ which are norms in L^*. Then by what we have proven above, the restriction map $\mathcal{A}[r_1'](L) \to \mathcal{A}[r_2'](L)$ is compact, and thus $\mathcal{A}[r_1](L) \to \mathcal{A}[r_2](L)$ which factors as $\mathcal{A}[r_1](L) \to \mathcal{A}[r_1'](L) \to \mathcal{A}[r_2'](L) \to \mathcal{A}[r_2](L)$ is also compact by Lemma 3.1.4.

The case of general R follows from the case $R = \mathbb{Q}_p$ by Lemma 3.1.27. □

6.1.3 Modules of Convergent Distributions

Definition 6.1.21 We denote by $\mathcal{D}[r](\mathbb{Q}_p)$ the continuous dual $\operatorname{Hom}_{\mathbb{Q}_p}(\mathcal{A}[r], \mathbb{Q}_p)$ of $\mathcal{A}[r](\mathbb{Q}_p)$. The space $\mathcal{D}[r](\mathbb{Q}_p)$ is a Banach module over $\mathbb{Q}_p$ for the norm

$$\|\mu\|_r = \sup_{f \in \mathcal{A}[r]} \frac{|\mu(f)|}{\|f\|_r}.$$

For R any $\mathbb{Q}_p$-Banach algebra, we define

$$\mathcal{D}[r](L) = \mathcal{D}[r](\mathbb{Q}_p) \hat{\otimes} R.$$

We insist that for R a general Banach $\mathbb{Q}_p$-algebra, we *do not define* $\mathcal{D}[r](R)$ as the dual of $\mathcal{A}[r](R)$.

Proposition 6.1.22 *The Banach L-modules* $\mathcal{D}[r](R)$ *are potentially orthonormalizable. The formation of* $\mathcal{D}[r](R)$ *commutes with base change. For* $r_1 > r_2 > 0$, *the natural restriction maps* $\mathcal{D}[r_2](R) \to \mathcal{D}[r_1](R)$ *are injective and compact.*

Proof The Banach $\mathcal{D}[r](\mathbb{Q}_p)$ is potentially orthonormalizable by Theorem 3.1.17. It is clear by definition that the formation of $\mathcal{D}[r](R)$ commutes with base change. The restriction maps $\mathcal{D}[r_2](\mathbb{Q}_p) \to \mathcal{D}[r_1](\mathbb{Q}_p)$ are the transpose of the restriction maps $\mathcal{A}[r_1](\mathbb{Q}_p) \to \mathcal{A}[r_2](\mathbb{Q}_p)$. They are injective since the maps

$\mathcal{A}[r_1](\mathbb{Q}_p) \to \mathcal{A}[r_2](\mathbb{Q}_p)$ have dense image. Note that the transpose of a compact map between $\mathbb{Q}_p$-Banach spaces is compact, since this is true for finite rank maps. Thus the maps $\mathcal{D}[r_2](\mathbb{Q}_p) \to \mathcal{D}[r_1](\mathbb{Q}_p)$ are compact by Lemma 6.1.20. For general L, the restriction maps $\mathcal{D}[r_2](R) \to \mathcal{D}[r_1](R)$ are defined by base change from the case $R = \mathbb{Q}_p$, so their injectivity and compactness follow (cf. Lemma 3.1.27). □

We note that the we have a natural injective norm-preserving map $\mathcal{D}[r](R) = \mathcal{D}[r](\mathbb{Q}_p)\hat{\otimes}R \to \mathrm{Hom}_R(\mathcal{A}[r](R), R)$ which identified $\mathcal{D}[r](R)$ as a closed subspace of $\mathrm{Hom}_R(\mathcal{A}[r](R), R)$. The following lemma is obvious from the definitions.

Lemma 6.1.23 *Let E be as in Proposition 6.1.14. Then through the isomorphisms (6.1.3) and (6.1.1)*

$$\mathrm{Hom}_R(\mathcal{A}[r](R), R) = \mathrm{Hom}_R(c_r(R)^E, R) = b_{\frac{1}{r}}(R)^E$$

the subspace $\mathcal{D}[r](R)$ of $\mathrm{Hom}_R(\mathcal{A}[r](R), R)$ is identified with the subspace $b'_{\frac{1}{r}}(R)^E$ of $b_{\frac{1}{r}}(R)^E$.

In particular, we see that $\mathcal{D}[r](R)$ is strictly smaller than the R-dual of $\mathcal{A}[r](R)$ when R is not finite-dimensional over $\mathbb{Q}_p$. This is the price to pay to have a $\mathcal{D}[r](R)$ whose formation commute with base change, which will be crucial for the application to the eigencurve in the next chapter.

6.2 Overconvergent Functions and Distributions

In this section as in the preceding one, R denotes a commutative $\mathbb{Q}_p$-Banach algebra with a norm denoted $|\ |$ extending the usual absolute value on $\mathbb{Q}_p$, so that $|p| = 1/p$.

6.2.1 *Semi-normic Modules and Fréchet Modules*

Let M be an R-module. A *semi-norm* on M is a map $p : M \to \mathbb{R}_{\geq 0}$ such that for every $x \in R$, $m, n \in M$, $p(m + m') \leq \max(p(m), p(n))$ and $p(xm) \leq |x|p(m)$. Note that one has automatically $p(0) = 0$ but there may be non-zero $m \in M$ such that $p(m) = 0$. Also if x is multiplicative, $p(xm) = |x|p(m)$ by the same proof as Exercise 3.1.1.

A family of semi-norms $(p_i)_{i\in I}$ on an R-module M defines a topology on M, for which a basis of neighborhood of an element $m_0 \in M$ are the sets $V_{S,\epsilon,m_0} := \{m \in M, \forall i \in S, p_i(m - m_0) < \epsilon\}$ where S runs among finite subsets of I and ϵ among positive real numbers. Addition and multiplication by a scalar in M are bi-continuous for this topology. We call a topology of this form *semi-normic*.

A *Fréchet module* M over R is an R-module M with a semi-normic topology which is metrizable and complete. Banach modules are examples of Fréchet modules.

Exercise 6.2.1 We define a *normable group* as a topological abelian group G whose topology can be defined by a ultrametric distance that is invariant by translation (that is $d(x+z, y+z) = d(x, y)$ for all $x, y, z \in G$).

1. Show that it amounts to the same to ask that G is the underlying topological group of a normed group in the sense of [27, §1.1], that is that the topology of G is defined by a norm $| \, | : G \to \mathbb{R}_+$ such that for all x, y in G, $|x| = 0$ if and only if $x = 0$; and $|x - y| \leq \max(|x|, |y|)$.
2. Let R be a Banach algebra over $\mathbb{Q}_p$ and M a semi-normic R-module. Show that if the topology of M is defined by countably many semi-norms, then the underlying topological abelian group M is a normable group, and in particular M is metrizable.

If $f : M \to N$ is a continuous linear map of Fréchet modules, then the open image theorem states that if f is surjective, it is open. The proof is the same as in the standard case, and can be found in [28, Théorème 1, Chapitre 1, §3.3].

Let M and N be R-module, and let p be a semi-norm on M and q a semi-norm on N. Then one defines a semi-norm $p \otimes q$ on $M \otimes_R N$ by setting $(p \otimes q)(z) = \inf_{z=\sum_{i=1}^r v_i \otimes w_i} \sup_{i \in \{1,\ldots,r\}} p(v_i)q(w_i)$. If M (resp. N) has a semi-normic topology defined by a family $(p_i)_{i \in I}$ (resp. $(q_j)_{j \in J}$) of semi-norms, then we define $M \hat{\otimes}_R N$ as the completion of $M \otimes_R N$ for the topology defined by the semi-norms $(p_i \otimes q_j)_{(i,j) \in I \times J}$, provided with the semi-normic topology defined by the same semi-norms. If $N = R'$ is a Banach R-algebra, then $M \hat{\otimes}_N R'$ is naturally a semi-normic R'-module, and if R'' is a Banach R'-algebra, there is a natural bi-continuous isomorphism of R''-module $(M \hat{\otimes}_R R') \hat{\otimes}_{R'} R'' = M \hat{\otimes}_R R''$.

We now give our attention to the case where $R = L$ is a finite extension of $\mathbb{Q}_p$ normed by its absolute value. In this case, the notion of *semi-normic topology* is the same as the notion of locally convex topology as explained in [105, §4]. More precisely, if $\mathcal{O}$ is the ring of integers of L, and V a vector space over L, a *lattice* Λ in V is an $\mathcal{O}$-submodule of V such that $L\Lambda = V$. A subset of V is *convex* if it is of the form $v + \Lambda$ where $v \in V$ and Λ is a lattice. A topology on V is *locally convex* if every point has a basis of neighborhoods that are convex. Given a family of lattices $(\Lambda_i)_{i \in I}$ such that for every $i, j \in I$, there exists $k \in I$ with $\Lambda_k \subset \Lambda_i \cap \Lambda_j$, and for all $i \in I$ and $x \in K^*$, there exists $j \in I$ such that $\Lambda_j \subset x\Lambda_i$, there is one and only one topology on V for which a basis of neighborhood of v are the $v + \Lambda_i$. This topology is locally convex, and all locally convex topologies arise this way.

The *gauge* of a lattice Λ is the semi-norm p_Λ defined by $p_\Lambda(v) = \inf_{x \in K, v \in x\Lambda} |x|$. If a locally convex topology on V is defined by a family of lattices Λ_i as above, then it is also defined by the semi-norms p_{Λ_i} and V is semi-normic. Conversely, given a family of semi-norms $(p_i)_{i \in I}$, the lattices of the form $V_{S,\epsilon,0}$ for $\epsilon > 0$ and $S \subset I$ finite form a family satisfying the above conditions which defines the same topology as the semi-norms p_i for $i \in I$.

If V is locally convex space over L, the dual $V^\vee$ of V is the space $\mathrm{Hom}_L(V, L)$ of all continuous linear maps $V \to L$. We provide $V^\vee$ with the so-called strong topology, defined as follows. Recall that a subset B of V is *bounded* if $p_i(B)$ is bounded above for every semi-norm p_i defining the topology of V. For every bounded subset B of V, we define a semi-norm p_B on $V^\vee$ by setting $p_B(l) = \sup_{b \in B} |l(b)|$. The strong topology on $V^\vee$ is the one defined by this family of semi-norms. We note that our notion of dual extends the one we have already used in the case of Banach spaces.

Exercise 6.2.2 Prove the theorem of Hahn-Banach: if V is a semi-normic space over a finite extension L of $\mathbb{Q}_p$, V_0 is a subspace of V, and l_0 is a continuous linear form on V_0, then l_0 may be extended into a continuous linear form on V.

Exercise 6.2.3 Let us recall that a continuous morphism $f : X \to Y$ between two normable groups is *strict* if the induced map $\bar{f} : X/\ker f \to \mathrm{Im}\, f$ is an homeomorphism, where $X/\ker f$ is given the quotient topology, and $\mathrm{Im}\, f$ the topology restricted from Y.

1. Show that if f is injective, it is strict if and only if $f : X \to \mathrm{Im}\, f$ is an homeomorphism.
2. Show that if f is surjective, it is strict if and only if it is open.
3. Show that if f is an isomorphism of groups, it is strict if and only if it as also an homeomorphism.
4. Let us denote by $| \, |$ two norms, one on X and one on Y defining the topologies of X and Y. Show that a morphism $f : X \to Y$ is strict if and only if for every real number $\epsilon > 0$, there exists a real number $\delta > 0$ such that for all $x \in X$, $|f(x)| \leq \delta$ implies $|x + k| \leq \epsilon$ for some $k \in \ker f$.
5. Let us assume that X and Y are Fréchet spaces over $\mathbb{Q}_p$. Then a continuous linear map $f : X \to Y$ is strict if and only if $f(X)$ is closed in Y.
6. Let R be a normed $\mathbb{Q}_p$-algebra, and let X and Y be R-modules with a Hausdorff topology defined by a countable family of R-modules semi-norms (hence X and Y are normable groups by question 2 of the above exercise). Let R' be a normed R-module which is R-flat. Let $f : X \to Y$ be a strict morphism of R-module. Then $f' = f \otimes 1 : X' = X \otimes_R R' \to Y' = Y \otimes_R R'$ is also strict. Here we provide X' (resp. Y') with the topology given by the family of semi-norms $p' = p \otimes | \, |_{R'}$ where p run among the family of semi-norms on X (resp. Y) and $| \, |_{R'}$ is the norm on R'.
7. If X and Y are two normed groups, $f : X \to Y$ is a strict continuous morphism, $\hat{X}$, $\hat{Y}$ are the completions of X and Y and $\hat{f} : \hat{X} \to \hat{Y}$ is the extension by continuity of f, then $\hat{f}$ is strict and one has $\ker \hat{f} = \widehat{\ker f}$ and $\mathrm{Im} \hat{f} = \widehat{\mathrm{Im} f}$. In particular, if $0 \to K \overset{g}{\to} X \overset{f}{\to} Y \to 0$ is an exact sequence, and if either f or g are strict, then $0 \to \hat{K} \overset{\hat{g}}{\to} \hat{X} \overset{\hat{f}}{\to} \hat{Y} \to 0$ is an exact sequence.

6.2.2 Modules of Overconvergent Functions and Distributions

Definition 6.2.4 For $r \geq 0$, we define the modules of overconvergent functions

$$\mathcal{A}^\dagger[r](R) = \varinjlim_{r'>r, r' \in p^{\mathbb{Q}}} \mathcal{A}[r'](R) = \bigcup_{r'>r} \mathcal{A}[r'](R).$$

The module $\mathcal{A}^\dagger[0](R)$ is simply the module of R-valued locally analytic functions on $\mathbb{Z}_p$: functions that around every point e admit a converging Taylor expansion on some ball of some positive radius r depending on e. The module $\mathcal{A}^\dagger[1](R)$ is the module of analytic functions on $\mathbb{Z}_p$ defined by a powers series that converges "a little bit more than necessary", that is on some ball of center 0 and radius $r > 1$.

We give $\mathcal{A}^\dagger[r](R)$ the *locally convex final topology* (see [105, §5.E]), that is the finest locally convex topology such that the natural morphism $\mathcal{A}[r'](L) \to \mathcal{A}^\dagger[r](L)$ for $r' > r$ are continuous. This makes $\mathcal{A}^\dagger[r](R)$ a semi-normic R-module. When $R = L$ is a field, the topology of $\mathcal{A}^\dagger[r](L)$ may be defined by the lattices Λ such that $\Lambda \cap \mathcal{A}[r'](L)$ is an open lattice of $\mathcal{A}[r'](L)$ for every $r' > r$.

Exercise 6.2.5 If M is a semi-normic R-module with topology defined by semi-norms $(p_i)_{i\in I}$, we denote by $c(M)$ the set of sequence $(m_n)_{n\in\mathbb{N}}$ of elements of M that goes to 0, that is such that $\lim_{n\to\infty} p_i(m_n) = 0$ for every $i \in I$. We provide $c(M)$ with the family of semi-norms $(q_i)_{i\in I}$ defined by $q_i((m_n)_{n\in\mathbb{N}}) = \sum_{n\in\mathbb{N}} p_i(m_n)$. This makes $c(M)$ a semi-normic R-module.

1. Prove that the sequences in $c(M)$ which are eventually 0 form a dense subset.
2. Prove that the natural morphism $c(R)\hat{\otimes}_R M \to c(M)$ is an isomorphism.

Definition 6.2.6 For any $r \geq 0$, we set

$$\mathcal{D}^\dagger[r](R) = \varprojlim_{r'>r,\ r' \in p^{\mathbb{Q}}} \mathcal{D}[r'](R).$$

In other words $\mathcal{D}^\dagger[r](R)$ is the intersection of the $\mathcal{D}[r'](R)$ for $r' > r$.

When $r = 0$, we shall write $\mathcal{D}^\dagger(R)$ instead of $\mathcal{D}^\dagger[0](R)$.

We give those modules the topology of the projective limit. In other words, the topology of $\mathcal{D}^\dagger[r](R)$ is defined by the family of norms $\|\mu\|_{r'}$ for all $r' > r, r' \in p^{\mathbb{Q}}$. It is clear that $\mathcal{D}^\dagger[r](R)$ is complete for that topology, and since the topology is defined by a countable family of semi-norms, it is also metrizable, hence a Fréchet R-module.

Proposition 6.2.7 *For any commutative $\mathbb{Q}_p$-Banach algebra R, we have a natural isomorphism of Fréchet R-modules $\mathcal{D}^\dagger[r](\mathbb{Q}_p)\hat{\otimes}R \to \mathcal{D}^\dagger[r](R)$.*

Proof The maps $\mathcal{D}^\dagger[r](\mathbb{Q}_p)\hat{\otimes}R \hookrightarrow \mathcal{D}[r'](\mathbb{Q}_p)\hat{\otimes}R$ for $r' > r$ induces an injective map $\mathcal{D}^\dagger[r](\mathbb{Q}_p)\hat{\otimes}R \hookrightarrow \varprojlim_{r'>r} \mathcal{D}[r'](\mathbb{Q}_p) = \mathcal{D}^\dagger[r](R)$. Since R is potentially orthonormalizable as a Banach space over $\mathbb{Q}_p$ (Theorem 3.1.17), we can choose an isomorphism $R \simeq c(\mathbb{Q}_p)$ and by the above observation, we only need to prove that the natural map $c(\mathcal{D}^\dagger[r](\mathbb{Q}_p)) \to \varprojlim_{r'>r} c(\mathcal{D}[r'](\mathbb{Q}_p)) = \cap_{r'>r} c(\mathcal{D}[r'](\mathbb{Q}_p))$ is an isomorphism, which is clear since a sequence (v_n) in $\mathcal{D}^\dagger[r]$ goes to 0 if and only if $\|v_n\|_{r'}$ goes to 0 for all $r' > r$. □

Corollary 6.2.8 *For any morphism $R \to R'$ of Banach algebras, we have a natural isomorphism of topological modules over R':*

$$\mathcal{D}^\dagger[r](R)\hat{\otimes}_R R' = \mathcal{D}^\dagger[r](R').$$

Proposition 6.2.9 *Assume that $R = L$ is a finite extension of $\mathbb{Q}_p$. Then $\mathcal{D}^\dagger[r](L)$ may be canonically identified, as a topological vector space, to the dual of $\mathcal{A}^\dagger[r](L)$.*

Proof Remember that $\mathcal{A}^\dagger[r](L) = \lim_{\substack{\leftarrow \\ r'>r}} \mathcal{A}[r'](L)$ by definition, where the transition maps are injective and compact (Lemma. 6.1.20), and that the dual of $\mathcal{A}[r'](L)$ with its operator norm (that is, its strong dual) is $\mathcal{D}[r'](L)$. We can therefore apply directly [105, Prop 16.10], which gives the result. □

6.2.3 *Integration of Functions Against Distributions*

Let L be a finite extension of $\mathbb{Q}_p$. If $f \in \mathcal{A}[r](L)$ or $f \in \mathcal{A}^\dagger[r](L)$, and $\mu \in \mathcal{D}[r](L)$ or $\mathcal{D}^\dagger[r](L)$, we have by definition in the first case, and by the above proposition in the second case, a scalar $\mu(f) \in L$. We will sometimes use the following longer but intuitive notation:

$$\int_{\mathbb{Z}_p} f(z) d\mu(z) := \mu(f).$$

Slightly more generally, for $a \in \mathbb{Z}_p$ and $n \in \mathbb{N}$, we set

$$\int_{a+p^n\mathbb{Z}_p} f(z) d\mu(z) := \int_{\mathbb{Z}_p} f(z) 1_{a+p^n\mathbb{Z}_p}(z) d\mu(z) = \mu(f 1_{a+p^n\mathbb{Z}_p}).$$

6.2.4 *Order of Growth of a Distribution*

In all this section, we assume that $R = L$ is a finite extension of $\mathbb{Q}_p$, normed by its absolute value. For simplicity, we shall write $\mathcal{D}^\dagger(L)$ instead of $D^\dagger[0](L)$.

Definition 6.2.10 Let $\nu \geq 0$ be a real number. We say that a distribution $\mu \in \mathcal{D}^\dagger(L)$ has *order* $\leq \nu$ if there exists a constant $C \in \mathbb{R}_+$ such that for all a in $\mathbb{Z}_p$, $k, n \in \mathbb{N}$, we have

$$\left| \int_{a+p^n\mathbb{Z}_p} (\frac{z-a}{p^n})^k d\mu(z) \right| < C p^{n\nu}. \qquad (6.2.1)$$

A distribution of order ≤ 0 is called *a measure* on $\mathbb{Z}_p$. A distribution which is of order $\leq \nu$ for some ν is called *a tempered distribution*.

Remark 6.2.11 Višik [124] and Colmez [42] use an interesting equivalent definition when ν is an integer: a distribution μ is of order $\nu \in \mathbb{N}$ if, as a continuous linear form on the space $\mathcal{A}^\dagger[0](L)$ of locally analytic functions on $\mathbb{Z}_p$, it extends to the larger space of class C^ν, defined as the space of functions that admits ν derivatives and such that the last derivative is Lipschitz. We shall not use this definition.

There is an easy characterization of the order of growth in terms of norms:

Lemma 6.2.12 *Let $\mu \in \mathcal{D}^\dagger(L)$ be a distribution. Then μ has order $\leq \nu$ if and only if there exists a real $D > 0$ such that $\|\mu\|_r \leq Dr^{-\nu}$ when $r \to 0^+$. Moreover, we can take $D = Cp$ if C is as in (6.2.1).*

Proof Note that $g_{a,n,k}(z) := (\frac{z-a}{p^n})^k 1_{a+p^n\mathbb{Z}_p}(z)$ satisfies $\|g_{a,n,k}\|_r = 1$ for $r = p^{-n}$. Thus if $\|\mu\|_r \leq Dr^{-\nu}$ for all r, we get for $r = p^{-n}$ that $|\mu(g_{a,n,k})| \leq Dp^{n\nu}$ and μ has order of growth $\leq \nu$.

For the converse, let μ have order of growth $\leq \nu$. Then there is a C such that $\left|\int_{a+p^n\mathbb{Z}_p} (\frac{z-a}{p^n})^k d\mu(z)\right| < Cp^{n\nu}$. Let $f \in \mathcal{A}[p^{-n}]$, and $e \in \mathbb{Z}_p$. Write the power series of f about e as $f(z) = \sum_{k=0}^{\infty} a_k(e)p^{-nk}(z-e)^k$. Then $\|f\|_{p^{-n}} = \sup_{e\in E} \sup_k |a_k(e)p^{-nk}|p^{-nk} = \sup_{e\in E} \sup_{k\in\mathbb{N}} |a_k(e)|$, where E is a finite subset of $\mathbb{Z}_p$ as in Proposition 6.1.14. Then

$$\mu(f) = \sum_{e\in E} \sum_{k=0}^{\infty} a_k(e) \int_{e+p^n\mathbb{Z}_p} (\frac{z-e}{p^n})^k d\mu(z)$$

so $|\mu(f)| \leq \|f\|_{p^{-n}} Cp^{n\nu}$.

Now if r is a positive real number, $r < 1$, there is an $n \in \mathbb{N}$ such that $p^{-n} < r \leq p^{-n+1}$, and we have $|\mu(f)| \leq \|f\|_{p^{-n}} Cp^{n\nu} \leq \|f\|_r Cpr^{-\nu}$, that is $\|\mu\|_r \leq Dr^{-\nu}$ with $D = Cp$. □

Theorem 6.2.13 (Višik, Amice-Vélu)

(i) *Let $\mu \in \mathcal{D}^\dagger(L)$ a distribution of order $\leq \nu$, and let N be an integer greater or equal to the integral part of ν. Then μ is uniquely determined by the linear forms for $a \in \mathbb{Z}_p$ and $n \in \mathbb{N}$:*

$$i_{\mu,a+p^n\mathbb{Z}_p} : \mathcal{P}_N(L) \to L$$
$$P \mapsto \int_{a+p^n\mathbb{Z}_p} P(z) d\mu(z).$$

Here as above, $\mathcal{P}_N(L)$ is the finite-dimensional space of polynomials of degree less than N over L.

(ii) *Conversely, suppose we are given, for every ball $a + p^n\mathbb{Z}_p$ in $\mathbb{Z}_p$, a linear form $i_{a+p^n\mathbb{Z}_p} : \mathcal{P}_N(L) \to L$ which satisfy the additivity relation (for all $a \in \mathbb{Z}_p$, $n \in \mathbb{N}$)*

$$i_{a+p^n\mathbb{Z}_p} = \sum_{i=0}^{p-1} i_{a+p^n i+p^{n+1}\mathbb{Z}_p}, \tag{6.2.2}$$

and such that there exist constants $C > 0$ and $\nu \geq 0$ such that for every $a \in \mathbb{Z}_p$, $k, n \in \mathbb{N}$, and $k \leq N$

$$|i_{a+p^n\mathbb{Z}_p}((\frac{z-a}{p^n})^k)| \leq Cp^{n\nu}. \tag{6.2.3}$$

Then there exists a unique distribution μ of order $\leq \nu$ such that $i_{\mu,a+p^n\mathbb{Z}_p} = i_{a+p^n\mathbb{Z}_p}$, and for any $n \in \mathbb{N}$ one has

$$\|\mu\|_{p^{-n}} \leq Cp^{n\nu}.$$

This theorem is not hard to prove, just somewhat computational. It says that a distribution on order ν is known when you know how it integrates polynomials of degree less or equal than $\lfloor \nu \rfloor$ on closed balls in $\mathbb{Z}_p$, and conversely, that any method to "integrate" those polynomials satisfying the obvious additivity relation and a growth condition can be extended in a method to integrate all locally analytic functions, in other words a distribution, which moreover satisfies the same growth condition. When $\nu = 0$, this takes the very intuitive form: a measure μ on $\mathbb{Z}_p$ is determined by the volume it gives to all balls $a + p^n\mathbb{Z}_p$. Conversely, if we have a way to attributes a volume to all ball $a + p^n\mathbb{Z}_p$ that is additive, and such that *the set of all such volumes is bounded*, then this defines a measure.

Proof We first prove (i). Fix a real $D > 0$ such that $\|\nu\|_r \leq Dr^{-\nu}$.

For f in $\mathcal{A}[r](L)$ for some $r > 0$, and $e \in \mathbb{Z}_p$, such that $f(z) = \sum_{i=0}^{\infty} a_i(e)(z-e)^i$, and for $n \geq 0$ such that $p^{-n} \leq r$, and $N \geq 0$ an integer, let us define $T_{f,e,n,N} \in \mathcal{P}_N(L)$ as the function with support $e + p^n\mathbb{Z}_p$ defined by

$$T_{f,e,n,N}(z) := \sum_{i=0}^{N} a_i(e)(z-e)^i.$$

That is, $T_{f,e,n,N}(z)$ is the degree N Taylor-approximation of f around e, restricted to $e + p^n\mathbb{Z}_p$. The following states how good is that approximation:

$$\|f 1_{e+p^n\mathbb{Z}_p} - T_{f,e,n,N}\|_{p^{-n}} \leq (p^n/r)^{N+1}\|f\|_r. \tag{6.2.4}$$

Indeed, $f1_{e+p^n\mathbb{Z}_p} - T_{f,e,n,N}$ has Taylor expansion about e as follows: $\sum_{i=0}^{N+1} a_i(e)(z-e)^i$; hence

$$\|f1_{e+p^n\mathbb{Z}_p} - T_{f,e,n,N}\|_{p^{-n}} = \sup_{i\geq N+1} |a_i(e)|p^{-ni} \leq (p^{-n}/r)^{N+1} \sup_{i\geq N+1} |a_i(e)|r^i$$
$$\leq (p^{-n}/r)^{N+1}\|f\|_r,$$

and (6.2.4) is proven.

Now let $\mu \in \mathcal{D}^\dagger(L)$, and $f \in \mathcal{A}[r](L)$. We consider the following approximation of $\mu(f)$, a kind of generalized Riemann sum, depending on the choice of an integer n such that $p^n \leq r$ and of a set of representative E of $\mathbb{Z}_p$ mod p^n

$$S_{\mu,f,n,N,E} := \sum_{e\in E} \mu(T_{f,e,n,N}) = \sum_{e\in E} i_{\mu,e+p^n\mathbb{Z}_p}(T_{f,e,n,N}) \in L$$

We have

$$|\mu(f) - S_{\mu,f,n,N,E}| = \left|\sum_{e\in E} \mu(f1_{e+p^n\mathbb{Z}_p} - T_{f,e,n,N})\right|$$
$$\leq (p^{-n}/r)^{N+1}\|\mu\|_{p^{-n}}\|f\|_r.$$
$$\leq D(p^{-n}/r)^{N+1}p^{n\nu}\|f\|_r$$

Since $\nu < N+1$, $|\mu(f)-S_{\mu,f,n,N,E}|$ goes to 0 for a fixed f, when n goes to infinity. It follows that $\mu(f)$ depends only on the linear forms $i_{\mu,e+p^n\mathbb{Z}_p}$, which proves (i).

Let us prove (ii). We begin by the following computation: For $n, n' \in \mathbb{N}$, $n' \geq n$ and $e, e' \in \mathbb{Z}_p$, $e' \in e + p^n\mathbb{Z}_p$ we have for all z such that $|z - e'| \leq p^{-n'}$

$$T_{f,e',n',N}(z) - T_{f,e,n,N}(z) = \sum_{i=0}^{N} a_i(e')(z-e')^i - \sum_{k=0}^{n} a_k(e)(z-e)^k$$
$$= \sum_{i=0}^{N} a_i(e')(z-e')^i - \sum_{k=0}^{N} a_k(e)\sum_{i=0}^{k} \frac{k!}{i!(k-i)!}(z-e')^i(e'-e)^{k-i}$$
$$= \sum_{i=0}^{N}(z-e')^i\left[a_i(e') - \sum_{k=i}^{n} \frac{k!}{(k-i)!i!}a_k(e)(e'-e)^{k-i}\right]$$
$$= \sum_{i=0}^{N}(z-e')^i E(i,n,e,e')$$

where $E(i, n, e, e')$ is an error term in L equal to the formula between bracket on the line above in L, and such that

$$|E(i, n, e, e')| \leq r^{-N-1}|e' - e|^{N+1-i}\|f\|_r \leq r^{-N-1}p^{-n(N+1-i)}\|f\|_r$$

using Scholium 6.1.15.

From a family of linear forms $i_{e+p^n\mathbb{Z}_p}$ as in the statement of the theorem, we shall construct $\mu \in \mathcal{D}^\dagger(L)$ by defining $\mu(f)$ as a limit of "Riemann sums" of the type considered above. Precisely we proceed as follows: Let $f \in \mathcal{A}[r](L)$ for some $r > 0$; for $n \in \mathbb{N}$ such that $p^{-n} \leq r$, and E a set of representative of $\mathbb{Z}_p$ mod p^n, define

$$S_{f,n,N,E} := \sum_{e \in E} i_{\mu,e+p^n\mathbb{Z}_p}(T_{f,e,n,N}).$$

For $n' \geq n$, E' any set of representative of $\mathbb{Z}_p$ mod $p^{n'}$, we compute

$$\begin{aligned}
|S_{f,n,N,E} - S_{f,n',N,E'}| &\leq \left|\sum_{e' \in E'} i_{e'+p^{n'}\mathbb{Z}_p}(T_{f,e,n,N}1_{e'+p^{n'}\mathbb{Z}_p} - T_{f,e',n',N})\right| \quad \text{(using (6.2.2))} \\
&\leq \left|\sum_{e' \in E'} i_{e'+p^{n'}\mathbb{Z}'_p}((z - e')^i E(i, n, e, e'))\right| \\
&\leq |E(i, n, e, e')|Cp^{n(\nu-i)} \quad \text{(using (6.2.3))} \\
&\leq Cr^{-N-1}\|f\|_r p^{n(\nu-(N+1))}
\end{aligned}$$

The above computation, using the hypothesis $\nu < N + 1$, shows that the sequence $n \mapsto S_{f,n,N,E}$, however we choose our systems of representatives E for each n, is a Cauchy sequence. We call $\mu(f)$ its limit, which does not depend on the choice of E. Thus we have constructed a linear form μ on $\mathcal{A}^\dagger[0](L)$ which satisfies by construction $i_{\mu,a+p^n\mathbb{Z}_p} = i_{a+p^n\mathbb{Z}_p}$.

Now fix an n, and let $f \in \mathcal{A}[p^{-n}](L)$ We have

$$\begin{aligned}
|S_{f,n,N,E}| &= |\sum_{e \in E} i_{e+p^n\mathbb{Z}_p}(\sum_{i=0}^{N} a_i(e)(z - e)^i)| \\
&\leq C \sup_{e \in E} |a_i(e)| p^{n\nu} p^{-ni} p^{n\nu} \quad \text{(using (6.2.3))} \\
&\leq C\|f\|_{p^{-n}} p^{n\nu}
\end{aligned}$$

For any $n' > n$, and E' system of representatives of $\mathbb{Z}_p$ modulo $p^{n'}$, it follows from the estimate of the above paragraph, applied to $r = p^{-n}$, that $|S_{f,n',N',E} - S_{f,n,N,E}| \leq C\|f\|_{p^{-n}} p^{n\nu}$. We thus get $|S_{f,n',N',E}| \leq C\|f\|_{p^{-n}} p^{n\nu}$ for all n', E', hence by passing to the limit,

$$|\mu(f)| \le C\|f\|_{p^{-n}} p^{n\nu}.$$

It follows that for every $n \ge 0$, the restriction of μ to $\mathcal{A}[p^{-n}](L)$ is a *continuous* linear form, of norm

$$\|\mu\|_{p^{-n}} \le Cp^{n\nu}.$$

That is, μ belongs to $\mathcal{D}[p^{-n}](L)$ for all n, hence to $\mathcal{D}(L)$. □

6.3 The Weight Space

The p-adic weight space $\mathcal{W}$ that we shall define below plays two different roles in this book, both very important. First its elements generalize the possible weights $k \in \mathbb{N}$ of a modular form, more precisely, the couples (k, ϵ) where ϵ is the p-part of the nebentypus of a modular form. Second it is the set in which we will draw the variable of our p-adic L-functions, in a way analogous to the fact that the variable of the complex L-function $L(f, s, \chi)$ is a couple (s, χ), where s is a complex variable and χ a Dirichlet character, or which is equivalent as we have seen in Sect. 5.4.1, a continuous character $\mathbb{A}_{\mathbb{Q}}^*/\mathbb{Q}^* \to \mathbb{C}^*$.

6.3.1 *Definition and Description of the Weight Space*

We use this analogy as a motivation for the following definition: a $\mathbb{C}_p$-valued weight is a continuous character $\kappa : \mathbb{A}_{\mathbb{Q}}^*/\mathbb{Q}_p^* \to \mathbb{C}_p^*$. Some remarks are in order: first, since $\mathbb{R}_+^*$ is connected, and $\mathbb{C}_p^*$ is totally disconnected, κ is trivial on $\mathbb{R}_+^*$ hence factors through a character $\kappa : \mathbb{A}_{\mathbb{Q}}^*/\mathbb{Q}^*\mathbb{R}_*^+ = \prod_l \mathbb{Z}_l^* \to \mathbb{C}_p^*$. Second:

Exercise 6.3.1 Show that a continuous morphism $\prod_l \mathbb{Z}_l^* \to \mathbb{C}_p^*$ is trivial on some open subgroup of $\prod_{l \ne p} \mathbb{Z}_l^*$.

Now, a basis of open subgroups of $\prod_{l \ne p} \mathbb{Z}_l^*$ is given by the

$$U_M := \{x \in \prod_{l \ne p} \mathbb{Z}_l^*,\ x_{l_k} \equiv 1 \pmod{{l_k}^{a_k}}\}$$

for M an integer prime to p, with a decomposition $M = \prod_{k=1}^r l_k^{a_k}$. Note that we have $\prod_{l \ne p} \mathbb{Z}_l^*/U_M = (\mathbb{Z}/M\mathbb{Z})^*$. Thus our character κ will factor through $\prod_l \mathbb{Z}_l^*/U_M = \mathbb{Z}_p^* \times (\mathbb{Z}/M\mathbb{Z})^*$. This motivates our definition:

Definition 6.3.2 Let M be an integer coprime to p. The p-adic *weight space of tame conductor* M is the rigid analytic space $\mathcal{W}_M$ over $\mathbb{Q}_p$ such that, for all commutative Banach $\mathbb{Q}_p$-algebra R,

$$\mathcal{W}_M(R) = \mathrm{Hom}(\mathbb{Z}_p^* \times (\mathbb{Z}/M\mathbb{Z})^*, R^*).$$

We write $\mathcal{W}$ instead of $\mathcal{W}_1$.

The uniqueness of the space $\mathcal{W}_M$ is clear from the definition and Yoneda's lemma, but less obvious is the implicit assertion of existence of a rigid analytic space $\mathcal{W}_M$ having the given functor of points. Let us prove it by describing explicitly this analytic structure:

First, set $q = p$ is p is odd and $q = 4$ if $p = 2$, and remember that that there is a canonical isomorphism $\mathbb{Z}_p^* = (\mathbb{Z}/q\mathbb{Z})^* \times (1 + q\mathbb{Z}_p)$, given by $x \mapsto (\tau(x), \langle x \rangle)$, where $\tau(x)$ is the Teichmüller lift of ($x \mod p$), that is the only $y \in \mathbb{Z}_p^*$ such that $y^{\phi(q)} = 1$ and $y \equiv x \pmod{q}$, and $\langle x \rangle = x/\tau(x)$. Remember also that the multiplicative group $1 + q\mathbb{Z}_p$ is isomorphic to the additive group $\mathbb{Z}_p$ (as follows: choose a generator γ of $1 + q\mathbb{Z}_p$ and send $a \in \mathbb{Z}_p$ to $\gamma^a := \exp_p(a \log_p(\gamma)) \in 1+q\mathbb{Z}_p$). Therefore, an element $\kappa \in \mathcal{W}_M(R)$ is given by a R-valued character κ_f on the finite group $(\mathbb{Z}/M\mathbb{Z})^* \times (\mathbb{Z}/q\mathbb{Z})^* = (\mathbb{Z}/Mq\mathbb{Z})^*$ and a character κ_p on $1+q\mathbb{Z}_p$. Such a character is of course completely defined by the image $\kappa_p(\gamma)$ of the chosen generator γ of $1+q\mathbb{Z}_p$. Since $\gamma^{p^n} \to 1$ when $n \to \infty$, we have $\kappa_p(\gamma)^{p^n} \to 1$, and this easily implies that $|\kappa_p(\gamma) - 1| < 1$. Conversely, by completeness, any element $x \in R^*$ satisfying $|x - 1| < 1$ defines a continuous character $\kappa_p : 1 + q\mathbb{Z}_p \to R^*$ sending γ to x.

The functor $R \mapsto \mathrm{Hom}((\mathbb{Z}/qM\mathbb{Z})^*, R^*)$ for all $\mathbb{Q}_p$-algebras R is obviously representable by a finite scheme (hence a finite rigid analytic space) over $\mathbb{Q}_p$, that we shall denote $((\mathbb{Z}/qM\mathbb{Z})^*)^\vee$. It has $\phi(qM)$ points over $\mathbb{C}_p$, but less over $\mathbb{Q}_p$, since not all characters must take values in $\mathbb{Q}_p$. When $M = 1$, the $\phi(q)$ points of this schemes are defined over $\mathbb{Q}_p$, and correspond to the power of the Teichmüller character τ.

The functor $R \mapsto \{x \in R^*, |x - 1| < 1\}$ for all complete $\mathbb{Q}_p$-algebras is represented by a rigid analytic space, the open ball $B(1, 1)$ of center 1 and radius 1.

Since $\mathcal{W}_M(R) = ((\mathbb{Z}/pM\mathbb{Z})^*)^\vee(R) \times B(1, 1)(R)$ functorially in R, by the map $\kappa \mapsto (\kappa_f, \kappa_p(\gamma))$, we have shown that $\mathcal{W}_M$ is representable by a rigid analytic scheme $((\mathbb{Z}/qM\mathbb{Z})^*)^\vee \times B(1, 1)$. Over $\mathbb{C}_p$, or even an extension of $\mathbb{Q}_p$ containing all $\phi(M)$-th roots of unity, it is the union of $\phi(qM)$ copies of the open unit ball of center 1 and radius 1.

For later use, let us record

Scholium 6.3.3 *Over $\mathbb{C}_p$, or over any extension of $\mathbb{Q}_p$ containing all $\phi(M)$-th roots of unity, $\mathcal{W}_M$ is a finite union of components $\mathcal{W}_{M,\kappa_f}$ indexed by the characters $\kappa_f : (\mathbb{Z}/qM\mathbb{Z})^* \to \mathbb{C}_p^*$. Each of these components is isomorphic to the open ball $B(1, 1)$. A generator γ of $1 + q\mathbb{Z}_p$ being fixed, an isomorphism $\mathcal{W}_{M,\kappa_f}(\mathbb{C}_p) \to B(1, 1)(\mathbb{C}_p)$ is given by $\kappa \mapsto \kappa(\gamma)$.*

6.3.2 Local Analyticity of Characters

It is easily seen that for $x \in R$, the power series $\log_p(1+x) := \sum_{n=1}^{\infty} \frac{(-1)^{n+1}x^n}{n}$ converges for $|x| < 1$. For $x, y \in R$, one has

$$\log_p((1+x)(1+y)) = \log_p(1+x)\log_p(1+y) \text{ if } |x| < 1, |y| < 1 \quad (6.3.1)$$

since the two sides are the same as formal power series in x and y, and both converge absolutely on the considered domain. On the smaller domain $|x| < p^{\frac{-1}{p-1}}$, the first term x in the power series defining $\log_p(x)$ is larger in absolute value than all the other terms. Thus one has

$$|\log_p(1+x)| = |x| \text{ if } |x| < p^{\frac{-1}{p-1}} \quad (6.3.2)$$

Similarly, it is easy to check that the power series $\exp_p(x) = \sum_{n=0}^{\infty} \frac{x^n}{n!}$ converges for $|x| < p^{\frac{-1}{p-1}}$, and that

$$\exp_p(x+y) = \exp_p(x)\exp_p(y) \text{ if } |x| < p^{\frac{-1}{p-1}}, |y| < p^{\frac{-1}{p-1}}. \quad (6.3.3)$$

Finally,

$$\exp_p(\log_p(x)) = x \text{ if } |x| < p^{\frac{-1}{p-1}} \quad (6.3.4)$$

as is seen by similar argument.

Theorem 6.3.4 *Let $\kappa : \mathbb{Z}_p^* \to R^*$ be a continuous character, that is an element of $\mathcal{W}(R)$. Extend κ into a function $\mathbb{Z}_p \to R$ by defining $\kappa(z) = 0$ for $z \in p\mathbb{Z}_p$. Then κ is locally analytic; that is, there exists $r > 0$ such that $\kappa \in \mathcal{A}[r](R)$.*

Proof Let $x = \kappa(\gamma) \in R^*$. We have $|x - 1| < 1$. Consider the power series

$$f(z) = \exp_p\left(\frac{\log_p(x)\log_p(z)}{\log_p(\gamma)}\right) \in R[[z]].$$

For $|z-1| < p^{\frac{-1}{p-1}}$, then $|\log_p(z)| = |z-1|$ by (6.3.2) and if z also satisfies $|z-1| < \frac{p^{\frac{-1}{p-1}}|\log_p(\gamma)|}{|\log_p(x)|}$, then the power series $f(z)$ converges. Calling r_0 the smaller of those two bound on $|z-1|$ (which is a positive real number depending on x), we see that for $|z-1| < r_0$, the series $f(z)$ converges and on the same domain, $f(zz') = f(z)f(z')$, by (6.3.2) and (6.3.3).

If N is a natural integer large enough so that $|x^{p^N}-1| < p^{\frac{-1}{p-1}}$ and $|\gamma^{p^N}-1| < r_0$ then f is defined at γ^{p^N} and one has

$$\begin{aligned} f(\gamma^{p^N}) &= \exp_p\left(\frac{\log_p(x)p^N\log_p(\gamma)}{\log_p(\gamma)}\right) \text{ (using (6.3.2))} \\ &= \exp_p(\log_p(x^{p^N})) \text{ (using (6.3.2) again)} \\ &= x^{p^N} \text{ (using (6.3.4))} \\ &= \kappa(\gamma^{p^N}) \end{aligned}$$

Hence κ and f agree at γ^{p^N}, and by multiplicativity and continuity, they agree on all p-adic integers z such that $|z-1| < |\gamma^{p^N}-1|$. This shows that $\kappa(z)$ is given by a converging power series on a sufficiently small disk centered at 1. Since κ is multiplicative, this is also true on a sufficiently small ball centered at any $e \in \mathbb{Z}_p^*$. And for $e \in p\mathbb{Z}_p$, the same is true since κ is 0 on $p\mathbb{Z}_p$. □

Definition 6.3.5 Let $\kappa \in \mathcal{W}(R)$. We denote by $r(\kappa) > 0$ the supremum of all real numbers r such that the function $z \mapsto \kappa(1+pz)$ belongs to $\mathcal{A}[r](R)$.

Exercise 6.3.6

1. Let x be a primitive p^n-root of 1 in $\mathbb{C}_p$, let κ_p be the character of $1+q\mathbb{Z}_p$ sending γ to x, and let $r_n = r(\kappa_p)$. Compute r_n.
2. Deduce that there is no universal $r > 0$ such that for all $\kappa \in \mathcal{W}(L)$, $\kappa(1+pz) \in \mathcal{A}[r](L)$.

6.3.3 Some Remarkable Elements in the Weight Space

Let $k \in \mathbb{Z}$. The map $z \mapsto z^k$, $\mathbb{Z}_p^* \to \mathbb{Q}_p^*$ is a continuous character, hence an element of $\mathcal{W}(\mathbb{Q}_p)$. We shall denote by k this element in $\mathcal{W}(\mathbb{Q}_p)$, and we call those elements the *integral points*, or *integral weights*. Hence we have an inclusion $\mathbb{Z} \subset \mathcal{W}(\mathbb{Q}_p)$. Note that $r(k) = +\infty$ (cf. Definition 6.3.5).

Lemma 6.3.7 *The set $\mathbb{Z}$ of integral weights is very Zariski-dense in $\mathcal{W}$ (cf. Definition 3.8.1). The p-adic topology on $\mathcal{W}$ induces by restriction the topology on $\mathbb{Z}$ for which a basis of neighborhood of $a \in \mathbb{Z}$ is given by the congruences classes of a modulo $p^n(p-1)$ for all n.*

Proof Any character κ_f of $\mathbb{Z}_p^*$ factoring through $(\mathbb{Z}/q\mathbb{Z})^*$ has the form $z \mapsto \tau(z)^{k_0}$ for some integer k_0 well determined modulo $\phi(q)$. The intersection of $\mathbb{Z}$ with the open ball $B(1,1)$ corresponding to κ_f is the set $\{\gamma^k,\ k \equiv k_0 \pmod{\phi(q)}\}$. In any ball of center γ^k will sit all the points $\gamma^{k+\phi(q)p^n}$ for n large enough, and those points, having an accumulation point in that ball, are Zariski-dense, which proves the first assertion. For two integers k, k' seen as characters to be close, we need their κ_f to be equal, so $k \equiv k' \pmod{\phi(q)}$ and γ^k to be close to $\gamma^{k'}$, which means k close to k' p-adically. This proves the second assertion. □

Now suppose we have chosen an embedding $\bar{\mathbb{Q}} \to \mathbb{C}_p$, and an embedding $\bar{\mathbb{Q}} \to \mathbb{C}$. Slightly more generally, let (k, ϵ) be a pair of an integer $k \in \mathbb{Z}$, and a Dirichlet character ϵ. The character ϵ takes values in $\mathbb{C}^*$, but those values are actually in $\bar{\mathbb{Q}}^*$, and well determined there by our choice of embedding, and we can see them as element of $\mathbb{C}_p^*$ by our other embedding. Write the conductor of ϵ as $p^\nu N$, where N is prime to p. If $N \mid M$, we can see ϵ as a character $\mathbb{Z}_p^* \times (Z/M\mathbb{Z})^* \to \mathbb{C}_p^*$ (factoring through the quotient $(\mathbb{Z}/p^\nu\mathbb{Z})^* \times (\mathbb{Z}/N\mathbb{Z})^*$). To (k, ϵ) we finally attach the character $z \mapsto z^k\epsilon(z)$ of $\mathbb{Z}_p^* \times (\mathbb{Z}/M\mathbb{Z})^*$, that we see as an element of $\mathcal{W}_M(\mathbb{C}_p)$. Those points are called the *arithmetic points* of $\mathcal{W}_M(\mathbb{C}_p)$.

6.3.4 The Functions $\log_p^{[k]}$ on the Weight Space

Definition 6.3.8 Let $k \geq 1$ be an integer. We define an analytic function on $\mathcal{W}$ by the formula

$$\log_p^{[k]}(\sigma) := \frac{\prod_{i=0}^{k-1} \log_p(\gamma^{-i}\sigma(\gamma))}{(\log_p \gamma)^k},$$

for $\sigma \in \mathcal{W}(\mathbb{C}_p)$, that is σ a continuous character $\mathbb{Z}_p^* \to \mathbb{C}_p^*$. Here γ is a generator of $1 + q\mathbb{Z}_p$.

Exercise 6.3.9

1. Show that the definition of $\log_p^{[k]}$ is independent of the choice of the generator γ.
2. Show that $\log_p^{[k]}$ vanishes at order 1 at the characters of the form $z^i\psi(z)$, where $0 \leq i \leq k-1$ is an integer and ψ is a character of finite order, and has no other zeros.
3. Show that $\log_p^{[k]} = \prod_{i=1}^{k} \log_p^{[1]}(\sigma/z^i)$
4. Show that for $\sigma \in \mathcal{W}(\mathbb{C}_p)$, the function $z \mapsto \frac{d^k\sigma}{dz^k} \cdot \frac{z^k}{\sigma(z)}$ is a constant on $\mathbb{Z}_p^*$, and that this constant is $\log_p^{[k]}(\sigma)$.

6.3.5 The Iwasawa Algebra and the Weight Space

We define the Iwasawa algebra, Λ, as the topological $\mathbb{Z}_p$-algebra

$$\Lambda := \mathbb{Z}_p[[\mathbb{Z}_p^*]] := \varprojlim \mathbb{Z}_p[\mathbb{Z}_p^*/(1 + p^n\mathbb{Z}_p)].$$

An element x in $\mathbb{Z}_p^*$ thus defines an element denoted $[x]$ in Λ.

We have a non-canonical isomorphism $\Lambda = \prod_{\kappa_f:(\mathbb{Z}/q\mathbb{Z})^*\to\mathbb{C}_p^*} \mathbb{Z}_p[[T]]$ sending $[\gamma]$, for γ a generator of $1+q\mathbb{Z}_p$ to $1+T$ diagonally.

There is a natural embedding $\Lambda \hookrightarrow \mathcal{O}(\mathcal{W})$ which sends an element $[x]$ for $x \in \mathbb{Z}_p^*$ to the function $\kappa \mapsto \kappa(x)$ for every $\kappa \in \mathcal{W}(\mathbb{C}_p)$ a continuous character $\mathbb{Z}_p^* \to \mathbb{C}_p^*$.

Exercise 6.3.10 Show that an element f of $\mathcal{O}(\mathcal{W})$ which satisfies $|f(w)| \leq 1$ for every $w \in \mathcal{W}(\mathbb{C}_p)$ belongs to Λ.

6.4 The Monoid $S_0(p)$ and Its Actions on Overconvergent Distributions

6.4.1 The Monoid $S_0(p)$

We define a monoid

$$S_0(p) = \left\{\gamma = \begin{pmatrix} a & b \\ c & d \end{pmatrix} \in M_2(\mathbb{Z}),\ p \nmid a,\ p \mid c,\ ad-bc \neq 0\right\}.$$

This is a submonoid of the monoid S defined in Sect. 5.1.1. We note that

$$S_0(p) \cap \mathrm{SL}_2(\mathbb{Z}) = \Gamma_0(p),$$

so a congruence subgroup Γ is contained in $S_0(p)$ if and only if it is contained in $\Gamma_0(p)$.

The monoid $S_0(p)$ acts on the open ball $B(0,p)$ of $\mathbb{C}_p$ on the right by $z_{|\gamma} = \frac{dz-b}{a-cz}$. Indeed, if $|z| < p$, $|cz| < 1$ and $|a-cz| = 1$, while $|dz-b| < p$.

Lemma 6.4.1 *If $\gamma \in S_0(p)$, then γ sends balls of radius r in $B(0,p)$ to balls of radius $|\det\gamma|\,r$,*

Proof Let $e \in B(0,p)$ and $f = e_{|\gamma} \in B(0,p)$. We want to prove that γ sends $B(e,r)$ (for any $r < p$) to $B(f, |\det\gamma|\,r)$. Since the matrices $\begin{pmatrix} 1 & e \\ 0 & 1 \end{pmatrix}$ and $\begin{pmatrix} 1 & f \\ 0 & 1 \end{pmatrix}$ have determinant 1 and act on $B(0,p)$ as translations, which do not change the radii of balls, one can replace γ by the matrix $\begin{pmatrix} 1 & -e \\ 0 & 1 \end{pmatrix}\gamma\begin{pmatrix} 1 & f \\ 0 & 1 \end{pmatrix}$, and in other words, assume that $e = f = 0$. In this case, $0_{|\gamma} = 0$, hence $\gamma = \begin{pmatrix} a & 0 \\ c & d \end{pmatrix}$, and $|\det\gamma| = |ad| = |d|$. The map $z \mapsto z_{|\gamma} = \frac{dz}{a-cz}$ sends the ball $B(0,r)$ onto $B(0,|d|r)$ since $|a-cz| = 1$ for all $z \in B(0,r)$. □

6.4.2 Actions of $S_0(p)$ on Functions and Distributions

Let R be a commutative Banach algebra over $\mathbb{Q}_p$, and let us choose $\kappa \in \mathcal{W}(R)$ a character $\mathbb{Z}_p^* \to R^*$.

For $f(z)$ an L-valued continuous function on $\mathbb{Z}_p$, and $\gamma \in S_0(p)$ we define a new function

$$(\gamma \cdot_\kappa f)(z) = \kappa(a - cz) f\left(\frac{dz - b}{a - cz}\right). \tag{6.4.1}$$

This formula should look familiar: when κ is an integral weight $k \geq 0$ (cf. Sect. 6.3.3), this is exactly the formula (5.1.6) defining the left-action of S on the space $\mathcal{P}_k(R)$. Note however that in our case, this formula makes sense only for $\gamma \in S_0(p)$, not $\gamma \in S$: the hypothesis $\gamma \in S_0(p)$ ensures that $a - cz \in \mathbb{Z}_p^*$, so $\kappa(a - cz)$ is defined, and also that $\frac{dz-b}{a-cz} \in \mathbb{Z}_p$, so that $f\left(\frac{dz-b}{a-cz}\right)$ makes sense. It is clear that we have defined, for every weight $\kappa \in \mathcal{W}(L)$, a left-action of $S_0(p)$ on the module of continuous R-valued functions on $\mathbb{Z}_p$.

Lemma 6.4.2 *Let $f \in \mathcal{A}[r](R)$, $\gamma \in S_0(p)$. Assume that $p^n \mid \det(\gamma)$, and let r' be such that $0 < r' < p$, and $r' \leq p^n r$. Then $f\left(\frac{dz-b}{a-cz}\right) \in \mathcal{A}[r'](R)$. Moreover if p^n divides exactly $\det(\gamma)$, and $p^n r < p$, then $\|f\|_r = \|f\left(\frac{dz-b}{a-cz}\right)\|_{p^n r}$.*

Proof The derivative of the map $g : z \mapsto \frac{dz-b}{a-cz}$ is $g'(z) = \frac{\det \gamma}{(a-cz)^2}$. For $z \in \mathbb{C}_p$, $|z| < p$, we have $|a - cz| = 1$, and we thus have $|g'(z)| \leq p^{-n}$. By Lemma 6.4.1, the map $z \mapsto g(z) = \frac{dz-b}{a-cz}$ sends a ball of radius r' (with $r' < p$) centered on an element e of $\mathbb{Z}_p$ into the ball of radius $p^{-n} r' \leq r$ and center $g(e) \in \mathbb{Z}_p$. Since g is analytic, this show that $f(g(z))$ is in $\mathcal{A}[p^n r]$ if f is in $\mathcal{A}[r]$.

If p^n divides r exactly, then g actually is a bijection from the closed ball of center e and radius $p^n r$ to the ball of center $g(e)$ and radius r. The assertion on the norms follows (by Exercise 6.1.18, question 2). □

Proposition and Definition 6.4.3 *Fix $\kappa \in \mathcal{W}(R)$. For any $0 < r < r(\kappa)$ (cf Definition. 6.3.5), the formula (6.4.1) defines a continuous left action of $S_0(p)$ on $\mathcal{A}[r](R)$, called the action of weight κ. If p^n divides exactly $\det \gamma$, and $p^n r < p$, then $\|\gamma \cdot_\kappa f\|_{p^n r} = \|f\|_r$. In particular, if $\det \gamma$ is prime to p, and $r < p$, then γ acts by isometry.*

Those actions are compatible with the restriction maps. Therefore, they define a left-action of weight κ on $\mathcal{A}^\dagger[r](R)$ for $0 \leq r < \kappa(w)$.

Proof By Definition 6.3.5, $\kappa(a-cz) = \kappa(a)\kappa(1-c/az)$ is in $\mathcal{A}[r](R)$ for $r < r(\kappa)$, since $p \mid c/a$ in $\mathbb{Z}_p$. Hence $\kappa(a - cz) f\left(\frac{dz-b}{a-cz}\right) \in \mathcal{A}[r](L)$, and we have defined an action on $\mathcal{A}[r](R)$. The other assertions follow from Lemma 6.4.2, noting that $|\kappa(a - cz)| = 1$ for all $z \in \mathbb{C}_p$, $|z| < p$. □

We now proceed to define a "dual" right-action of weight κ on $\mathcal{D}[r](R)$, when $r < r(\kappa)$. The difficulty comes from the fact that for R infinite-dimensional over $\mathbb{Q}_p$, $\mathcal{D}[r](R)$ is not the R-dual of $\mathcal{A}[r](R)$, but a proper sub-module of it.

On $\mathcal{A}[r](R)^\vee = \mathrm{Hom}_R(\mathcal{A}[r](R), R)$, we can define the dual action of the action on $\mathcal{A}[r]$ by

$$\mu_{|_\kappa \gamma}(f) = \mu(\gamma \cdot_\kappa f).$$

Lemma 6.4.4 *Let $\mu \in \mathcal{D}[r](R) \subset \mathcal{A}[r](R)^\vee$. Then $\mu_{|_\kappa \gamma} \in \mathcal{D}[r](R)$.*

Proof Fix $\gamma \in S_0(p)$. The map $\tau : \mathcal{A}[r](R) \to \mathcal{A}[r](R)$, $f \mapsto \gamma \cdot_\kappa f$ can be decomposed as $\tau_2 \circ \tau_1$, where $\tau_1(f) = f\left(\frac{dz-b}{a-cz}\right)$ and $\tau_2(f) = \kappa(a - cz)f(z)$. We only need to show is that ${}^t\tau$ stabilizes the sub-module $\mathcal{D}[r](R)$ of $\mathcal{A}[r](R)^\vee$, and for this we only need to show that ${}^t\tau^1$ and ${}^t\tau^2$ stabilizes this sub-module. Note that τ_1 clearly comes by extension of scalars from the map $\mathcal{A}[r](\mathbb{Q}_p) \to \mathcal{A}[r](\mathbb{Q}_p)$, $f \mapsto f\left(\frac{dz-b}{a-cz}\right)$, while τ_2 is the multiplication by an element of $\mathcal{A}[r](R)$. Thus the lemma follows from Propositions 6.1.8 and 6.1.9 via the translation given by Lemma 6.1.23. □

Hence, we have defined a right action of $S_0(p)$ on $\mathcal{D}[r](R)$ of weight κ, if $r < r(\kappa)$.

Proposition 6.4.5 *This action is continuous. Assume $r < p$. Let $\mu \in \mathcal{D}[r](R)$, $\gamma \in S_0(p)$ such that $p^n \mid \gamma$. Then $\mu_{|_\kappa}(\gamma) \in \mathcal{D}[r/p^n](R)$. If p^n divides exactly $\det \gamma$, then moreover $\|\mu_{|_\kappa \gamma}\|_{r/p^n} = \|\mu\|_r$. In particular, if $\det \gamma$ is prime to p, then γ acts by isometry.*

Proof Since the injection $\mathcal{D}[r](R) \subset \mathcal{A}[r](R)^\vee$ is norm-preserving, this follows from the analog result for $\mathcal{A}[r]$ (Proposition 6.4.3) except for the fact that $\mu_{|_\kappa}(\gamma) \in \mathcal{D}[r/p^n](R)$ which is proven using the same method as in the above lemma. □

Since the action of weight κ on the $\mathcal{D}[r](R)$ for various $r < r(\kappa)$ are compatible, we can use them to define an action of weight κ on $\mathcal{D}^\dagger[r](R)$ for $0 \le r < r(\kappa)$.

We put an index κ when we think of $\mathcal{A}[r](R)$, $\mathcal{A}^\dagger[r](R)$, $\mathcal{D}[r](R)$, $\mathcal{D}^\dagger[r](R)$ as provided with that action: $\mathcal{A}_\kappa[r](R)$, $\mathcal{A}^\dagger_\kappa[r](R)$, $\mathcal{D}_\kappa[r](R)$, $\mathcal{D}^\dagger[r]_\kappa(R)$. As before, $\mathcal{D}^\dagger_\kappa(R)$ means $\mathcal{D}^\dagger_\kappa[0](R)$.

6.4.3 *The Module of Locally Constant Polynomials and Its Dual*

In this Sect. 6.4.3, we assume (just for the sake of simplicity) that $R = L$ is a finite extension of $\mathbb{Q}_p$, normed by its absolute value.

Definition 6.4.6 Let $k \geq 0$ be an integer. For $r > 0$, we shall denote by $\mathcal{P}_k[r](L)$ the subspace of $\mathcal{A}[r](L)$ of functions which on each closed ball $\bar{B}(a, r)$ in $\mathbb{Z}_p$ are polynomials of degree at most k. We shall set $\mathcal{V}_k[r](L) = \mathrm{Hom}_L(\mathcal{P}_k[r](L), L)$.

Lemma 6.4.7 *For $r > 0$, the L-vector spaces $\mathcal{P}_k[r](L)$ and $\mathcal{V}_k[r](L)$ are free of finite rank and their formation commutes with any base change. If we put $\mathcal{A}[r](L)$ its action of weight k, then $\mathcal{P}_k[r](L)$ is $S_0(p)$-stable in $\mathcal{A}_k[r](L)$. Hence we have a dual surjective map $\mathcal{D}_k[r](L) \to \mathcal{P}_k[r](L)$*

Proof This is clear. □

We provide $\mathcal{P}_k[r](L)$ with its left-action of $S_0(p)$ induced by the action of weight k on $\mathcal{A}[r](L)$, and $\mathcal{V}_k[r](L)$ with the dual right-action.

Lemma 6.4.8 *For all $r > 0$, one has a trivial $S_0(p)$-equivariant injective map $\mathcal{P}_k(L) \to \mathcal{P}_k[r](L)$, and a dual trivial $S_0(p)$-equivariant surjective map $\mathcal{V}_k(L) \to \mathcal{V}_k[r](L)$. These maps are isomorphisms if $r \geq 1$.*

Proof A polynomial with coefficients in L and degree at most k defines a function $\mathbb{Z}_p \to L$ that lies in $\mathcal{P}_k[r](L)$ for all r. Thus we get a map $\mathcal{P}_k(L) \to \mathcal{P}_k(r)(L)$, which is clearly injective. If $r \geq 1$, a function in $\mathcal{P}_k[r](L)$ coincides with a polynomial of degree at most k on $\bar{B}(0, 1) \supset \mathbb{Z}_p$, which proves the surjectivity of $\mathcal{P}_k(L) \to \mathcal{P}_k[r](L)$. The fact that this map is $S_0(p)$-equivariant follows from the fact that as we have already noticed, the action of $S_0(p)$ on $\mathcal{A}_k[r](L)$ and on $\mathcal{P}_k[r]$ are given by the same formulas. The assertion for $\mathcal{V}_k$ follows. □

Recall that $\mathcal{P}_k(L)$ and $\mathcal{V}_k(L)$ (with their action of $S \subset S_0(p)$) were defined in Sect. 5.1.2.

Definition 6.4.9 For any $r > 0$, we denote by ρ_k the $S_0(p)$-equivariant surjective maps $\mathcal{D}_k[r](L) \to \mathcal{V}_k[L]$ obtained by composing the morphism $\mathcal{D}_k[r](L) \to \mathcal{V}_k[r](L)$ and $\mathcal{V}_k[r](L) \to \mathcal{V}_k(L)$ of the two preceding lemmas. For $r \geq 0$, we also denote by ρ_k the map $\mathcal{D}_k^\dagger[r](L) \to \mathcal{V}_k(L)$ obtained by composing $\mathcal{D}_k^\dagger[r](L) \hookrightarrow \mathcal{D}_k[r'](L)$ for some $r' > r$ with $\rho_k : \mathcal{D}_k[r'](L) \to \mathcal{V}_k(L)$.

Exercise 6.4.10 Show that the maps $\rho_k : \mathcal{D}_k^\dagger[r](L) \to \mathcal{V}_k(L)$ commute with the restriction maps $\mathcal{D}_k[r](L) \to \mathcal{D}_k[r'](L)$ for $r' > r$. Show that $\rho_k : \mathcal{D}_k^\dagger[r](L) \to \mathcal{V}_k(L)$ is independent of the choice of r' used to define it, and is surjective.

6.4.4 The Fundamental Exact Sequence for Overconvergent Functions

In this section, we keep assuming that $R = L$ is a finite extension of $\mathbb{Q}_p$, normed by its absolute value.

Let $k \in \mathbb{N}$. Let us consider the map $(\frac{d}{dz})^{k+1} : \mathcal{A}[r](L) \to \mathcal{A}[r](L)$ or $\mathcal{A}^\dagger[r](L) \to \mathcal{A}^\dagger[r](L)$. The kernel of this map is obviously the finite-dimensional subspace $\mathcal{P}_k[r](L)$ defined in the preceding Sect. 6.4.3.

Lemma 6.4.11 *The map $(\frac{d}{dz})^{k+1} : \mathcal{A}^\dagger[r](L) \to \mathcal{A}^\dagger[r](L)$ is surjective, continuous and open.*

Proof Let E be a finite set as in Proposition 6.1.14. If $f \in \mathcal{A}^\dagger[r](L)$, then at any point $e \in E$, $f(z)$ has a Taylor expansion of the form $f(z) = \sum_{n\geq 0} a_n(e)(z-e)^n$, with $|a_n(e)|(r')^n \to 0$ for some $r' > r$. Then

$$f = (\frac{dg}{dz})^{k+1}, \text{ with } g(z) = \sum_{n\geq k+1} \frac{a_{n-k-1}(e)}{n(n-1)\dots(n-k)}(z-e)^n.$$

For any r'' such that $r < r'' < r'$, $|\frac{a_{n-k-1}(e)}{n(n-1)\dots(n-k)}|r''^n \to 0$. Hence $g \in \mathcal{A}^\dagger[r]$. This proves the surjectivity. We observe also that there is a constant $C(r', r'') > 0$ depending on r' and r'', but not on f, such that

$$\|g\|_{r''} \leq C\|f\|_{r'} \tag{6.4.2}$$

Since the map $(\frac{d}{dz})^{k+1} : \mathcal{A}[r'](L) \to \mathcal{A}[r'](L)$ is clearly continuous for all r', and so is the post-composition $(\frac{d}{dz})^{k+1} : \mathcal{A}[r'] \to \mathcal{A}^\dagger[r]$ of this map with the continuous inclusion $\mathcal{A}[r'](L) \to \mathcal{A}^\dagger[r](L)$. Hence the inductive limit of those maps, $(\frac{d}{dz})^{k+1} : \mathcal{A}^\dagger[r](L) \to \mathcal{A}^\dagger[r](L)$ is also continuous by Schneider [105, Lemma 5.1.i].

To prove that $(\frac{d}{dz})^{k+1}$ is open, we need only to prove that it sends any open lattice on a neighborhood of 0. An open lattice in $\mathcal{A}^\dagger[r]$ is by definition a lattice Λ such that $\Lambda \cap \mathcal{A}[r''](L)$ is open in $\mathcal{A}[r''](L)$ for all $r'' > r$. Hence $\Lambda \cap \mathcal{A}[r''](L)$ contains an open ball of center 0 and some radius R (depending on r'') in $\mathcal{A}[r''](L)$. But by the first paragraph of this proof, and in particular by (6.4.2), the image of Λ by $(\frac{d}{dz})^{k+1}$ contains an open ball of radius R/C in $\mathcal{A}[r'](L)$ for all $r' > r''$, hence is open. □

Exercise 6.4.12 Prove that the map $(\frac{d}{dz})^{k+1} : \mathcal{A}[r](L) \to \mathcal{A}[r](L)$ is not surjective. (This is the main reason we need to work with the Fréchet spaces $\mathcal{A}^\dagger[r](L)$, rather than with the simpler Banach spaces $\mathcal{A}[r](L)$.)

We have established the existence of an exact sequence:

$$0 \to \mathcal{P}_k[r](L) \to \mathcal{A}^\dagger[r](L) \overset{(\frac{d}{dz})^{k+1}}{\longrightarrow} \mathcal{A}^\dagger[r](L) \to 0 \tag{6.4.3}$$

If put $\mathcal{A}^\dagger[r](L)$ its weight k action of $S_0(p)$, then the natural inclusion $\mathcal{P}_k[r](L) \hookrightarrow \mathcal{A}_k^\dagger[r](L)$ is $S_0(p)$-equivariant. Can we make the second arrow of the exact sequence, $(\frac{d}{dz})^{k+1}$, equivariant as well?

Lemma 6.4.13 *The exact sequence*

$$0 \to \mathcal{P}_k(L) \to \mathcal{A}_k^\dagger(L)[r] \overset{(\frac{d}{dz})^{k+1}}{\longrightarrow} \mathcal{A}_{-2-k}^\dagger(L)[r](k+1) \to 0$$

is $S_0(p)$-equivariant, where the $(k+1)$ means, as above, that the action of $S_0(p)$ is twisted by $\det^{k+1}$.

Proof We need to prove that for all $f \in \mathcal{A}^\dagger[r]$, and $\gamma \in S_0(p)$, one has

$$\frac{d^{k+1}}{dz^{k+1}}\left((a-cz)^k f\left(\frac{dz-b}{a-cz}\right)\right) = \det\gamma^{k+1}(a-cz)^{-2-k}\frac{d^{k+1}f}{dz^{k+1}}\left(\frac{dz-b}{a-cz}\right). \tag{6.4.4}$$

We prove (6.4.4) by induction on k, the formula being clearly true for $k = 0$. Assuming the formula is true for $k' < k$ (for all f), we compute using the product rule

$$\frac{d^{k+1}}{dz^{k+1}}\left((a-cz)^k f\left(\frac{dz-b}{a-cz}\right)\right) = \frac{d^k}{dz^k}\left(-ck(a-cz)^{k-1} f\left(\frac{dz-b}{a-cz}\right)\right) + \frac{d^k}{dz^k}\left((\det\gamma)(a-cz)^{k-2} f'\left(\frac{dz-b}{a-cz}\right)\right)$$

By induction, using (6.4.4) for $k' = k-1$, the first term of the sum is

$$\frac{d^k}{dz^k}\left(-ck(a-cz)^{k-1} f\left(\frac{dz-b}{a-cz}\right)\right) = -ck(a-cz)^{-1-k}(\det\gamma)^k\frac{d^k f}{dz^k}\left(\frac{dz-b}{a-cz}\right).$$

By induction also, using (6.4.4) for $k' = k-2$ and f replaced by f', the second term is

$$\begin{aligned}\frac{d^k}{dz^k}\left((\det\gamma)(a-cz)^{k-2} f'\left(\frac{dz-b}{a-cz}\right)\right) &= \frac{d}{dz}\left(\det\gamma^k(a-cz)^{-k}\frac{d^k}{dz^k}\left(f\left(\frac{dz-b}{a-cz}\right)\right)\right)\\ &= ck\det\gamma^k(a-cz)^{-k-1}\frac{d^k f}{dz^k}\left(\frac{dz-b}{a-cz}\right)\\ &+ \det\gamma^{k+1}(a-cz)^{-2-k}\frac{d^{k+1}f}{dz^{k+1}}\left(\frac{dz-b}{a-cz}\right)\end{aligned}$$

Adding the two terms, we find (6.4.4). □

6.4.5 The Fundamental Exact Sequence for Overconvergent Distributions

We still assume that L is a finite extension of $\mathbb{Q}_p$. In this case, the strong dual of $\mathcal{A}^\dagger[r](L)$ is $\mathcal{D}^\dagger[r](L)$ (Proposition 6.2.9). We call $\theta_k : \mathcal{D}^\dagger[r](L) \to \mathcal{D}^\dagger[r](L)$ the transpose of the map $\frac{d^{k+1}}{dz^{k+1}} : \mathcal{A}^\dagger[r](L) \to \mathcal{A}^\dagger[r](L)$. The transpose of the inclusion map $\mathcal{P}_k[r](L) \hookrightarrow \mathcal{A}^\dagger[r](L)$ has already been given a name: it is $\rho_k : \mathcal{D}^\dagger[r](L) \to \mathcal{V}_k[r](L)$. Since the functor "strong dual" is contravariant, applied to the exact sequence (6.4.3), it gives a complex

$$0 \longrightarrow \mathcal{D}^\dagger_{-2-k}[r](L)(k+1) \xrightarrow{\theta_k} \mathcal{D}^\dagger_k[r](L) \xrightarrow{\rho_k} \mathcal{V}_k[r](L) \longrightarrow 0.$$

Theorem 6.4.14 *The complex of Frechet vector spaces*

$$0 \longrightarrow \mathcal{D}^\dagger_{-2-k}[r](L)(k+1) \xrightarrow{\theta_k} \mathcal{D}^\dagger_k[r](L) \xrightarrow{\rho_k} \mathcal{V}_k[r](L) \longrightarrow 0$$

is an exact sequence, and it is $S_0(p)$-equivariant.

Proof The injectivity of θ_k follows trivially from the surjectivity of $\frac{d^{k+1}}{dz^{k+1}}$. The surjectivity of ρ_k follows from Hahn-Banach's theorem, see Exercise 6.2.2. Let us prove exactness in the middle. Let $\mu \in \mathcal{D}^\dagger[r](L)$. If $\rho_k(\mu) = 0$, then $\mu(\mathcal{P}_k[r](L)) = 0$, hence $\mu = \mu' \circ \frac{d^{k+1}}{dz^{k+1}}$ for a unique linear form μ' on $\mathcal{A}^\dagger[r](L)$. The form μ' is continuous because $\frac{d^{k+1}}{dz^{k+1}}$ is open. Hence $\mu' \in \mathcal{D}^\dagger[r](L)$ and $\theta_k(\mu') = \mu$. □

Exercise 6.4.15 Why is the $(k+1)$ not becoming a $-1-k$ when we dualize?

Exercise 6.4.16 Show that the map $\Theta_k : \mathcal{D}^\dagger[r](L) \to \mathcal{D}^\dagger[r](L)$ is strict (cf. Exercise 6.2.3).

6.5 Rigid Analytic and Overconvergent Modular Symbols

In all this section we fix an integer N such that $p \nmid N$ and a congruence subgroup Γ such that $\Gamma_1(N) \subset \Gamma \subset \Gamma_0(N)$.

6.5.1 Definitions and Compactness of U_p

In all this §, R is a commutative Banach algebra over $\mathbb{Q}_p$, and we fix a weight $\kappa \in \mathcal{W}(R)$.

Definition 6.5.1 For R a commutative Banach algebra over $\mathbb{Q}_p$, $\kappa \in \mathcal{W}(R)$, and $0 < r < r(\kappa)$ (resp. $0 \leq r < r(\kappa)$) we define the space of *rigid analytic modular symbols* (resp. *overconvergent modular symbols*) of weight κ as $\mathrm{Symb}_\Gamma(\mathcal{D}_\kappa[r](R))$ (resp. $\mathrm{Symb}_\Gamma(\mathcal{D}^\dagger_\kappa[r](R))$).

The module $\mathrm{Symb}_\Gamma(\mathcal{D}_\kappa[r](R))$ is naturally a Banach R-module for the norm $\|\Phi\|_r = \sup_{D\in\Delta_0} \|\Phi(D)\|_r$ for $\Phi \in \mathrm{Symb}_\Gamma(\mathcal{D}_\kappa[r](R))$, which makes sense because of

Lemma 6.5.2 *The set of* $\|\Phi(D)\|_r$ *for* $D \in \Delta_0$ *is bounded.*

Proof By Manin's lemma (Exercise 4.1.1), we can choose a finite family of generators $(D_i)_{i\in I}$ of Δ_0 under $\mathbb{Z}[\Gamma]$. That is, any $D \in \Delta_0$ can be written $\sum_{i\in I, j\in J_i} n_{i,j}\gamma_{i,j} \cdot D_i$ where I and the J_i are finite sets, and for $j \in J_i$, $\gamma_{i,j} \in \Gamma$ and $n_{i,j} \in \mathbb{Z}$. It is clear that $\|\Phi(D)\|_r \leq \sup_{i,j} \|\Phi(D_i)_{|\gamma_{i,j}}\|_r = \sup_{i\in I} \|\Phi(D_i)\|_r$ since the $\gamma_{i,j}$ act by isometries (Proposition 6.4.5). The result follows since I is finite. □

From this proof we have learned more than just what we wanted to show:

Scholium 6.5.3 *We can choose divisors* $D_i \in \Delta_0$*, for* $i \in I$*,* I *finite, such that the map* $\Phi \mapsto (\Phi(D_i))_{i\in I}$, $\mathrm{Symb}_\Gamma(\mathcal{D}_\kappa[r](R)) \to \mathcal{D}[r](R)^I$ *is a closed isometric embedding.*

Since $S_0(p)$ acts on $\mathcal{D}_\kappa[r](R)$ and $\mathcal{D}^\dagger_\kappa[r](R)$, we get an action of $\mathcal{H}(S_0(p), \Gamma)$ on $\mathrm{Symb}_\Gamma(\mathcal{D}_\kappa[r](R))$ and $\mathrm{Symb}_\Gamma(\mathcal{D}^\dagger_\kappa[r](R))$. The Hecke operators $T_l = [\Gamma\begin{pmatrix}1 & 0\\ 0 & l\end{pmatrix}\Gamma]$ for $l \nmid N$, $U_l = [\Gamma\begin{pmatrix}1 & 0\\ 0 & p\end{pmatrix}\Gamma]$ for $l \mid N$, and the Diamond operators $\langle a\rangle$ for $a \in (\mathbb{Z}/N\mathbb{Z})^*$ belong to $\mathcal{H}(S_0(p), \Gamma)$, and we denote, as above, by $\mathcal{H}$ the commutative subring they generate.

Lemma 6.5.4 *The Hecke operators* T_l, U_l, $\langle a\rangle \in \mathcal{H}$ *are bounded on* $\mathrm{Symb}_\Gamma(\mathcal{D}_\kappa[r](R))$ *and have norm at most one. For* U_p*, we have the following more precise result: if* $\Phi \in \mathrm{Symb}_\Gamma(\mathcal{D}_\kappa[r](R))$*, then* $\Phi_{|U_p} \in \mathrm{Symb}_\Gamma(\mathcal{D}_\kappa[r/p](R))$ *and*

$$\|\Phi_{|U_p}\|_r \leq \|\Phi_{|U_p}\|_{r/p} \leq \|\Phi\|_r.$$

Proof If T is any Hecke operator, we have by definition $\Phi_{|T}(D) = \sum \phi(D_{|\gamma})_{|\gamma}$, where the sum is a finite sum for some $\gamma \in S_0(p)$. But if p^n divides exactly $\det\gamma$, we have by Proposition 6.4.5:

$$\|\Phi(D_{|\gamma})_\gamma\|_{r/p^n} = \|\Phi(D_{|\gamma})\|_r \leq \|\Phi\|_r.$$

For T an Hecke operator distinct of U_p, all the γ's intervening in the sum have determinant 1, and the result is clear. For $T = U_p$, writing $\gamma_a = \begin{pmatrix} 1 & a \\ 0 & p \end{pmatrix}$, we get

$$\begin{aligned}\|\Phi_{|U_p}(D)\|_{r/p} &= \|\sum_{a=0}^{p-1} \phi(D_{|\gamma_a})_{|\gamma_a}\|_{r/p} \\ &\leq \|\Phi(D_{|\gamma_a})\|_r \\ &\leq \|\Phi\|_r\end{aligned}$$

hence the second inequality in the displayed formula. The first inequality is obvious. □

Remark 6.5.5 It might seem at first glance that in the above proof, the inequality $\|\phi(D_\gamma)\|_{p^{-n}r} \leq \|\phi(D_\gamma)\|_r$ is gross when $n \geq 1$ and can be improved. But this is not true, as for the constant function 1 for example, $\|1\|_{p^{-n}r} = \|1\|_r$. Thus, the upper bound we get for the norm of U_p for example, namely 1, cannot be improved, at least this way. This estimate will play a fundamental role in the proof of Stevens' control theorem below.

Theorem 6.5.6 (Stevens) *Let $0 < r < \min(r(\kappa), p)$. Let ν be a real number.*

(i) *The operator U_p acts compactly on $Symb_\Gamma(\mathcal{D}_\kappa[r](R))$.*
(ii) *The module $Symb_\Gamma(\mathcal{D}_\kappa[r](R))^{\leq\nu}$ is projective of finite rank.*
(iii) *The natural maps (that is, induced by the transposed restriction maps)*

$$Symb_\Gamma(\mathcal{D}_\kappa[r_1](R))^{\leq\nu} \to Symb_\Gamma(\mathcal{D}_\kappa[r_2](R))^{\leq\nu}$$

are isomorphisms for any $0 < r_1 \leq r_2 < \min(r(\kappa), p)$.

Proof Let $\Phi \in \mathrm{Symb}_\Gamma(\mathcal{D}_\kappa[r](R))$ and $D \in \Delta_0$. We have

$$\Phi_{|U_p}(D) = \sum_{a=0}^{p-1} \Phi(\beta(a,p)\cdot D)_{|_\kappa \beta(a,p)}$$

where $\beta(a,p) = \begin{pmatrix} 1 & a \\ 0 & p \end{pmatrix}$. Since $\det(\beta_{a,p}) = p$, it follows from Proposition 6.4.5 that $\mu_{|_\kappa \beta(a,p)} \in \mathcal{D}[r/p](R)$ for every distribution $\mu \in \mathcal{D}[r](R)$. Hence $\Phi_{|U_p}$ actually belongs to $\mathrm{Symb}_\Gamma(\mathcal{D}_\kappa[r/p](R))$. In other words, the map U_p factors through:

$$\mathrm{Symb}_\Gamma(\mathcal{D}_\kappa[r](R)) \to \mathrm{Symb}_\Gamma(\mathcal{D}_\kappa[r/p](R)) \stackrel{\text{restriction}}{\longrightarrow} \mathrm{Symb}_\Gamma(\mathcal{D}_\kappa[r](R)). \tag{6.5.1}$$

Therefore, to prove (i) it suffices to prove that the second map is compact. But by Scholium 6.5.3, this map is the restriction to a closed subspace of the restriction map $\mathcal{D}[r'](R)^I \to \mathcal{D}[r](R)^I$ (for I a finite set), which is compact by Lemma 6.1.22.

The assertion (ii) follows from (i) by applying Proposition 3.4.2.

To prove (iii) we only need to prove that for r, r' such that $0 < r \le r' \le rp$, and $r' < p$, the restriction map $\mathrm{Symb}_\Gamma(\mathcal{D}_\kappa[r])^{\le\nu} \to \mathrm{Symb}_\Gamma(\mathcal{D}_\kappa[r'])^{\le\nu}$ is an isomorphism. The injectivity follows from the injectivity of $\mathcal{D}[r'] \to \mathcal{D}[r]$ and the left exactness of Symb_Γ. Since any vector in $\mathrm{Symb}_\Gamma(\mathcal{D}_\kappa[r'])^{\le\nu}$ is in the image of U_p, the surjectivity follows from (6.5.1). □

Exercise 6.5.7 Show that U_p is injective on $\mathrm{Symb}_\Gamma(\mathcal{D}_\kappa^\dagger(L))$.

6.5.2 *Space of Overconvergent Modular Symbols of Finite Slope*

In all this subsection, L will be a finite extension of $\mathbb{Q}_p$, and we shall work with various vector spaces over L, all provided with an operator U_p. If V is a Banach space over L (hence potentially orthonormalizable), and U_p is compact, we have already defined for each real ν, *the part of slope* $\le \nu$ of V, denoted $V^{\le\nu}$. We shall need to extend the definition to vector spaces V that are not Banach (e.g. to Fréchet spaces), and we do this as follows:

Definition 6.5.8 Let ν be a real number.

(i) We say that a monic polynomial $P(X) \in L[X]$ has slope $\le \nu$ (resp. $< \nu$) if all its roots in some extension of L have p-adic valuations $\le \nu$ (resp. $< \nu$)
(ii) If V is any L-vector space with a linear operator U_p acting on it, we define the subspace of vectors of slope $\le \nu$ (resp. $< \nu$) than ν, denoted by $V^{\le\nu}$ (resp. $V^{<\nu}$), as the sum of the subspaces $\ker P(U_p)$ where P runs among monic polynomials of slope $\le \nu$ (resp. $< \nu$).

When V is a Banach space and U_p a compact operator, this new definition of $V^{\le\nu}$ coincides with the one given in Definition 3.4.1 in view of Lemma 3.4.4. In this case $V^{\le\nu}$ is finite dimensional by Proposition 3.4.2. This new definition allows us to talk of $\mathrm{Symb}_\Gamma(\mathcal{D}_k^\dagger(L))^{\le\nu}$ or $\mathrm{Symb}_\Gamma(\mathcal{D}_k^\dagger(L))^{<\nu}$.

Lemma 6.5.9 *The natural maps*

$$Symb_\Gamma(\mathcal{D}_k^\dagger[r_1](L))^{\le\nu} \to Symb_\Gamma(\mathcal{D}_k[r_2](K))^{\le\nu}$$

are isomorphisms for any $0 \le r_1 < r_2 \le \min(r(\kappa), p)$.

Proof This follows immediately from (iii) of Theorem 6.5.6. □

Lemma 6.5.10 *The associations* $V \mapsto V^{\leq \nu}$ *and* $V \mapsto V^{<\nu}$ *define left exact functors from the category of* L*-vector spaces with an endomorphism* U_p *to the category of* L*-vector spaces.*

Proof This is obvious. □

Exercise 6.5.11 Prove that in general, the functor $V \mapsto V^{\leq \nu}$ is not right exact.

The following proposition gives a control on how the norms behave in the equality of vector spaces $\mathrm{Symb}_\Gamma(\mathcal{D}_\kappa[r_1](L))^{\leq \nu} = \mathrm{Symb}_\Gamma(\mathcal{D}_\kappa[r_2](L))^{\leq \nu}$. It shall not be used before Chap. 8.

Proposition 6.5.12 *For any* $0 < r_1 < r_2 \leq \min(r(\kappa), p)$ *as in (iii), there exists a real constant* D*,* $0 < D < 1$ *depending on* r_1 *and* r_2 *and of* ν*,* ***but not of*** κ*, such that if* $\Phi \in \mathrm{Symb}_\Gamma(\mathcal{D}_\kappa[r_1](L))^{\leq \nu} = \mathrm{Symb}_\Gamma(\mathcal{D}_\kappa[r_2](L))^{\leq \nu}$ *we have*

$$D\|\Phi\|_{r_1} \leq \|\Phi\|_{r_2} \leq \|\Phi\|_{r_1}$$

Proof We obviously have $\|\Phi\|_{r_2} \leq \|\phi\|_{r_1}$. By an easy induction we may assume that Φ is an eigenvector for U_p, that is that $\Phi_{|U_p} = \lambda U_p$ with $v_p(\lambda) \leq \nu$, in other words with $|\lambda^{-1}| \leq p^\nu$. Choose an integer n large enough so that $r_2/p^n \leq r_1$. We have

$$\begin{aligned}
\|\Phi\|_{r_1} &\leq \|\Phi\|_{r_2/p^n} \\
&= \|\lambda^{-n}\Phi_{|U_p^n}\|_{r_2/p^n} \\
&= \|\lambda^{-n}\Phi\|_{r_2} \quad \text{(by Proposition 6.5.4)} \\
&= p^{n\nu}\|\Phi\|_{r_2}.
\end{aligned}$$

Hence $D\|\Phi\|_{r_1} \leq \|\Phi\|_{r_2}$ with $D = p^{n\nu}$, and clearly D does not depend of κ. □

Let us also note a result similar to but simpler than (iii) Theorem 6.5.6.

Proposition 6.5.13 *For all* $r < p$*, and all real* ν*, the natural map*

$$\mathrm{Symb}_\Gamma(\mathcal{V}_k(L))^{\leq \nu} \to \mathrm{Symb}_\Gamma(\mathcal{V}_k[r](L))^{\leq \nu}$$

is an isomorphism.

Proof For $r \geq 1$ this is clear because $\mathcal{V}_k[r](L) = \mathcal{V}_k(L)$. Otherwise, it suffices to prove that for r, r' such that $0 < r \leq r' \leq rp$, and $r' < p$, the restriction map $\mathrm{Symb}_\Gamma(\mathcal{V}_k[r])^{\leq \nu} \to \mathrm{Symb}_\Gamma(\mathcal{V}_k[r'])^{\leq \nu}$ is an isomorphism, which is proved exactly as (iii) of the above proposition. □

Let us finish this section by another simple and nice result of Stevens relating the slope and the rate of growth:

Proposition 6.5.14 (Stevens) *Let* $\Phi \in \mathrm{Symb}_\Gamma(\mathcal{D}^\dagger(L))^{\leq \nu}$ *be a modular symbol. Then for all* $D \in \Delta_0$*, the distribution* $\Phi(D)$ *has order of growth* $\leq \nu$*.*

Proof We may assume that $U_p\Phi = \alpha\Phi$ for some $\alpha \in L^*$. Then $|\alpha|^n \|\Phi\|_r = \|U_p^n \Phi\|_r = \|\Phi\|_{r/p^n}$ so $\|\Phi\|_{r/p^n} = O(p^{-nv_p(\alpha)})$ from which we deduce $\|\Phi\|_{r'} = O(r'^{-v_p(\alpha)})$. The result then follows from Lemma 6.2.12. □

6.5.3 Computation of an H_0

In this section, L is a finite extension of $\mathbb{Q}_p$.

Lemma 6.5.15 *Let Δ be the operator of discrete differentiation on $\mathcal{A}^\dagger[r](L)$:*

$$\Delta f := f(z+1) - f(z).$$

The sequence

$$0 \to L \to \mathcal{A}^\dagger[r](L) \xrightarrow{\Delta} \mathcal{A}^\dagger[r](L) \to 0$$

is exact, and Δ is continuous and open.

Proof The exactness in the middle expresses the fact that if $\Delta f = 0$, i.e. $f(z+1) = f(z)$ for all $z \in \mathbb{Z}_p$, then f is constant; this is clear as we deduce $f(z+n) = f(z)$ for all $n \in \mathbb{Z}$ by induction, and then $f(z+a) = f(z)$ for all $a \in \mathbb{Z}_p$ by continuity, therefore f constant on $\mathbb{Z}_p$. The exactness at the right, that is the surjectivity of Δ is a little harder, and quite analog to Lemma 6.4.11: if $f \in \mathcal{A}^\dagger[r](L)$, then it is in $\mathcal{A}[r'](L)$ for some $r' > r$, we can write formally for any $e \in \mathbb{Z}_p$, $f(z) = \sum_{n\geq 0} b_n(e)(z-e)^{(n)}$ where $x^{(n)} := x(x-1)\dots(x-n+1)$. This is easily seen from the writing $f(z) = \sum_{n\geq 0} a_n(e)(z-e)^n$ using that there is a triangular matrix with entries in $\mathbb{Z}_p$ and diagonal terms equal to 1 that transforms the basis $((z-e)^n)_n$ of the space of polynomials into the basis $((z-e)^{(n)})_n$, and one sees the same way that for any ρ, $|b_n(e)|\rho^n$ goes to 0 if and only if $|a_n(e)|\rho^n$ goes to 0. In particular, the series $\sum_{n\geq 0} b_n(e)(z-e)^{(n)}$ converges to $f(z)$ for $|z-e| \leq r'$. Let us now define g by the formula (for $e \in \mathbb{Z}_p$) $g(z) = \sum_{n=1}^{\infty} \frac{b_{n-1}(e)(z-e)^{(n)}}{n}$. It is immediately seen that this series converges for $|z-e| < r''$ where r'' is any real such that $r'' < r'$, and for a given z is independent on e. Hence, choosing an r'' such that $r < r'' < r'$, we have defines a $g \in \mathcal{A}^\dagger[r''](L) \subset \mathcal{A}^\dagger[r](L)$ and we see easily that $\Delta g = f$ because $\Delta(z-e)^{(n)} = (z-e)^{(n-1)}$. This proves that Δ is surjective. The same argument as in Lemma 6.4.11 also proves that Δ is continuous and open. □

Lemma 6.5.16 *For R any commutative Banach algebra over $\mathbb{Q}_p$, we have an exact sequence*

$$0 \longrightarrow \mathcal{D}_k^\dagger[r](R) \xrightarrow{\Delta^*} \mathcal{D}_k^\dagger[r](R) \xrightarrow{\mu \mapsto \mu(1)} R \longrightarrow 0,$$

where the map Δ^*, transpose of the map Δ defined above, is strict (cf. Exercise 6.2.3).

Proof To obtain the lemma in the case where $R = L$ is a finite extension of $\mathbb{Q}_p$, we just dualize the exact sequence of the above lemma (cf. the proof of Theorem 6.4.14 to see why we still get an exact sequence), and we note that Δ^* is strict since it is a morphism of Banach space with closed image (Exercise 6.2.3, question 5). For general R, we tensorize the exact sequence for $\mathbb{Q}_p$ by R and complete: the sequence remains exact and Δ^* remains strict by questions 6 and 7 of Exercise 6.2.3. □

Lemma 6.5.17 (Pollack-Stevens) *For every $k \in \mathbb{Z}$, every $r > 0$, $H_0(\Gamma, \mathcal{D}_k^\dagger[r](L)) = 0$, excepted for $k = 0$ where $H_0(\Gamma, \mathcal{D}_0^\dagger[r](L)) = L$.*

Proof Recall that by definition $H_0(\Gamma, V) = V/IV$, where I is the ideal in $\mathbb{Z}[G]$ generated by the element $\gamma - 1$ for $\gamma \in \Gamma$. Let $g = \begin{pmatrix} 1 & 1 \\ 0 & 1 \end{pmatrix} \in \Gamma$. Then g acts on $\mathcal{D}_k^\dagger[r](L)$ as Δ. From the exact sequence of the above lemma, we see that $\mu \mapsto \mu(1)$ defines an isomorphism $\mathcal{D}_k^\dagger[r](L)/(g-1)\mathcal{D}_k^\dagger[r](L) \simeq L$.

If $k = 0$, one sees at once that for all $\mu \in \mathcal{D}_k^\dagger[r](L)$ and all $\gamma \in \Gamma$, $\mu_{|(\gamma-1)}(1) = \mu(1) - \mu(1) = 0$, hence $\mathcal{D}_k^\dagger[r](L)/I\mathcal{D}_k^\dagger[r](L) \simeq L$.

Now assume that $k \neq 0$. Let us consider the distribution δ_0' in $\mathcal{D}_k^\dagger[r](L)$ which sends $f \in \mathcal{A}_k^\dagger[r](L)$ to $f'(0)$. For $\gamma \in \Gamma$, let $\mu_\gamma = (\delta_0')_{|(\gamma-1)} \in I\mathcal{D}_k^\dagger[r]$. We compute $\mu_\gamma(1) = \delta_0'((a-cz)^k - 1) = kca^{k-1}$. Hence the image of μ_γ in L is non-zero since $k \neq 0$ provided we chose a γ such that $ac \neq 0$, which is always possible. Hence when $k \neq 0$, $\mathcal{D}_k^\dagger[r]/(g-1)\mathcal{D}_k^\dagger[r] = 0$. □

6.5.4 The Fundamental Exact Sequence for Modular Symbols

Proposition 6.5.18 (Pollack-Stevens) *Let $k > 0$ be an integer, L a finite extension of $\mathbb{Q}_p$, $r > 0$. We have an exact sequence, compatible with the Hecke operators*

$$0 \to \mathrm{Symb}_\Gamma(\mathcal{D}_{-2-k}^\dagger[r](L))(k+1) \to \mathrm{Symb}_\Gamma(\mathcal{D}_k^\dagger[r](L)) \stackrel{\rho_k}{\to} \mathrm{Symb}_\Gamma(\mathcal{V}_k[r](L)) \to 0,$$

where the $(k+1)$ after $\mathrm{Symb}_\Gamma(\mathcal{D}_{-2-k}^\dagger[r])$ means that the action of an Hecke operator $[\Gamma s \Gamma]$ for $s \in S_0(p)$ is $(\det s)^{k+1}$ times what it would be otherwise, and ρ_k is the map induced by the map $\mathcal{D}_k^\dagger[r](L) \to \mathcal{V}_k(L)$ dual of the natural inclusion $\mathcal{P}_k(L) \subset \mathcal{A}_k[r](L)$.

Let $\nu > 0$. Assume $r < p$. We have an exact sequence, compatible with the Hecke operators:

$$0 \to \mathrm{Symb}_\Gamma(\mathcal{D}_{-2-k}^\dagger[r](L))(k+1)^{\leq \nu} \to \mathrm{Symb}_\Gamma(\mathcal{D}_k^\dagger[r](L))^{\leq \nu} \stackrel{\rho_k}{\to} \mathrm{Symb}_\Gamma(\mathcal{V}_k(L))^{\leq \nu} \to 0,$$

and similarly with $\leq \nu$ replaced by $< \nu$.

Proof We apply the long exact sequence of cohomology with compact support to the fundamental exact sequence, obtaining

$$0 \longrightarrow \mathrm{Symb}_\Gamma(\mathcal{D}^\dagger_{-2-k}[r](L))(k+1) \longrightarrow \mathrm{Symb}_\Gamma(\mathcal{D}^\dagger[r]_k(L))$$
$$\longrightarrow \mathrm{Symb}_\Gamma(\mathcal{V}_k(L)) \longrightarrow H^2_c(\Gamma, \mathcal{D}^\dagger_{-2-k}(L)(k+1))$$

The last term is by Poincaré's duality (forgetting the Hecke action) $H_0(\Gamma, \mathcal{D}^\dagger_{-2-k}[r](L))$, so Lemma 6.5.17 allows us to complete the proof of the first exact sequence.

Applying the left-exact functor $V \mapsto V^{\le \nu}$ (Lemma 6.5.10(i)) to that exact sequence, we get

$$0 \to \mathrm{Symb}_\Gamma(\mathcal{D}^\dagger_{-2-k}[r](L))(k+1)^{\le\nu} \to \mathrm{Symb}_\Gamma(\mathcal{D}^\dagger_k[r](L))^{\le\nu} \to \mathrm{Symb}_\Gamma(\mathcal{V}_k[r](L))^{\le\nu}.$$

Let us prove that the last map $\mathrm{Symb}_\Gamma(\mathcal{D}^\dagger_k[r](L))^{\le\nu} \to \mathrm{Symb}_\Gamma(\mathcal{V}_k[r](L))^{\le\nu}$. is surjective. Let $w \in \mathrm{Symb}_\Gamma(\mathcal{V}_k[r](L))^{\le\nu}$. Then w is in the image of some $v \in \mathrm{Symb}_\Gamma(\mathcal{D}^\dagger_k[r](L))$, and for any $r' > r$, such a v belongs to the Banach space $\mathrm{Symb}_\Gamma(\mathcal{D}_k[r'](L))$ on which U_p acts compactly. Then Lemma 6.5.10(iii) tells us that there is a $v' \in \mathrm{Symb}_\Gamma(\mathcal{D}_k[r'](L))^{\le\nu}$ whose image by the natural map followed by the restriction map is $w \in \mathrm{Symb}_\Gamma(\mathcal{V}_k[r](L))$. By Theorem 6.5.6, $v' \in \mathrm{Symb}_\Gamma(\mathcal{D}^\dagger_k[r](L))^{\le\nu}$. This proves the desired surjectivity. To conclude the proof of the proposition one applies Lemma 6.5.13, which allows us to replace $\mathrm{Symb}_\Gamma(\mathcal{V}_k[r](L))^{\le\nu}$ by $\mathrm{Symb}_\Gamma(\mathcal{V}_k(L))^{\le\nu}$. □

6.5.5 Stevens's Control Theorem

Theorem 6.5.19 *Let L be any finite extension of* $\mathbb{Q}_p$.

(i) *[Stevens's control theorem] The natural map*

$$\rho_k : \mathit{Symb}_\Gamma(\mathcal{D}_k(L))^{<k+1} \longrightarrow \mathit{Symb}_\Gamma(\mathcal{V}_k(L))^{<k+1}$$

is an isomorphism.

(ii) *If we provide both* $\mathit{Symb}_\Gamma(\mathcal{D}_k(L))^{<k+1}$ *and* $\mathit{Symb}_\Gamma(\mathcal{V}_k(L))^{<k+1}$ *of the norm* $\| \,\|_1$, *then* ρ_k *is an isometry.*

Proof We shall give two different proofs of the important result (i). The first one, which is the proof outlined by Stevens in his preprint [120] and completed by Pollack-Stevens [100] is sleek and short, but does not give (ii). From

Proposition 6.5.18, we have an exact sequence

$$0 \to (\mathrm{Symb}_\Gamma(\mathcal{D}_{-2-k}(L))(k+1))^{<k+1} \to \mathrm{Symb}_\Gamma(\mathcal{D}_k(L))^{<k+1} \to \mathrm{Symb}_\Gamma(\mathcal{V}_k(L))^{<(k+1)} \to 0,$$

We can rewrite the first term as $\mathrm{Symb}_\Gamma(\mathcal{D}_{-2-k}(L))^{<0}$, but since U_p acts on $\mathrm{Symb}_\Gamma(\mathcal{D}^\dagger_{-2-k}(L))$ with norm at most 1 (Lemma 6.5.4), this is 0. The result (i) follows.

The second proof is essentially the method of Višik (cf. [124] or [92]), and will give (ii) as well as (i). Let $\phi \in \mathrm{Symb}_\Gamma(\mathcal{V}_k(L))^{<k+1}$. We shall prove that there is a unique $\Phi \in \mathrm{Symb}_\Gamma(\mathcal{D}_k(L))^{<k+1}$ such that $\rho_k(\Phi) = \phi$, and moreover that $\|\Phi\|_1 = \|\phi\|_1$. We may clearly assume that ϕ is a generalized eigenvector of U_p, then by induction a true eigenvector of U_p, of eigenvalues $\lambda \in L$.

First we prove the uniqueness of Φ. Let $D \in \Delta_0$. Write $\mu = \Phi(D)$. Then μ is a distribution of order $\leq \nu$. Moreover for every $n \in \mathbb{N}$, $a \in \mathbb{Z}_p$, and $0 \leq j \leq k$ we have $\mu((z-a)/p^n)^j 1_{a+p^n\mathbb{Z}_p}) = \lambda^{-n}\Phi(D_{|U_p^n})(z^j) = \lambda^{-n}\phi(D_{|U_p^n})(z^j)$. Hence all the values $\mu((z-a)/p^n)^j 1_{a+p^n\mathbb{Z}_p})$ are determined by ϕ, hence μ itself is determined by ϕ by Theorem 6.2.13(i), and so is Φ.

Now we prove the existence of Φ, as follows. For $D \in \Delta_0$ we define $\Phi(D)$ as the unique distribution μ of slope $\leq \nu$ such that for every $n \in \mathbb{N}$, $a \in \mathbb{Z}_p$, and $0 \leq j \leq k = \mu((z-a)/p^n)^j 1_{a+p^n\mathbb{Z}_p}) = \lambda^{-n}\phi(D_{|U_p^n})(z^k)$. Such a distribution exists by Theorem 6.2.13(ii), because the additivity (6.2.2) hypothesis follows from the fact $\phi_{|U_p} = \lambda U_p$ and hypothesis (6.2.3) with $C = \|\phi\|_1$ follows from $|\lambda| \geq p^{-\nu}$; moreover it satisfies $\|\mu\|_1 \leq \|\phi\|_1$. By uniqueness of that μ, it follows easily that Φ is a modular symbol, which clearly satisfies $\rho_k(\Phi) = \phi$ and $\|\Phi\|_1 \leq \|\phi\|_1$. The other equality $\|\phi\|_1 \leq \|\Phi\|_1$ follows from the fact that ρ_k has norm at most one since the natural map $\mathcal{D}[1](L) \to \mathcal{V}_k(L)$ has norm one by the theorem of Hahn-Banach, as it is the dual map of the inclusion map $\mathcal{P}_k(L) \to \mathcal{A}[1](L)$ which is clearly of norm one. □

Exercise 6.5.20 According to Exercise 5.3.37, for $\Gamma = \Gamma_0(N) \cap \Gamma_1(p)$, the boundary modular symbols $\phi_{k,u,v} \in \mathrm{Symb}_\Gamma(\mathcal{V}_k(L))$ with u/v a p-integer belongs to the space $\mathrm{Symb}_\Gamma(\mathcal{V}_k(L))^{k+1}$, hence have a unique preimage by ρ_k, denoted $\Phi_{k,u,v} \in \mathrm{Symb}_\Gamma(\mathcal{D}_k^\dagger)$. Show that $\Phi_{k,u,v}$ has the following description: $\Phi_{k,u,v} \in \mathrm{Hom}(\Delta, \mathcal{V}_k)$ is supported in the Γ-orbit of u/v and such that for $\gamma = \begin{pmatrix} a & b \\ c & d \end{pmatrix} \in \Gamma$, $f \in \mathcal{A}_k^\dagger(L)$

$$\Phi_{k,u,v}\left(\gamma \cdot \frac{u}{v}\right)(f(z)) = f\left(\gamma \cdot \frac{u}{v}\right)(cu+dv)^k. \tag{6.5.2}$$

6.6 The Mellin Transform

6.6.1 The Real Mellin Transform

The real Mellin transform takes a function $g(y)$ of one positive real variable y, with rapid decay near 0 and ∞, to the analytic function of one complex variable

$$s \mapsto M(g)(s) = \int_{\mathbb{R}_+^*} g(y) y^s \frac{dy}{y}.$$

For instance, the L-function of a modular form $f(z)$ is, up to a simple factor $\frac{\Gamma(s)}{(2\pi)^s}$, the Mellin transform of $f(iy)$, see Proposition 5.4.1.

We can see the map $y \mapsto y^s$ as a character of $\mathbb{R}_+^*$, and all continuous characters $\mathbb{R}_+^* \to \mathbb{C}^*$ are of this form. Thus, we can consider the Mellin transform of f, $M(f)$, as a function of a continuous character $\mathbb{R}_+^* \to \mathbb{C}^*$, rather than as a function of s. Moreover $f(iy)\ \frac{dy}{y}$ is a measure on $\mathbb{R}_+^*$. Thus we can see the Mellin transform as a map $\mu \mapsto M(\mu)$ from the set of measures μ on $\mathbb{R}_+^*$ with rapid decay with values in the space of analytic functions on the set of characters $\mathrm{Hom}(\mathbb{R}_+^*, \mathbb{C}^*)$. Actually, since the characters $y \mapsto y^s$ are C^∞, we can even define the Mellin transform $M(\mu)$ of a distribution μ with rapid decay at 0 and ∞.

Exercise 6.6.1 The measure $\frac{dy}{y(e^y-1)}$ is not of rapid decay at 0, but show that you can still define its Mellin transform for $\mathrm{Re}\, s > 1$. Show that its Mellin transform is $\Gamma(s)\zeta(s)$.

6.6.2 The p-Adic Mellin Transform

By analogy, we now define the p-adic Mellin transform over $\mathbb{Q}_p$.

Definition 6.6.2 Let us denote by $\mathcal{R}$ the $\mathbb{Q}_p$-space of analytic functions on the weight space $\mathcal{W}$.

The weight space $\mathcal{W}$ is a disjoint union of finitely many balls $B(1, 1)$ (see Scholium 6.3.3). For any such ball, and any positive number $r < 1$, $r \in p^{\mathbb{Q}}$, an analytic function f attains its maximum on the smaller ball $B(1, r)$, and this maximum defines a semi-norm on $\mathcal{R}$. Provided with the topology defined by all these semi-norms, $\mathcal{R}$ is a Fréchet space.

If R is a Banach algebra, and $\sigma : \mathbb{Z}_p^* \to R^*$ is a continuous character, that is an element of $\mathcal{W}(R)$, then σ extended by 0 on $p\mathbb{Z}_p$, belongs to the set of function $\mathcal{A}[r](R)$ for some $r > 0$. A measure $\mu \in \mathcal{D}^\dagger(\mathbb{Q}_p)$ belongs to $\mathcal{D}[r](\mathbb{Q}_p)$ and defines a measure $(\mu \otimes 1)$ in $\mathcal{D}[r](R)$. This measure can be applied to σ giving a value $(\mu \otimes 1)(\sigma) \in R$. We denote this value by $\int_{\mathbb{Z}_p^*} \sigma(z) d\mu(z)$.

Proposition and Definition 6.6.3 *For any overconvergent distribution $\mu \in \mathcal{D}^\dagger(\mathbb{Q}_p)$, there exists a unique analytic function $M(\mu) \in \mathcal{R}$ such that for every finite extension L of $\mathbb{Q}_p$, and every $\sigma \in \mathcal{W}(L) = \mathrm{Hom}(\mathbb{Z}_p^*, L^*)$,*

$$M(\mu)(\sigma) = \int_{\mathbb{Z}_p^*} \sigma(z) d\mu(z) \in L^*. \tag{6.6.1}$$

The function $M(\mu)$ is called the p-adic Mellin transform of the distribution μ.

Proof The uniqueness of $M(\mu)$ is clear since a function on a reduced rigid analytic space over $\mathbb{Q}_p$ is defined by its value at all L-points for L running among finite extension of $\mathbb{Q}_p$.

For the existence, let us choose an admissible covering of $\mathcal{W}$ by $\mathbb{Q}_p$-affinoids $U_i = \mathrm{Spec}\, R_i$. The natural inclusion $U_i \to \mathcal{W}$ defines a character $\sigma_i : \mathbb{Z}_p^* \to R_i^*$. Thus $\int_{\mathbb{Z}_p^*} \sigma_i(z) d\mu(z)$ is a rigid analytic function $M_i(\mu)$ on U_i, which clearly satisfies (6.6.1) for every $\sigma \in U_i(L)$, $L/\mathbb{Q}_p$ finite extension. Those functions $M_i(\mu)$ on U_i satisfies the gluing condition and therefore defines a function $M(\mu)$ on $\mathcal{W}$, that is an element of $\mathcal{R}$ which satisfies (6.6.1). □

Exercise 6.6.4 Show that $M : \mu \mapsto M(\mu)$, $\mathcal{D}^\dagger(\mathbb{Q}_p) \to \mathcal{R}$ is continuous with respect to the Fréchet topologies. For those not familiar with Fréchet topologies, what you need to show is that for every $r' < 1$, there exist an $r > 0$, and a $C > 0$, such that for all $\mu \in \mathcal{D}^\dagger(\mathbb{Q}_p)$,

$$\sup_{|x-1| \le r'} |M_\mu(x)| < C \|\mu\|_r.$$

6.6.3 Properties of the p-Adic Mellin Transform

For U an open subset of $\mathbb{Z}_p$, and μ a distribution over $\mathbb{Z}_p$, we define the restriction $\mu_{|U}$ of μ to U as the distribution $\mu_{|U}(f) = \mu(f 1_U)$. We say that μ has *support in* U if $\mu = \mu_{|U}$. Obviously, the Mellin transform M_μ depends only of the restriction $\mu_{|\mathbb{Z}_p^*}$ of μ to $\mathbb{Z}_p^*$. We call $\mathcal{D}_U^\dagger(\mathbb{Q}_p)$ the subspace of distributions in $\mathcal{D}^\dagger(\mathbb{Q}_p)$ with support in U. It is clearly closed, hence inherits from $\mathcal{D}^\dagger(\mathbb{Q}_p)$ a structure of Fréchet space.

Definition 6.6.5 Let $\nu > 0$ be a real. We say that an analytic function $h(x) = \sum_{n \ge 0} a_n (x-1)^n$ converging on the open ball $|x - 1| < 1$ *has order* $\le \nu$ if $|a_n| = O(n^\nu)$. We say that a function on $\mathcal{W}$ *has order* $\le \nu$ if its restriction to the $p - 1$ connected components of $\mathcal{W}$ have order $\le \nu$.

Exercise 6.6.6 Show that a function $h(x)$ as in the definition has order 0 if and only if it is bounded on $B(1, 1)$. Show that this is equivalent to saying that $h(x)$ has finitely many zeros on $B(1, 1)$ or is identically 0.

Exercise 6.6.7 Show that $\log(1+x)$ has order 1.

Exercise 6.6.8 Assume that $\nu \in \mathbb{N}$. Show that $h(x)$ has order $\leq \nu$ is equivalent to $\sup_{|x|<r} |h(x)| = O(\sup_{|x|<r} |\log_p(1+x)^\nu|)$.

Exercise 6.6.9 Give an example of function of order $1/2$.

Theorem 6.6.10 (Višik, Amice-Vélu)

(i) *The application $\mu \mapsto M(\mu)$ is an isomorphism of Fréchet spaces from the subspace $\mathcal{D}^\dagger(\mathbb{Q}_p)_{\mathbb{Z}_p^*}$ of $\mathcal{D}^\dagger(\mathbb{Q}_p)$ of distributions with support of $\mathbb{Z}_p^*$ onto $\mathcal{R}$.*

(ii) *Let $\mu \in \mathcal{D}^\dagger(\mathbb{Q}_p)$, and $\nu \geq 0$ a real number. Then μ has order $\leq \nu$ if and only if $M(\mu)$ has order $\leq \nu$.*

Proof See e.g. [42]. □

Proposition 6.6.11 *If μ and μ' are two distributions, such that $\mu' = \Theta_{k+1}\mu$, then*

$$M(\mu')(\sigma) = \log_p^{[k+1]}(\sigma) M(\mu)(\sigma/z^{k+1})$$

for $\sigma \in \mathcal{W}(\mathbb{C}_p)$.

Proof By induction, using point c. of Exercise 6.3.9, we may assume $k = 0$. Let us write $\sigma(z)$ as $\psi(\tau(z))\sigma_p(z)$, where ψ is a character of $(\mathbb{Z}/q\mathbb{Z})^*$, and σ_p is the character of the cyclic group $1+q\mathbb{Z}_p$ sending a generator γ to $x := \sigma(\gamma)$. One then has

$$\begin{aligned} M(\mu')(\sigma) &= M(\mu)\left(\frac{d\sigma(z)}{dz}\right) \\ &= M(\mu)\left(\frac{d\left(\psi(\tau(z))\sigma_p\left(\frac{z}{\tau(z)}\right)\right)}{dz}\right) \\ &= M(\mu)\left(\frac{\psi(\tau(z))}{\tau(z)}\frac{d\sigma_o}{dz}\left(\frac{z}{\tau(z)}\right)\right) \quad \text{since } \tau(z) \text{ is locally constant} \end{aligned}$$

But for $z \in 1 + q\mathbb{Z}_p$, $\sigma_p(z) = \exp(\log(z)\log(x)/\log(\gamma))$, so $\frac{d\sigma_p}{dz}(z) = \frac{\log_p(x)}{z\log_p(\gamma)}\sigma_p(z)$. Therefore

$$\begin{aligned} M(\mu')(\sigma) &= M(\mu)\left(\frac{\psi(\tau(z))}{\tau(z)}\frac{\log_p(x)}{(z/\tau(z))\log_p(\gamma)}\sigma_p(z)\right) \\ &= \log_p(x)/\log_p(\gamma) M(\mu)(\sigma/z) \\ &= \log_p^{[1]}(\sigma) M(\mu)(\sigma/z). \end{aligned}$$

□

6.6.4 The Mellin Transform over a Banach Algebra

Now, let R be any Banach algebra over $\mathbb{Q}_p$. Recall that $\mathcal{D}^\dagger(R) = \mathcal{D}^\dagger(\mathbb{Q}_p)\hat{\otimes}_{\mathbb{Q}_p} R$ by Proposition 6.2.7.

Definition 6.6.12 The Mellin transform over L is the L-linear map

$$M_R : \mathcal{D}^\dagger(R) = \mathcal{D}^\dagger(\mathbb{Q}_p)\hat{\otimes}_{\mathbb{Q}_p} R \to \mathcal{R}\hat{\otimes}_{\mathbb{Q}_p} R$$

defined by $M_R = M\hat{\otimes}\mathrm{Id}_R$.

If R is an affinoid algebra over $\mathbb{Q}_p$, then an element of $\mathcal{R}\hat{\otimes}_{\mathbb{Q}_p} R$ is just a function on the rigid space $\mathcal{W} \times \mathrm{Sp}\, R$ analytic in the first variable and locally analytic in the second. This is using this Mellin transform over R that we shall construct two-variables p-adic L-function: one variable in the weight space $\mathcal{W}$ and one variable in the rigid analytic space $\mathrm{Sp}\, R$, which in the application will be an open set in an eigenvariety: see Chap. 8.

6.7 Applications to the p-Adic L-functions of Non-critical Slope Modular Forms

As usual, $N \geq 1$ is an integer, and we denote by $\mathcal{H}$ the Hecke algebra generated by the $T_l, l \nmid N, U_l, l \mid N$, and $\langle a \rangle, a \in (\mathbb{Z}/N\mathbb{Z})^*$.

Let $f(z) = \sum_{n\geq 1} a_n q^n$, $q = e^{2i\pi z}$ be a cuspidal form of weight $k+2$, level $\Gamma_1(N)$ and character ϵ, which is an eigenform for all Hecke operators $T \in \mathcal{H}$.

The idea of Stevens's method to construct a p-adic L-function attached to f is to see f as a symbol in $\mathrm{Symb}_{\Gamma_1(N)}(\mathcal{V}_k)$ by Manin-Shokurov's theorem and to lift it uniquely as a symbol in $\mathrm{Symb}_{\Gamma_1(N)}(\mathcal{D}_k^\dagger)$ by Stevens control theorem (Theorem 6.5.19). In order to do the latter, we apparently need two things: that the level $\Gamma_1(N)$ be a subgroup of $\Gamma_0(p)$ (that is that $p \mid N$), otherwise $\mathrm{Symb}_{\Gamma_1(N)}(\mathcal{D}_k^\dagger)$ is not even defined; and that the slope of our modular symbol be strictly less than $k+1$. The second condition is an intrinsic limit of the method—see next chapter for how it can be removed. The first condition $p \mid N$ is quite restrictive, but fortunately, there is a way to make it true when it is not to begin with. Actually, and interestingly, there is not one way, but two ways to do so.

6.7.1 Refinements

We start with an f as above, and we assume that we are in the case where p is prime to N. An easy way to remedy to this problem is to consider $f(z)$ as a form of level

$\Gamma := \Gamma_1(N) \cap \Gamma_0(p)$ (and same weight $k+2$), which it is. But as is well known, the form $f(pz)$ is also a form of level Γ (and same weight).

Exercise 6.7.1 Check this fact.

The forms $f(z)$ and $f(pz)$ have the inconvenience of not being eigenvectors for U_p. But if we call α and β the two roots of the equations

$$X^2 - a_p X + p^{k+1}\epsilon(p), \tag{6.7.1}$$

and if we set

$$f_\alpha(z) = f(z) - \beta f(pz) \tag{6.7.2}$$

$$f_\beta(z) = f(z) - \alpha f(pz), \tag{6.7.3}$$

then one has:

Lemma 6.7.2 *The two forms f_α and f_β are eigenforms for U_p of eigenvalues α and β respectively.*

Proof Let V_p be the operator $f(z) \mapsto f(pz)$. One has $U_p V_p = \mathrm{Id}$ and $T_p = U_p + p^{k+1}\langle p\rangle V_p$ (see Sect. 2.6.4). Therefore $U_p f_\alpha = U_p(f - \beta V_f f) = U_p f - \beta f = a_p f - \epsilon(p)p^{k+1}V_p f - \beta f = \alpha f - \alpha\beta V_p f = \alpha f_\alpha$. □

The forms f_α and f_β are also eigenforms for all the other Hecke operators, with the same eigenvalues as f, and for the Diamond operators: their nebentypus is ϵ seen as a non-primitive character of $(\mathbb{Z}/Np\mathbb{Z})^*$—that is, it is trivial at p.

Definition 6.7.3 The forms f_α and f_β are called the *refinements*[4] of f at p.

Remark 6.7.4 It is conjectured, but not known excepted in weight 2 (that is when $k = 0$), that $\alpha \neq \beta$. See [39].

Exercise 6.7.5 If f is a newform then $f(z)$ and $f(pz)$ form a basis of the space of all forms of weight $k+2$, level $\Gamma_1(N) \cap \Gamma_0(p)$, nebentypus ϵ, and with the same eigenvalues as f for operators T_l, for l prime to Np and U_l for $l \neq p$. If $\alpha \neq \beta$, then f_α and f_β are a basis of that space, in which U_p is diagonal. If $\alpha = \beta$, then U_p is not diagonal.

We note that the complex numbers α and β are algebraic by definitions, and actually integral. Choosing an embedding $\bar{\mathbb{Q}} \hookrightarrow \mathbb{C}$ (which determines α and β as elements of $\bar{\mathbb{Q}}$) and then of $\bar{\mathbb{Q}} \to \mathbb{C}_p$ (which allows us to see α and β as elements

[4] Another terminology calls them *p-stabilizations*.

of $\mathbb{C}_p$), we have by looking at Eq. (6.7.1)

$$v_p(\alpha) + v_p(\beta) = k + 1 \tag{6.7.4}$$

$$v_p(\alpha) \geq 0 \tag{6.7.5}$$

$$v_p(\beta) \geq 0 \tag{6.7.6}$$

Hence $0 \leq v_p(\alpha), v_p(\beta) \leq k + 1$. We also have $v_p(a_p) \geq \min(v_p(\alpha), v_p(\beta))$ with equality when $v_p(\alpha) \neq v_p(\beta)$. In particular, if $v_p(\alpha) = 0$, then $v_p(\beta) = k + 1 \neq 0$ and $v_p(a_p) = 0$.

Definition 6.7.6 We say that f is *p-ordinary* if $v_p(a_p) = 0$. If it is, then $v_p(\alpha) = 0$ and $v_p(\beta) = k+1$ up to exchanging α and β, and we call f_α the *ordinary refinement*, and f_β the *critical slope refinement*.

6.7.2 Construction of the p-Adic L-functions

We shall deal at the same time with the cases $p \nmid N$ and $p \mid N$. In both cases, let $\Gamma = \Gamma_1(N) \cap \Gamma_0(p)$ which of course is simply $\Gamma_1(N)$ in the case $p|N$. Let us call $g = f_\alpha$ if $p \nmid N$, where α is one of the root of (6.7.1) with $v_p(\alpha) < k + 1$. We have two choices of such α when f is not p-ordinary (and $\alpha \neq \beta$), but only one when f is p-ordinary. If $p \mid N$, then we set $g = f$. The U_p-eigenvalue of g is $\alpha := a_p$ and we shall assume that $v_p(a_p) < k + 1$.

To summarize, in any case we end up with a normalized cuspidal eigenform g of weight $k+2$, level Γ containing $\Gamma_0(p)$, nebentypus ϵ, and such that $U_p g = \alpha g$ with $\alpha < k + 1$. We now proceed to the construction. Remember that we have chosen embeddings $\bar{\mathbb{Q}} \to \bar{\mathbb{C}}$ and $\bar{\mathbb{Q}} \to \bar{\mathbb{Q}}_p$.

Step 1 To the cuspidal form g we attach the two modular symbols ϕ_g^+/Ω_g^+ and ϕ_g^-/Ω_g^- in $\mathrm{Symb}_\Gamma(\mathcal{V}_k(K_g))$. (Here $K_g \subset \bar{\mathbb{Q}} \subset \mathbb{C}$ is the number field generated by the eigenvalues of g; the symbols $\phi_g^\pm$ were defined above, in Theorem 5.3.19; the periods $\Omega_g^\pm$ are defined in Definition 5.4.9.)

Let L be the finite extension of $\mathbb{Q}_p$ generated in $\bar{\mathbb{Q}}_p$ by the image of K_g by our embedding $\bar{\mathbb{Q}} \hookrightarrow \bar{\mathbb{Q}}_p$. Then we can see $\phi_g^\pm/\Omega_g^\pm$ as elements of $\mathrm{Symb}_\Gamma^\pm(\mathcal{V}_k(L))$.

Note that the periods $\Omega_g^\pm$ and the modular symbols $\phi_g^\pm/\Omega_g^\pm$ are well-defined only up to multiplication by an element of K_g^*. However, as explained in Remark 5.4.10 we can do better if we need to. For instance, if $\mathfrak{p}$ is the place of K_g above p defined by our fixed embedding $K_g \subset L$ (hence L is the completion of K_g at $\mathfrak{p}$), and if $\mathfrak{p}$ is principal, we can require our periods $\Omega^\pm$ to be chosen such that $\phi_g^\pm/\Omega_g^\pm$ are $\mathfrak{p}$-integral modular symbols and not divisible by $\mathfrak{p}$. This defines our modular symbols up to a $\mathfrak{p}$-unit. Even if $\mathfrak{p}$ is not principal, we can do the same if we allow the periods $\Omega^\pm$ to be taken in a suitable finite extension of K_g, or in L.

Step 2 The modular symbols $\phi_g^{\pm}/\Omega_g^{\pm}$ have the same eigenvalues for the Hecke operators as g, in particular they have U_p-eigenvalue α. Since $v_p(\alpha) < k+1$, we have

$$\phi_g^{\pm}/\Omega_g^{\pm} \in \mathrm{Symb}_{\Gamma}^{\pm}(\mathcal{V}_k(L))^{<k+1}$$

Therefore, applying Steven's control theorem (Theorem 6.5.19), we see that there exist unique symbols

$$\Phi_g^{\pm} \in \mathrm{Symb}_{\Gamma}^{\pm}(\mathcal{D}_k(L))$$

such that

$$\rho_k(\Phi_g^{\pm}) = \phi_g^{\pm}/\Omega_g^{\pm} \tag{6.7.7}$$

Step 3 We define a distribution

$$\mu_g^{\pm} = \Phi_g^{\pm}(\{\infty\} - \{0\}) \in \mathcal{D}^{\dagger}(L) \tag{6.7.8}$$

Step 4 One sees easily that $M_{\mu_g^{\pm}}(\sigma) = 0$ whenever $\sigma(-1) \neq \pm 1$. We thus **define** the p-adic L-function of g on $\mathcal{W}^+$) by

$$L_p^+(g, \sigma) = M_{\mu_g^+}(\sigma), \quad \forall \sigma \in \mathcal{W}^+(\mathbb{C}_p)$$

he p-adic L-function of g on $\mathcal{W}^-$) by

$$L_p^+(g, \sigma) = M_{\mu_g^-}(\sigma), \quad \forall \sigma \in \mathcal{W}^-(\mathbb{C}_p)$$

The construction is over.

Remark 6.7.7

(i) The construction is canonical, except for the normalization of the two periods $\Omega_g^{\pm}$. Thus, the functions L_p^+ and L_p^- are each well-defined only up to multiplication by an element in L^*.

(ii) Without the hypothesis $v_p(\alpha) < k+1$, it is still possible to perform each step of the construction, but we don't know in step 2 that the symbols $\Phi_g^{\pm}$ are unique. Instead, we have a whole finite-dimensional space of possible symbols (of unknown dimension), hence a finite-dimensional vector space of possible p-adic L-functions, with no way (without further insight) to know which one is better.

6.7.3 Computation of the p-Adic L-functions at Special Characters

Now we want to compute some of the values of the p-adic L-function we just defined. For that we need to be able to compute some integrals $\mu_g^{\pm}(h) = \int h(z)d\mu^{\pm}(z)dz$.

As a warm-up, we begin with the case $h(z) = z^j$ for $0 \le j \le k$.

$$\begin{aligned}\mu_g^{\pm}(z^j) &= \Phi_g^{\pm}(\{\infty\} - \{0\})(z^j)\\ &= \frac{1}{\Omega_g^{\pm}}\phi_g^{\pm}(\{\infty\} - \{0\})(z^j) \quad \text{since } \phi_g^{\pm}/\Omega_g^{\pm} = \rho_k(\Phi_g^{\pm})\\ &= \frac{\Gamma(j+1)}{(2\pi)^{j+1}\Omega_g^{\pm}}L(f, j+1) \text{ or } 0 \quad \text{according to } \pm = (-1)^i \text{ or not}\end{aligned}$$

The last line was computed using the definition of $\phi_g^{\pm}$ and formula (5.4.1) of Proposition 5.4.1.

Exercise 6.7.8 Where did we use that $0 \le j \le k$ in this computation?

Now, let's try something a little bit more complicated: let $n \ge 0$, $a \in \{0, 1, \dots, p^n - 1\}$, and as above $0 \le j \le k$. We want to compute $\mu_g^{\pm}(z^j 1_{a+p^n\mathbb{Z}_p}(z))$.

$$\begin{aligned}\mu_g^{\pm}(z^j 1_{a+p^n\mathbb{Z}_p}(z)) &= \Phi_g^{\pm}(\{\infty\} - \{0\})(z^j 1_{a+p^n\mathbb{Z}_p}(z))\\ &= \alpha^{-n}(\Phi_g^{\pm})_{|U_p^n}(\{\infty\} - \{0\})(z^j 1_{a+p^n\mathbb{Z}_p}(z)) \quad (\text{since } U_p^n\Phi_g^{\pm} = \alpha^n\Phi_g^{\pm})\\ &= \alpha^{-n}\sum_{b=1-p^n}^{0} \Phi_g^{\pm}(\{\infty\} - \{b/p^n\})((p^n z - b)^j 1_{a+p^n\mathbb{Z}_p}(p^n z - b))\\ &\qquad (\text{by definition of the action of } U_p^n, \text{ using } \Gamma\begin{pmatrix}1 & 0\\ 0 & p^n\end{pmatrix}\Gamma = \coprod_{b=1-p^n}^{0}\Gamma\begin{pmatrix}1 & b\\ 0 & p^n\end{pmatrix})\\ &= \alpha^{-n}\Phi_g^{\pm}(\{\infty\} - \{a/p^n\})((a + p^n z)^j) \quad (\text{only the fittest term survives: } b = -a)\\ &= \frac{\alpha^{-n}}{\Omega_g^{\pm}}\phi_g^{\pm}(\{\infty\} - \{a/p^n\})((a + p^n z)^j) \quad (\text{since } \phi_g^{\pm}/\Omega_g^{\pm} = \rho_k(\Phi_g^{\pm}))\\ &= \frac{\alpha^{-n}p^{nj}}{\Omega_g^{\pm}}\phi_g^{\pm}(\{\infty\} - \{a/p^n\})((\frac{a}{p^n} + z)^j)\end{aligned}$$

Now if χ is a Dirichlet character of conductor p^n, with $n \geq 1$ and $\chi(-1)(-1)^j = \pm 1$, and if we see χ as a character of $\mathbb{Z}_p^*$,

$$\begin{aligned}
L_p^{\pm}(g, \chi z^j) &= \sum_{a=0}^{p^n-1} \chi(a) \mu_g^{\pm}(z^j 1_{a+p^n\mathbb{Z}_p}(z)) \text{ (by def., using that } \chi(a) = 0 \text{ if } p|a) \\
&= \frac{\alpha^{-n} p^{nj}}{\Omega_g^{\pm}} \sum_{a=0}^{p^n-1} \chi(a) \Phi_g^{\pm}(\{\infty\} - \{a/p^n\})(\frac{a}{p^n} + z)^j) \\
&= \frac{\alpha^{-n} p^{nj}}{\Omega_g^{\pm}} \frac{j! \tau(\chi)}{(-2i\pi)^{j+1}} L(g, \chi^{-1}, j+1) \quad \text{by (5.4.6)} \\
&= \frac{p^{n(j+1)} j!}{\alpha^n (-2i\pi)^{j+1} \tau(\chi^{-1}) \Omega_g^{\pm}} L(g, \chi^{-1}, j+1) \text{ (since } \tau(\chi)\tau(\chi^{-1}) = p^n)
\end{aligned}$$

And we have computed our p-adic L-function (at special characters) in terms of the complex L-function.

The computation when χ is of conductor $p^0 = 1$, that is when χ is trivial, is a little different (we still assume that $(-1)^j = \pm 1$): for $0 \leq j \leq k$,

$$\begin{aligned}
L_p^{\pm}(g, z^j) &= \Phi_g^{\pm}(\{\infty\} - \{0\})(z^j 1_{\mathbb{Z}_p^*}(z)) \\
&= \Phi_g^{\pm}(\{\infty\} - \{0\})(z^j) - \Phi_g^{\pm}(\{\infty\} - \{0\})(z^j 1_{p\mathbb{Z}_p}(z)) \\
&= (1 - \alpha^{-1} p^j) \Phi_g^{\pm}(\{\infty\} - \{0\})(z^j) \\
&= (1 - \alpha^{-1} p^j) \frac{\Gamma(j+1)}{(2\pi)^{j+1} \Omega_g^{\pm}} L(g, j+1)
\end{aligned}$$

To summarize:

Theorem 6.7.9 (Mazur–Swinnerton-Dyer, Manin, Višik, Amice-Vélu) *Let g be a modular cuspidal normalized eigenform of weight $k+2$, character ϵ, level $\Gamma_1(N)$. Assume that $p \mid N$ and that $U_p g = \alpha g$ with $v_p(g) < k+1$. Then for any value of the sign $\pm$, there exists a unique function $L_p^{\pm}(g, \sigma)$ that enjoys the two following properties:*

(interpolation) *For any finite image character $\chi : \mathbb{Z}_p^* \to \mathbb{C}_p^*$ of conductor p^n, and any integer j such that $0 \leq j \leq k$ and $\chi(-1)(-1)^j = \pm$,*

$$L_p^{\pm}(g, \chi z^j) = e_p(g, \chi, j) \frac{p^{n(j+1)} j!}{\alpha^n (-2i\pi)^{j+1} \tau(\chi^{-1}) \Omega_g^{\pm}} L(g, \chi^{-1}, j+1),$$

where $e_p(g, \chi, j) = 1$ if χ is non-trivial and $e_p(g, 1, j) = (1 - \alpha^{-1} p^j)$.

(growth rate) *The order of growth of $L_p^{\pm}(g, \sigma)$ is $\leq v_p(\alpha)$.*

Proof We have just constructed a p-adic L-function $L_p(g)$ that satisfies the given interpolation property. The growth rate of $\mu_g^{\pm}$ is $\leq v_p(\alpha)$ by Proposition 6.5.14, so the growth rate of $L_p(g,\sigma)$ is $\leq v_p(\alpha)$ by Theorem 6.6.10(ii). This shows the existence part of the theorem.

By Theorem 6.6.10(ii), the uniqueness of $L_p(g,\sigma)$ is equivalent to the uniqueness of $\mu_g^{\pm}$ satisfying the corresponding property (i) and (ii). Property (i) means that we can compute $\mu_g^{\pm}(P)$ for any polynomial P of degree less or equal to k. Since the order of growth of $\mu_g^{\pm}$ is $\leq v_p(\alpha) < k+1$, the uniqueness follows from Theorem 6.2.13(ii). □

The result is not completely satisfying when we started with an eigenform f of level $\Gamma_1(N)$ such that $p \nmid N$. We know that we can apply the above theorem to $g = f_\alpha$ or $g = f_\beta$ (provided g is not of critical slope). But we want to express the interpolation property in terms of f, not of the auxiliary function g.

Observe that $L(f(pz), s) = p^{-s}L(f,s)$ (this is best seen on the integral representation), hence

$$\begin{aligned} L(f_\alpha, s) &= (1-\beta p^{-s})L(f,s) \\ &= (1-\alpha^{-1}\epsilon(p)p^{k+1-s})L(f,s) \text{ using } \alpha\beta = p^{k+1}\epsilon(p) \end{aligned}$$

For non trivial χ, applying the same result to the form $f_{\chi^{-1}}$ we get

$$L(f_\alpha, \chi^{-1}, s) = (1-\alpha^{-1}\epsilon(p)\chi^{-1}(p)p^{k+1-s})L(f,\chi^{-1},s) = L(f,\chi^{-1},s)$$

since $\chi^{-1}(p) = 0$. From this, we get

Corollary 6.7.10 (Same as Above, Plus Mazur-Tate-Teitelbaum) *Let f be a modular cuspidal normalized eigenform of weight $k+2$, character ϵ, level $\Gamma_1(N)$. Assume that N is prime to p. Let α, β be the two roots of $X^2 - a_pX + p^{k+1}\epsilon(p)$, and choose one, say α, with $v_p(\alpha) < k+1$. There for any choice of the sign $\pm$, there exists a unique function $L_p^{\pm}(f_\alpha,\sigma)$ that enjoys the two following properties:*

(interpolation) *For any finite image character $\chi : \mathbb{Z}_p^* \to \mathbb{C}_p^*$ of conductor p^n, and any integer j such that $0 \leq j \leq k$ and $\chi(-1)(-1)^j = \pm$,*

$$L_p^{\pm}(f_\alpha, \chi z^j) = e_p(f,\alpha,\chi,j)\frac{p^{n(j+1)}j!}{\alpha^n(-2i\pi)^{j+1}\tau(\chi^{-1})\Omega_g^{\pm}}L(f,\chi^{-1},j+1),$$

where $e_p(f,\alpha,\chi,j) = 1$ if χ is non trivial and $e_p(f,\alpha,1,j) = (1-\alpha^{-1}\epsilon(p)p^{k-j})(1-\alpha^{-1}p^j)$

(growth rate) *The order of growth of $L_p(f_\alpha,\sigma)$ is $\leq v_p(\alpha)$.*

Remark 6.7.11

(i) The construction of the p-adic L-function of a modular form we have given (or for that matter, any other known construction) is quite indirect in the sense that it uses complex transcendental method. One could expect that to construct from an algebraic object (the sequence of coefficients a_n of f) a p-adic object $L_p(f_\alpha, \sigma)$, it would suffice to use purely algebraic and p-adic means. But we used the modular symbols $\phi_{g^\pm}$, which is defined by complex transcendental means from g.

(ii) It is expected, but not known in general, that the order of growth of $L_p(f_\alpha, \sigma)$ is exactly $v_p(\alpha)$. The only case where this is known is when $a_p = 0$ (and $p \nmid N$), in which case both $L_p^\pm(f_\alpha, \sigma)$ and $L_p^\pm(f_\beta, \sigma)$ have the expected order $(k-1)/2$ (see [98]). Also, it is known (*loc. cit.*) and due to Mazur that at least one of the functions $L_p^\pm(f_\alpha, \sigma)$ and $L_p^\pm(f_\beta, \sigma)$ has positive order.

(iii) We know little on the values of L_p at characters σ not of the form χz^k for $0 \le j \le k$, or for that matter on the derivatives of L_p at characters of that form. There are important conjectures predicting those values, and relating it to arithmetic invariants of f.

(iv) Fix an M prime to p. One can extend the p-adic L-function of f_α form a function on $\mathcal{W}$ to a function on $\mathcal{W}_M$ (with essentially no change in the proof). The formulas are essentially the same, excepted that now χ is interpreted as a Dirichlet character of conductor $p^n M$ (and accordingly the p^n in the formula has to be replaced by $p^n M$), and that the *p-adic multiplier* $e_p(f, \alpha, \chi, j)$ is defined by a slightly more complicated formula. See [92] for the precise formulas.

(v) The question of whether the p-adic multiplier $e_p(f, \alpha, \chi, j)$ is 0, and what consequences this vanishing has when it occurs, is discussed extensively in [92].

(vi) There is another completely independent way to construct the p-adic L-function of a modular form: it is by using Perrin-Riou's method applied to Kato's Euler system. We shall not discuss this point view here. See [44] and [45] and the references there for more information.

(vii) The p-adic L-functions we constructed are never 0. This is easy to see when the form g has weight $k+2 > 2$, since then the interpolation property gives, for χ a non-trivial Dirichlet character $L_p^\pm(g, \chi z^k) = CL(g, \chi^{-1}, k+1)$ where C is a non-zero constant. If $k+2 > 2$, that is $k > 0$, then $k+1 \ge k/2 + 3/2$ so $k+1$ is either in the interior or on the boundary (when $k=1$, i.e. in weight 3) of the domain of convergence of the Dirichlet Series of $L(g, \chi^{-1}, s)$. In both case, it is known that this functions does not vanish at $k+1$, because of the Euler product description in the first case, of the Jacquet-Shalika generalization of the theorem of Hadamard and De La Vallée-Poussin in the second case. Proving that the p-adic L-function does not vanish for a form of weight 2 is harder, and requires a difficult theorem of Rohrlich.

(viii) When $v_p(\alpha) = k+1$, we can still construct a p-adic L-function satisfying (i) and (ii) by choosing for $\Phi_g^\pm$ any lift of $\phi_g^\pm/\Omega_g^\pm$, but then the construction is not canonical, and (i) and (ii) does not determine uniquely the p-adic L-function. It is not possible to remediate this lack of uniqueness (there definitely are several functions satisfying (i) and (ii)), but it is possible to modify the construction so as to get a *canonical, well-defined* p-adic L-function, and to state properties that characterize it uniquely. See Chap. 8.

6.8 Notes and References

Classical modular symbols appear in [88], in connection with the study of the arithmetic properties of the special values of a modular forms and with the construction of its p-adic L-function. They are studied systematically in [115].

Part III
The Eigencurve and its p-Adic L-Functions

Chapter 7
The Eigencurve of Modular Symbols

Stevens's overconvergent modular symbols can replace Coleman's overconvergent modular forms in the construction of the eigencurve. This was outlined by Stevens himself in a talk in Paris in 2000, but he never publicly released his personal notes on the subject. A very quick sketch of his argument is given in [78] or in [96] (a few lines in each cases) and a detailed construction of the local pieces of Stevens's eigencurve is to be found in [8] of which we give in Sect. 7.1 a global version. We then (Sect. 7.2) use Chenevier's comparison theorem (see Sect. 3.8) to prove that this eigencurve is essentially the same as the Coleman-Mazur eigencurve. Nevertheless, the construction with modular symbols is essential to our purpose of constructing and studying p-adic L-functions in family, as it provides over the eigencurve a natural module of distribution-valued modular symbols which is closely related to L-functions. In the next Sect. 7.3, we offer a detailed discussion of the points of the eigencurve that correspond to classical modular forms. Those points, and their neighborhoods in the eigencurve, are the main objects of interest for arithmetical applications. Section 7.4 introduces and studies one of the last remaining important characters of our intrigue, the family of Galois representations that the eigencurve carry. Finally, using this family and Galois cohomology arguments, we study the local geometry of the eigencurve at classical points, focussing on questions of smoothness and étaleness over the weight space.

7.1 Construction of the Eigencurve Using Rigid Analytic Modular Symbols

From now on, we will fix an integer N, a prime p not dividing N, and we sate

$$\Gamma = \Gamma_1(N) \cap \Gamma_0(p).$$

We will construct in this Sect. 7.1 *the p-adic eigencurve of tame level N*.

J. Bellaïche, *The Eigenbook*, Pathways in Mathematics,
https://doi.org/10.1007/978-3-030-77263-5_7

We call $\mathcal{H}_0$ the commutative ring generated by symbols called T_l, for l not dividing Np, U_p, and $\langle a \rangle$ for $a \in (\mathbb{Z}/N\mathbb{Z})^*$. Thus $\mathcal{H}_0$ is the ring called $\mathcal{H}_0$ or $\mathcal{H}_0(N)$ we introduced in Sect. 2.6.4 and used in the preceding chapters, except that the symbol T_p is now written U_p. (We could actually work with the full Hecke algebra $\mathcal{H}$, including the U_l's for $l \nmid Np$; this would lead to a different, non-reduced, eigencurve.)

We shall also need the Hecke algebra $\mathcal{H}_0^p$: it is the polynomial ring in the T_l, $l \nmid Np$, and $\langle a \rangle$, $a \in (\mathbb{Z}/N\mathbb{Z})^*$, that is the same as $\mathcal{H}_0$ without U_p (or what was called $\mathcal{H}_0(Np)$ in Sect. 2.6.4).

7.1.1 *Overconvergent Modular Symbols Over an Admissible Open Affinoid of the Weight Space*

In all this section, let $W = \operatorname{Sp} R$ be an admissible open affinoid of the weight space $\mathcal{W}$. So R is an affinoid algebra over $\mathbb{Q}_p$, in particular a noetherian commutative $\mathbb{Q}_p$-Banach algebra. The inclusion map $W = \operatorname{Sp} R \in \mathcal{W}$ is a point in $\mathcal{W}(R)$, hence by definition of $\mathcal{W}$ a continuous group homomorphism

$$K : \mathbb{Z}_p^* \to R^*.$$

The map K should be thought of as a family of characters of $\mathbb{Z}_p^*$ parametrized by the weights $w \in W$. That is, for L a finite extension of $\mathbb{Q}_p$, if w is a point of $\mathcal{W}$ that belongs to $W(L) = \operatorname{Hom}_{\text{cont,ring}}(R, L^*)$, then the character $\kappa : \mathbb{Z}_p^* \to L^*$ attached to w is precisely the composition

$$\kappa : \mathbb{Z}_p^* \stackrel{K}{\to} R^* \stackrel{w}{\to} L^*. \tag{7.1.1}$$

We have defined in Sects. 6.1.2, 6.1.3, 6.2.2 Banach R-modules $\mathcal{A}[r](R)$, $\mathcal{D}[r](R)$ for $r > 0$, and R-modules $\mathcal{A}^\dagger[r](R)$, $\mathcal{D}^\dagger[r](R)$ for $r \geq 0$. Let us recall (Proposition 6.1.22 and Corollary 6.2.8) that the formation of the Banach module $\mathcal{D}[r](R)$ and of the Fréchet module $\mathcal{D}^\dagger[r](R)$ commutes with base changes $R \to R'$, where R' is any $\mathbb{Q}_p$-Banach algebra (in particular, when R' is an affinoid algebra and $\operatorname{Sp} R'$ an affinoid subdomain of $\operatorname{Sp} R = W$), in the sense that $\mathcal{D}^\dagger[r](R)\hat{\otimes}_R R' = \mathcal{D}^\dagger[r](R')$ and $\mathcal{D}^\dagger[r](R)\hat{\otimes}_R R' = \mathcal{D}^\dagger[r](R')$

Lemma 7.1.1 *If L is any finite extension of $\mathbb{Q}_p$, and if $w \in W(L)$, then we have a natural isomorphism*

$$\mathcal{D}^\dagger[r](R) \otimes_{R,w} L = \mathcal{D}^\dagger[r](L).$$

Proof This is a special case of Corollary 6.2.8 using the easy observation that since L is finite over R, our $\hat{\otimes}_R L$ functor is just $\otimes_R L$ $\square$

Now, by the work done in Sect. 6.4, we can use the weight $K : \mathbb{Z}_p^* \to R^*$ to define a weight-K action of $S_0(p)$ on $\mathcal{A}[r](R)$, $\mathcal{D}[r](R)$, for all r such that $0 < r < r(K)$, $\mathcal{A}^\dagger[r](R)$ and $\mathcal{D}^\dagger[r](R)$, for all r such that $0 \le r < r(K)$, where $r(K)$ is a positive real number depending only on K, that is only on W, defined in Definition 6.3.5. We shall write below $r(W)$ instead of $r(K)$. When provided with these actions, these modules are denoted by $\mathcal{A}_K[r](R)$, $\mathcal{D}_K[r](R)$, $\mathcal{A}_K^\dagger[r](R)$, $\mathcal{D}_K^\dagger[r](R)$.

Lemma 7.1.2 *Let L be a finite extension of $\mathbb{Q}_p$, $w \in W(L) = \mathrm{Hom}(R, L)$ and $\kappa : \mathbb{Z}_p^* \to L^*$ the corresponding character. The natural isomorphism $\mathcal{D}_K^\dagger[r](R) \otimes_{R,w} L = \mathcal{D}_\kappa^\dagger[r](L)$ is compatible with the $S_0(p)$-actions (and similarly for the other modules $\mathcal{D}_K[r](R)$ etc.)*

Proof This is clear from the definitions of those isomorphisms and from (7.1.1). □

Hence one should think of the R-module $\mathcal{D}_K^\dagger[r](R)$ with its $S_0(p)$-action as the family of $S_0(p)$-modules $\mathcal{D}_\kappa^\dagger[r]$ parametrized by points $w \in W$, with κ the character corresponding to w.

We can form modules of modular symbols, $\mathrm{Symb}_\Gamma(\mathcal{D}_K[r](R))$ which is a Banach R-module or $\mathrm{Symb}_\Gamma(\mathcal{D}_K^\dagger[r](R))$.

There is one technical point that must be addressed before we carry in this context the work done over a field in Sect. 6.5.2:

From now on we shall make the following assumption on R:

Hypothesis 7.1.3 *There is a multiplicative element π in R which has the highest norm < 1, and if $R_0 = \{x \in R, |x| \le 1\}$, then $\tilde{R} = R_0/\pi R_0$ is a PID.*

We note that this hypothesis is satisfied for instance when R is a Tate algebra in one variable over a finite extension L of $\mathbb{Q}_p$, $R = L\langle T\rangle$. Indeed, in this case, if $\mathcal{O}_L$ is the ring of integers of L and k_L the residue field, we can take π to be the uniformizer and we see that $\tilde{R} = k_L[T]$ is a PID.

Lemma 7.1.4 *If R satisfies Hypothesis 7.1.3, the R-module $\mathrm{Symb}_\Gamma(\mathcal{D}_K[r](R))$ is potentially orthonormalizable.*

Proof This follows from scholium 6.5.3 and from the following abstract lemma. □

Lemma 7.1.5 *Assume that R satisfies Hypothesis 7.1.3. Let M be any potentially orthonormalizable Banach R-module, and N be any closed sub-module of M. Then N is potentially orthonormalizable.*

Proof Note that our assumption implies that R satisfies Hypothesis 3.1.15. Changing the norm of M we may assume that M is orthonormalizable, in particular $|M| \subset |R|$. Let M_0 and N_0 be the closed unit balls in M and N. We have $M_0 \subset N_0$, and $\pi M_0 \cap N_0 = \pi N_0$, since being in the left hand side means being in N and having norm < 1, which is the same sing as being in πN_0. Hence we have an inclusion of

$\tilde{R}$-modules $N_0/\pi N_0 \hookrightarrow M_0/\pi M_0$. By Lemma 3.1.16, $M_0/\pi M_0$ is free, and so $N_0/\pi N_0$ is free too because a sub-module of a free module over a PID is free, and N is orthonormalizable by Lemma 3.1.16 again. □

Exercise 7.1.6 Is $\mathrm{Symb}_\Gamma(\mathcal{D}_K[r](R))$ potentially orthonormalizable for all $\mathbb{Q}_p$-Banach algebra R?

By Theorem 6.5.6(i), the operator U_p acts compactly on the Banach module $\mathrm{Symb}_\Gamma(\mathcal{D}_K[r](R))$ for $0 < r < \min(r(K), p)$. If R satisfies Hypothesis 7.1.3, this Banach module is potentially orthonormalizable and thus U_p has a Fredholm's determinant $F(T) = \det(1 - TU_p) \in R\{\{T\}\}$ (cf. Sect. 3.1), which moreover is independent of r by Theorem 6.5.6(iii), hence depends only on $W = \mathrm{Sp}\, R$. If ν is a real number that is *adapted* to $F(T)$ (Definition 3.3.10), we shall simply say that ν is adapted to W. In this case we can defined the sub-module $\mathrm{Symb}_\Gamma(\mathcal{D}_K[r](R))^{\leq \nu}$ of $\mathrm{Symb}_\Gamma(\mathcal{D}_K[r](R))$, which are finite and projective over R, and independent of r.

The next two subsections are aimed at proving two results concerning the commutation of the formation of $\mathrm{Symb}_\Gamma(\mathcal{D}_K[r](R))$ with base change.

7.1.2 The Restriction Theorem

The aim of this subsection is to prove the following result.

Theorem 7.1.7 *Let R be a Tate algebra of dimension 1. Let $W' = \mathrm{Sp}\, R'$ be an affinoid subdomain of $W = \mathrm{Sp}\, R$. Then for every r such that $0 < r \leq r(W)$, the two Banach-modules with action of $\mathcal{H}_0$,*

$$\mathrm{Symb}_\Gamma(\mathcal{D}_K[r](R))\hat{\otimes}_R R' \quad \text{and} \quad \mathrm{Symb}_\Gamma(\mathcal{D}_K[r](R'))$$

are linked.[1]

First note that by definition $r(W') \geq r(W)$, so $\mathrm{Symb}_\Gamma(\mathcal{D}_K[r](R'))$ is defined.

Let us recall (cf. Lemma 6.5.2 and above) that if V is an R-module with an R-linear $S_0(p)$-action, and if q is a norm, or a semi-norm, of R-module on V which is preserved by Γ, then $\mathrm{Symb}_\Gamma(V)$ inherits from q a norm of R-module, or a semi-norm, that we shall still denote by q. If V is a topological module whose topology is defined by a family of norms or semi-norms q_i as above, then we consider $\mathrm{Symb}_\Gamma(V)$ as provided with the topology defined by the norms q_i. Let us define the *difference operator*: $\Delta : V \to V,\ v \mapsto v_{|\begin{pmatrix} 1 & 1 \\ 0 & 1 \end{pmatrix}} - v$.

[1]For the definition of *linked*, see Sect. 3.5.

Lemma 7.1.8 (Pollack-Stevens) *There exists an integer $t > 0$, an integer $s \geq 0$, and elements $\gamma_i \in \Gamma$ for $i = 1, \dots, t$, all depending only on Γ, such that if $n = t+s+1$, and $\mu_V^* : V^n \to V$ is defined by $(v_i) \mapsto \sum_{i=1}^t (v_i)_{|\gamma_i - \mathrm{Id}} + \sum_{i=t+1}^{t+s} v_i + \Delta v_n$, then we have an isomorphism*

$$Symb_\Gamma(V) = \ker(V^n \xrightarrow{\mu_V^*} V)$$

that respects the topology of $Symb_\Gamma(V)$ when we have put one on it.

The proof is a refined version of the argument based on Manin's lemma used in the proof of Lemma 4.1.5, obtained by choosing artfully a system of generators of Δ_0 as $\mathbb{Z}[\Gamma]$-module: see [99, Theorem 2.9].

Lemma 7.1.9 *If V is an R-module topologized by a family of semi-norms, and V is Hausdorff, we have a natural injective isometric $\mathcal{H}_0$-compatible R-linear morphism*

$$\widehat{Symb_\Gamma(V)} \hookrightarrow Symb_\Gamma(\hat{V}).$$

If $\mu_V^* : V \to V$ is strict (cf. Exercise 6.2.3), then the displayed morphism is an isomorphism.

Proof If V is as in the statement, we have

$$\mathrm{Symb}_\Gamma(\hat{V}) = \ker(\hat{V}^n \xrightarrow{\mu_{\hat{V}*}} \hat{V})$$

and it is clear from the description of μ_V^* that $\mu_{\hat{V}}^*$ is just the extension by continuity of μ_V^*. Hence from the injective map $V \hookrightarrow \hat{V}$ an injective map $\mathrm{Symb}_\Gamma(V) \hookrightarrow \mathrm{Symb}_\Gamma(\hat{V})$ inducing by the universal property of completion a map

$$\widehat{\mathrm{Symb}_\Gamma(V)} \hookrightarrow \mathrm{Symb}_\Gamma(\hat{V}).$$

This proves the first half of the lemma. The second half then follows from question 7 of Exercise 6.2.3, or equivalently [27, Prop. 1.1.9/5]. □

Lemma 7.1.10 *The map μ_V^* is strict when $V = \mathcal{D}_K^\dagger[r](R)$.*

Proof In this case, the map μ_V^* sends $(v_i)_{i=1,\dots n}$ to

$$(v_i) \mapsto \sum_{i=1}^{t} (v_i)_{|_K \gamma_i - \mathrm{Id}} + \sum_{i=t+1}^{t+s} v_i + \Delta^* v_n$$

where $\Delta^* v_n = (v_n)_{|\begin{pmatrix} 1 & 1 \\ 0 & 1 \end{pmatrix}} - v_n$ is as in Lemma 6.5.16. In particular, by this lemma, Δ^* is strict, has closed image, and $V/\Delta^*(V)$ is identified with R by the map which evaluates a distribution v in V on the constant function 1.

We claim that μ_V^* has closed image in V. By the above, it suffices to show that the image of $\text{Im}\mu_V^*$ in $V/\Delta^*(V) = R$ is closed. But since μ_V^* is R-linear, that image is a submodule of R, that is an ideal of R, and it is well-known that every ideal of a Tate algebra is closed.

The Fréchet modules V^n and V over R can be seen as Fréchet spaces over $\mathbb{Q}_p$. It suffices to keep the same semi-norms, and to forget the scalar multiplication by elements of $R - \mathbb{Q}_p$. The topologies of course are not changed. The map μ_V^* is just a $\mathbb{Q}_p$-linear space between two Fréchet spaces, and has closed image. It is therefore strict by question 5 of Exercise 6.2.3. □

Lemma 7.1.11 *We have an isometric natural isomorphism of normed R-modules, respecting the action of* $\mathcal{H}_0$

$$Symb_\Gamma(\mathcal{D}_K[r](R)) \otimes_R R' = Symb_\Gamma(\mathcal{D}_K[r](R) \otimes_R R')$$

and a natural isomorphism respecting the norms and the action of $\mathcal{H}_0$

$$Symb_\Gamma(\mathcal{D}_K^\dagger[r](R)) \otimes_R R' = Symb_\Gamma(\mathcal{D}_K^\dagger[r](R) \otimes_R R')$$

Proof By Conrad [48, page 13], R' is R-flat. Hence the lemma follows from Lemma 4.1.5. □

Proposition 7.1.12 *We have*

$$Symb_\Gamma(\mathcal{D}_K^\dagger[r](R')) = Symb_\Gamma(\mathcal{D}_K^\dagger[r](R))\hat{\otimes}_R R'$$

Proof One has the following sequence of isomorphisms of R'-module with action of $\mathcal{H}_0$:

$$\begin{aligned}
\text{Symb}_\Gamma(\mathcal{D}_K^\dagger[r](R')) &= \text{Symb}_\Gamma(\mathcal{D}_K^\dagger[r](R)\hat{\otimes}_R R') \text{ (by Corollary 6.2.8)}\\
&= \text{Symb}_\Gamma(\overline{\mathcal{D}_K^\dagger[r](R) \otimes_R R'}) \text{ (by def. of } \hat{\otimes})\\
&= \overline{\text{Symb}_\Gamma(\mathcal{D}_K^\dagger[r](R) \otimes_R R')} \text{ (by Lemmas 7.1.9 and 7.1.10).}\\
&= \overline{\text{Symb}_\Gamma(\mathcal{D}_K^\dagger[r](R)) \otimes_R R'} \text{ (by Lemma 7.1.11)}\\
&= \text{Symb}_\Gamma(\mathcal{D}_K^\dagger[r](R))\hat{\otimes}_R R' \text{ (by def. of } \hat{\otimes}).
\end{aligned}$$

□

We are now ready to prove Theorem 7.1.7. Observe that one has the following isomorphisms and embeddings of R'-modules with action of $\mathcal{H}_0$:

$$\begin{aligned}
\text{Symb}_\Gamma(\mathcal{D}_K[r](R))\hat{\otimes}_R R' &= \overline{\text{Symb}_\Gamma(\mathcal{D}_K[r](R)) \otimes_R R'} \text{ (by def. of } \hat{\otimes})\\
&\hookrightarrow \text{Symb}_\Gamma(\overline{\mathcal{D}_K[r](R) \otimes_R R'}) \text{ (by Lemma 7.1.9)}
\end{aligned}$$

$$= \mathrm{Symb}_\Gamma(D_K[r](R)\hat{\otimes}_R R') \text{ (by def. of } \hat{\otimes})$$
$$= \mathrm{Symb}_\Gamma(D_K[r](R')) \text{ (by Proposition 6.1.22).}$$

On the other hand, choosing an r' and r'' such that $r < r' < r'' < r(W)$, and remembering that $\mathcal{D}_k[r] \subset \mathcal{D}_k[r'] \subset \mathcal{D}_k[r'']$, one has

$$\begin{aligned}\mathrm{Symb}_\Gamma(D_K[r](R')) &\hookrightarrow \mathrm{Symb}_\Gamma(\mathcal{D}_K^\dagger[r'](R')) \text{ (since } \mathrm{Symb}_\Gamma \text{ is left-exact)}\\ &= \overline{\mathrm{Symb}_\Gamma(\mathcal{D}_K^\dagger[r'](R)) \otimes_R R'} \text{ (by Prop 7.1.12)}\\ &\hookrightarrow \overline{\mathrm{Symb}_\Gamma(\mathcal{D}_K[r''](R)) \otimes_R R'} \text{ (since } \mathrm{Symb}_\Gamma \text{ is left-exact)}\\ &= \mathrm{Symb}_\Gamma(\mathcal{D}_K[r''](R))\hat{\otimes}_R R' \text{ (by def. of } \hat{\otimes}).\end{aligned}$$

Finally we note that $\mathrm{Symb}_\Gamma(\mathcal{D}_K[r''](R))\hat{\otimes}_R R'$ and $\mathrm{Symb}_\Gamma(\mathcal{D}_K[r](R))\hat{\otimes}_R R'$ are linked by Theorem. 6.5.6, and thus $\mathrm{Symb}_\Gamma(D_K[r](R'))$, which sits in between is also linked to them by Lemma 3.5.3. Theorem 7.1.7 is now proved.

Remark 7.1.13 I do not know if μ_V^* is strict when V is the Banach space $\mathcal{D}_k[r](\mathbb{Q}_p)$. Of course, it is the case if $s > 0$ because then μ_V^* is surjective and open, but for a torsion-free Γ, one has $s = 0$ unfortunately. When $s = 0$, the fact that $\Delta : \mathcal{D}_k[r](\mathbb{Q}_p) \to \mathcal{D}_k[r](\mathbb{Q}_p)$ is not strict suggests that μ_V^* may not be strict either.

This explains the indirectness of the proof of Theorem 7.1.7. Indeed, while this theorem is concerned with modular symbols valued in the Banach modules $\mathcal{D}_k[r](R)$ and $\mathcal{D}_k[r](R')$, we are forced to work with their Fréchet cousins $\mathcal{D}_k^\dagger[r](R)$ and $\mathcal{D}_k^\dagger[r](R')$, for which μ_V^* is strict, because it is not true that the functor *completion* commutes with taking the kernel of a continuous morphism of Banach or Fréchet space, when the morphism is not supposed to be strict (see Exercise 7.1.14 below). Otherwise, the theorem would be proved exactly as Lemma 4.1.5.

Exercise 7.1.14 Let $R = \mathbb{Q}_p\langle T\rangle$ be the Tate algebra in one variable, so $\mathrm{Sp}\, R$ is the closed ball of center 0 and radius 1. Let M be an orthonormalizable Banach module over R of orthonormal basis $(e_i)_{i\geq 1}$, and define a continuous linear map $u : M \to M$ by $u(e_1) = \sum_{i=1}^\infty p^i T^i e_i$ and $u(e_i) = p^{i-1}e_{i-1}$ for $i > 1$. Show that u is injective. Show that if $\mathrm{Sp}\, R'$ is the ball of center 0 and radius $r < 1$, $r \in p^{\mathbb{Q}}$, then $u \otimes 1 : M \otimes_R R' \to M \otimes_R R'$ is injective, but its completion $u\hat{\otimes}1 : M\hat{\otimes}_R R' \to M\hat{\otimes}_R R'$ is not injective. (This example is due to V. Lafforgue, cf. [35, footnote 1]).

7.1.3 The Specialization Theorem

In all this section, $W = \mathrm{Sp}\, R$ is an admissible affinoid open subset of $\mathcal{W}$.

Lemma 7.1.15 *Let* $0 \le r < r(K)$. *If* $0 \notin W(\mathbb{Q}_p) = Sp\, R(\mathbb{Q}_p) = \mathrm{Hom}(R, \mathbb{Q}_p)$, *then* $H_0(\Gamma, \mathcal{D}^\dagger[r](R)) = 0$. *If* $0 \in W(\mathbb{Q}_p) = \mathrm{Hom}(R, \mathbb{Q}_p)$, *then we have a natural isomorphism of* $\mathbb{Q}_p$*-vector spaces* $H_0(\Gamma, \mathcal{D}^\dagger[r](R)) = \mathbb{Q}_p$. *This isomorphism is an isomorphism of R-module if we give the RHS an R-module structure using the morphism* $R \to \mathbb{Q}_p$ *corresponding to the character* 0.

Proof Recall that the sequence of R-modules

$$0 \longrightarrow \mathcal{D}^\dagger[r](R) \overset{\Delta}{\longrightarrow} \mathcal{D}^\dagger[r](R) \overset{\rho}{\longrightarrow} R \longrightarrow 0 \tag{7.1.2}$$

is exact by Lemma 6.5.16. We need to compute $H^0(\Gamma, \mathcal{D}_K^\dagger[r](R)) = \mathcal{D}_K^\dagger[r](R)/I\mathcal{D}_K^\dagger[r](R)$ where I is the augmentation ideal of $\mathbb{Z}[\Gamma]$. We proceed as in [100, Lemma 5.2], but "in family".

The ideal I contains $(g-1)$ with $g = \begin{pmatrix} 1 & 1 \\ 0 & 1 \end{pmatrix}$, which acts as Δ. Thus $I\mathcal{D}^\dagger[r](R)$ contains $\ker \rho$, and we have an exact sequence

$$0 \to \rho(I\mathcal{D}^\dagger(R)) \to R \to H_0(\Gamma, \mathcal{D}^\dagger[r](R)) \to 0. \tag{7.1.3}$$

We now construct two explicit elements in $\rho(I\mathcal{D}^\dagger(R))$.

First, set $\gamma = \begin{pmatrix} 1+pN & pN \\ -pN & 1-pN \end{pmatrix} \in \Gamma_1(Np) \subset \Gamma$. Let δ_1 be the distribution that sends f to $f(1)$. We compute

$$\begin{aligned} \rho((\delta_1)_{|\gamma-1}) &= \delta_1(\tilde{K}(1+pN+pNz)-1) \\ &= \tilde{K}(1+2pN)-1 \in \rho(I\mathcal{D}^\dagger(R)) \end{aligned}$$

Second, let a be an integer such that $a \pmod p$ is a generator of $(\mathbb{Z}/q\mathbb{Z})^*$, and $a \equiv 1 \pmod N$ (remember that $q = p$ if p is odd, $q = 4$ if $p = 2$). By an easy application of Bezout's lemma there exists integers b, c, d such that $\gamma := \begin{pmatrix} a & b \\ c & d \end{pmatrix}$ is in Γ. Let δ_0 be the distribution $f \mapsto f(0)$. Then we compute:

$$\begin{aligned} \rho((\delta_0)_{|\gamma-1}) &= \delta_0(\tilde{K}(a+cz)-1) \\ &= \tilde{K}(a)-1 \in \rho(I\mathcal{D}^\dagger(R)) \end{aligned}$$

From the exact sequence (7.1.3) we deduce the existence of a surjective map

$$R/(\tilde{K}(1+2pN)-1, \tilde{K}(a)-1) \to H_0(\Gamma, \mathcal{D}^\dagger[r](R)). \tag{7.1.4}$$

Since $1+2pN$ generates $1+q\mathbb{Z}_p$, and a generates $(\mathbb{Z}/q\mathbb{Z})^*$, we see that $R/(\tilde{K}(1+2pN)-1, \tilde{K}(a)-1)$ is either $\mathbb{Q}_p$ or 0, according to whether $0 \in W$ or not, and that in the former case, the map $R \to R/(\tilde{K}(1+2pN)-1, \tilde{K}(a)-1) = \mathbb{Q}_p$ is the one corresponding to the trivial character 0. In this case, we have a natural map

$$H_0(\Gamma, \mathcal{D}^\dagger[r](R)) \to H_0(\Gamma, \mathcal{D}_0^\dagger[r]) \tag{7.1.5}$$

surjective by right exactness of H_0, hence $H_0(\Gamma, \mathcal{D}^\dagger[r](R))$ has dimension at least 1 over $\mathbb{Q}_p$, since by Lemma 6.5.17, $H_0(\Gamma, \mathcal{D}_0^\dagger[r])$ has dimension 1. This shows that the surjective maps (7.1.4) and (7.1.5) are both isomorphisms and that $H_0(\Gamma, \mathcal{D}^\dagger[r](R))$ has dimension exactly 1 over $\mathbb{Q}_p$. □

Theorem 7.1.16 *Let $W = Sp\, R$ be a nice affinoid of $\mathcal{W}$. Let L be a finite extension of $\mathbb{Q}_p$. Let $w \in W(L) = \mathrm{Hom}(R, L)$ which by definition is a character $\mathbb{Z}_p^* \to L^*$. Let $0 \leq r < r(K)$ be a real number. Then there is a natural injective morphism of L-vector spaces, compatible with the action of $\mathcal{H}_0$*

$$Symb_\Gamma(\mathcal{D}^\dagger[r](R)) \otimes_{R,w} \mathbb{Q}_p \hookrightarrow Symb_\Gamma(\mathcal{D}_w^\dagger[r]). \tag{7.1.6}$$

This map is surjective excepted when $w = 0$. In this case, the cokernel is a space of dimension 1.

Proof Lemma 7.1.1 may be reformulated as the exact sequence

$$0 \longrightarrow \mathcal{D}^\dagger[r](R) \xrightarrow{\times u} \mathcal{D}^\dagger[r](R) \xrightarrow{w} \mathcal{D}_w^\dagger[r] \longrightarrow 0.$$

where u is a generator of the ideal $\ker w$ in R. The long exact sequence of cohomology with compact support attached to this short exact sequence is

$$\begin{aligned} 0 \to H_c^1(\Gamma, \mathcal{D}^\dagger[r](R)) &\overset{\times u}{\to} H_c^1(\Gamma, \mathcal{D}^\dagger[r](R)) \to H_c^1(\Gamma, D_w^\dagger[r]) \\ \to H_c^2(\Gamma, \mathcal{D}^\dagger[r](R)) &\overset{\times u}{\to} H_c^2(\Gamma, \mathcal{D}^\dagger[r](R)) \end{aligned}$$

Using Poincaré duality for Γ, which is functorial and compatible with the Hecke operators in $\mathcal{H}_0$ we see that the last morphism is the same as $H_0(\Gamma, \mathcal{D}^\dagger[r](R)) \overset{\times u}{\to} H_0(\Gamma, \mathcal{D}^\dagger[r](R))$. The kernel A of this map is, by the lemma above, 0 excepted when κ is trivial in which case it is $\mathbb{Q}_p$. Using the functorial isomorphism $\mathrm{Symb}_\Gamma(V) = H_c^1(Y(\Gamma), V)$ of Ash-Stevens (Theorem 4.4.1), we get an exact sequence

$$0 \to \mathrm{Symb}_\Gamma(\mathcal{D}^\dagger[r](R)) \overset{\times u}{\to} \mathrm{Symb}_\Gamma(\mathcal{D}^\dagger[r](R)) \to \mathrm{Symb}_\Gamma(\mathcal{D}_w^\dagger[r]) \to A \to 0,$$

and the result follows. □

Lemma and Definition 7.1.17 *The image of the map (7.1.6) is independent of the choice of the affinoid W containing w. We call it $Symb_\Gamma(\mathcal{D}_w^\dagger[r])_g$.*

Proof There is nothing to prove in the case $w \neq 0$. When $w = 0$, consider the image for an affinoid W and for another affinoid W'. If $W \subset W'$, the image for the former is included into the image of the latter, and since both have codimension 1 in $\mathrm{Symb}_\Gamma(\mathcal{D}^\dagger_w[r])$, they are equal. In general, consider a affinoid $W'' \subset W \cap W'$. □

The g should make think of "global", as $\mathrm{Symb}^\pm_\Gamma(\mathcal{D}_w)_g$ is the space of modular symbols in $\mathrm{Symb}^\pm_\Gamma(\mathcal{D}_w)$ that comes by specialization from "global" modular symbols, i.e. defined over W.

We want to determine the action of $\mathcal{H}_0$ and ι on the line $\mathrm{Symb}_\Gamma(\mathcal{D}^\dagger_0[r])/\mathrm{Symb}_\Gamma(\mathcal{D}^\dagger_0[r])_g$. We shall not be able to do that before the next section, but as a preparation:

Definition 7.1.18 The *system of eigenvalues of* E_2^{crit} is the character $\mathcal{H}_0 \to L$ (L any ring) that sends T_l to $1 + l$ (for $l \nmid Np$), U_p to p, $\langle a \rangle$ to 1.

Lemma 7.1.19 *There exists a non-zero boundary modular symbol* $\Phi_{E_2^{\mathrm{crit}}}$ *in* $BSymb_\Gamma(\mathcal{D}_0[r](\mathbb{Q}_p)) \subset Symb_\Gamma(\mathcal{D}_0[r](\mathbb{Q}_p))$ *(for any* $r > 0$*) which is an eigenvector for* $\mathcal{H}_0$ *with the system of eigenvalues of* E_2^{crit} *and an eigenvector for* ι *with sign* -1.

Proof We can assume $\Gamma = \Gamma_0(p)$. We will define $\Phi_{E_2^{\mathrm{crit}}}$ as the restriction to Δ_0 of an element Φ of $\mathrm{Hom}(\Delta, \mathcal{D}_0[r](\mathbb{Q}_p))^\Gamma$. Since every cusp is $\Gamma_0(p)$-equivalent to 0 or ∞, in order to define Φ it is sufficient to give two distributions $\Phi(0)$ and $\Phi(\infty)$ in $\mathcal{D}_0[r]$ invariant by the stabilizers of 0 and ∞ in $\Gamma_0(p)$ respectively. We set $\Phi(\infty) = 0$, and $\Phi(0)(f) = f'(0)$ for all $f \in \mathcal{A}[r](\mathbb{Q}_p)$. Checking the required invariance of the latter distribution amounts to checking that $f(z)$ and $f(z/(1+pz))$ have the same derivative at 0 for every test function $f \in \mathcal{A}[r](\mathbb{Q}_p)$, which is obvious. To see that Φ is an eigenvector for $\mathcal{H}_0$ we compute for example:

$$\Phi_{|U_p}(0) = \sum_{a=0}^{p-1} \Phi(a/p)_{\left|\begin{pmatrix} 1 & a \\ 0 & p \end{pmatrix}\right.}$$

and in this sum, only the term $a = 0$ matters since for $a \neq 0$, a/p is the $\Gamma_0(p)$-class of ∞. So $\Phi_{|U_p}(0)(f(z)) = \left(\frac{df(pz)}{dz}\right)_{|z=0} = pf'(0) = p\Phi(0)(f)$. On the other hand $\Phi_{|U_p}(\infty) = 0$ since $\begin{pmatrix} 1 & a \\ 0 & p \end{pmatrix} \cdot \infty = \infty$ for all a. Hence we see that $\Phi_{|U_p} = p\Phi$. A similar computation, left to the reader, shows that $\Phi_{|T_l} = (1+l)\Phi$, and that $\Phi_{|\langle a \rangle} = \Phi$. Finally, to compute the sign of Φ, we just observe that

$$\Phi_{\begin{pmatrix} 1 & 0 \\ 0 & -1 \end{pmatrix}}(0)(f(z)) = \Phi(0)(f(-z)) = \left(\frac{df(-z)}{dz}\right)_{|z=0} = -f'(0),$$

and $\Phi_{\begin{pmatrix} 1 & 0 \\ 0 & -1 \end{pmatrix}}(\infty) = 0$ so $\Phi_{|\iota} = -\Phi$. □

7.1.4 Construction

We fix as above $\Gamma = \Gamma_1(N) \cap \Gamma_0(p)$ (with $p \nmid N$).

We are ready to apply the eigenvariety machine in order to construct the p-adic eigencurve of modular symbols (of tame level N). Actually we shall not construct only one eigencurve, but two, denoted $\mathcal{C}^+$, $\mathcal{C}^-$. To unify the two construction, let $\pm$ mean either $+$ or $-$.

We describe the eigenvariety data that will feed the machine.

(ED1) For the Hecke ring we take the ring $\mathcal{H}_0$ generated by symbols T_l $(l \nmid Np)$, U_p and the diamond operators $\langle a \rangle$ for $a \in (\mathbb{Z}/p\mathbb{Z})^*$.

(ED2) For $\mathcal{W}$ we take the usual weight space defined in Sect. 6.3. For $\mathfrak{C}$ we take the following admissible covering: each connected component of $\mathcal{W}$ is isomorphic, after the choice of a generator of $1+q\mathbb{Z}_p$, to the open ball of center 1 and radius 1: we define $\mathfrak{C}$ as the set of closed balls (in those open balls) of center an integer and radius $\rho \in p^{\mathbb{Q}}$, $\rho < 1$. Note that if $W \in \mathfrak{C}$, $W = \operatorname{Sp} R$, then R satisfies Hypothesis 7.1.3.

(ED3) If $W = \operatorname{Sp} R$ is in $\mathfrak{C}$, we choose an r such that $0 < r < \min(r(W), p)$ and

$$M_W := \operatorname{Symb}^{\pm}_{\Gamma}(\mathcal{D}_K[r](R)).$$

Note that M_W is a direct summand of $\operatorname{Symb}^{\pm}_{\Gamma}(\mathcal{D}_K[r](R))$ hence satisfies property (Pr) by Lemma 7.1.4. We have a natural action of the Hecke operators $\psi : \mathcal{H}_0 \to \operatorname{End}_R(M_W)$. The choice of r does not matter since $\operatorname{Symb}^{\pm}_{\Gamma}(\mathcal{D}_K[r](R))$ and $\operatorname{Symb}^{\pm}_{\Gamma}(\mathcal{D}_K[r'](R))$ are linked by Theorem 6.5.6(iii).

We check the required conditions:

(ED1) The action of U_p on M_W is compact by Theorem 6.5.6(i).

(ED2) If $W' = \operatorname{Sp} R' \subset W = \operatorname{Sp} R$, the module $M_W \hat{\otimes}_R R'$ and $M_{W'}$ are linked by the restriction theorem (Theorem 7.1.7).

Hence the eigenvariety machine provides us with an eigenvariety $\mathcal{C}^{\pm}$, which is an equidimensional of dimension 1, separated, rigid analytic space over $\mathbb{Q}_p$ with

(i) A rigid $\mathbb{Q}_p$-analytic map $\kappa : \mathcal{C}^{\pm} \to \mathcal{W}$, which is locally finite and flat.
(ii) Analytic functions T_l (for $l \nmid N$), U_p and $\langle a \rangle$ on $\mathcal{C}^{\pm}$ (that is, the analytic function T_l is defined as the image of $T_l \in \mathcal{H}_0$ by the map $\psi : \mathcal{H}_0 \to \mathcal{O}(\mathcal{C}^{\pm})$).

Note that the flatness of κ does not follow from general properties of eigenvariety, but follows in our case from the fact for $W = \operatorname{Spec} R$ in $\mathfrak{C}$, R is PID, hence any torsion free module is flat over R, in particular the algebra $\mathcal{T}_{W,\nu}$, so $\kappa : \mathcal{C}_{W,\nu} \to W$. Since flat is a local property, the flatness of κ follows.

Lemma 7.1.20 *The functions T_l (for $l \nmid N$), U_p and $\langle a \rangle$ on $\mathcal{C}^{\pm}$ are power-bounded.*

Proof It is enough to prove that for any affinoid of the form $\mathcal{C}_{W,\nu} = \operatorname{Sp} \mathcal{T}_{W,\nu}$ of $\mathcal{C}^{\pm}$, the image of the operators T_l, U_p, etc. in $\mathcal{T}_{W,\nu}$ for one norm of algebra on this Banach algebra is bounded by 1. Let us chose the operator norm on $\mathcal{T}_{W,\nu} \in \mathrm{End}_R(M_W^{\leq \nu})$ attached to the norm $\| \|_r$ on $M_W^{\leq \nu} = \mathrm{Symb}_\Gamma(\mathcal{D}^*[r](R))$. This is Lemma 6.5.4. □

7.2 Comparison with the Coleman-Mazur Eigencurve

We keep the same notation and hypotheses as above. We have constructed two eigencurves $\mathcal{C}^+$ and $\mathcal{C}^-$ with modular symbols and we want to compare them with the classical eigencurves $\mathcal{C}$ (the full eigencurve of Coleman-Mazur and Buzzard) and $\mathcal{C}^0$ (the *cuspidal eigencurve*, which has been widely considered in the literature, but, it seems, never properly defined).

For this we need to explain how these first eigencurves are constructed.

7.2.1 The Coleman-Mazur Full Eigencurve $\mathcal{C}$

The Coleman-Mazur eigencurve can be constructed with the same data (ED1), that is the ring $\mathcal{H}_0$, and (ED2) that is the same weight space $\mathcal{W}$ with the same admissible covering $\mathfrak{C}$ as above, but uses different data (ED3), that we now describe.

Let $X(\Gamma)$ be the completed modular curve of level $\Gamma = \Gamma_1(N) \cap \Gamma_0(p)$. Let us recall [53, page 236] that it is the proper flat scheme of relative dimension 1 over $\mathbb{Z}$ (or algebraic stack in the case $N \leq 2$) representing the functor which to a scheme S attaches the set of isomorphism classes of triplets (E, H, α), where E is a generalized elliptic curve[2] over S, H a locally free rank p subgroup scheme of E/S which meets every irreducible component of every geometric fiber of E/S, and α a structure[3] of level $\Gamma_1(N)$. Let us denote by $X(\Gamma)_{\mathbb{Q}_p}$ the extension of scalars of $X(\Gamma)$ to $\mathbb{Q}_p$, and also the rigid analytic space attached to it. The cusps of $X(\Gamma)_{\mathbb{Q}_p}(\mathbb{C}_p)$ are the points (E, α, H) with $E = \mathrm{Tate}(q^N)$[4] (the only points with E non-smooth). They are in a $\mathrm{SL}_2(\mathbb{Z})$-equivariant bijection with $\mathbb{P}^1(\mathbb{Q})/\Gamma$, and there are (for $N \geq 3$) $\mu_\infty = \frac{1}{2}\sum_{d|N} \phi(d)\phi(N/d)$ of them. The point $(\mathrm{Tate}(q^N), \mu_p, \alpha_{\mathrm{can}})$ where α_{can} is the natural level-N structure on $\mathrm{Tate}(q^N)$ is called the *infinity* cusp, denoted by ∞.

For v be a rational number such that $0 \leq v < p/(p+1)$, Coleman, following earlier work of Katz, defined [38, §B2] an affinoid subdomain $X(\Gamma, v)_{\mathrm{can}}$, of

---

[2]For the definition of a generalized elliptic curve, see [53, Chapter II]. Let us just recall that the smooth locus E^{reg} of a generalized elliptic curve is a group scheme over S, that E is smooth if and only if it is an elliptic curve over S, and that non-smooth generalized elliptic curves over fields correspond to cusps in the moduli schemes of generalized elliptic curves.

[3]The definition of such a structure can be found in [77] but ww will not use it here.

[4]For the definition of the generalized elliptic curve $\mathrm{Tate}(q^N)$, see [53].

$X(\Gamma)_{\mathbb{Q}_p}$, as follows (in the case $p \geq 5$; we refer the reader to [38, §B2] for the general case). Note that by properness, the $\mathbb{C}_p$-points of $X(\Gamma)_{\mathbb{Q}_p}$ are its $\mathcal{O}_{\mathbb{C}_p}$-points.

First we define the affinoid subdomain $X(\Gamma, v)$ of $X(\Gamma)_{\mathbb{Q}_p}$ whose points over $\mathbb{C}_p$ are the triples (E, H, α) where E is a generalized elliptic scheme over $\mathcal{O}_{\mathbb{C}_p}$ such that for η a generator of $\Omega_{E/S}$, $|E_{p-1}(E, \eta)| \geq p^{-v}$. Here E_{p-1} is the usual Eisenstein series of level 1 and weight $p - 1$.[5] For a point (E, H, α) of $X(\Gamma, v)$, under our hypotheses $v < p/(p-1)$, a theorem of Lubin (see [77, Theorem 3.1]) defines a specific locally free rank p subgroup scheme H_{can} of E called the *canonical subgroup*. (In the case where E is an ordinary elliptic curve or E is a non-smooth generalized elliptic curve, then H modulo p^n is isomorphic to μ_p.) We define $X(\Gamma, v)_{\text{can}}$ as the affinoid subdomain of $X(\Gamma, v)$ of points (E, H, α). It is not hard to see that $X(\Gamma, v)_{\text{can}}$ is disjoint from its complement in $X(\Gamma, v)$, and that $X(\Gamma, v)_{\text{can}}$ contains as cusps exactly the cusps of $X(\Gamma)$ which are in the $\Gamma_1(N)$-orbit of ∞ (that is half of them).

Let R be an affinoid $\mathbb{Q}_p$-algebra. Coleman and Mazur [40, page 46] define the Banach R-module of v-overconvergent modular forms $M^\dagger[v](\Gamma, R)$ (denoted $M_U^\dagger(N)(r)$ *loc. cit.*) as the module of R-valued analytic functions on $X(\Gamma, v)_{\text{can}}$. From the definition it is clear that these modules are potentially orthonormalizable Banach R-modukles, and that their formations commutes with base change (i.e. there is a canonical isomorphism $M^\dagger[v](\Gamma, R)\hat{\otimes}_R R' = M^\dagger[v](\Gamma, R')$).

If k is a positive integer, and $R = L$ any finite extension of $\mathbb{Q}_p$, then we see the space of classical modular form $M_k(\Gamma, L)$ as a sub-space of $M^\dagger[v](\Gamma, R)$ by sending f to $f/E_{p-1}^{k/(p-1)}$.

Theorem 7.2.1 (Coleman) *For every affinoid algebra R and every weight $\kappa \in \mathcal{W}(R)$, and for every v sufficiently small with respect to (κ, R), there is an action (called the weight κ action) of $\mathcal{H}_0$ on $M^\dagger[v](\Gamma, R)$.*

For this action, the operator U_p is compact.

This action is compatible with base change, in the sense that if R' is a Banach R-algebra, and κ' is κ seen as an element of $\mathcal{W}(R')$, then v is sufficiently small w.r.t. (R', κ) for the weight-κ-action of $\mathcal{H}_0$ on $M^\dagger[v](\Gamma, R')$ to be defined, and the isomorphism $M^\dagger[v](\Gamma, R)\hat{\otimes}_R R' = M^\dagger[v](\Gamma, R')$ is compatible with weight κ $\mathcal{H}_0$-actions.

Moreover, if $k \geq 2$ is an integer, the inclusion $M_k(\Gamma, L) \to M^\dagger[v](\Gamma, L)$ is $\mathcal{H}_0$-equivariant if we give the target its weight $(k-2)$-action.

Of course, the $k - 2$ appearing at the end of the statement is a pure matter of convention. We could write a k instead, but choosing $k - 2$ will simplify the comparison with modular symbols below.

[5] To understand this definition, recall that E_{p-1} lifts the Hasse-invariant modulo p. Thus $\tilde{X}(\Gamma, 0)$ consist of all points (E, H, α) such that $|E_{p-1}(E, \eta)| = 1$, that is such that either E is a non-smooth generalized elliptic curve (i.e. (E, α, H) is a cusp of $X(\Gamma)_{\mathbb{Q}_p}$), or E is a true elliptic curve and the reduction of E mod p is ordinary. If $v > 0$, then $\tilde{X}(\Gamma, v)$ is a slightly larger affinoid containing also points (E, H, α) where E has supersingular reduction.

The weight-κ action is constructed, and the theorem is proved in [38]. The construction uses the q-expansion of modular forms and the fact that E_{p-1} has q-expansion equal to 1. A more conceptual construction of the weight-κ action, making the theorem obvious, has been given independently in [2] and [97].

We can now describe the data (ED3) for the Coleman-Mazur eigencurve: For every $W = \mathrm{Sp}\, R \subset \mathcal{W}$ in $\mathfrak{C}$, we let $\kappa : \mathbb{Z}_p \to R^*$ be the canonical character, we choose a v small enough, and we set $M_R = M^\dagger[v](\Gamma, R)$ with its weight-κ action. Note that the theorem implies that these data satisfies condition (EC1) and (EC2) of the eigenvariety machine. The eigenvariety defined by this machine is by definition the Coleman-Mazur full eigencurve $\mathcal{C}$.

Not also that the last sentence of the theorem can be rephrased as: one has $\mathcal{H}_0$-equivariant specialization isomorphisms, for $k \in \mathbb{Z} \subset W(\mathbb{Q}_p)$

$$M^\dagger[r](\Gamma, R) \otimes_{R,k} \mathbb{Q}_p = M^\dagger[r]_{k+2}(\Gamma, \mathbb{Q}_p). \tag{7.2.1}$$

For the comparison of $\mathcal{C}$ with $\mathcal{C}^+$ and $\mathcal{C}^-$ We shall need in addition the Coleman's control's theorem:

$$M_{k+2}(\Gamma, \mathbb{Q}_p)^{<k+1} = M^\dagger_{k+2}(\Gamma, \mathbb{Q}_p)^{<k+1}. \tag{7.2.2}$$

This is the main result of [37].

7.2.2 The Cuspidal Eigencurve

Since $M^\dagger[v](R)$ is the space of R-valued functions on $X(\Gamma, v)^{\mathrm{can}}$, and since this rigid analytic space over $\mathbb{Q}_p$ has $\mu_\infty/2 = 1/4 \sum_{d|N} \phi(d)\phi(N/d)$ cusps (if $N \geq 3$), there is a map "evaluations at cusps"

$$\mathrm{ev}_R : M^\dagger[v](R) \to R^{\mu_\infty/2}.$$

We define

$$S^\dagger[v](R) := \ker \mathrm{ev}_R \tag{7.2.3}$$

and we call this module the modules of *cuspidal v-overconvergent forms* on R. For any character $\kappa : \mathbb{Z}_p^* \to R^*$, it follows from Coleman's construction of the weight-κ action of $\mathcal{H}_0$ that $S^\dagger[v](R)$ is $\mathcal{H}_0$-equiavariant. Note that because the formation of ev_R evidently commutes with base change, and because $\mathrm{ev}_{\mathbb{Q}_p}$ is surjective, ev_R is surjective for all R, and the formation of $S^\dagger[v](R)$, which a direct-summend sumodule of $M^\dagger[v](R)$, also commutes with any base change.

Remark 7.2.2 Here is a subtle but important point to keep in mind. If $f \in M_{k+2}(\Gamma, \mathbb{Q}_p)$ is a classical modular form, we can see f as an overconvergent modular form $f \in M^\dagger[v](\mathbb{Q}_p)$. But saying that $f \in S^\dagger[v](\mathbb{Q}_p)$ (i.e. saying that f is *cuspidal as an overconvergent modular form*) **does not imply** that $f \in S_{k+2}(\Gamma)$ (i.e. that f is *cuspidal as a classical form*. The reason for this is that the requirement for $f \in S^\dagger[v](\mathbb{Q}_p)$ is the vanishing of f at all cusps of $X(\Gamma, v)^{\mathrm{can}}$, but they constitute only half the cusps of $X(\Gamma)$.

For instance, if $E_{k,\mathrm{crit}} \in M_k(\Gamma_0(p))$ is the critical-slope refinement $E_k(z) - E_k(pz)$ of the Eisenstien series E_k (see Sect. 6.7.1), it is clear that $E_{k,\mathrm{crit}}$ vanishes at the cusp ∞, hence is cuspidal *as an overconvergent modular form*, but clearly this form is not cuspidal in the classical sense (and in fact it does not vanish at the cusp 0).

We now define a new eigenvariety data, with the same data (ED1) and (ED2) as above, and for (ED3), we take $M_R = S^\dagger[v](R)$ for a v chosen sufficiently small with respect to the canonical character $\kappa : \mathbb{Z}_p \to R^*$. Note that in particular,

$$S^\dagger[r](\Gamma, R) \otimes_{R,k} \mathbb{Q}_p = S^\dagger[r]_{k+2}(\Gamma, \mathbb{Q}_p) \tag{7.2.4}$$

We note that the M_R are potentially orthonormalizable Banach R-modules by Lemma 7.1.5. Conditions (EC1) and (EC2) are thus obviously satisfied. The eigenvariety machine produces an eigencurve, called *the cuspidal eigencurve* $\mathcal{C}_0$.

To compare $\mathcal{C}_0$ with our other eigencurves, we shall need the following version of Coleman's control theorem : one has $\mathcal{H}_0$-equivariant isomorphisms, for $k \in \mathbb{Z} \subset W(\mathbb{Q}_p)$

$$S_{k+2}(\Gamma, \mathbb{Q}_p)^{<k+1} = S^\dagger_{k+2}(\Gamma, \mathbb{Q}_p)^{<k+1} \tag{7.2.5}$$

This can be deduced from (7.2.2) as follows: if $f \in S^\dagger_{k+2}(\Gamma, \mathbb{Q}_p)^{<k+1}$, we need to prove that $f \in S_{k+2}(\Gamma, \mathbb{Q}_p)$. We may assume that f is an eigenform for $\mathcal{H}_0$. By (7.2.2), $f \in M_{k+2}(\Gamma, \mathbb{Q}_p)$ and by (7.2.3), $f_{|\gamma}(\infty) = 0$ for $\gamma \in \Gamma_1(N)$. Since f is an eigenform, it is either a cuspidal form or an Eisenstein series. If $f \in \mathcal{E}_{k+2}(\Gamma, \mathbb{Q}_p)$, then by Exercise 2.6.15, the eigenvalue of U_p on f is p^{k+1} times a root of unity, contradicting the hypothesis. Therefore, $f \in S_{k+2}(\Gamma, \mathbb{Q}_p)$ as wanted.

7.2.3 *Applications of Chenevier's Comparison Theorem*

Theorem 7.2.3

(i) *There exist unique closed immersions*

$$\mathcal{C}^0 \hookrightarrow \mathcal{C}^\pm \hookrightarrow \mathcal{C}$$

which are compatible with the weight morphisms κ to $\mathcal{W}$ and the maps $\mathcal{H}_0 \to \mathcal{O}(\mathcal{C}^0), \mathcal{O}(\mathcal{C}^\pm), \mathcal{O}(\mathcal{C})$. Moreover, these eigencurves are all reduced, and $\mathcal{C} = \mathcal{C}^+ \cup \mathcal{C}^-$.

(ii) *For $k \in \mathbb{Z} \subset \mathcal{W}(\mathbb{Q}_p)$, L a finite extension of $\mathbb{Q}_p$ there exist $\mathcal{H}_0$-injections between the $\mathcal{H}_0$-modules*

$$S^\dagger_{k+2}(\Gamma)(L)^{ss,\#} \subset Symb^\pm_\Gamma(\mathcal{D}_k(L))^{ss,\#}_g \subset M^\dagger_{k+2}(\Gamma)^{ss,\#},$$

where ss means semi-simplification as an $\mathcal{H}_0$-module. One has $Symb^\pm_\Gamma(\mathcal{D}_k(L))_g = Symb^\pm_\Gamma(\mathcal{D}_k(L))$ if $k \neq 0$ and $Symb^\pm_\Gamma(\mathcal{D}_0(L))/Symb^\pm_\Gamma(\mathcal{D}_0(L))_g$ is a line on which $\mathcal{H}_0$ acts by the system of eigenvalues of E_2^{crit} and ι acts by -1.

(iii) *A system of $\mathcal{H}_0$-eigenvalues of finite slope (that is, with a non-zero U_p-eigenvalue) appears in $Symb_\Gamma(\mathcal{D}_k)$ if and only if it appears on $M^\dagger_{k+2}(\Gamma)$ except when $N = 1$ (that is $\Gamma = \Gamma_0(p)$) for the system of eigenvalues of E_2^{crit} which appears in $Symb_{\Gamma_0(p)}(\mathcal{D}_0)$ but not in $M^\dagger_2(\Gamma_0(p))$.*

Proof We shall apply Theorem 3.8.10 repeatedly.

For $x \in \mathcal{W}(L)$, we shall denote by $(M^0_x)^\#$, resp. $(M^\pm_x)^\#$, resp. $(M_x)^\#$, the finite slope part of the fiber at x of the eigenvariety data used to construct $\mathcal{C}^0$, $\mathcal{C}^\pm$, $\mathcal{C}$ respectively. Hence we have, if $x = k \in \mathbb{Z}$, $(M^0_k)^\# = S^\dagger_{k+2}(\Gamma, \mathbb{Q}_p)^\#$ and $(M_k)^\# = M^\dagger_{k+2}(\Gamma, \mathbb{Q}_p)^\#$ by (7.2.1) and (7.2.4), while $(M^\pm_k)^\# = \mathrm{Symb}^\pm_\Gamma(\mathcal{D}_k(\mathbb{Q}_p))_g$ by definition.

We know define classical structures for the eigenvariety data used to construct $\mathcal{C}^0$, $\mathcal{C}^\pm$, $\mathcal{C}^-$ (See Sect. 3.8.1.) In all cases we take for $X = \mathbb{N} \subset \mathcal{W}$ (CSD1) the set of integers $k \geq 0$. It is very Zariski-dense. For $x = k \in \mathbb{N}$, we take for $(M^0_k)^{\mathrm{cl}}$, resp. $(M_k)^{\mathrm{cl}}$ the finite dimensional $\mathcal{H}_0$-modules $S_{k+2}(\Gamma, \mathbb{Q}_p)^\#$, resp. $M_{k+2}(\Gamma, \mathbb{Q}_p)^\#$, while for $(M^\pm_k)^{\mathrm{cl}}$ we take $\mathrm{Symb}^\pm_\Gamma(\mathcal{V}_k(\mathbb{Q}_p))^\#$ which defines the data (CSD2). To check that condition (CSC) is satisfied, fix a $\nu \in \mathbb{R}$. The set of $k \in X$ such that there exists an $\mathcal{H}_0$-isomorphism $(M^*_k)^{\mathrm{cl},\leq\nu} \simeq (M^*_k)^{\leq\nu}$ contains all the k such that $k+1 > \nu$ by Coleman's or by Stevens' control theorem, hence condition (CSC2) is clearly satisfied in all cases.

Having defined those classical structures, we check the hypothesis of Theorem 3.8.10: We observe that we have $\mathcal{H}_0$-equivariant map $S_{k+2}(\Gamma, \mathbb{Q}_p) \hookrightarrow \mathrm{Symb}^\pm_\Gamma(\mathcal{V}_k(\mathbb{Q}_p)) \hookrightarrow M_{k+2}(\Gamma, \mathbb{Q}_p)$. Hence four applications of Theorem 3.8.10 shows that there are unique closed immersions

$$\mathcal{C}^0 \hookrightarrow \mathcal{C}^\pm \hookrightarrow \mathcal{C}$$

compatible with the map to $\mathcal{W}$ and from $\mathcal{H}_0$. Moreover it gives us an $\mathcal{H}_0$-equivariant inclusion

$$S_{k+2}(\Gamma)(L)^{\mathrm{ss},\#} \subset \mathrm{Symb}^\pm_\Gamma(\mathcal{D}_k(L))^{\mathrm{ss},\#}_g \subset M^\dagger_{k+2}(\Gamma)^{\mathrm{ss},\#}.$$

That $\mathcal{C}$ is the union of $\mathcal{C}^+$ and $\mathcal{C}^-$ is a consequence of Exercise 3.7.2 since we have $M_{k+2}(\Gamma, \mathbb{Q}_p) \hookrightarrow \mathrm{Symb}_\Gamma^+(\mathcal{V}_k(\mathbb{Q}_p)) \oplus \mathrm{Symb}_\Gamma^-(\mathcal{V}_k(\mathbb{Q}_p)) = \mathrm{Symb}_\Gamma(\mathcal{V}_k(\mathbb{Q}_p))$.

We know that $\mathrm{Symb}_\Gamma^\pm(\mathcal{D}_k(L))_g^{\mathrm{ss},\#}$ is the same as $\mathrm{Symb}_\Gamma^\pm(\mathcal{D}_k(L))^{\mathrm{ss},\#}$ except perhaps for $k = 0$ and at most one value of the sign $\pm$, where it might be of codimension one. Let us assume that $N = 1$, that is $\Gamma = \Gamma_0(p)$ for a minute: we know by Lemma 7.1.19 that the system of eigenvalues of E_2^{crit} appears in $\mathrm{Symb}_{\Gamma_0(p)}^-(\mathcal{D}_0(L))^{\mathrm{ss},\#}$, but it does not appear in $\mathrm{Symb}_{\Gamma_0(p)}^-(\mathcal{D}_0(L))_g^{\mathrm{ss},\#} \subset M_2^\dagger(\Gamma_0(p))^{\mathrm{ss},\#}$ since "E_2^{crit} is not overconvergent": this is the main result of [41], see also the appendix of [33]. Hence, we see that $\mathrm{Symb}_{\Gamma_0(p)}(\mathcal{D}_0(L))/\mathrm{Symb}_{\Gamma_0(p)}(\mathcal{D}_0(L))_g$ is a line on which $\mathcal{H}_0$ acts by the system of eigenvalues of E_2^{crit} and ι acts by -1. Now, for $N > 1$, we have a natural inclusion $\mathrm{Symb}_{\Gamma_0(p)}(\mathcal{D}_0(L))/\mathrm{Symb}_{\Gamma_0(p)}(\mathcal{D}_0(L))_g \subset \mathrm{Symb}_\Gamma(\mathcal{D}_0(L))/\mathrm{Symb}_\Gamma(\mathcal{D}_0(L))_g$ which is an equality since both spaces have dimension 1. Note however that the system of eigenvalues od E_2^{crit} does appear in $\mathrm{Symb}_\Gamma^\pm(\mathcal{D}_0(L))$ for both values of $\pm$ when $N > 1$ because for any prime factor l of N the critical refinement $E_{2,l}^{\mathrm{crit}}$ of the new exceptional Eisenstein series $E_{2,l}$ belongs to $S_2^\dagger(\Gamma, L)^{\mathrm{ss},\#} \subset \mathrm{Symb}_\Gamma^\pm(\mathcal{D}_0(L))$. This completes the proof of (ii) and (iii).

It remains to show that the eigencurves are reduced. Let us prove this for $\mathcal{C}$, the other cases being similar. Let us check the hypothesis of Theorem 3.8.8. Fix $\nu \in \mathbb{R}$. For $k \in X$ such that $k > 2\nu - 1$, we claim that $M_{k+2}(\Gamma, \mathbb{Q}_p)^{\leq \nu}$ is semi-simple as an $\mathcal{H}_0$-module. All the Hecke operators in $\mathcal{H}_0$ excepted U_p are normal, hence semi-simple. It is enough to prove that U_p acts semi-simply on every generalized $\mathcal{H}_0$-eigenspace of $M_{k+2}(\Gamma)^{\leq \nu}$. For a generalized eigenspace which is new at p, this results from Atkin-Lehner's theory (see Sect. 2.6.4). Consider therefore a generalized eigenspace $M_{k+2}(\Gamma)_{(f)}^{\leq \nu}$ which is old at p, corresponding to a newform f for $\Gamma_1(N')$ for some N' dividing N (possibly $N' = N$), of T_p-eigenvalue a_p and nebentypus ϵ. Since $X^2 - a_p X + p^{k+1}\epsilon(p)$ has a unique root α such that $v_p(\alpha) \leq \nu$ (because $\nu < (k+1)/2$), we have that $g \mapsto g_\alpha$ is an isomorphism $M_{k+2}(\Gamma_1(N))_{(f)} \simeq M_{k+2}(\Gamma)_{(f)}^{\leq \nu}$, and that U_p acts by the scalar α on our generalized eigenspace. This complete the proof of the claim. Since the set of $k > 2\nu - 1$ is obviously Zariski-dense in any ball of center k' and some radius, we can apply Theorem 3.8.8 which tells us that $\mathcal{C}$ is reduced. □

Remark 7.2.4 As we saw in the proof, the spaces of classical modular symbols $\mathrm{Symb}_\Gamma(\mathcal{V}_k(L))$ tend to be semi-simple as $\mathcal{H}_0$-module, except perhaps for the operator U_p on the subspace of old symbols of slope $(k+1)/2$—and even this exception is conjectured not to happen (and known not to happen when $k = 2$).

In contrast, the space of overconvergent modular symbols $\mathrm{Symb}_\Gamma^\pm(\mathcal{D}_k(L))^{\leq \nu}$ are in general not semi-simple as $\mathcal{H}_0$-module (except of course if $\nu < k+1$ since in this case they are isomorphic to submodules of the module of classical modular symbols). This non semi-simplicity happens already for $\nu = k+1$, as we shall see.

Similar remarks can be made with modular symbols replaced by modular forms.

It is therefore a natural question whether we can remove the semi-simplification in (ii) of the theorem, that is if

$$S^{\dagger}_{k+2}(\Gamma)(L)^{\#} \subset \mathrm{Symb}^{\pm}_{\Gamma}(\mathcal{D}_k(L))^{\#}_{g} \subset M^{\dagger}_{k+2}(\Gamma)^{\#},$$

as $\mathcal{H}_0$-module, and not only after semi-simplification (Fabrizio Andreatta, reporting on a joint work with Glenn Stevens and Adrian Iovita in Goa in August 2010 stated this question as a conjecture). We shall see below that this question has a positive answer at most classical points (of slope $k+1$, otherwise this is trivial). However, in general, the answer to this question is no. For a counter-example (at a point corresponding to a classical CM form of weight one), see the author's *CM modules on eigenvarieties and p-adic L-function*, in preparation.

Another natural question in the same vein is whether we can remove the "finite slope" assumption in statement (ii) and (iii) of the theorem. Again, the answer is no. Indeed, by Exercise 6.5.7, $\mathrm{Symb}^{\pm}_{\Gamma}(\mathcal{D}_k(L))$ contains no vector of infinite slope, that is killed by U_p. However, it is well-known that $S^{\dagger}_{k+2}(\Gamma)(L)$ does contain non-zero vectors killed by U_p: take a cuspidal classical new form of weight k for $\Gamma_0(p^n)$ with $n > 1$. One has $U_p f = 0$ (for example from Lemma 7.3.7(ii) below.) Such a form can be interpreted as an overconvergent form in $S^{\dagger}_{k+2}(\Gamma_0(p))(L)$ because the canonical subgroup of order p^n extends into the supersingular annulus.

7.3 Points of the Eigencurve

We keep the notations of the preceding section.

7.3.1 *Interpretations of the Points as Systems of Eigenvalues of Overconvergent Modular Symbols*

Let $x \in \mathcal{C}^{\pm}(L)$, L any finite extension of $\mathbb{Q}_p$. To x we can attach a character $\lambda_x = \mathcal{H}_0 \to L$, $\lambda_x(T) = \psi(T)(x)$ for $T \in \mathcal{H}_0$, in other words a *system of eigenvalues of* $\mathcal{H}_0$.

Theorem 7.3.1 *Let w be a point of $\mathcal{W}$, L' a finite extension of the field $L(w)$. Then the map $x \mapsto \lambda_x$ is a bijection from the set of L'-points x of $\mathcal{W}$ such that $\kappa(x) = w$ onto the set of all systems of eigenvalues appearing in a space $\mathrm{Symb}^{\pm}_{\Gamma}(\mathcal{D}_w(L))^{\#}$ (excepted, when $w = 0$ and $\pm 1 = -1$, the system of eigenvalues of E_2^{crit}).*

Proof This is just Theorem 3.7.1 using (ii) of Theorem 7.2.3. □

7.3.2 Very Classical Points

Definition 7.3.2 A point x in $\mathcal{C}^{\pm}$ of field of definition $L(x)$ is called *very classical*[6] of weight $k \in \mathbb{N}$ if $\kappa(x) = k$ and the system of eigenvalues λ_x appears in $\mathrm{Symb}_{\Gamma}^{\pm}(\mathcal{V}_k(L(x)))$.

We observe that for the classical structure on $\mathcal{C}^{\pm}$ introduced during the proof of (i) of Theorem 7.2.3, the classical points in the sense of Sect. 3.8.1 are the points we call now *very classical*. Hence, by Proposition 3.8.6:

Proposition 7.3.3 *The set of very classical points in $\mathcal{C}^{\pm}$ is very Zariski-dense.*

We shall denote by λ_x^p the restriction of λ_x to $\mathcal{H}_0^p$ (that is forgetting U_p). We define the *minimal level* of x as the minimal level $M_0 \mid Np$ of λ_x^p in $\mathrm{Symb}_{\Gamma}^{\pm}(\mathcal{V}_k(L))$, when λ_x^p is not the system of eigenvalues of E_2, and by convention as being p when λ_x^p is the system of eigenvalues of E_2.

In particular, we say that x is *p-new* if that minimal level is divisible by p, *p-old* if it is not.

Lemma 7.3.4 *If x is a p-new very classical point with $\kappa(x) = k \in \mathbb{Z}$ then $v_p(U_p(x)) = k/2$.*

Proof By Theorem 5.3.35, the system of eigenvalues λ_x appear in $M_{k+2}(\Gamma, L)$ but not in $M_{k+2}(\Gamma_1(N), L)$. This system is either cuspidal or Eisenstein. If it is cuspidal, the result follows from [93, Theorem 4.6.17/2]. If it is Eisenstein, then according to the classification of new Eisenstein series (see Sect. 2.6.4) the system has to be the one of the exceptional Eisenstein series $E_{2,p}$ (since the p-valuation of the level of normal new Eisenstein form $E_{k,\psi,\tau}$ with trivial nebentypus at a prime p is always even, because the level of the form is the product of the conductors of the two primitive Dirichlet characters ψ and τ, while the nebentypus is $\psi\tau$.) In this case, we have $v_p(U_p(x)) = 0$ (exceptional new Eisenstein series are ordinary), and $k = 0$, hence the result. □

Proposition 7.3.5 *The set of very classical p-old points in $\mathcal{C}^{\pm}$ is very Zariski-dense.*

Proof This follows also from Proposition 3.8.6, just replacing in the classical structure the space $\mathrm{Symb}_{\Gamma}(\mathcal{V}_k(\mathbb{Q}_p))^{\#}$ by its subspace $\mathrm{Symb}_{\Gamma}(\mathcal{V}_k(\mathbb{Q}_p))^{\#,p-\mathrm{old}}$. The preceding lemma ensures that condition (CSC) is still satisfied. □

7.3.3 Classical Points

Definition 7.3.6 Let L be a complete field containing $\mathbb{Q}_p$. An L-point x in $\mathcal{C}^{\pm}(L)$ is *classical* if the system of eigenvalues λ_x appears in $\mathrm{Symb}_{\Gamma_1(M)}^{\pm}(\mathcal{V}_k(L))$ for some $k \in \mathbb{N}$, $M \in \mathbb{N}$, $M \geq 1$.

[6] French *très classique*. This terminology seems to be due to Chenevier.

It amounts to the same to requiring that the system of eigenvalues λ_x appears in the space of modular forms $M_{k+2}(\Gamma_1(M), L)$. It is not hard to see, using Theorem 5.3.35 and the Hecke estimate on coefficients of modular forms, that the integer k is uniquely determined by x. We shall call it the *weight* of the classical point x, but not without warning the reader that it is not always true, as we shall see, that $\kappa(x) = k$. Also, as for very classical points, we define the minimal level M_0 of x as the minimal level of the system of eigenvalues λ_x^p of $\mathcal{H}_0^p$ in $\mathrm{Symb}^{\pm}_{\Gamma_1(M)}(\mathcal{V}_k(L))$. Writing $M_0 = N_0 p^{t_0}$ with $p \nmid N_0$, we call N_0 the *minimal tame level* and p^{t_0} the *minimal wild level*.

While the very classical points appear on the eigencurve, in a sense, *by construction*, it is not clear a priori that there are any classical points on the eigencurve beyond the very classical points. The aim of the next section is to produce a good supply of classical points. Later, with help of the Galois representations, we shall prove that this supply together with the very classical points exhaust all classical points.

7.3.4 Hida Classical Points

Remember that for any congruence subgroup Γ containing $\Gamma_0(p)$, U_p denotes the Hecke operator $[\Gamma\begin{pmatrix}1 & 0\\ 0 & p\end{pmatrix}\Gamma]$.

Lemma 7.3.7 *Let L be a finite extension of $\mathbb{Q}_p$, W an L-vector space with a structure of right $S_0(p)$-module.*

(i) *If $w \in W^{\Gamma_0(p^t)\cap\Gamma_1(N)}$ with $p \nmid N$, for some $t \geq 1$, and $U_p w = bw$ with $b \neq 0$, then $w \in W^{\Gamma_0(p)\cap\Gamma_1(N)}$.*
(ii) *Let $w \in W^{\Gamma_1(p^t N)} - W^{\Gamma_1(p^{t-1}M)}$ with $p \nmid N$, for some $t \geq 1$. Assume that $U_p w = bW$, and that the diamond operators $\langle a\rangle$ for $a \in (\mathbb{Z}/p^t\mathbb{Z})^*$ act on w through a character ϵ of $(\mathbb{Z}/p^t\mathbb{Z})^*$. Then $b \neq 0$ if and only if ϵ is primitive.*

Proof An easy matrix computation (cf. e.g. [93, Lemma 4.5.11]) show that for $t > 1$,

$$(\Gamma_0(p^t)\cap\Gamma_1(N))\begin{pmatrix}1 & 0\\ 0 & p\end{pmatrix}(\Gamma_0(p^t)\cap\Gamma_1(N)) = (\Gamma_0(p^t)\cap\Gamma_1(N))\begin{pmatrix}1 & 0\\ 0 & p\end{pmatrix}(\Gamma_0(p^{t-1})\cap\Gamma_1(N)).$$

Since the first double class describes the action of U_p, one sees that $bw = U_p w \in W^{\Gamma_0(p^{t-1})\cap\Gamma_1(N)}$, hence w is in this space, so (i) follows by induction on t.

The proof of (ii) is similar, see [93, theorem 4.6.17] for details. □

Now let us fix an integer $k \geq 0$, an integer $t \geq 1$, and a Dirichlet character $\epsilon : (\mathbb{Z}/p^t\mathbb{Z})^* \to L^*$, where L is some finite extension of $\mathbb{Q}_p$.

We see ϵ as an element of $\mathcal{W}(L)$ by precomposing it with the canonical surjection $\mathbb{Z}_p^* \to (\mathbb{Z}/p^t\mathbb{Z})^*$. Let us call $w \in \mathcal{W}(L)$ the character $z \mapsto z^k\epsilon(k)$. We clearly have $r(\epsilon) = r(w) = p^{-\nu}$ (see Definition 6.3.5) if p^ν is the conductor of ϵ. For $0 < r < r(\epsilon)$, consider the inclusion $\mathcal{P}_k[r](L) \subset \mathcal{A}_w[r](L)$.

Lemma 7.3.8 *The left-action of $S_0(p)$ on $\mathcal{A}_w[r]$ leaves stable $\mathcal{P}_k[r]$*

Proof For $\gamma = \begin{pmatrix} a & b \\ c & d \end{pmatrix} \in S_0(p)$, and P in $\mathcal{P}_k[r]$, $P_{|_w\gamma}(z) = \epsilon(a - cz)P_{|_k\gamma}(z)$ and $\epsilon(a-cz) = \epsilon(a)\epsilon(1-\frac{c}{a}z)$ is constant on every closed ball of radius r since $p|\frac{c}{a}$. □

We denote by $\mathcal{P}_w[r](L)$ the space $\mathcal{P}_k[r](L)$ with the action of $S_0(p)$ induced by its action on $\mathcal{A}_w[r]$, and by $\mathcal{V}_w[r](L)$ its L-dual with its dual right $S_0(p)$-action. Hence we get a surjective $S_0(p)$-equivariant map $\rho_w : \mathcal{D}_w[r](L) \to \mathcal{V}_w[r](L)$.

The results of the following proposition are proved exactly as in the case $\epsilon = 1$ (cf. Proposition 6.5.18 and Theorem 6.5.19). We leave the details to the reader.

Proposition 7.3.9 *The map*

$$\rho_w : Symb_\Gamma(\mathcal{D}_w[r](L))^\# \to Symb_\Gamma(\mathcal{V}_w[r](L))^\#$$

is surjective, and induces an isomorphism

$$\rho_w : Symb_\Gamma(\mathcal{D}_w[r](L))^{<k+1} \to Symb_\Gamma(\mathcal{V}_w[r](L))^{<k+1}.$$

Proposition 7.3.10 *We have a natural isomorphism, compatible with the action of $\mathcal{H}_0$*

$$Symb_\Gamma(\mathcal{V}_w[r](L))^\# = Symb_{\Gamma_1(Np^t)}(\mathcal{V}_k(L))[\epsilon]^\#.$$

Proof First, by Lemma 7.3.7(i), we have

$$\mathrm{Symb}_\Gamma(\mathcal{V}_w[r](L))^\# = \mathrm{Symb}_{\Gamma_0(p^t)\cap\Gamma_1(N)}(\mathcal{V}_w[r](L))^\#.$$

Next, observe that the action of $\Gamma_1(Np^t)$ on $\mathcal{V}_w[r](L)$ is the same as its action on $\mathcal{V}_k[r](L)$ since $\epsilon(a - cz) = 1$ for $|z| \leq 1$ and $a \equiv 1 \pmod{p^t}$, $p^t \mid c$. Hence an obvious inclusion map

$$\mathrm{Symb}_{\Gamma_0(p^t)\cap\Gamma_1(N)}(\mathcal{V}_w[r](L)) \subset \mathrm{Symb}_{\Gamma_1(Np^t)}(\mathcal{V}_k[r](L)).$$

The right hand side has an action of the diamonds operators $\langle a\rangle$ for $a \in (\mathbb{Z}/p^t\mathbb{Z})^*$, and one sees easily that the above inclusion induces an equality

$$\mathrm{Symb}_{\Gamma_0(p^t)\cap\Gamma_1(N)}(\mathcal{V}_w[r](L)) = \mathrm{Symb}_{\Gamma_1(Np^t)}(\mathcal{V}_k[r](L))[\epsilon]. \qquad (7.3.1)$$

Finally, $\mathrm{Symb}_{\Gamma_1(Np^t)}(\mathcal{V}_k[r](L))^\# = \mathrm{Symb}_{\Gamma_1(Np^t)}(\mathcal{V}_k(L))^\#$ by a trivial generalization of Proposition 6.5.13. □

Theorem 7.3.11 *There exists a natural $\mathcal{H}_0$-equivariant surjective map*

$$Symb_\Gamma(\mathcal{D}_w(L))^\# \to Symb_{\Gamma_1(Np^t)}(\mathcal{V}_k(L))[\epsilon]^\#$$

which induces an isomorphism

$$Symb_\Gamma(\mathcal{D}_w(L))^{<k+1} \to Symb_{\Gamma_1(Np^t)}(\mathcal{V}_k(L))[\epsilon]^{<k+1}$$

Proof This results from the two propositions above, using that $\mathrm{Symb}_\Gamma(\mathcal{D}_w[r](L))^\# = \mathrm{Symb}_\Gamma(\mathcal{D}_w(L))^\#$ (Theorem 6.5.6). □

Proposition and Definition 7.3.12 *Let L be a finite extension of $\mathbb{Q}_p$. For $t \geq 1$, and ϵ a primitive character of $(\mathbb{Z}/p^t\mathbb{Z})^* \to L^*$, every system of $\mathcal{H}_0$-eigenvalues λ appearing in $Symb_{\Gamma_1(Np^t)}(\mathcal{V}_k(L))[\epsilon]^\#$ defines a unique point $x \in \mathcal{C}(L)$ such that $\lambda_x = \lambda$ and $\kappa(x)$ is the character $z \mapsto z^k\epsilon(z)$. This point x is classical of weight k and level Np^t. We shall call those classical points Hida classical points.*

Proof The first assertion follows from Theorem. 7.3.11. The rest is obvious. □

In other words a point of the eigencurve is a Hida classical point if $\kappa(x)$ has the form $z \mapsto z^k\epsilon(z)$ for a (clearly unique) integer k and a (clearly unique) non-trivial finite order character ϵ, and its system of eigenvalues λ_x appears in $\mathrm{Symb}_{\Gamma_1(Np^t)}(\mathcal{V}_k(L))[\epsilon]^\#$ for k and ϵ determined by $\kappa(x)$ as above, and for the integer $t \geq 1$ such that p^t is the conductor of ϵ.

Remark 7.3.13 The reader should note that for a Hida classical point x of weight k and level Np^t, one does **not** have $\kappa(x) = k$. Instead, $\kappa(x)$ is the character $z \mapsto z^k\epsilon(z)$, where ϵ is a primitive character of conductor p^t. In particular, a Hida classical point is not very classical.

Proposition 7.3.14 *Let $x \in \mathcal{C}(L)$ be a point such that $\kappa(x)$ is of the form $z \mapsto z^k\epsilon(z)$ with ϵ a finite order character, and assume that $v_p(U_p(x)) < k+1$. Then x is either a very classical point or a Hida classical point. In particular, it is a classical point.*

Proof If $\epsilon = 1$, then we already know that x is very classical by Theorem 6.5.19. If $\epsilon \neq 1$, it is Hida classical by Theorem 7.3.11. □

Remark 7.3.15 At this stage, it is not clear that every classical point is either a Hida classical point or a very classical point. We don't even know that there are no classical points whose minimal tame level is not a divisor of N, or such that $\kappa(x)$ is not of the form $z \mapsto z^k\epsilon(z)$ for $k \in \mathbb{N}$ and ϵ a finite order character. With the help of the family of Galois representations carried by the eigencurve, it will be easy to prove these results: see Theorem 7.4.6 below.

7.4 The Family of Galois Representations Carried by the Eigencurve

7.4.1 Construction of the Family of Galois Representations

For the definition and basic properties of pseudorepresentations, see Sect. 2.6.7

Note that the map $\kappa : \mathcal{C}^\pm \to \mathcal{W}$ determines by definition of $\mathcal{W}$ a continuous morphism of groups $\mathbb{Z}_p^* \to \mathcal{O}(\mathcal{C}^\pm)^*$, that we shall also denote by κ.

Theorem 7.4.1 *There exists a unique continuous two-dimensional pseudorepresentation (τ, δ) of $G_{\mathbb{Q},Np}$ with values in $\mathcal{O}(\mathcal{C}^\pm)$, such that for every l prime to Np,*

$$\tau(\mathrm{Frob}_l) = T_l.$$

Moreover we have $\tau(c) = 0$, where c is a complex conjugation in $G_{\mathbb{Q},Np}$, and δ factors through the quotient

$$\mathrm{Gal}(\mathbb{Q}(\zeta_{Np^\infty})/\mathbb{Q}) = \mathbb{Z}_p^* \times (\mathbb{Z}/N\mathbb{Z})^*$$

of $G_{Q,Np}$. Its restriction to $\mathbb{Z}_p^*$ is the map $t \mapsto t\kappa(t)$, $\mathbb{Z}_p^* \to \mathcal{O}(\mathcal{C}^\pm)^*$ and its restriction to $(\mathbb{Z}/N\mathbb{Z})^*$ sends a to $\langle a\rangle \in \mathcal{O}(\mathcal{C}^\pm)^*$.

Proof We shall prove Theorem 7.4.1 by an interpolation argument due to Chenevier. Suppose we have a Zariski-dense set of points $Z \in \mathcal{C}^\pm$ such that for $z \in Z$ there exists a Galois representation $\rho_z : G_{\mathbb{Q},Np} \to \mathrm{GL}_2(\bar{\mathbb{Q}}_p)$ satisfying $\mathrm{tr}\,\rho_z(\mathrm{Frob}_l) = T_l(z)$ for all l not dividing Np. Let $\mathcal{O}(\mathcal{C}^\pm)^0$ be the set of power-bounded elements in $\mathcal{O}(\mathcal{C}^\pm)$. Consider the map

$$\mathrm{ev}_Z : \mathcal{O}(\mathcal{C}^\pm)^0 \xrightarrow{\prod_{z\in Z}\mathrm{ev}_z} \prod_{z\in Z}\bar{\mathbb{Q}}_p.$$

This is obviously a continuous morphism of algebras for the product topology on the target; it is injective by density of Z and reducedness of $\mathcal{C}$; and since $\mathcal{O}(\mathcal{C}^\pm)^0$ is compact by Proposition 3.7.9, the range of ev_Z is compact (hence closed in $\prod_{z\in Z}\bar{\mathbb{Q}}_p$) and ev_Z is an homeomorphism of its source over its image.

Now consider the continuous map

$$\tau_Z : G_{\mathbb{Q},Np} \xrightarrow{\prod_{z\in Z}\mathrm{tr}\,\rho_z} \prod_{z\in Z}\bar{\mathbb{Q}}_p.$$

We have $\tau_Z(\mathrm{Frob}_l) = (T_l(z))_{z\in Z} = \mathrm{ev}_Z(T_l)$ for $l \nmid Np$, and since $T_l \in \mathcal{O}(\mathcal{C}^\pm)^0$ (Lemma 7.1.20), $\tau_Z(\mathrm{Frob}_l)$ lies in the image of ev_Z. Since the Frob_l's are dense in $G_{\mathbb{Q},Np}$, τ_Z is continuous, and the image of ev_Z is closed, we deduce that

$\tau_Z(G_{\mathbb{Q},Np})$ entirely lies in the image of ev_Z. We can therefore define $\tau = \mathrm{ev}_Z^{-1} \circ \tau_Z : G_{\mathbb{Q},Np} \to \mathcal{T}^{\pm}$, which is a continuous map. Moreover $\mathrm{ev}_Z(\tau(\mathrm{Frob}_l)) = \tau_Z(\mathrm{Frob}_l) = (T_l(z))_{z\in Z}$, while $\mathrm{ev}_Z(T_l) = (T_l(z))_{z\in Z}$. Hence $\tau(\mathrm{Frob}_l) = T_l$ by injectivity of ev_Z.

To apply the above, take for Z the set of very classical points in $\mathcal{C}^{\pm}$. The existence of a suitable ρ_z for $z \in Z$ is known by Eichler-Shimura (in weight $k+2=2$) and Deligne (in weight $k+2>2$). Indeed, if $z \in Z$, the character $\kappa(z)$ is an integral character $w \mapsto w^{k_z}$, and z corresponds to a classical modular symbol ϕ_z of weight k_z (see Theorem 7.3.1), hence to a modular form f_z of weight k_z+2 and level $\Gamma_0(p) \cap \Gamma_1(N)$, which is an eigenform for $\mathcal{H}_0$ which eigenvalue for $T \in \mathcal{H}_0$ the value $T(z)$ of the function T on $\mathcal{C}^{\pm}$. The Eichler-Shimura relation gives $\mathrm{tr}\,\rho_z(\mathrm{Frob}_l) = T_l(z)$. The density of Z follows from Proposition 7.3.3.

Since the ρ_z are odd, $\mathrm{tr}\,\rho_z(c) = 0$ for $z \in Z$, and by density $\tau(c) = 0$.

By Eichler-Shimura relations we have $\det \rho_z = \omega_p^{k+1}\epsilon_z(\omega_N)$, where $\omega_p : G_{\mathbb{Q},Np} \to \mathrm{Gal}(\mathbb{Q}(\mu_p^{\infty})/\mathbb{Q}) = \mathbb{Z}_p^*$ is the cyclotomic character, and ω_N is the natural map $G_{\mathbb{Q},Np} \to \mathrm{Gal}(\mathbb{Q}(\mu_M)/\mathbb{Q}) = (\mathbb{Z}/M\mathbb{Z})^*$. Hence, δ_z factors through $G_{\mathbb{Q},Np} \stackrel{\omega_p\times\omega_N}{\longrightarrow} \mathbb{Z}_p^* \times (\mathbb{Z}/N\mathbb{Z})^*$ and is the map $t \mapsto \kappa(t)(z)\,t$ on $\mathbb{Z}_p^*$ and $a \mapsto \langle a\rangle(z)$ on $(\mathbb{Z}/N\mathbb{Z})^*(z)$. So if we define $\delta : G_{\mathbb{Q},Np} \to \mathcal{O}(\mathcal{C}^{\pm})^*$ as in the statement in the theorem, it is clear that $\delta(z) = \delta_z$ for every $z \in Z$.

To check that (τ, δ) is a pseudorepresentation of dimension 2, it suffices to check the four conditions of Definition 2.6.23, but since Z is Zariski-dense, and τ, δ are analytic (that is, in $\mathcal{O}(\mathcal{C}^{\pm})$), it suffices to prove that those relations holds for $(\tau(z), \delta(z))$ at each point $z \in Z$, and these relation holds since by Chebotarev $\tau(z) = \tau_z = \mathrm{tr}\,\rho_z$, $\delta(z) = \delta_z = \det \rho_z$ and $(\mathrm{tr}\,\rho_z, \det \rho_z)$ is indeed a pseudorepresentation.

Finally, the uniqueness assertion is clear since by Chebotarev τ is determined by the condition put on it, and since 2 is invertible in $\mathcal{O}(\mathcal{C}^{\pm})$, $\delta(g) = (\tau(g)^2 - \tau(g^2))/2$ for every $g \in G_{\mathbb{Q},Np}$, and δ is determined by τ. □

Corollary 7.4.2 *If L is any finite extension of $\mathbb{Q}_p$, and every $x \in \mathcal{C}^{\pm}(L)$, there exists a unique continuous semi-simple Galois representation*

$$\rho_x : G_{\mathbb{Q},Np} \to \mathrm{GL}_2(L)$$

such that for every prime l not dividing Np,

$$\mathrm{tr}\,\rho_x(\mathrm{Frob}_l) = T_l(x).$$

One has $\mathrm{tr}\,\rho_x(c) = 0$. *Moreover* $\det \rho_x$ *factors through the quotient*

$$\mathrm{Gal}(\mathbb{Q}(\zeta_{Np^{\infty}})/\mathbb{Q}) = \mathbb{Z}_p^* \times (\mathbb{Z}/N\mathbb{Z})^*$$

of $G_{Q,Np}$. Its restriction to $\mathbb{Z}_p^*$ is the map $w \mapsto z\kappa_x(w)$, $\mathbb{Z}_p^* \to L(x)^*$ and its restriction to $(\mathbb{Z}/N\mathbb{Z})^*$ sends a to $\langle a\rangle(x) \in L(x)^*$.

Proof If we post-compose τ and δ with the *evaluation at* x morphism $\mathrm{ev}_x : \mathcal{O}(\mathcal{C}^{\pm}) \to L$, we get a continuous pseudorepresentation of dimension 2: $(\tau_x = \mathrm{ev}_x \circ \tau, \delta_x = \mathrm{ev}_x \circ \delta)$ of $G_{\mathbb{Q},Np}$ with values in L.

By Theorem 2.6.24, over a suitable finite extension L' of L, τ_x and δ_x are the trace and determinant of a unique (up to isomorphism) semi-simple continuous representation

$$\rho_{x,L'} : G_{\mathbb{Q},Np} \to \mathrm{GL}_2(L').$$

Let us assume first that $\rho_{x,L'}$ is absolutely irreducible. By the general theory of descent of representations, since $\mathrm{tr}\,\rho_{x,L'}(G_{\mathbb{Q},Np}) \subset L$, there exists an L-algebra D of dimension 4, with center L, satisfying $D \otimes L' \simeq M_2(L')$ (hence defining a morphism of group $D^* \hookrightarrow M_2(L')^* = \mathrm{GL}_2(L')$) and a continuous morphism of groups $\rho_x : G_{\mathbb{Q},Np} \to D^*$ such that the composition $G_{\mathbb{Q},Np} \stackrel{\rho_x}{\to} D^* \stackrel{i}{\to} \mathrm{GL}_2(L')$ is isomorphic to the representation $\rho_{x,L'}$. Moreover the algebra D is either $M_2(L)$ or is a division algebra. Let us prove that D cannot be a division algebra. If it was, the element $\rho_x(c) \in D^*$ would satisfy $\rho_x(c)^2 = 1$, hence $\rho_x(c) = \pm 1$ in D contradicting $\mathrm{tr}\,\rho_x(c) = 0$. Hence $D = M_2(L)$, and $\rho_x : G_{\mathbb{Q},Np} \to \mathrm{GL}_2(L)$ is the continuous representation we were looking for.

If $\rho_{x,L'}$ is absolutely reducible, then up to replacing L' by a finite extension, we may assume that $\rho_{x.L'}$ is the sum of two characters. In particular $\rho_{x,L'}(G_{\mathbb{Q},Np})$ is commutative. The element $\rho_{x,L'}(c)$ has square 1 and trace 0. Hence up to changing $\rho_{x,L'}$ by a conjugate we can assume that $\rho_{x,L'}(c) = \begin{pmatrix} 1 & 0 \\ 0 & -1 \end{pmatrix}$. Let $g \in G_{\mathbb{Q},Np}$. Since the matrix $\rho_{x,L'}(g)$ commutes with $\begin{pmatrix} 1 & 0 \\ 0 & -1 \end{pmatrix}$ it is diagonal, hence of the form $\begin{pmatrix} a & 0 \\ 0 & d \end{pmatrix}$ with $a, b \in L'$. Since $\mathrm{tr}\,\rho_{x,L'}(g) \in L$ and $\mathrm{tr}\,\rho_{x,L'}(gc) \in L$, we obtain $a + c \in L$ and $a - d \in L$, which implies that a and d are in L. Hence χ_1 and χ_2 take values in L^*, and $\rho_x = \chi_1 \oplus \chi_2$ satisfies the required properties. □

7.4.2 *Local Properties at $l \neq p$ of the Family of Galois Representations*

We shall prove some important local properties of the Galois representations ρ_x. Before stating them, a few notations and reminders:

For any prime l, we call $G_{\mathbb{Q}_l}$ the absolute Galois group of $\mathbb{Q}_l$. We have a natural $G_{\mathbb{Q},Np}$-conjugacy class of maps $i_l : G_{\mathbb{Q}_l} \to G_{\mathbb{Q},Np}$ whose image is a decomposition group at l and for ρ a representation of $G_{\mathbb{Q},Np}$ we denote by $\rho_{|G_{\mathbb{Q}_l}}$ the composition $\rho \circ i_l$ which is a representation of $G_{\mathbb{Q}_l}$ well-defined up to isomorphism.

Let us recall the notion of tame conductor $N(\rho)$ of a representation ρ, due to Serre (using work of Swan and Grothendieck): for l a prime different from p, we set

$$n_l(\rho_{|G_{\mathbb{Q}_l}}) = \dim \rho - \dim \rho^{I_l} + \mathrm{sw}(\mathrm{gr}\rho).$$

Here I_l is the inertia subgroup of $G_{\mathbb{Q}_l}$, $\mathrm{gr}\rho$ is the representation of I_l on the sum of the graded pieces of $\rho_{|G_{\mathbb{Q}_l}}$ with respect to the Grothendieck filtration, and sw is the Swan conductor of that representation of I_l. Remember that I_l acts on $\mathrm{gr}\rho$ through a finite quotient. Finally, we define the tame conductor by:

$$N(\rho) = \prod_{l \neq p} l^{n_l(\rho_{|G_{\mathbb{Q}_l}})}.$$

We recall from the theory of p-adic families of $G_{\mathbb{Q}_l}$-representations in [13] the following result:

Lemma 7.4.3 *Let l be a prime distinct from p. Let $T : G_{\mathbb{Q}_l} \to \mathcal{O}(X)$ be a continuous 2-dimensional pseudocharacter, where X is a rigid analytic space over $\mathbb{Q}_p$. For $y \in X$, let $\rho_y : G_{\mathbb{Q}_l} \to \mathrm{GL}_2(\bar{\mathbb{Q}}_p)$ be the unique semi-simple representation of trace T_y. Let $x \in X$. There is a Zariski-open subset U of X containing x, such that for every $y \in U$, $n_l(\rho_y) \geq n_l(\rho_x)$. If ρ_x is not isomorphic to an unramified twist of $1 \oplus \omega_p$, then there is a Zariski-open subset U of X containing x, $n_l(\rho_y) = n_l(\rho_x)$.*

Proof By [13, Lemma 7.8.11] we are reduced to the case where there is a continuous representation $\rho : G_{\mathbb{Q}_l} \to \mathrm{GL}_2(R)$ whose trace is equal to the restriction of T to U. In this case, since $\mathrm{gr}\rho$ factors through a finite quotient of G, it is a locally constant representation and so is its Swan conductor. Thus we only have to prove that $\dim \rho_y^{I_l} \leq \dim \rho_x^{I_l}$ for $y \in U$ and $\dim \rho_y^{I_l} = \dim \rho_x^{I_l}$ if ρ_x is not isomorphic to an unramified twist of $1 \oplus \omega_p$. But this follows from [13, §7.8.3]. □

Proposition 7.4.4 *Let f be a modular form, eigenform for $\mathcal{H}_0$, and assume f is not an exceptional Eisenstein series. If $N_0 p^t$ (with N_0 prime to p, $t \geq 0$) is the minimal level of f, one has $N(\rho_f) = N_0$.*

Proof Let f' be the new form attached to f, so f' is of level $N_0 p^t$. For a cuspidal form f', the result is a famous theorem of Carayol (cf. [31] and [85, Lemma 4.1]). If $f = E_{k+2,\tau,\psi}$ with τ, ψ primitive of conductor Q and R, and minimal level QR, the Galois representation ρ_f is $\tau \oplus \psi\omega_p^{k+1}$ so $N(\rho_f)$ is the prime-to-p part of $QR = N_0 p^t$. □

Corollary 7.4.5 *If x is a point of the eigencurve $\mathcal{C}^{\pm}$, the tame conductor $N(\rho_x)$ of ρ_x divides N.*

Proof This follows from the preceding proposition and Lemma 7.4.3. □

Theorem 7.4.6 *The only classical points on $\mathcal{C}^{\pm}$ are the very classical points and the Hida classical points.*

Proof If x is a classical point of weight k and minimal level $N_0 p^t$ with N_0 not divisible by p, then by definition ρ_x is the representation attached to the system of eigenvalues λ_x which appears in $\mathrm{Symb}_{\Gamma_1(N_0 p^t)}(\mathcal{V}_k(L))$ and is new. Since the diamond operators $\langle a \rangle$ for $a \in (\mathbb{Z}/p^t\mathbb{Z})^*$ commutes with $\mathcal{H}_0$, λ_x must appear in $\mathrm{Symb}_{\Gamma_1(N_0 p^t)}(\mathcal{V}_k(L))[\epsilon]$ for some character ϵ of $(\mathbb{Z}/p^t\mathbb{Z})^*$. Hence the restriction to $\mathbb{Z}_p^*$ of the determinant of ρ_x is $z \mapsto z^{k+1}\epsilon(z)$. By Corollary 7.4.2, $\kappa(x)(z) = z^k\epsilon(z)$, and by Corollary 7.4.5, $N_0 \mid N$.

Now we distinguish two cases: the first case is when $t = 0$ or $t = 1$, and the character ϵ is trivial. In this case λ_x appears in $\mathrm{Symb}_{\Gamma_0(p)\cap\Gamma_1(N_0)}(\mathcal{V}_k(L))$, hence in $\mathrm{Symb}_\Gamma(\mathcal{V}_k(L))$ and we have $\kappa(x) = k$ by the above paragraph. Hence x is a very classical point (p-new if $t = 1$ and p-old if $t = 0$). The second case is when $t \geq 2$ or ϵ is non-trivial. Since $t = 0$ implies obviously $\epsilon = 1$, this means either that $t = 1$ and ϵ non-trivial, or $t \geq 2$. Note that in both case, ϵ is primitive: for $t = 1$, any non-trivial character if $(\mathbb{Z}/p^t\mathbb{Z})^*$ is primitive, and for $t > 1$, this follows from Lemma 7.3.7(ii). Summarizing, λ_x appears in $\mathrm{Symb}_{\Gamma_1(Np^t)}(\mathcal{V}_k(L))[\epsilon]$ and $\kappa(x)$ is the character $z \mapsto z^k\epsilon(z)$ with ϵ primitive: this is exactly the definition of x being Hida-classical. □

Definition 7.4.7 Let x be a classical point on the eigencurve $\mathcal{C}^{\pm}$. We shall say that x is *normal* if we have both:

(N1) The system of eigenvalues $\lambda_{x,0}$ of $\mathcal{H}_0$ is not the system of eigenvalues of E_2.
(N2) The system of eigenvalues $\lambda_{x,0}$ of $\mathcal{H}_0$ is not the system of eigenvalues of a new Eisenstein series $E_{2,\psi,\tau}$ of weight 2 with τ and ψ primitive Dirichlet characters of conductor Q and R, and there is a prime factor l of N such that such that $\tau_{|G_{\mathbb{Q}_l}} = \psi_{|G_{\mathbb{Q}_l}}$.

Note that all cuspidal classical points, and all classical points corresponding to an Eisenstein series of weight > 2, or of weight 2 but with square-free level, are normal.

Lemma 7.4.8 *Let x be classical point on the eigencurve of tame minimal level N_0. If x is normal, then in a sufficiently small neighborhood of x all classical points have minimal level N_0.*

Proof If x is normal, there exists by Atkin-Lehner's theory a new form f of tame level N_0 whose system of eigenvalues for $\mathcal{H}_0$ is λ_x, and $\rho_x = \rho_f$. For l a prime different from p, we need to show that $n_l(\rho_y)$ is constant on a neighborhood of x. There is nothing to prove for $l \nmid N$. For $l \mid N$, the result follows from Lemma 7.4.3, excepted when $(\rho_f)_{|G_{\mathbb{Q}_l}}$ is $1 \oplus \omega_p$ up to an unramified twist (and in particular, ρ_f is unramified at ℓ and $\rho_f(\mathrm{Frob}_\ell)$ has two eigenvalues α and αl.) But we claim that that this cannot happen if f is normal. If f is cuspidal, and ρ_f unramified at l, then the two eigenvalues of $\rho_f(\mathrm{Frob}_\ell)$ are known to have the same complex absolute value, which is not the case here. If f is Eisenstein, then $(\rho_f)_{|G_{\mathbb{Q}_l}}$ is $1 \oplus \omega_p$ up to an unramified twist only if f is exceptional, excluded by (N1), or if $f = E_{2,\tau,\psi}$, with $\tau_{|G_{\mathbb{Q}_l}} = \psi_{|G_{\mathbb{Q}_l}}$ and those two characters are unramified, excluded by (N2). □

7.4.3 *Local Properties at p of the Family of Galois Representations*

In this subsection and in the rest of the chapter, we suppose the reader somewh familiar with p-adic Hodge theory, or at least to know the basic properties of Se weights and of the functor D_{crys}. A good introduction to this subject can be foun in the notes of the Clay Hawaii Summer School, in particular [11] and [24].

Lemma 7.4.9 *For* $x \in \mathcal{C}^{\pm}$, *the Hodge-Tate-Sen weights of* $(\rho_x)_{|G_{\mathbb{Q}_p}}$ *are* 0 *an* $-d\kappa(x) - 1$, *where* $d\kappa(x)$ *is the derivative at* 1 *of the character* $\kappa(x) : \mathbb{Z}_p^* \to L^*$.

Proof If x is very classical and $\kappa(x) = k \in \mathbb{N}$, the Galois representation ρ_x is th one attached to a modular form of weight $k+2$, hence the Hodge-Tate weights c $(\rho_x)_{|G_{\mathbb{Q}_p}}$ are (according to our normalization) 0 and $-k-1$, that is 0 and $-d\kappa(x)+$ since $\kappa(x) : z \mapsto z^k$ has derivative k at 1.

Lemma 7.4.10 *For* $x \in \mathcal{C}^{\pm}$, *the space* $D_{crys}((\rho_x)_{|G_{\mathbb{Q}_p}})^{\phi=U_p(x)}$ *has dimension a least* 1 *over* L.

Proof If x is very classical and p-old, then $(\rho_x)_{|G_{\mathbb{Q}_p}}$ is crystalline and its ϕ eigenvalues are $U_p(x)$ and $p^{k+1}U_p(x)^{-1}$. Then the result follows from Kisin' theorem [13, Theorem 3.3.3(i)], [79].

Exercise 7.4.11 Assume that $N = l$ is a prime. By Theorem 7.2.3 and its proo there is a classical point x in $\mathcal{C}^{\pm}$ such that λ_x is the system of eigenvalues of E_2^{cri} Show that there exists an affinoid neighborhood of x in $\mathcal{C}^{\pm}$ in which every classic point y has minimal tame level l.

7.5 The Ordinary Locus

Definition 7.5.1 The *ordinary locus* $\mathcal{C}^{\pm}_{\text{ord}}$ of the eigencurve $\mathcal{C}^{\pm}$ is the locus of point x such that $v_p(U_p(x)) = 0$.

Proposition 7.5.2 *The ordinary locus* $\mathcal{C}^{\pm}_{\text{ord}}$ *is a union of connected components o* $\mathcal{C}^{\pm}$. *In particular, it is equidimensional of dimension* 1. *The very classical points ar very Zariski-dense in it. The restriction* $\kappa : \mathcal{C}^{\pm}_{\text{ord}} \to \mathcal{W}$ *of the weight map* κ *is finite*

Proof Recall that $U_p \in \mathcal{O}(\mathcal{C}^{\pm})$ is bounded by 1, and that the ball of function bounded by 1 in $\mathcal{O}(\mathcal{C}^{\pm})$ is compact (Proposition 3.7.9). Thus every subsequenc of the sequence $U_p^{n!}$ has a converging subsequence, and all those convergin subsequence have the same limit since these limits obviously assume the value on x such that $|U_p(x)| = 1$ and 0 elsewhere. It follows that $e = \lim_{n\to\infty} U_p^{n!}$ exist in $\mathcal{O}(\mathcal{C})$ and take the value 1 on $\mathcal{C}^{\pm}_{\text{ord}}$, 0 elsewhere. In particular e is an idempoten in $\mathcal{O}(\mathcal{C}^{\pm})$, and $\mathcal{C}^{\pm}_{\text{ord}}$ is defined by $e = 1$, hence is a union of connected components The assertion about density follows. For the finiteness, see [34, Prop. 6.11].

Remark 7.5.3 Well before the eigencurve was constructed, Hida had defined the so-called Hida-family, as some eigenalgebras over the Iwasawa algebra Λ of function bounded by 1 in $\mathcal{O}(\mathcal{W})$. One can show that $\mathcal{C}^{\pm}_{\mathrm{ord}}$ is the base change to $\mathcal{W}$ of the corresponding Hida family.

Here we see the trade-off between Hida families and eigenvarieties. Hida families only contains the ordinary modular forms, but they are defined over the Iwasawa algebra, which allows ones to reason modulo p when we need to.

Lemma 7.5.4 *If x is a very classical point of weight k such that $v_p(U_p(x)) = 0$ or $v_p(U_p(x)) = k+1$, then $(\rho_x)_{|G_{\mathbb{Q}_p}}$ is an extension of an unramified character χ_1 by an Hodge-Tate character χ_2 of weight $-k-1$.*

Proof In both cases, x is old, and corresponds to a refinement of an ordinary modular form. The result is thus known by a deep result of Hida and Wiles, cf. [125]. □

Proposition 7.5.5 *If $x \in \mathcal{C}^{\pm}_{\mathrm{ord}}$, then $(\rho_x)_{|G_{\mathbb{Q}_p}}$ is an extension of an unramified character χ_1 (in particular with Hodge-Tate weight 0) by a character χ_2 whose Hodge-Tate weight is $-d\kappa(x)$.*

Proof For x very classical on $\mathcal{C}^{\pm}_{\mathrm{ord}}$, this is the preceding lemma, and one concludes by Proposition 7.5.2. □

7.6 Local Geometry of the Eigencurve

7.6.1 Clean Neighborhoods

Let x be a point of the eigencurve $\mathcal{C}^{\pm}$ of field of definition L. We want to study the geometry of the eigencurve near x. In order to avoid the minor but annoying complications related to rationality questions, we shall make a base change to L. That is, we consider in this section both the eigencurve and the weight space as rigid spaces over L instead of $\mathbb{Q}_p$ (without changing notations). In particular, all affinoids of $\mathcal{C}^{\pm}$ and $\mathcal{W}$ will be L-affinoids without further notice. Let us call $w = \kappa(x)$, which is a closed point of $\mathcal{W}$ of field of definition L. After we make this base change, saying that κ is étale at x is the same as saying that it is an isomorphism on its image over any small enough neighborhood of x.

Lemma and Definition 7.6.1 *There exists a basis of affinoid admissible neighborhoods U of $x \in \mathcal{C}^{\pm}$ such that*

(a) *There exist an admissible open affinoid $W = \mathrm{Sp}\, R$ of $\mathcal{W}$ (so R is an affinoid L-algebra), and $\nu \in \mathbb{R}$ adapted to W, such that U is the connected component of x in $\mathcal{C}^{\pm}_{W,\nu}$.*
(b) *The point x is the only point in U above w.*
(c) *The map $\kappa : U \to W$ is étale at every point of U except perhaps x.*

We shall call such a neighborhood a *clean* neighborhood of x.

Proof By construction, the $\mathcal{C}^{\pm}_{W,\nu}$ are a basis of affinoid neighborhoods of x in $\mathcal{C}^{\pm}$. Fix such a neighborhood. Choose two admissible open disjoint subsets U_1 and U_2 of $\mathcal{C}^{\pm}_{W,\nu}$ such that U_1 contains x and U_2 contains every points above w excepted x (this is possible since $\kappa^{-1}(w) \cap \mathcal{C}^{\pm}_{W,\nu}$ is a finite set). Since $\kappa : \mathcal{C}^{\pm}_{W,\nu} \to W$ is finite flat, the $\kappa^{-1}(W')$ for $w \in W' \subset W$ form a basis of neighborhoods of $\kappa^{-1}(w)$ [18, 2.1.6], so it is possible to find a $W' \in \mathfrak{C}$ such that $\kappa^{-1}(W') \subset U_1 \cup U_2$. For such a W', the connected component U of x in $\mathcal{C}^{\pm}_{W'\nu} = \kappa^{-1}(W')$ is contained in U_1, hence contains no point in $\kappa^{-1}(w)$ other than x. By the openness (already for the Zariski topology) of the étale locus of a finite flat map, if W' is small enough then $\kappa : U \to W'$ is étale excepted perhaps at x. □

If U is a clean neighborhood of x, it follows from the definition that U is a connected component of a local piece $\mathcal{C}^{\pm}_{W,\nu} = \operatorname{Sp} \mathcal{T}^{\pm}_{W,\nu}$ where $\mathcal{T}^{\pm}_{W,\nu}$ is the eigenalgebra corresponding to the action of $\mathcal{H}$ on the finite projective R-module (actually free since R is a PID) $\operatorname{Symb}^{\pm}_{\Gamma}(\mathcal{D}_K[r])^{\leq \nu}$ which is defined for $r > 0$ small enough, and independent of r. That is to say, there is an idempotent $\varepsilon \in \mathcal{T}^{\pm}_{W,\nu}$ such that U is defined in $\mathcal{C}^{\pm}_{W,\nu}$ by the equation $\varepsilon = 1$, and $U = \operatorname{Sp}(\varepsilon \mathcal{T}^{\pm}_{W,\nu})$. Also, the module $\varepsilon \operatorname{Symb}^{\pm}_{\Gamma}(\mathcal{D}_K[r])^{\leq \nu}$ is a direct summand of $\operatorname{Symb}^{\pm}_{\Gamma}(\mathcal{D}_K[r])^{\leq \nu}$, hence is finite projective (free) over R, and is stable by $\mathcal{H}$ since $\varepsilon \in \mathcal{T}^{\pm}_{W,\nu}$. Clearly, $\varepsilon \mathcal{T}^{\pm}_{W,\nu}$ is the eigenalgebra defined by the action of $\mathcal{H}$ on the module $\varepsilon \operatorname{Symb}^{\pm}_{\Gamma}(\mathcal{D}_K[r])^{\leq \nu}$.

Since the geometry of clean neighborhoods of x will be the main object of study in this section, we shall introduce shorter notations for the objects of the above paragraph. For U a clean neighborhood of x:

(i) We shall write $M := \epsilon \operatorname{Symb}^{\pm}_{\Gamma}(\mathcal{D}_K[r])^{\leq \nu}$. It is a finite free module over R. We call d its rank.
(ii) We shall write $\mathcal{T}$ for the eigenalgebra defined by the action of $\mathcal{H}$ on M. It is a finite free module over R. We call e its rank. We have $U = \operatorname{Sp} \mathcal{T}$.
(iii) We still write $\kappa : U \to W$ for the restriction of $\kappa : \mathcal{C}^{\pm} \to \mathcal{W}$ to U. It is a finite flat map of degree e.
(iv) For $w' \in W(L)$ we write $M_{w'} := M \otimes_{R,w'} L$ and $\mathcal{T}_{w'} = \mathcal{T} \otimes_{R,w'} L$. The space $M_{w'}$ is of dimension d, independently of w', equal to the rank of M and gets an action of $\mathcal{H}$. There is a surjective morphism with nilpotent kernel from $\mathcal{T}_{w'}$ to the eigenalgebra defined by that action (see Proposition 2.4.1), and this morphism is an isomorphism when $w' \neq w$ by Proposition 2.4.3 since κ is étale at any point over w'. (N.B. At w, the algebra $\mathcal{T}_w$ may be strictly larger that the eigenalgebra on M_w).

A simple but important observation is that $\mathcal{T}_w$ is a local L-algebra (since x is the only point above w in U) and that, as an $\mathcal{H}$-module

$$M_w = \operatorname{Symb}^{\pm}_{\Gamma}(\mathcal{D}_w)_{(x)},$$

where the subscript $_{(x)}$ indicates as usual that the we take the generalized eigenspace of $\mathcal{H}$ for the system of eigenvalues λ_x in the space $\mathrm{Symb}^{\pm}_{\Gamma}(\mathcal{D}^{\dagger}_{w}(L))$ (or what amounts to the same, in the space $\mathrm{Symb}^{\pm}_{\Gamma}(\mathcal{D}_w[r](L))$.) This observation is just a special case of Theorem 2.5.9.

Lemma 7.6.2 *If x is a point such that $\kappa(x)$ has the form $z \mapsto z^k\epsilon(z)$ for some $k \geq 0$ and some finite order character ϵ of conductor p^t (e.g. a classical point, cf. Theorem 7.4.6) which has minimal tame level N_0 (so $N_0 \mid N$). If $N_0 < N$, also assume that x is normal. Then $d = \sigma(N/N_0)e$ where $\sigma(n)$ is the number of positive divisors of a positive integer n.*

Proof By assumption, the weight $w = \kappa(x)$ has the form $z \mapsto z^k\epsilon(z)$ for some $k \geq 0$ and some finite order character ϵ of conductor p^t. We can choose a non-negative integer k' such that $k' > 2\nu - 1$, $k' > \nu - 1$, $k' \neq k$ and the character $w'(z) = z^{k'}\epsilon(z)$ is in W, and moreover such that all points x' above w' have minimal tame level N_0 (by Lemma 7.4.8 and the assumption of normality if $N_0 \neq N$, and by Lemma 7.4.3 if $N_0 = N$.).

In particular $M_{w'} = \varepsilon\mathrm{Symb}^{\pm}_{\Gamma}(\mathcal{D}_{w'}(L))^{\leq\nu} = \varepsilon\mathrm{Symb}^{\pm}_{\Gamma}(\mathcal{V}_{w'}(L))^{\leq\nu}$ by Stevens's control theorem (Theorem 6.5.19 and Theorem 7.3.11). Since κ is étale over w', after extending the field of scalars L if necessary, we have $\mathcal{T}_{w'} = L^e$ as L-algebra, which means that the eigenalgebra of the action of $\mathcal{H}$ on $M_{w'}$ is L^e. Therefore, $M_{w'}$ is the direct sum of the eigenspaces $M_{w'}(x') = \mathrm{Symb}^{\pm}_{\Gamma}(\mathcal{V}_{w'}(L))[x']$ for x' running among the e points of the fiber of κ at w'. It therefore suffices to prove that $\dim_L \mathrm{Symb}^{\pm}_{\Gamma}(\mathcal{V}_{w'}(L))(x') = \sigma(N/N_0)$ for each such x'. This is done in the following lemma (of which the assumption on $\lambda(U_p)$ is satisfied because $\nu < (k'+1)/2$). □

Lemma 7.6.3 *Let $w \in W(L)$ be of the form $w(z) = z^k\epsilon(z)$ with $k \geq 0$ and ϵ a finite order character of $\mathbb{Z}_p^*$. Let λ be an $\mathcal{H}$-system of eigenvalues appearing in $\mathrm{Symb}^{\pm}_{\Gamma}(\mathcal{V}_w(L))$, which is not the system of eigenvalues of E_2, and let N_0 be its minimal tame level. If $\epsilon = 1$, assume moreover that $\lambda(U_p)^2 \neq p^{k+1}\lambda(\langle p\rangle)$. Then the $\mathcal{H}$-eigenspace $\mathrm{Symb}^{\pm}_{\Gamma}(\mathcal{V}_w(L))[\lambda]$ and generalized eigenspace $\mathrm{Symb}^{\pm}_{\Gamma}(\mathcal{V}_w(L))_{(\lambda)}$ are equal and both have dimension $\sigma(N/N_0)$.*

Proof By Theorem 5.3.35, this amounts to proving that in the case where $\epsilon = 1$,

$$\dim_L M_{k+2}(\Gamma, L)_{(\lambda)} = \dim_L M_{k+2}(\Gamma, L)[\lambda] = \sigma(N/N_0), \tag{7.6.1}$$

and in the case where $\epsilon \neq 1$, using also Proposition 7.3.10, and denoting the conductor of ϵ by p^t,

$$\dim_L M_{k+2}(\Gamma_1(Np^t), L)[\epsilon][\lambda] = \dim_L M_{k+2}(\Gamma_1(Np^t), L)[\epsilon]_{(\lambda)} = \sigma(N/N_0). \tag{7.6.2}$$

We first prove (7.6.2), by applying Atkin-Lehner theory, cf. Sect. 2.6.4, for the level Np^t. As in *loc. cit.*, we denote by $\mathcal{H}_0(Np^t)$ the polynomial ring generated by the T_l for $l \nmid N$ and the diamond operators a for $\langle a\rangle$ for $a \in (\mathbb{Z}/p^tN\mathbb{Z})^*$. Thus

$\mathcal{H}_0(Np^t)$ contains the polynomial ring we currently denote by $\mathcal{H}_0^p$ and in addition variables corresponding to the diamond operators $\langle a \rangle$ for $a \in (\mathbb{Z}/p^t\mathbb{Z})^*$. Thus, ϵ and the restriction λ_0 of λ from $\mathcal{H}_0$ to $\mathcal{H}_0^p$ together define a character $\lambda_0' : \mathcal{H}_0(Np^t) \to L$. The minimal level of this character is $N_0 p^t$ since we already know its minimal tame level is N_0 and its minimal wild level is p^t because that's the conductor of ϵ. Hence

$$\dim M_{k+2}(\Gamma_1(Np^t), L)[\lambda_0'] = \sigma(Np^t/N_0 p^t) = \sigma(N/N_0)$$

by Atkin-Lehner theory. Finally, we note that

$$\dim M_{k+2}(\Gamma_1(Np^t), L)[\epsilon][\lambda] = \dim M_{k+2}(\Gamma_1(Np^t), L)[\lambda_0'].$$

Indeed, the difference between the two eigenspaces is that in the LHS the operator U_p is included, but not in the RHS. But since λ is new at p, including U_p does not change the eigenspace by Atkin-Lehner theory. Moreover, for the same reason, U_p acts semi-simply on $\dim M_{k+2}(\Gamma_1(Np^t), L)[\epsilon]_{(\lambda)}$, and since all the other operator in $\mathcal{H}_0$ are semi-simple, we see that this generalized eigenspace is actually equal to the corresponding eigenspace.

We now turn to (7.6.1). The case where λ is p-new is done exactly as (7.6.2), so let us assume that λ is p-old. In this case, the minimal level of x is N_0, hence, if λ_0 denotes the restriction of the system of eigenvalues x to $\mathcal{H}_0$, we have $\dim M_{k+2}(\Gamma_1(N), L)[\lambda_0] = \sigma(N/N_0)$ and $\dim M_{k'+2}(\Gamma, L)[\lambda_0] = \sigma(pN/N_0) = 2\sigma(N/N_0)$ by Atkin-Lehner's theory. Now consider the two *refinement* maps

$$M_{k+2}(\Gamma_1(N), L)[\lambda_0] \to M_{k+2}(\Gamma, L)[\lambda_0],$$

defined by

$$f(z) \mapsto f(z) - \frac{p^{k+1}\lambda(\langle p \rangle)}{\lambda(U_p)} f(pz)$$

and

$$f(z) \mapsto f(z) - \lambda(U_p) f(pz).$$

It is easy to see that these maps are injective (e.g. by looking at q-expansions). By Lemma 6.7.2, the image of the first map is included in $M_{k+2}(\Gamma, L)[\lambda]$ (that is the $\mathcal{H}$-eigenspace for λ), hence in $M_{k+2}(\Gamma, L)_{(\lambda)}$, and the image of the second is included in $M_{k+2}(\Gamma, L)[\lambda']$, hence in $M_{k+2}(\Gamma, L)_{(\lambda')}$ where λ' is the system of eigenvalues of $\mathcal{H}_0$ that restricts to λ_0 on $\mathcal{H}_0^p$ and sends U_p to $\frac{p^{k+1}\lambda(\langle p \rangle)}{\lambda(U_p)}$. Since those two generalized eigenspaces are in direct sum because of our hypothesis $\lambda(U_p) \neq \frac{p^{k+1}\lambda(\langle p\rangle)}{\lambda(U_p)}$, the sum of their dimension is at most $2\sigma(N/N_0)$, but as we just saw each of those spaces has dimension at least $\sigma(N/N_0)$. Hence they both have dimension $\sigma(N/N_0)$, and are equal to the corresponding eigenspaces, which proves (7.6.1), hence the lemma. □

7.6.2 Étaleness of the Eigencurve at Non-critical Slope Classical Points

Theorem 7.6.4 *Let x be a normal classical point on $\mathcal{C}^{\pm}$ of weight k, and such that $U_p(x)^2 \neq p^{k+1}\langle p\rangle(x)$ If x is of* non-critical slope, *that is if $v_p(U_p(x)) < k+1$, then κ is étale at x. In particular, $\mathcal{C}^{\pm}$ is smooth at x.*

Proof We choose a clean neighborhood on x and keep all the notations used above. Let N_0 be the minimal tame level of x and write $w = \kappa(x)$. We have $d = \dim M_w = \dim \mathrm{Symb}_\Gamma^{\pm}(\mathcal{D}_w)_{(x)}$ The latter is, by Stevens' control theorem (using that x is of non-critical slope), $\dim \mathrm{Symb}_\Gamma^{\pm}(\mathcal{V}_w)_{(x)}$. Hence, $d = \sigma(N/N_0)$ by Lemma 7.6.3. But by Lemma 7.6.2, $d = e\sigma(N/N_0)$. Hence $e = 1$. □

Remark 7.6.5 In this theorem, $v_p(U_p(x)) < k+1$ is the serious hypothesis, while x normal and $U_p(x)^2 \neq p^{k+1}\langle p\rangle(x)$ are the technical hypotheses. We discuss here whether those technical hypotheses are really necessary.

The hypothesis $U_p(x)^2 \neq p^{k+1}\langle p\rangle(x)$ when x is very classical is necessary because when $U_p(x)^2 = p^{k+1}\langle p\rangle(x)$, the eigenalgebra defined by the action of $\mathcal{H}$ on M_w is non semi-simple (cf. Exercise 2.6.21), hence κ is not étale at x. On the other hand, it is conjectured, as already noted, that $U_p(x)^2 = p^{k+1}\langle p\rangle(x)$ never happens for a classical point, and this is known for $k = 0$.

About the hypothesis that x is normal, there is one noteworthy abnormal case where the conclusion of Theorem 7.6.4 still holds and can be proved with only a slight modification of the proof: it is when $N = 1$ and x is the point corresponding to the ordinary Eisenstein series $E_{2,p}$ (actually, this point lies on the ordinary Eisenstein line in the eigencurve of tame level 1, which is isomorphic to the weight space $\mathcal{W}$ through the weight map κ.) I do not know if the eigencurve is étale at the other abnormal classical points x of non-critical slope, but it is unlikely.

Remark 7.6.6 The same argument using Coleman's control theorem instead of Stevens's shows that the eigencurve of modular forms $\mathcal{C}$ as well is smooth at x. As a consequence, the injection $\mathcal{C}^{\pm} \hookrightarrow \mathcal{C}$ is an isomorphism in some neighborhood of x.

7.6.3 Geometry of the Eigencurve at Critical Slope Very Classical Points

In this section, we study normal very classical points x in $\mathcal{C}^{\pm}$ of weight $\kappa(x) = k \in \mathbb{N}$ that are of *critical slope* (that is, $v_p(U_p(x)) = k+1$). Note that by Lemma 7.3.4 those points are always p-old.

Lemma and Definition 7.6.7 *Let x be a normal very classical point of weight k and critical slope. There exists a unique (up to isomorphism) Galois representation $\rho_x^P : G_{\mathbb{Q},Np} \to \mathrm{GL}_2(\bar{\mathbb{Q}}_p)$ such that*

(i) *The representation ρ_x^P satisfies the Eichler-Shimura relation* $\mathrm{tr}\,(\rho_f(\mathrm{Frob}_l)) = T_l(x)$ *for all l prime to Np.*
(ii) *The restriction $(\rho_x^P)_{|G_{\mathbb{Q}_p}}$ is crystalline at p.*
(iii) *The representation ρ_x^P is indecomposable.*

Moreover, the Hodge-Tate weights of $(\rho_x)_{|G_{\mathbb{Q}_p}}$ are 0 and $-k-1$ and $U_p(x)$ is an eigenvalue of the crystalline Frobenius φ on $D_{crys}((\rho_f)_{|G_{\mathbb{Q}_p}})$.

We call ρ_x^P the preferred Galois representation attached to x.

Proof When x is cuspidal, ρ_x is irreducible, hence satisfies (iii). It also satisfies (i), (ii) (since x is p-old), and the "moreover" by the known property of Galois representations attached to modular forms. Any representation that satisfies (i) has its semi-simplification isomorphic to $\rho_x^{\mathrm{ss}} = \rho_x$, hence is isomorphic to ρ_x, and the uniqueness follows.

When x is Eisenstein, it is attached to a new Eisenstein series $E_{k+2,\psi,\tau}$ and we have $\rho_x = \tau\omega_p^{k+1} \oplus \psi$. A representation of ρ_x^P satisfying (i) and (iii) is either a non-trivial extension (in the category of $G_{\mathbb{Q},Np}$-representations) of $\tau\omega_p^{k+1}$ by ψ or a non-trivial extension of ψ by $\tau\omega_p^{k+1}$. We will show that there is one and only one (up to isomorphism of $G_{\mathbb{Q},Np}$-representations) non-trivial extension of ψ by $\tau\omega_p^{k+1}$ and none of $\tau\omega_p^{k+1}$ by ψ satisfying (ii).

First consider a non-trivial extension of ψ by $\tau\omega_p^{k+1}$ and twist it by ψ^{-1}. It becomes a non-trivial extension V of 1 by $\psi^{-1}\tau\omega_p^{k+1}$. Property (ii) is equivalent to $V_{|G_{\mathbb{Q}_p}}$ being crystalline (using that τ and ψ are crystalline). Such extensions V are parametrized by the Bloch-Kato style Selmer group,

$$\ker\left(H^1(G_{\mathbb{Q},Np}, \psi^{-1}\tau\omega_p^{k+1}) \to H^1_{/f}(G_{\mathbb{Q}_p}, \psi^{-1}\tau\omega_p^{k+1})\right), \tag{7.6.3}$$

where the notation $H^1_{/f}$ means H^1/H^1_f. We claim that this space has dimension 1. This space is actually the same as

$$H^1_f(\mathbb{Q}, \psi^{-1}\tau\omega_p^{k+1}) = \ker\left(H^1(G_{\mathbb{Q},Np}, \psi^{-1}\tau\omega_p^{k+1}) \to \prod_{l|Np} H^1_{/f}(G_{\mathbb{Q}_l}, \psi^{-1}\tau\omega_p^{k+1})\right)$$

because for $l \mid N$ we have $H^1_{/f}(G_{\mathbb{Q}_l}, \psi^{-1}\tau\omega_p^{k+1}) = 0$ excepted if $\psi^{-1}\tau\omega_p^{k+1} = \omega_p$ on $G_{\mathbb{Q}_l}$, which means $\psi = \tau$ on $G_{\mathbb{Q}_l}$ and $k = 0$, which is excluded by out hypothesis that x is normal. To conclude this case, we observe that the dimension $H^1_f(\mathbb{Q}, \epsilon\omega_p^n)$ is known (essentially by the work of Soulé, cf. [11] for details) to be 1 when $n \geq 1$, $\epsilon(-1)(-1)^n = -1$, and $\epsilon\omega_p^n \neq \omega_p$, and 0 in all other cases. So $H^1_f(\mathbb{Q}, \psi^{-1}\tau\omega_p^{k+1})$ has dimension 1, because $\psi^{-1}\tau\omega_p^{k+1} = \omega_p$ implies that x is abnormal, as either N is different from 1 and any prime factor l of N satisfies (N2) or $N = 1$, which implies $\psi = \tau = 1$ and $k = 0$ is excluded by (N1). Hence we have proved the existence and uniqueness of an extension of ψ by $\tau\omega_p^{k+1}$ satisfying the required conditions.

For the extensions in the other direction, we are reduced to compute, by similar argument, $H^1_f(\mathbb{Q}, \psi\tau^{-1}\omega_p^{-k-1})$ which is 0 by the result of Soulé quoted above. □

Definition 7.6.8 We say that a point x on $\mathcal{C}^{\pm}$ with $\kappa(x) = k \in \mathbb{N}$ has a *companion point* y if there exists a point $y \in \mathcal{C}^{\pm(-1)^{k+1}}$ such that $\kappa(y) = -2 - k$, $T_l(y) = l^{-k-1}T_l(x)$ for all $l \nmid N$, $U_p(y) = p^{-k-1}U_p(x)$ and $\langle a\rangle(y) = \langle a\rangle(x)$ for $a \in (\mathbb{Z}/N\mathbb{Z})^*$. The point y, if it exists, is necessarily unique and is called the *companion* of x.

Remark 7.6.9 Our definition of a companion is slightly different from the standard one, which would be phased exactly the same but for $\mathcal{C}^{\pm}$ and $\mathcal{C}^{\pm(-1)^{k+1}}$ both replaced by the full Coleman-Mazur-Buzzard eigencurve $\mathcal{C}$. Our definition is clearly more natural in our context of modular symbols, the requirement that the companion lies on the eigencurve of sign $\pm(-1)^{k+1}$ being the counterpart for the ι-involution of the twist by l^{-k-1} for the T_l-operators. But we claim that our definition is the right one even beyond the context of modular symbols, because it removes the difference of behavior with respect to the existence of a companion that exist between cuspidal form and Eisenstein series with the standard definition, cf. for instance [37, §7].

Theorem 7.6.10 *Let x be a very classical point on $\mathcal{C}^{\pm}$ of weight k. We assume that x is not abnormal, and that it is of critical slope, that is $v_p(U_p(x)) = k + 1$. The following are equivalent:*

(a) *The weight map κ is étale at x.*
(b) *The map $\rho_k : \mathrm{Symb}^{\pm}_{\Gamma}(\mathcal{D}_k)_{(x)} \to \mathrm{Symb}^{\pm}_{\Gamma}(\mathcal{V}_k)_{(x)}$ is an isomorphism.*
(c) *The point x has no companion point y on the eigencurve $\mathcal{C}^{\pm(-1)^{k+1}}$.*
(d) *The restriction $(\rho^P_x)_{|G_{\mathbb{Q}_p}}$ is not the direct sum of two characters.*

Theorem 7.6.11 *Let x be as above, and if x is cuspidal, assume moreover than $H^1_g(\mathbb{Q}, \mathrm{ad}\rho_x) = 0$. Then the eigencurve $\mathcal{C}^{\pm}$ is smooth at x. Moreover the injection $\mathcal{C}^{\pm} \hookrightarrow \mathcal{C}$ is an isomorphism in some neighborhood of x.*

We shall prove together Theorems 7.6.10 and 7.6.11.

Remark 7.6.12 Excepted for the restriction to forms that are refinement of forms of level prime to p, the equivalence between (c) and (d) above is the famous result of Breuil and Emerton [23] (which was before a conjecture of Gross) on the equivalence between having a companion form and having a Galois representations that splits at p. The proof we give here is completely different (and arguably, more elementary) than the one given in [23]. This proof is due to John Bergdall, see [17].

Proof We write by N_0 the minimal tame level of x. It is also the minimal level of x since as we have already observed, a critical point x is p-old.

Equivalence Between (a) and (b) Using notations and results of Sect. 7.6.1, we have $d = e\sigma(N/N_0)$, $\dim \mathrm{Symb}^{\pm}_{\Gamma}(\mathcal{V}_k)_{(x)} = \sigma(N/N_0)$, and $d = \mathrm{Symb}^{\pm}_{\Gamma}(\mathcal{D}_k)_{(x)}$. The map ρ_k is surjective, so it is an isomorphism if and only if the dimension of its source and target are the same, that is if and only if $d = \sigma(N/N_0)$, which is equivalent to $e = 1$.

Equivalence Between (b) and (c) By Proposition 6.5.18, we have an $\mathcal{H}$-equivariant exact sequence

$$0 \to \mathrm{Symb}^{\pm(-1)^{k+1}}_{\Gamma}(\mathcal{D}_{-2-k}(L))_{(y)}(k+1) \to \mathrm{Symb}^{\pm}_{\Gamma}(\mathcal{D}_k(L))_{(x)} \to \mathrm{Symb}^{\pm}_{\Gamma}(\mathcal{V}_k(L))_{(x)} \to 0$$

where we denote by y the system of $\mathcal{H}$-eigenvalues described in the theorem. Hence (b) is equivalent to $\mathrm{Symb}^{\pm(-1)^{k+1}}_{\Gamma}(\mathcal{D}_{-2-k}(L))_{(y)} = 0$ which by Theorem 7.3.1 is equivalent to the fact that x has no companion point y.

Proof that (d) Implies (c) When x Is Cuspidal Assume non-(c), so x has a companion point y. Since $U_p(y) = p^{-k-1}U_p(x)$, we have $v_p(U_p(y)) = 0$, so y is in the ordinary locus (Definition 7.5.1). Hence $(\rho_y)_{|G_{\mathbb{Q}_p}}$ is an extension of a character of Hodge-Tate weight 0 by a character of Hodge-Tate $k+1$. Since $\rho_x = \rho_y(k+1)$ by Cebotarev's density theorem, we deduce that $(\rho_x)_{|G_{\mathbb{Q}_p}}$ is an extension of a character of Hodge-Tate weight $-k-1$ by a character of Hodge-Tate weight 0. On the other hand, by Lemma 7.5.4, $(\rho_x)_{|G_{\mathbb{Q}_p}}$ is an extension of a character of Hodge-Tate weight 0 by a character of Hodge-Tate weight $-k-1$. Hence $(\rho_x)_{|G_{\mathbb{Q}_p}}$ is the sum of two characters, hence non-(d).

Proof of Theorem 7.6.11 When x Is Cuspidal Since x is cuspidal, the Galois representation ρ_x of $G_{\mathbb{Q},Np}$ is absolutely irreducible. Consider the pseudorepresentation $(\tau_{\mathbb{T}}, \delta_{\mathbb{T}}) : G_{\mathbb{Q},Np} \to \mathbb{T}$, where $\mathbb{T}$ is the completed local ring of the eigencurve of modular form $\mathcal{C}$ at the point x, and where $\tau_{\mathcal{T}}$, $\delta_{\mathbb{T}}$ are the composition of τ and δ with the restriction map $\mathcal{O}(\mathcal{C}) \to \mathbb{T}$. By Theorem 2.6.24, $\tau_{\mathbb{T}}$ and $\delta_{\mathbb{T}}$ are the trace and determinant of a unique continuous representation $\rho : G_{\mathbb{Q},Np} \to \mathrm{GL}_2(\mathbb{T})$ whose residual representation is ρ_x.

We consider the following deformation problem: for all Artinian local algebra A with residue field L, we define $D(A)$ as the set of strict isomorphism classes of representations $\rho_A : G_{\mathbb{Q}} \to \mathrm{GL}_2(A)$ deforming ρ_x such that

(i) The restriction $(\rho_A)_{|G_{\mathbb{Q}_p}}$ has a constant weight equal to 0;
(ii) the A-module $D_{\mathrm{crys}}((\rho_A)_{|G_{\mathbb{Q}_p}})^{\varphi=\tilde{\beta}}$ is free of rank one for some $\tilde{\beta}$ lifting β;
(iii) for $l \mid N$, the restriction $(\rho_A)_{|I_l}$ to the inertia subgroup at l is constant.

By [79, Prop. 8.7] (using the fact that $\beta \neq \alpha$ and that $0 \neq k+1$), and [13, Prop. 7.6.3(i)] (for condition (iii)), we see that D is pro-representable, say by a complete Noetherian local ring R.

We claim that $\rho \otimes \mathbb{T}/I$ is, for all cofinite length ideal I in $\mathbb{T}$, an element of $D(\mathbb{T}/I)$: property (i) follows from Sen's theory, see e.g. [13, Lemma 4.3.3(i)]; property (ii) is Kisin's theorem (see [79] or [13, §3.3.3]). For property (iii), let $l \mid N$. Let N_x and $N_{s(x)}^{\mathrm{gen}}$ be the special monodromy operator of $\rho_{|I_l}$ at x and the generic monodromy operator of $T_{|I_l}$ at a component $s(x)$ of the eigencurve through x (cf. [13, definition 7.8.16]). To prove (iii), by [13, Prop. 7.8.19 and Lemma 7.8.17] it is sufficient to prove that $N_x \sim N_{s(x)}^{\mathrm{gen}}$ (see [13, §7.8.1] for the definition of the equivalence relation $\sim$ and the pre-order $\prec$ on the set of nilpotent matrices). Assume first that the cuspidal modular form f attached to x is not special at l, so the monodromy operator N_x of $(\rho_x)_{|I_l}$ is trivial. Then in a neighborhood of x, there is a dense set of classical points that are newforms of the same level (Lemma 7.4.8) and same nebentypus, hence that are not special either, and thus have a trivial monodromy operator. Hence by [13, Prop. 7.8.19(2)], $N_x = N_{s(x)}^{\mathrm{gen}} = 0$. If f is special, then N_x is non-zero, but since by [13, Prop. 7.8.19(3)] $N_x \prec N_{s(x)}^{\mathrm{gen}}$ and we are in dimension 2, $N_x \sim N_{s(x)}^{\mathrm{gen}}$ in this case as well. This completes the proof of the claim.

Thus, ρ defines a morphism of algebras $R \to \mathbb{T}$. A standard argument shows that this morphism is surjective. Since $\mathbb{T}$ has Krull dimension 1, if we prove that the tangent space of R has dimension at most 1, it would follow that the map $R \to \mathbb{T}$ is an isomorphism and that R is a regular ring of dimension 1. This would prove that $\mathcal{C}$ is smooth at x.

The tangent space of R, $t_D := D(L[\varepsilon])$, lies inside the tangent space of the deformation ring of ρ_x without local condition, which is canonically identified with $H^1(G_{\mathbb{Q}}, \mathrm{ad}\rho_f)$. Since $H^1_f(G_{\mathbb{Q}}, \mathrm{ad}\rho_x) = H^1_g(G_{\mathbb{Q}}, \mathrm{ad}\rho_x) = 0$ by hypothesis, this space injects into $\prod_{l \mid Np} H^1_{/f}(G_{\mathbb{Q}_l}, \mathrm{ad}\rho_x)$ (where $H^1_{/f}$ means H^1/H^1_f) and so does t_D. The image of t_D in $H^1_{/f}(G_{\mathbb{Q}_l}, \mathrm{ad}\rho_x)$ is 0 for $l \mid N$ by (iii). Hence t_D injects in $H^1_{/f}(G_{\mathbb{Q}_p}, \mathrm{ad}\rho_x)$.

If $(\rho_x)_{|G_{\mathbb{Q}_p}}$ is not the direct sum of two characters, that is if (d) holds, then by what we have already proved (a) holds, that is κ is étale at x which in particular is smooth. So let us assume henceforth that $(\rho_x)_{|G_{\mathbb{Q}_p}} = \chi_1 \oplus \chi_2$. Both χ_1 and χ_2 are crystalline, and say χ_1 has weight 0 while χ_2 has weight $k+1$. Since $v_p(U_p(x)) = k+1$, $U_p(x)$ is the eigenvalue of the crystalline Frobenius on $D_{\mathrm{crys}}(\chi_2)$. Then we compute:

$$H^1_{/f}(G_{\mathbb{Q}_p}, \mathrm{ad}\rho_f) = H^1_{/f}(G_{\mathbb{Q}_p}, \chi_1\chi_1^{-1}) \oplus H^1_{/f}(G_{\mathbb{Q}_p}, \chi_2\chi_1^{-1}) \oplus H^1_{/f}(G_{\mathbb{Q}_p}, \chi_1\chi_2^{-1}) \oplus H^1_{/f}(G_{\mathbb{Q}_p}, \chi_2\chi_2^{-1}).$$

The condition on $D_{\mathrm{crys}}(-)^{\varphi=\tilde{\beta}}$ implies that the image of t_d in the third and fourth factors are 0 (the third factor is 0 anyway). In particular any deformation of ρ_x in $D(L[\varepsilon])$ is triangular with diagonal terms a deformation $\tilde{\chi}_1$ of χ_1 and the constant deformation χ_2 of χ_2, and the Sen weights of this deformation are thus the Sen weight of $\tilde{\chi}_1$ and $k+1$. The condition on the constant weight 0 in our deformation problem D thus implies that $\tilde{\chi}_1$ has constant weight 0, hence that it is constant: thus, the image of t_D in the first factor is 0. Therefore t_D injects in $H^1_{/f}(G_{\mathbb{Q}_p}, \chi_2\chi_1^{-1})$. Since $\chi_2\chi_1^{-1}$ is not trivial (its Hodge-Tate weight is not 0), local Tate duality and Euler characteristic formula implies that $H^1(G_{\mathbb{Q}_p}, \chi_2\chi_1^{-1})$ has dimension 1, excepted if $\chi_2\chi_1^{-1}$ is the cyclotomic character of $G_{\mathbb{Q}_p}$, in which case $H^1(G_{\mathbb{Q}_p}, \chi_2\chi_1^{-1})$ has dimension 2 but $H^1_{/f}(G_{\mathbb{Q}_p}, \chi_2\chi_1^{-1})$ has dimension 1. Therefore, t_D has dimension at most one, which is what remained to check for the proof that $\mathcal{C}$ is smooth at x, of dimension 1.

Now since $x \in \mathcal{C}^{\pm}$ and $\mathcal{C}^{\pm}$ is equidimensional of dimension 1, it follows that the inclusion $\mathcal{C}^{\pm} \hookrightarrow \mathcal{C}$ is a local isomorphism near x, and therefore that $\mathcal{C}^{\pm}$ is also smooth at x.

Proof that (a) Implies (d) When x Is Cuspidal Assume non-(d), that is $(\rho_x)_{|G_{\mathbb{Q}_p}}$ is the sum of two characters. Then as we have seen in the above proof, any deformation in $D(L[\varepsilon])$ has a constant Sen weight $k+1$. Since $R = \mathbb{T}$, this means that one Sen weight of the restriction to any tangent vector at x of $\mathcal{C}^{\pm}$ is constant $k+1$. In other words, the tangent map of κ maps any tangent vector at x to 0, that is, κ is not étale at x, which is non-(a).

We have thus finished the proof of both theorems when x is cuspidal. When x is Eisenstein, it remains to prove that (d) is equivalent to (a), (b), and (c), and that $\mathcal{C}^{\pm}$ is smooth at x. With our hypotheses that x is normal, the proof is a straightforward generalization of [12]. We quickly remind the main steps for the convenience of the reader.

Proof of Theorem 7.6.11 When x Is Eisenstein
Then the system of $\mathcal{H}_0$-eigenvalues $\lambda_{x,0}$ is the same as the one of a unique new Eisenstein series $E_{2,\psi,\tau}$. Let $K = \mathrm{Frac}(\mathbb{T})$ which is a finite product of fields. Then there is a representation $\rho : G_{\mathbb{Q}_{Np}} \to \mathrm{GL}_2(K)$ with $\mathrm{tr}\,\rho = T$ and $\rho(c) = \begin{pmatrix} -1 & 0 \\ 0 & 1 \end{pmatrix}$ (use the argument given after Theorem 7.4.1). If $\psi(c) = -1$ we conjugate ρ by $\begin{pmatrix} 0 & 1 \\ 1 & 0 \end{pmatrix}$ so that we have in any case:

$$\rho(c) = \begin{pmatrix} \tau(c)\omega_p^{k+1}(c) & 0 \\ 0 & \psi(c) \end{pmatrix}.$$

Writing

$$\rho(g) = \begin{pmatrix} a(g) & b(g) \\ c(g) & d(g) \end{pmatrix}$$

we define B (resp. C, resp. B_p, resp. C_p) as the $\mathbb{T}$-submodule of K generated by the $b(g)$, $g \in G_{\mathbb{Q},Np}$ (resp. $c(g)$, $g \in G_{\mathbb{Q},Np}$, resp. $b(g)$, $g \in G_{\mathbb{Q}_p}$, resp. $c(g)$, $g \in G_{\mathbb{Q}_p}$). Then we have natural injective L-linear maps ι_B (resp. etc.) that fits into commutative diagrams (cf. [13, §1.7] or [12]):

$$\begin{array}{ccc} (B/\mathfrak{m}B)^* & \stackrel{\iota_B}{\longrightarrow} & \mathrm{Ext}^1_{G_{\mathbb{Q},Np}}(\psi, \tau\omega_p^{k+1}) \\ \downarrow & & \downarrow \\ (B_p/\mathfrak{m}B_p)^* & \stackrel{\iota_{B_p}}{\longrightarrow} & \mathrm{Ext}^1_{G_{\mathbb{Q}_p}}(\psi, \tau\omega_p^{k+1}) \end{array} \tag{7.6.4}$$

$$\begin{array}{ccc} (C/\mathfrak{m}C)^* & \stackrel{\iota_C}{\longrightarrow} & \mathrm{Ext}^1_{G_{\mathbb{Q},Np}}(\tau\omega_p^{k+1}, \psi) \ , \\ \downarrow & & \downarrow \\ (C_p/\mathfrak{m}C_p)^* & \stackrel{\iota_{C_p}}{\longrightarrow} & \mathrm{Ext}^1_{G_{\mathbb{Q}_p}}(\tau\omega_p^{k+1}, \psi) \end{array} \tag{7.6.5}$$

where the left vertical maps are dual of the inclusion $B_p \subset B$ and $C_p \subset C$, and the top vertical maps are restriction maps. Moreover, by Kisin's lemma the image of ι_B lies in the subspace of $\mathrm{Ext}^1_{G_{\mathbb{Q},Np}}(\psi, \tau\omega_p^{k+1})$ parametrizing extensions that are crystalline at p. As seen is the proof of Lemma 7.6.7, this subspace has dimension 1. It follows that $B/\mathfrak{m}B$ has dimension at most 1, and by Nakayama that B is a principal ideal. Also, from the proof of Lemma 7.6.7 we get that the restriction map $\mathrm{Ext}^1_{G_{\mathbb{Q},Np}}(\tau\omega_p^{k+1}, \psi) \to \mathrm{Ext}^1_{G_{\mathbb{Q}_p}}(\tau\omega_p^{k+1}, \psi)$ is an isomorphism between spaces of dimension 1 and we conclude as above that C and C_p are principal, and even that they are equal provided that $C_p \neq 0$.

The ideal BC (resp. B_pC_p) is the reducibility ideal of T (resp. $T_{|G_{\mathbb{Q}_p}}$) as defined in [13, chapter 1] or [12]. It is proved in [12] that B_pC_p is the ideal of the schematic fiber of κ at k and that BC is the maximal ideal $\mathfrak{m}$ of $\mathbb{T}$. By the above $\mathfrak{m}$ is principal, which shows that $\mathbb{T}$ is a discrete valuation ring, hence Theorem 7.6.11 in the Eisenstein case.

Equivalence Between (a) and (d) in the Eisenstein Case We keep the notations of the above paragraph. Since $BC = \mathfrak{m}$ and B_pC_p is the ideal of the fiber of κ, we see that (a) is equivalent to $B_pC_p = BC$. By what we have seen, $C_p = C$ (C being clearly non-zero since $BC = \mathfrak{m}$), so the later is equivalent to $B_p = B$. By the commutative diagram (7.6.4) this is equivalent to restriction map $\mathrm{Ext}^1_{G_{\mathbb{Q},Np}}(\psi, \tau\omega_p^{k+1}) \to \mathrm{Ext}^1_{G_{\mathbb{Q}_p}}(\psi, \tau\omega_p^{k+1})$ being injective, which by construction of ρ_x^P is clearly equivalent to (d).

□

7.6.4 Critical Slope Eigenforms and Points on $\mathcal{C}^{\pm}$

Proposition 7.6.13 *Let g be modular form, of weight $k+2$, level $\Gamma = \Gamma_1(N) \cap \Gamma_0(p)$, which is new at N. We assume that g is an eigenform of $\mathcal{H}_0$. Then the unique point x_g of $\mathcal{C}$ corresponding to g (Theorem 7.3.1) belong to both $\mathcal{C}^+$ and $\mathcal{C}^-$. If g is cuspidal, then x_g is a very classical point of both $\mathcal{C}^+$ and $\mathcal{C}^-$. If g is Eisenstein, then x_g is a very classical point of $\mathcal{C}^{\epsilon(g)}$ (where $\epsilon(g)$ is the sign of the Eisenstein series g, defined in Definition 5.3.33) but not of $\mathcal{C}^{-\epsilon(g)}$.*

Before beginning the proof we observe that a form g as in the statement is necessarily old at p, hence is the refinement $g(z) = f_\beta(z) = f(z) - \alpha f(pz)$ of a newform f of level $\Gamma_1(N)$ which is ordinary at p (that is, its coefficient a_p is a p-adic unit). Here α and β are the roots of $X^2 - a_p X + p^{k+1}\epsilon(p)$ of p-valuation 0 and $k+1$ respectively.

Proof By hypothesis, the system of $\mathcal{H}_0$-eigenvalues of g appears in $M_{k+2}(\Gamma)$. By Theorem 5.3.35, if g is cuspidal it appears in $\mathrm{Symb}^{\pm}_{\Gamma}(\mathcal{V}_k)$ for both values of the sign $\pm$, while if g is Eisenstein, it appears in $\mathrm{Symb}^{\epsilon(p)}_{\Gamma}(\mathcal{V}_k)$ but not in $\mathrm{Symb}^{-\epsilon(p)}_{\Gamma}(\mathcal{V}_k)$. Hence when g is cuspidal, the point x_g belongs to $\mathcal{C}^+$ and $\mathcal{C}^-$ and is very classical in both.

If g is Eisenstein, this implies that x^g belongs to $\mathcal{C}^{\epsilon}(g)$ and is very classical in it, and that x_g if it belongs to $\mathcal{C}^{-\epsilon(g)}$ is not very classical (and by Theorem 7.4.6, is not classical either.) It thus remains only to prove that x_g belong to $\mathcal{C}^{-\epsilon(g)}$. By Exercise 2.6.15, $g \in S^{\dagger}(\Gamma)$, hence x_g belongs to the cuspidal eigencurve $\mathcal{C}_0$, and since by Theorem 7.2.3, $\mathcal{C}^0 \subset \mathcal{C}^{\pm}$ in $\mathcal{C}$ for both values of the sign $\pm$, x_g belongs to both $\mathcal{C}^+$ and $\mathcal{C}^-$. □

Corollary 7.6.14 *If g is as in the proposition, the $\mathcal{H}_0$-modules $\mathrm{Symb}^{+}_{\Gamma}(\mathcal{D}_k)_{(x)}$ and $\mathrm{Symb}^{-}_{\Gamma}(\mathcal{D}_k)_{(x)}$ are isomorphic and their dimension as vector spaces, called e, is the degree of both weight maps $\mathcal{C}^+ \to \mathcal{W}$ and $\mathcal{C}^- \to \mathcal{W}$ at x_g.*

Proof Since $\mathcal{C}^+$ and $\mathcal{C}^-$ are isomorphic in a neighborhood of x_g, and g is very classical in at least one of these curve, the preceding theorem applies and gives the corollary. □

We know want to review what is known or conjectured about the local degree e of the weight map at x_g for the different type of modular eigenform g as in the proposition.

There are three cases to consider:

When f is a cuspidal non CM new form of weight k, ordinary at p, **it is conjectured** but not known in general, that $e = 1$, that is **that κ is étale** at x_g. Indeed, it is conjectured,[7] that the restriction at p of the Galois representation $\rho_g = \rho_{f_\beta}$, which is reducible singe g is ordinary, is always non-split, which is assertion (d) of Theorem 7.6.10.

[7]This was at first a question of Greenberg, now become a folklore conjecture. See [61].

When f is an Eisenstein form, f is always ordinary at p, and **it is again conjectured**, as a part of Jansen's conjecture (cf. [74, Question 2, page 349] and [11, §5.2]) that $e = 1$, i.e. **that κ is étale** at x_g. See [12] for details.

When f is cuspidal CM, it has CM by a well-determined imaginary quadratic field K (i.e., K is the unique quadratic field such that $(\rho_f)_{|G_K}$ is the sum of two characters) and f is ordinary at p if and only if p splits in K. In this case, it is known that $e \geq 2$, in other words, κ **is never étale** at x_g.

Exercise 7.6.15 Prove this result, as follows.

1. Prove that $(\rho_f)_{|G_K}$ is the direct sum of two distinct characters, χ and χ'
   We use a notation of the proof of Theorem 7.6.10, namely the pseudo-representation $(\tau_{\mathbb{T}}, \delta_{\mathbb{T}}) : G_{\mathbb{Q},Np} \to \mathbb{T}$, where $\mathbb{T}$ is the completed local ring of the eigencurve of modular form $\mathcal{C}$ at the point x_g. Modulo the maximal ideal $\mathfrak{m}$ of $\mathbb{T}$, this pseudo-representation is the one attached to ρ_g, and by a, its restriction to G_K is the sum of two characters
2. Prove that the reduction modulo $\mathfrak{m}^2$ of $(\tau_{\mathbb{T}}, \delta_{\mathbb{T}})$, when restricted to G_K, is the sum of two continuous characters $G_K \to (\mathbb{T}/\mathfrak{m}^2)^*$.
3. Deduce (using [12, §5.4]) that the map κ is not étale at y.

It is also a consequence of Jansen's conjecture that $e = 2$ in this case. For more detail about this case, see [9].

7.6.5 Complements on the Geometry of the Eigencurve at Classical Points

In the preceding subsection, we have focused on the normal very classical case in order to avoid too many complications or variants. Here we give similar results concerning the remaining cases, with very rough sketches of proof.

There are two noteworthy abnormal cases where the same (and actually even stronger) results hold, with some modification of the proof:

Proposition 7.6.16 *Assume that x is a very classical point of weight $\kappa(x) = 0$, of critical slope $v_p(U_p(x)) = 1$, and such that the system of eigenvalues of x restricted to $\mathcal{H}_0$, $\lambda_{x,0} : \mathcal{H}_0 \to L$ is either the system of Eigenvalues of E_2 or the system of eigenvalues of $E_{2,\tau,\tau}$ where τ is a non-trivial Dirichlet character (note that such very classical points x exist on $\mathcal{C}^{\pm}$ if and only if $\pm = +1$). Then the weight map κ is étale at x, and in particular the conclusions of Lemma 7.6.7, Theorems 7.6.11, and 7.6.10, hold. Moreover, the assertions (a) to (d) of Theorem 7.6.10 are true.*

Proposition 7.6.17 *Assume that the tame level is a prime power $N = l^{\nu}$, and that x is a very classical point of weight $\kappa(x) = 0$, of critical slope $v_p(U_p(x)) = 1$. Assume that the system of eigenvalues of x restricted to $\mathcal{H}_0$, $\lambda_{x,0} : \mathcal{H}_0 \to L$ is either the system of eigenvalues of $E_{2,l}$ or the system of eigenvalues of $E_{2,\tau,\tau}$ where τ is a primitive non-trivial Dirichlet character, necessarily of conductor a power* *of l. (Note that such very classical points x exists on $\mathcal{C}^{\pm}$ if and only if $\pm = +1$). Then the conclusions of Lemma 7.6.7, Theorem 7.6.11, and Theorem 7.6.10, hold. Moreover, the condition (a) to (d) in Theorem 7.6.10 are true.*

Proof In the proof of Lemma 7.6.7, the only thing that changes is the proof that the space (7.6.3) has dimension 1. Namely, we now have to prove that

$$\ker\left(H^1(G_{\mathbb{Q},Np}, \omega_p) \to H^1_{/f}(G_{\mathbb{Q}_p}, \omega_p)\right)$$

is of dimension 1. By Kummer theory, this dimension of this space is the rank of the group of N-units in $\mathbb{Q}$, so is 1 if (and only if) N is a prime power. For the extensions in the other direction, nothing changes.

In the proof of the theorems, we see that nothing changes until the computation of the rank of ι_B, but the result is the same (that is 1) by the above paragraph. The computation of the rank of ι_C is not changed. Hence the theorems hold. Assertion (d) of Theorem 7.6.10 hold since a N-unit of $\mathbb{Q}$ of infinite order obviously is also of infinite order when seen in $\mathbb{Q}_p$. □

Remark 7.6.18 For the other abnormal very classical critical-slope points, Lemma 7.6.7 is false, as there are in this case several independent non-isomorphic extension of ψ by $\tau\omega_p$ in the category of $G_{\mathbb{Q},Np}$-representation that are crystalline at p. I do not know if Theorem 7.6.10 and 7.6.11 hold in this case, but it is unlikely. If $\mathcal{C}^{\pm}$ was smooth at x in this case, then Ribet's lemma would single out a canonical non-trivial extension of ψ by $\tau\omega_p$ as above, and I don't see why any of those extensions would deserve such a privilege.

Concerning classical points that are not very classical, one has

Theorem 7.6.19 *Let x be a normal classical point. If x is cuspidal, assume than $H^1_g(\mathbb{Q}, \mathrm{ad}\rho_x) = 0$. Then the eigencurve $\mathcal{C}^{\pm}$ is smooth at x.*

However, the details of the proof, as well as the statement of Theorem 7.6.10 needs to be slightly modified. For instance, the preferred representation should not be required to be crystalline, but rather to have two crystalline periods after twists by specific characters. We leave these changes to the interested reader.

There are also points $x \in \mathcal{C}^{\pm}$ such that λ_x is the system of eigenvalues of a classical modular eigenform f of weight 1. Those points are not classical in our sense, since their system of eigenvalues are not seen in spaces of classical modular symbols, but could reasonably be called classical. Concerning these points, one has:

Theorem 7.6.20 (Bellaïche-Dimitrov) *Let $x \in \mathcal{C}^{\pm}$ be a point corresponding to a refinement of a classical cuspidal modular eigenform f of weight 1 which is regular[8] at p. Then $\mathcal{C}^{\pm}$ is smooth at x. Moreover $\mathcal{C}^{\pm}$ is étale over $\mathcal{W}$ at x if and only if f is not a CM form for a quadratic extension of $\mathbb{Q}$ in which p splits.*

[8]A form f of level prime to p is *regular at p* if its two p-adic refinements f_α and f_β are distinct.

The proof is completely different from the proof of Theorems 7.6.10 and 7.6.19. It uses transcendence theory in the form of Baker's theorem on linear independence of logarithms of algebraic numbers. See [15] for details.

On the other hand, Betina et al. [22] have recently proven that if f is the unique p-refinement of a classical cuspidal Eisenstein series of weight 1 which is irregular at p, then f appears on the cuspidal eigencurve $\mathcal{C}_0$, which is smooth at f, but f belongs to three smooth components of $\mathcal{C}$, that cross normally (i.e. like the three axes in the 3-dimensional space). In particular the singularity of $\mathcal{C}$ at f is not Gorenstein, Similarly, Betina and Dimitrov [20] have proven that if f is the unique p-refinement of a classical cuspidal CM forms of weight 1 which is irregular at p. then $\mathcal{C}_0$ has either three of four irreducible components at f, and in any case a non-Gorenstein singularity at this point.

7.7 Global Properties of the Eigencurve

7.7.1 *Integrality of Fredholm Determinants and Integral Models of the Eigencurves*

Consider one of the eigencurve we constructed in this chapter, $\mathcal{C}^0$, $\mathcal{C}^+$, $\mathcal{C}^-$ or $\mathcal{C}$. Let us call M_W the Banach modules used in the construction of the eigencurve. As already observed at the beginning of the proof of Theorem 3.8.10, for every $h \in \mathcal{H}$, there is a globally defined Fredholm determinant $\det(1 - T\psi(hU_p)) \in \mathcal{O}(\mathcal{W})\{\{T\}\} = \sum_{n=0}^{\infty} a_n T^n$ which glues the Fredholm determinant $\det(1 - T\psi_W(hU_p)) \in \mathcal{O}(W)\{\{T\}\}$ for $W \in \mathfrak{C}$.

Proposition 7.7.1 *The coefficients of the Fredholm determinant* $\det(1 - T\psi(hU_p)) \in \mathcal{O}(\mathcal{W})[[T]]$ *belong to the Iwasawa subalgebra* Λ *of* $\mathcal{O}(\mathcal{W})$ *(see Sect. 6.3.5).*

Proof By definition, h is a R-linear combination of operators T_l, U_p, and $\langle a \rangle$. By Lemma 6.5.4, for any $w \in \mathcal{W}(\mathbb{C}_p)$ the operator hU_p acting on the Banach space M_w has norm ≤ 1. Therefore, the coefficients of its Fredholm determinant, $a_n(w)$ have absolute values ≤ 1. By Exercise 6.3.10, it follows that $a_n \in \Lambda$ for every n. □

This simple observation suggests that the eigencurve $\mathcal{C}$ (or $\mathcal{C}^{\pm}$, or $\mathcal{C}^0$) has an "integral model", or is, in some sense, "defined over Λ". Actually, a precise sense has been given to this phrase, which has recently been proved. Namely, it is possible to extend the construction of the eigencurve to get an adic space (in the sense of Huber) which is locally finite and torsion-free on $\mathrm{spa}(\Lambda, \Lambda)^{an}$ and which reduces to the eigencurve $\mathcal{C}$ over the locus $p \neq 0$ of $\mathrm{spa}(\Lambda, \Lambda)^{an}$ (which is naturally identified to $\mathcal{W}$). See [3, Theorem 6.3] and [75, §6.1].

7.7.2 Valuative Criterion of Properness

The eigencurve $\mathcal{C}$ (or $\mathcal{C}^{\pm}$, or $\mathcal{C}^0$) is not proper over $\mathcal{W}$, since it is not of finite type. However, it is possible to prove the following version of the valuative criterion of properness, which was conjectured by Buzzard and Calegari [30].

Theorem 7.7.2 *Let D be an open disk in the weight space, and D^* the same disk deprived of its center. Then any section $s^* : D^* \to \mathcal{C}$ of κ extends to a section $s : D \to \mathcal{C}$.*

This theorem was first proved by Diao and Liu [57]. An other proof has recently been given by Ye [128]. Using this result, Hattori and Newton have proved that every irreducible component of the eigencurve of finite degree is finite over the weight space: see [67].

7.7.3 Open Questions

There are still many open questions on the global geometry of the eigencurve. Some are standing since the first construction by Coleman and Mazur:

Are there infinitely many irreducible components in $\mathcal{C}$? This question seems as wide open as when it was first asked.

Is any of its components, not contained in the ordinary locus $\mathcal{C}^0$, of finite degree over $\mathcal{W}$? The answer is conjecturally yes, see [75].

Remember the isomorphism between the weight space $\mathcal{W}$ and the disjoint union of $p-1$ copies of the ball $B(0, 1)$. Consider, for a positive real $r < 1$, the "boundary" of the weight space B_r, namely the admissible open subspace $\mathcal{W}$ consisting of the points of all the components $B(0, 1)$ such that $r < |z| < 1$. Is it true that the eigencurve becomes étale over the weight space after restriction to B_r, for r sufficiently close to 1? This is a conjecture of Buzzard. Recent and important progress have been made in its direction: see [84].

7.8 Notes and References

The construction of the eigencurve given in Sect. 7.1 and its comparison to the Coleman-Mazur eigencurve Sect. 7.2 are based on the author's paper [8], but extends it from the local to the global case. That is, the global eigencurve is constructed here, instead of just the local pieces $\mathcal{C}^{\pm}_{W,\nu}$ as in [8], which requires proving the delicate restriction theorem (Theorem 7.1.7). As said in the introduction, the whole idea on the construction of the eigencurve with modular symbols valued in family of distributions is based on a talk by Glenn Stevens (known to the author through hearsay), and an important part of the technical machinery necessary to actually make the construction work comes from the paper by Pollack and Stevens on modular symbols with value distributions (not families thereof) [99], whose results were exposed and expanded in the preceding chapter (cf. Chap. 6).

The material on points on the eigencurve (Sect. 7.3) is standard. The idea that classical point, with a non-trivial nebentypus at p, appear on family interpolating very classical points with trivial nebentypus at p, seems originally due to Hida. The construction of the Galois representation on $\mathcal{C}^{\pm}$ is due to Chenevier (though it was known before due to Coleman and Mazur, with a more indirect proof) and its local study follows small parts of [13]. The study of the geometry of the eigenvariety is based on work by the author, either alone [8] and [9], or with Chenevier [12], and with Dimitrov [15].

Chapter 8
p-Adic L-Functions on the Eigencurve

In Chap. 5, we defined, using the method of Manin, Višik and Amice-Velu as reinterpreted by Mazur-Tate-Teitelbaum and Stevens) the p-adic L-function of a classical cuspidal eigenform of non-critical slope of weight ≥ 2.

In this chapter, using the eigencurve, we extend this construction to most overconvergent eigenforms, including the classical eigenforms of critical slope (cuspidal or Eisenstein), as well as (most) classical forms of weight 1. We then prove that these p-adic L-function can be put together to define a family of L-functions on the eigencurve (sometimes called a 2-variable p-adic L-function.)

8.1 Good Points and p-Adic L-Functions

In this section, we fix a prime p, an integer N not divisible by p, and a choice of sign $\pm$. We denote by Γ the congruence subgroup $\Gamma_0(p) \cap \Gamma_1(N)$, $\mathcal{H}_0$ the Hecke algebra generated by the T_l's for $l \nmid Np$, U_p, and $\langle a \rangle$ for $a \in (\mathbb{Z}/N\mathbb{Z})^*$. As in the preceding chapter, $\mathcal{C}^\pm$ denotes the p-adic eigencurve of modular symbols of tame level N and sign $\pm$.

8.1.1 Good Points on the Eigencurve

Lemma 8.1.1 *Let L be a finite extension of $\mathbb{Q}_p$, x be a point on $\mathcal{C}^\pm(L)$, with $\kappa(x) = w \in \mathcal{W}$. The following statements are equivalent:*

(i) *The dual of the generalized eigenspace $\mathrm{Symb}^\pm_\Gamma(\mathcal{D}^\dagger_w(L))^\vee_{(x)}$ is free of rank 1 over the algebra $\mathcal{T}_x$ of the schematic fiber of κ at x.*
(ii) *For any clean neighborhood U of x, the module $M^\vee$ is flat of rank one over $\mathcal{T}$ at x (we use the notations M and $\mathcal{T}$ attached to U defined in Sect. 7.6.1).*

J. Bellaïche, *The Eigenbook*, Pathways in Mathematics,
https://doi.org/10.1007/978-3-030-77263-5_8

(iii) *For any sufficiently small clean neighborhood U of x, the module $M^\vee$ is free of rank one over $\mathcal{T}$ at x.*

Proof The equivalence between (ii) and (iii) is clear. If U is a clean neighborhood, x is the only point of U above w, so $\mathcal{T}_x$ is the fiber $\mathcal{T}_w$ of $\mathcal{T}$ at w and the fiber M_w of M at w is $\mathrm{Symb}_\Gamma^\pm(\mathcal{D}_w^\dagger(L))_{(x)}$ (cf. Sect. 7.6.1), hence $\mathrm{Symb}_\Gamma^\pm(\mathcal{D}_w^\dagger(L))_{(x)}^\vee$ is the fiber of $M^\vee$ at w. Since $M^\vee$ and $\mathcal{T}$ are finite flat over R, Nakayama's lemma implies that (i) is equivalent to (ii). □

Definition 8.1.2 A point $x \in \mathcal{C}^\pm$ satisfying the conditions of the above lemma is called a *good* point.

If x is a good point, there exists a clean neighborhood of x where all points are good (by (iii) of Lemma 8.1.1).

Proposition 8.1.3 *If x is a good point, $\kappa(x) = w$, then the eigenspace $\mathrm{Symb}_\Gamma^\pm(\mathcal{D}_w^\dagger(L))[x]$ has dimension 1 over L.*

Proof Let $\mathfrak{m}_x$ be the maximal ideal of the local algebra $\mathcal{T}_x$. We need to prove that $\mathrm{Symb}_\Gamma^\pm(\mathcal{D}_w^\dagger(L))[\mathfrak{m}_x]$ has dimension 1. This space is dual of $\mathrm{Symb}_\Gamma^\pm(\mathcal{D}_w^\dagger(L))^\vee/\mathfrak{m}_x\mathrm{Symb}_\Gamma^\pm(\mathcal{D}_w^\dagger(L))^\vee$. Since $\mathrm{Symb}_\Gamma^\pm(\mathcal{D}_w^\dagger(L))^\vee$ is free of rank one over $\mathcal{T}_x$, this space has dimension one. □

Exercise 8.1.4 Let x be a classical normal good point on $\mathcal{C}^\pm$. Show that the minimal tame level of x is N.

Theorem 8.1.5 *Let x be a point of $\mathcal{C}^\pm$ such that the weight $w := \kappa(x)$ has the form $w : z \mapsto w(z)z^k\epsilon(z)$ for some $k \geq 0$ and some finite order character ϵ of conductor p^t. If x has tame level N and $\mathcal{C}^\pm$ is smooth at x, then $\mathcal{C}^\pm$ is a good point.*

Moreover, if $\mathcal{O}_{\mathcal{W},w}$ is the local ring of $\mathcal{W}^\pm$ at w, u a uniformizer of this discrete valuation ring, and $\mathcal{O}_{\mathcal{C}^\pm,x}$ the local ring of $\mathcal{C}^\pm$ at x, then there is an isomorphism of $\mathcal{O}_{\mathcal{W},w}$-algebras

$$\mathcal{O}_{\mathcal{W},w}[t]/(t^e - u) = \mathcal{O}_{\mathcal{C}^\pm,x},$$

where e is the index of ramification of κ at x. This isomorphism induces an isomorphism $\mathcal{T}_x = L[t]/t^e$.

Proof Choose a clean neighborhood U of x, and define R, M, $\mathcal{T}$ as usual. The localization $\mathcal{O}_{\mathcal{C},x}$ at x is by assumption a PID. By Exercise 2.3.2, the localization of the finite modules M and $M^\vee$ at x are torsion-free over $\mathcal{O}_{\mathcal{C},x}$, hence are free. Hence by shrinking U if necessary we may assume that $M^\vee$ is free over $\mathcal{T}$. The rank of $M^\vee$ over R is equal to the rank of M, namely d, while the rank of $\mathcal{T}$ is e, and by Lemma 7.6.2 and the hypothesis that x has tame level N, $d = e$. Hence $M^\vee$ is of rank one over $\mathcal{T}$, and hence x is a good point.

The second set of assertions is clear since $\mathcal{O}_{\mathcal{C},x}/\mathcal{O}_{\mathcal{W},w}$ is an extension of DVRs. □

8.1.2 The p-Adic L-Function of a Good Point

Definition 8.1.6 Let x be a good point of the eigencurve $\mathcal{C}^{\pm}$, of field of definition L. We define the modular symbol $\Phi_x^{\pm}$ as a generator of the one-dimensional space $\mathrm{Symb}_{\Gamma}^{\pm}(\mathcal{D}_w^{\dagger}(L))[x]$. We define the *L-distribution* of x, $\mu_x^{\pm}$, as $\Phi_x^{\pm}(\{\infty\} - \{0\})$ and the p-adic L-function of x as the Mellin transform $L_p^{\pm}(x, \sigma)$ of $\mu_x^{\pm}$.

Thus $\sigma \mapsto L_p^{\pm}(x, \sigma)$ is an analytic function $\mathcal{W}^{\pm}(\mathbb{C}_p) \to \mathbb{C}_p$ defined on half of the weight space. Obviously the symbol $\Phi_x^{\pm}$, the L-distribution $\mu_x^{\pm}$ and the L-function $L_p^{\pm}(x, \sigma)$ are only defined up to a scalar in L^*.

Remark 8.1.7 The modular symbol $\Phi_x^{\pm}$ is, by definition, non-zero. But this does not imply that the L-distribution $\mu_x^{\pm}$ and the p-adic L-function $L_p^{\pm}(x, \sigma)$ it defines are not zero. In fact, we will see below that for points x corresponding to ordinary Eisenstein series, the p-adic L-function is always 0.

Proposition 8.1.8 *The p-adic L-function $L_p^{\pm}(x, \sigma)$ is a function of order (see Definition 6.2.12) at most $v_p(U_p(x))$.*

Proof The distribution $\mu_x^{\pm}$ has order $\leq v_p(U_p(x))$ by Proposition 6.5.14. The proposition then follows from Theorem 6.6.10(b). □

8.1.3 Companion Points and p-Adic L-Functions

Proposition 8.1.9 *Let $x \in \mathcal{C}^{\pm}$ be a good normal point of weight $\kappa(x) = k \in \mathbb{N}$. Assume that x has a companion point $y \in \mathcal{C}^{\pm(-1)^k}$ (see Definition 7.6.8). Then the point y (of weight $-2-k$) is also a good point, and one has, for $\sigma \in \mathcal{W}^{\pm}(\mathbb{C}_p)$:*

$$L_p^{\pm}(x, \sigma) = \log_p^{[k+1]}(\sigma) L_p^{\pm(-1)^{k+1}}(y, \sigma/z^{k+1}), \tag{8.1.1}$$

up to multiplication by a non-zero constant.

Proof By Proposition 6.5.18, we have an $\mathcal{H}_0$-equivariant injection

$$\mathrm{Symb}_{\Gamma}^{\pm(-1)^{k+1}}(\mathcal{D}_{-2-k}(L))_{(y)}(k+1) \hookrightarrow \mathrm{Symb}_{\Gamma}^{\pm}(\mathcal{D}_k(L))_{(x)} \tag{8.1.2}$$

hence an $\mathcal{H}_0$-equivariant surjection

$$\mathrm{Symb}_{\Gamma}^{\pm(-1)^{k+1}}(\mathcal{D}_{-2-k}(L))_{(x)}^{\vee} \to \mathrm{Symb}_{\Gamma}^{\pm}(\mathcal{D}_k(L))_{(y)}(k+1)^{\vee}.$$

Since x is a good point, the LHS is a free module of rank 1 over $\mathcal{T}_x$, and in particular, is a faithful module over $\mathcal{T}_x$. It follows that the eigenalgebra generated by $\mathcal{H}_0$ in

$\mathrm{End}(\mathrm{Symb}_\Gamma^{\pm(-1)^{k+1}}(\mathcal{D}_{-2-k}(L))^\vee_{(x)})$ is exactly $\mathcal{T}_x$. Thus, there is an ideal I in the local algebra $\mathcal{T}_x$ such that

(i) $\mathrm{Symb}_\Gamma^{\pm}(\mathcal{D}_k(L))_{(y)}(k+1)^\vee = \mathrm{Symb}_\Gamma^{\pm(-1)^{k+1}}(\mathcal{D}_{-2-k}(L))^\vee_{(x)} \otimes_{\mathcal{T}_x} \mathcal{T}_x/I$;
(ii) the eigenalgebra generated by $\mathcal{H}_0$ over $\mathrm{Symb}_\Gamma^{\pm}(\mathcal{D}_k(L))_{(y)}(k+1)^\vee$ is $\mathcal{T}_x/I$;
(iii) $\mathrm{Symb}_\Gamma^{\pm}(\mathcal{D}_k(L))_{(y)}(k+1)^\vee$ is free of rank one over $\mathcal{T}_x/I$;

Therefore the action of $\mathcal{T}_y$ on $\mathrm{Symb}_\Gamma^{\pm}(\mathcal{D}_k(L))_{(y)}(k+1)^\vee$ factors through $\mathcal{T}_x/I$.

To conclude that y is good, we therefore just need to show that $\mathcal{T}_y$ and $\mathcal{T}_x/I$ have the same dimension over L. Choosing a clean neighborhood of y and using the notation d and e as in Sect. 7.6.1, we have by definition $e = \dim_L \mathcal{T}_y$ and $d = \dim_L \mathrm{Symb}_\Gamma^{\pm}(\mathcal{D}_k(L))^\vee_{(y)} = \dim \mathrm{Symb}_\Gamma^{\pm}(\mathcal{D}_k(L))_{(y)} = \dim \mathcal{T}_x/I$. So we need to prove that $d = e$. Now the representation ρ_y is just the representation ρ_x up to unramified twist, and it has therefore the same tame conductor, namely N (Exercise 8.1.4). Lemma 7.6.2 thus gives $d = e$.

In follows from (8.1.2) that $\Phi_x^{\pm} = \Theta_{k+1}\Phi_y^{\pm(-1)^{k+1}}$ (up to a scalar in L^*), hence $\mu_x^{\pm} = \Theta_{k+1}\mu_y^{\pm(-1)^{k+1}}$. Equation (8.1.1) follows from Proposition 6.6.11. □

8.2 p-Adic L-Functions of an Overconvergent Eigenform

In this section, we fix a prime p, an integer N not divisible by p. Write as usual $\Gamma = \Gamma_1(N) \cap \Gamma_0(p)$. We denote as usual by $\mathcal{C}$ and $\mathcal{C}_0$ the full and cuspidal p-adic eigencurve of tame level N.

8.2.1 Definition

For L a finite extension of $\mathbb{Q}_p$, and for $\kappa \in \mathcal{W}(L)$ an element of the weight space, let $g \in M_\kappa^\dagger(\Gamma, L)$ be an overconvergent $\mathcal{H}_0$-eigenform, which defines a point x in the full eigencurve $\mathcal{C}$ of Coleman-Mazur.

Since $\mathcal{C} = \mathcal{C}^+ \cup \mathcal{C}^-$, the point x belongs to $\mathcal{C}^+$, or $\mathcal{C}^-$, or both. Note that when $x \in \mathcal{C}_0$, x belong to both $\mathcal{C}^+$ and $\mathcal{C}^-$.

Definition 8.2.1 For every value of the sign $\pm$ such that $x \in \mathcal{C}^{\pm}$ and x is good in $\mathcal{C}^{\pm}$, we define the p-adic L-function of g on $\mathcal{W}^{\pm}(\mathbb{C}_p)$,

$$\sigma \mapsto L_p^{\pm}(g, \sigma)$$

as the p-adic L function $L_p^{\pm}(x, \sigma)$ of the point x of $\mathcal{C}^{\pm}$.

Thus g may have one p-adic L-function defined on half the weight space, or two p-adic L-functions defined on both halves of the weight space, each of them is only well defined up to a scalar in L^*.

8.2.2 Classical Cuspidal Eigenforms of Non-critical Slope

Proposition 8.2.2 *Let $g \in S_{k+2}(\Gamma, L)^{<k+1}$ be a classical cuspidal $\mathcal{H}_0$-eigenform of non-critical slope and let x be the point of $\mathcal{C}$ corresponding to g. Then x belongs to $\mathcal{C}^+$ and $\mathcal{C}^-$ and is a good points of both eigencurve.*

Proof Since g is cuspidal, x belongs to $\mathcal{C}^0$ hence to both $\mathcal{C}^+$ and $\mathcal{C}^-$, and the point x is normal. Let λ be the $\mathcal{H}_0$ system of eigenvalue of g. In the case $\lambda(U_p)^2 \neq p^{k+1}\lambda(\langle p\rangle)$, Theorem 7.6.4 shows that the eigencurve $\mathcal{C}^+$ or $\mathcal{C}^-$ is étale at x so in each case, a sufficiently small clean neighborhood $\mathcal{T} = R$ and $M^\vee$, which is free over R is free over $\mathcal{T}$, and x is good.

In the case $\lambda(U_p)^2 \neq p^{k+1} = \lambda(\langle p\rangle)$ (which is conjectured never to happen), $S_{k+2}(\Gamma, L)_{(g)}$ has dimension 2 but $S_{k+2}(\Gamma, L)[g]$ has dimension 1 (Exercise 6.7.5). By Eichler-Shimura and Stevens's control theorem, $\mathrm{Symb}_\Gamma^\pm(\mathcal{D}_k^\dagger(L))_{(g)}$ has dimension 2 and $\mathrm{Symb}_\Gamma^\pm(\mathcal{D}_k^\dagger(L))[g]$ has dimension 1, for both choice of the sign $\pm$. In a clean neighborhood, we thus have $d = 2$, and by Lemma 7.6.2, $e = d = 2$. The fiber $\mathcal{T}_x$ thus has dimension 2. Since the action of $\mathcal{T}_x$ on the 2-dimensional space $\mathrm{Symb}_\Gamma^\pm(\mathcal{D}_k^\dagger(L))_{(g)}^\vee$ factors through the action of the Hecke algebra, which is non semi-simple, we see that $\mathrm{Symb}_\Gamma^\pm(\mathcal{D}_k^\dagger(L))_{(g)}^\vee$ is free of rank one on $\mathcal{T}_x$, which shows that x is good.

□

For $g \in S_{k+2}(\Gamma, L)^{<k+1}$ be a classical cuspidal $\mathcal{H}_0$-eigenform of non-critical slope, we can thus define, according to Definition 8.2.1 both p-adic L-functions $L_p^+(g, \sigma)$ and $L_p^-(g, \sigma)$, each up to a multiplicative constant in L^*.

However, we had already defined two p-adic L-functions attached to g in Sect. 6.7.

Let us recall how it was done. By Eichler-Shimura, for each choice of the sign $\pm$, there is an element $\phi_g^\pm$ in $\mathrm{Symb}_\Gamma^\pm(\mathcal{V}_k(L))$ corresponding to g. The element $\phi_g^\pm$ is well-defined up to a scalar in L^*, but by requiring it lies on the lattice $\mathrm{Symb}_\Gamma^\pm(\mathcal{V}_k(\mathcal{O}_L))$ of $\mathrm{Symb}_\Gamma^\pm(\mathcal{V}_k(L))$, and not on the sub-lattice $\mathfrak{m}_L \mathrm{Symb}_\Gamma^\pm(\mathcal{V}_k(\mathcal{O}_L))$ (where $\mathfrak{m}_L$ is the maximal ideal of $\mathcal{O}_L$), the element $\phi_g^\pm$ is then well-defined up to an element of $\mathcal{O}_L^*$.

The morphism $\rho_k^* : \mathrm{Symb}_\Gamma^\pm(\mathcal{D}_k^\dagger(L))^{<k+1} \to \mathrm{Symb}_\Gamma^\pm(\mathcal{V}_k(L))^{k+1}$ is an isomorphism, and we called $\Phi_g^\pm$ the pre-image of $\phi_g^\pm$ by ρ_k^*; we then defined the p-adic distribution of g, $\mu_g^\pm$, as $\Phi_g^\pm(\{\infty\} - \{0\})$, and the p-adic L-function of g, $L_p^\pm(g, \sigma)$ as the Mellin transform of $\mu_g^\pm$. Thus $\mu_g^\pm$ and $L^\pm(g, \sigma)$ are well defined up to multiplication by an element of $\mathcal{O}_L^*$.

On the other hand, our new definition requires to choose a generator $\Phi_x^\pm$ of $\mathrm{Symb}_\Gamma^\pm(\mathcal{D}_k^\dagger(L))[x]$ and define the distribution $\mu_x^\pm$ as $\Phi_g^\pm(\{\infty\}-\{0\})$ and the L-function as the Mellin transform of $\mu_x^\pm$. These are defined up to an element of L^*.

Proposition 8.2.3 *The new definitions given here of the p-adic L-functions of g agree, up to multiplication by a non-zero scalar in L^*, with that of Sect. 6.7.*

Proof For a given choice of the sign $\pm$, let x be the point corresponding to g on $\mathcal{C}^\pm$. Since g of tame critical N, the space $\mathrm{Symb}_\Gamma^\pm(\mathcal{V}_k(L))[x]$ has dimension 1 and so does $\mathrm{Symb}_\Gamma^\pm(\mathcal{D}_k^\dagger(L))[x]$. The element $\phi_g^\pm$ by construction lies in the former space, hence $\Phi_g^\pm$ lies in the latter. It is thus equal, up to a scalar in L^* to the generator $\Phi_x^\pm$ of $\mathrm{Symb}_\Gamma^\pm(\mathcal{D}_k^\dagger(L))[x]$. □

8.2.3 Ordinary Eisenstein Eigenforms

Theorem 8.2.4 *Let $k \geq 0$ be an integer, Q and R be two integers such that $QR = N$, τ and ψ Dirichlet characters of conductors Q and R such that $\tau(-1)\psi(-1) = (-1)^k$, L a finite extension of $\mathbb{Q}_p$ containing the image of τ and ψ. Let $g = E_{k+2,\tau,\psi,\mathrm{ord}} \in \mathcal{E}_k(\Gamma, L)$ be the ordinary refinement (see Exercise 2.6.15) of the new Eisenstein series $E_{k,\tau,\psi}$.*

Then the point $x \in \mathcal{C}$ corresponding to $\mathcal{C}$ belongs to $\mathcal{C}^{\tau(-1)}$ but not $\mathcal{C}^{-\tau(-1)}$. It is a good point of $\mathcal{C}^{\tau(-1)}$. One can thus define a distribution $\mu_g^{\tau(-1)}$ and half a p-adic function of g, $L_p^{\tau(-1)}(g,\sigma)$ for $\sigma \in \mathcal{W}^{\tau(-1)}$. The distribution $\mu_g^{\tau(-1)}$ has support contained in $\{0\}$, and the L-function $L_p^{\tau(-1)}(g,\sigma)$ is zero or all $\sigma \in \mathcal{W}^{\tau(-1)}$.

Proof According to Eichler-Shimura (Theorem 5.3.35) and Proposition 5.3.34 the system of eigenvalues of g appears once in $\mathrm{Symb}_\Gamma^{\tau(-1)}(\mathcal{V}_k(L))$, and not in $\mathrm{Symb}_\Gamma^{-\tau(-1)}(\mathcal{V}_k(L))$. Since this system has slope 0, by Stevens's control theorem, the system of eigenvalues of g appears once in $\mathrm{Symb}_\Gamma^{\tau(-1)}(\mathcal{D}_k^\dagger(L))$, and not in $\mathrm{Symb}_\Gamma^{-\tau(-1)}(\mathcal{D}_k(L))$. Hence by Theorem 7.3.1, x is in $\mathcal{C}^{\tau(-1)}$ but not in $\mathcal{C}^{-\tau(-1)}$. The fact that x is good in $\mathcal{C}^{-1}$ is proved as in the easy case of the above theorem.

We prove the assertion about $\mu_g^{\tau(-1)}$ and $L^{\tau(-1)}(g,\sigma)$ only in the case k even, the case k odd needing only some minor modifications (see [14] for details). In this case, let us write by $\Phi_x^{\tau(-1)}$ the overconvergent corresponding to x in $\mathrm{Symb}_\Gamma^{\tau(-1)}(\mathcal{D}_k^\dagger(L))$ and by $\phi_x^{\tau(-1)} \in \mathrm{Symb}_\Gamma^{\tau(-1)}(\mathcal{V}_k(L))$ its image by ρ_k. Since $\phi_{x,\tau(-1)}$ is ordinary, it is a linear combination of classical boundary symbols of the form $\phi_{k,u,v}$, with u, v relatively prime integers such that u/v a p-integer (cf. question 3 of Exercise 5.3.37.) By Exercise 6.5.20, $\Phi_x^{\tau(-1)}$ is a linear combination of $\Phi_{k,u,v}$ with u, v as above.

But for $f \in \mathcal{A}_k^\dagger(L)$,

$$\Phi_{k,u,v}(\{\infty\}-\{0\})(f)=\Phi_{k.u,v}(\{\infty\})(f)-\Phi_{k,u,v}(\{0\})(f)=-\Phi_{k,u,v}(\{0\})(f)$$

since $\Phi_{k,u,v}$ is supported in the Γ-orbit of the p-integer u/v, which does not contain ∞. By applying (6.5.2) to a matrix $\gamma=\begin{pmatrix}-v & u\\ c & d\end{pmatrix}\in\Gamma$ (the coefficients c and d being chosen using a Bézout relation), we see that $\Phi_{k,u,v}(\{0\})(f)=\Phi_{k,u,v}\left(\left\{\gamma\cdot\frac{u}{v}\right\}\right)(f)=f(0)(cu+dv)^k=f(0)$ since $1=\det\gamma=ad+bc$. Therefore $\Phi_{k,u,v}(\{\infty\}-\{0\})$ is the negative of the Dirac measure at 0, and $\mu_x^{\tau(-1)}=\Phi_x^{\tau(-1)}(\{\infty\}-\{0\})$ is a multiple of the Dirac measure at 0. Its restriction to $\mathbb{Z}_p^*$, and in particular its Mellin transform $L_p^{\tau(-1)}(\sigma)$ is thus zero. □

8.2.4 Classical Eigenforms of Critical Slope

One of the most important consequence of the construction of p-adic L-function using the eigencurve is the possibility to define p-adic L-functions for classical eigenforms of critical slope.

Theorem 8.2.5 *Let $k\geq 0$ be an integer. Let L be a finite extension of $\mathbb{Q}_p$ and $g\in M_{k+2}(\Gamma,L)$ a modular form, of tame level N, eigenvector for $\mathcal{H}_0$, with critical slope $k+1$. Then g is the critical refinement, f_β, of a modular newform $f\in M_{k+2}(\Gamma_1(N),L)$. Assume that f is normal.*

The point x of $\mathcal{C}$ corresponding to g is in both $\mathcal{C}^+$ and $\mathcal{C}^-$ and smooth and good in both eigencurves, and κ has the same index of ramification e at x on $\mathcal{C}^+$ and $\mathcal{C}^-$. We can define both p-adic L-functions $L^\pm(g,\sigma)$ on $\mathcal{W}^\pm$. These p-adic L-functions have order $\leq k+1$ and satisfy the following interpolation property:

For any finite image character $\chi:\mathbb{Z}_p^*\to\mathbb{C}_p^*$ of conductor p^n, and any integer j such that $0\leq j\leq k$ and $\chi(-1)(-1)^j=\pm$,

$$L_p^\pm(f_\beta,\chi z^j)=\begin{cases}\frac{e_p(f,\alpha,\chi,j)p^{n(j+1)}j!}{\beta^n(-2i\pi)^{j+1}\tau(\chi^{-1})\Omega_g^\pm}L(f,\chi^{-1},j+1) & \text{if } e=1 \text{ and } f \text{ is cuspidal}\\ 0 & \text{otherwise}\end{cases}$$

where $e_p(f,\beta,\chi,j)=1$ if χ is non trivial and $e_p(f,\beta,1,j)=(1-\beta^{-1}\epsilon(p)p^{k-j})(1-\beta^{-1}p^j)$.

If f is a cuspidal form with CM by K, let ψ be the character $G_K\to L^*$ such that $\rho_f=\mathrm{Ind}_{G_K}^{G_\mathbb{Q}}\psi$, and let us denote by $\tau\to L_K(\tau)$, where τ runs amongst characters $G_K\to\mathbb{C}_p^*$ the Katz p-adic L-function (defined implicitly in [77] but explicitly in [63], denoted there $L_p(0,\psi)$), for all $\sigma\in\mathcal{W}^\pm(\mathbb{C}_p)$

$$L_p^\pm(f_\beta,\sigma)=\log_p^{[k+1]}(\sigma)L_K(\psi\sigma_K^{-1})\qquad(8.2.1)$$

If f is the Eisenstein series $E_{k,\tau,\psi}$, then, on $\mathcal{W}^{\pm}(\mathbb{C}_p)$

$$L_p(E_{k,\tau,\psi,\mathrm{crit}},\sigma) = \begin{cases} \log_p^{[k+1]}(\sigma)L_p(\tau,\sigma z)L_p(\psi,\sigma z^{-k}) & \text{if } \pm 1 = \tau(-1)(-1)^{k+1}. \\ 0 & \text{if } \pm 1 = \tau(-1)(-1)^{k} \end{cases}$$

where $L_p(\tau,-)$ (resp. $L_p(\psi,-)$) are the p-adic Dirichlet L-function of the Dirichlet character τ (resp. ψ).

We note that, contrary to the situation of non-critical slope cuspidal forms, the interpolation property does not by itself characterizes the p-adic L-functions, since their slope may be $k+1$.

Proof The assertion of smoothness of $\mathcal{C}^{\pm}$ at x, as well as the assertion of the degree e was proved in Proposition 7.6.13. The assertions of goodness follows in view of Theorem 8.1.5.

Also by this theorem, we know that $\mathcal{T}_x \simeq L[t]/t^e$. Consider the map ρ_k : $\mathrm{Symb}_\Gamma(\mathcal{D}_k^\dagger(L))_{(x)} \to \mathrm{Symb}_\Gamma(\mathcal{V}_k(L))_{(x)}$. The right hand side is a free module of rank 1 over $\mathcal{T}_x$, hence of dimension e over L.

The left hand side has dimension 1 if f is cuspidal. In this case, since ρ_k is Hecke-equivariant, the kernel of ρ_k is the hyperplane $t\mathrm{Symb}_\Gamma^{\pm}(\mathcal{D}_k^\dagger(L))_{(x)}$. The symbol $\Phi_x^{\pm}$ is by definition a non-zero Hecke eigenvector in $\mathrm{Symb}_\Gamma^{\pm}(\mathcal{D}_k^\dagger(L))_{(x)}$, hence it lies in $t^{e-1}\mathrm{Symb}_\Gamma^{\pm}(\mathcal{D}_k^\dagger(L))_{(x)}$. We therefore conclude that $\rho_k(\Phi_x^{\pm}) = 0$ if and only if $e = 1$ in the cuspidal case, and the interpolation formula thus follows in the cuspidal case exactly as in Sect. 6.7.3. In case $e > 1$, Theorem 7.6.10 shows that x has a companion point y of the eigencurve. By Proposition 8.1.9, $L_p^{\pm}(f_\beta,\sigma) = \log_p^{[k+1]}(\sigma)L_p^{\pm(-1)^k}(y,\sigma)$. Moreover, in the case where f has CM, the point y can easily be identified, as an ordinary CM point on the eigencurve of negative weight, and the function $L_p^{\pm(-1)^k}(y,\sigma)$ can be evaluated in term of Katz p-adic L-function. This leads to formula (8.2.1). We refer the reader to [9] for the details, which are not hard.

We know turn to the case where f is an Eisenstein series, say $f = E_{k,\tau,\psi}$.

On $\mathcal{C}^{-\tau(-1)}$ the point x is not very classical, and one has $\mathrm{Symb}_\Gamma^{\tau(-1)}(\mathcal{V}_k(L))_{(x)} = 0$. It follows that ρ_k is 0, hence its kernel is non trivial, which means that x has a companion point y. Again, this point can be easily identified: it corresponds to an ordinary Eisenstein series of negative weight. Like ordinary Eisenstein series of positive weight (that is, classical ordinary Eisenstein series, see Theorem 8.2.4), these series are easily seen to have a p-adic L-function equal to zero (see [14, §5.2 and §5.3]). It follows then from Proposition 8.1.9 that $L_p^{-\tau(-1)}(g,\sigma) = 0$.

On $\mathcal{C}^{\tau(1)}$, the point x is very classical, so Theorem 7.6.10 does apply, and x has a companion point if and only if $e > 1$. The existence of the factorization of $L_p^{\tau(-1)}(g,\sigma)$, even in the case $e = 1$, is proved in [14]. □

8.2.5 *Classical Eigenforms of Weight* 1

A classical modular form of weight 1 corresponds to no classical modular symbol, and the traditional method to attach to them p-adic L-function fails. However, the eigencurve method allows to define a p-adic L-function in this arithmetically important case too.

Let $f \in S_1(\Gamma_1(N), \mathbb{C})$ be an new form of nebentypus ϵ. Let us call a_p the eigenvalue of U_p. Let us factor $X^2 + a_pX + \epsilon(p)$ as $(X - \alpha)(X - \beta)$. Here α and β are both p-adic units, hence of p-adic valuations 0. The refinements f_α and f_β are thus both ordinary and critical slope. Recall that f is *regular* if $\alpha \neq \beta$. In this case we have assume that f is regular.

Theorem 8.2.6 *Let x be the point corresponding of $\mathcal{C}$ to f_α (or f_β). Then x belongs to $\mathcal{C}^+$ and $\mathcal{C}^-$ and is a good point there. In particular, we can define the p-adic L-function of f_α, $L_p(f_\alpha, \sigma)$ on both $\mathcal{W}^+$ and $\mathcal{W}^-$.*

Proof As recalled in Theorem 7.6.20, the smoothness of the point x in $\mathcal{C}^+$ and $\mathcal{C}^-$ is proved in [15]. The fact that x is a good point follows from Theoram 8.1.5. □

For generalizations, see for instance [19] and [55]. For applications, see for instance [51].

8.3 The 2-Variable p-Adic L-Function

In this section, given a good point $x \in \mathcal{C}^\pm$, we show that on a suitable affinoid neighborhood U of x we can define a two-variable p-adic L-function, $\tilde{L}_p(y, \sigma)$, analytic on $U \times \mathcal{W}^\pm$, well defined up to an element of $\mathcal{O}(U)^*$. This function interpolates the one-variable p-adic L-functions defined above (Definition 8.1.6) at every point y on U (which is always good):

$$\forall \sigma \in \mathcal{W}^\pm, \ \tilde{L}_p(y, \sigma) = c(y)L_p(y, \sigma),$$

where $c(y) \in L(y)^*$, $L(y)$ being the field of definition of y.

Note that here, for $y \in U$, the functions $\sigma \to L_p(y, \sigma)$ are themselves defined up to multiplication by a scalar of $L(y)^*$, so $c(y)$ is completely undetermined. However, in the special cases where y corresponds to a cuspidal modular eigenform g of non-critical slope, then the function $L_p(y, \sigma) = L_p^\pm(g, \sigma)$ is defined only up to a scalar in $L(y)^*$ of absolute value 1. Hence, the absolute value $|c(y)|$ is well-defined. An important complement of the preceding interpolation property is the fact that for y running among all such points in U, the normalization constant $c(y)$ stays bounded above, and below from zero.

Those results follows from taking the Mellin transform on the distributions considered in the following theorem.

Theorem 8.3.1 *Let $x \in \mathcal{C}^{\pm}$ be a good point. For $U = Sp\,\mathcal{T}$ a sufficiently small clean neighborhood of x, there exists a distribution $\tilde{\mu} \in \mathcal{D}^{\dagger}(\mathcal{T})$, well-defined up to multiplication by an element in $\mathcal{O}(U)^* = \mathcal{T}^*$, such that for all $y \in U$, of field of definition L,*

$$ev_y(\tilde{\mu}) = \mu_y, \tag{8.3.1}$$

where $\mu_y \in D^{\dagger}(L)$ is the distribution attached to y as in Definition 8.1.6, well-defined up to an element of L^*, and the equality is up to that indeterminacy.

Moreover, a choice of $\tilde{\mu}$ as above being fixed, there exists two real constants $0 < c < C$ such that if y is a point of U with $\kappa(y) = k \in \mathbb{Z}$ with the same system of eigenvalues as a cuspidal eigenform $g \in S_{k+2}(\Gamma, L)$ of non-critical slope, then we have $ev_y(\tilde{\mu}) = c(y)\mu_g^{\pm}$ where $c(y)$ is an element of L^* so that $c < |c(y)| < C$.

Proof We first explain the construction of the distribution $\tilde{\mu} \in \mathcal{D}^{\dagger}(\mathcal{T})$, well defined up to multiplication by an element of $\mathcal{T}^*$, whose Mellin's transform is the desired 2-variable p-adic L-function.

Since x is good, choosing the clean neighborhood U small enough ensures that $M^{\vee}$ is free of rank one over $\mathcal{T}$, and that M is free of finite rank over R. We recall that $M = \varepsilon \mathrm{Symb}_{\Gamma}(\mathcal{D}_K(R))^{\leq \nu}$ for some real ν, and some idempotent ε. We **choose a $\mathcal{T}$-generator** u of $M^{\vee}$. Thus u is well-defined up to multiplication by an element of $\mathcal{T}^*$. The choice of u define an isomorphism $\mathcal{T} \to M^{\vee}$ of $\mathcal{T}$-modules, $1 \mapsto u$, whose inverse we will denote by τ_u.

Hence we can consider the following sequence of isomorphisms:

$$\mathrm{Hom}_R(M, \mathcal{D}^{\dagger}(R)) \xrightarrow{\sim} M^{\vee} \otimes_R \mathcal{D}^{\dagger}(R) \xrightarrow[\tau_u \otimes \mathrm{Id}]{\sim} \mathcal{T} \otimes_R \mathcal{D}^{\dagger}(R) \xrightarrow{\sim} \mathcal{D}^{\dagger}(\mathcal{T})$$

Here the first isomorphism is just the obvious one, and the last one is Corollary 6.2.8 (remember that $\mathcal{T}$ and M are both free of finite type over R).

The module $\mathrm{Hom}_R(M, \mathcal{D}^{\dagger}(R))$ has a canonical element, noted $\tilde{e}$. It is the morphism that sends a modular symbol $\Phi \in M$ to

$$\tilde{e}(\Phi) := \Phi(\{\infty\} - \{0\}).$$

We define $\tilde{\mu} \in D^{\dagger}(\mathcal{T})$ as the image of $\tilde{e}$ by the sequence of isomorphisms above. Note that $\tilde{\mu}$ is well-defined only up to an element of $\mathcal{T}^*$ since the middle isomorphism is.

Now we prove the specialization formula (8.3.1).

If y is a point of U, of ideal $\mathfrak{p}_y$, $L = \mathcal{T}/\mathfrak{p}_y$ the field of definition of y.

Let $w \in (\mathrm{Spec}\, R)(L)$ the weight of y. Write $M_w = M \otimes_{\mathcal{T},w} L$ and $(M^{\vee})_w = M^{\vee} \otimes_{\mathcal{T},w} L$. Since M are locally free of finite type over R, we have a natural isomorphism $(M_w)^{\vee} = (M^{\vee})_w$, and since $M^{\vee}$ is free of rank one over $\mathcal{T}$ generated by u, $(M^{\vee})_w$ is free of rank one over $\mathcal{T}_w = \mathcal{T} \otimes_{R,w} L$, the algebra of the schematic fiber of U at w, and the image u_w of $u \in (M^{\vee})_w$ is a generator of $(M^{\vee})_w$, defining an isomorphism $\tau_{u_w} : (M^\vee)_w \to \mathcal{T}_w$. Hence the commutative diagram, where the second line is just the first line tensorized by L.

$$\begin{array}{ccccccc}
\mathrm{Hom}_R(M, \mathcal{D}^\dagger(R)) & \xrightarrow{\sim} & M^\vee \otimes_R \mathcal{D}^\dagger(R) & \xrightarrow[\tau_u \otimes \mathrm{Id}]{\sim} & \mathcal{T} \otimes_R \mathcal{D}^\dagger(R) & \xrightarrow{\sim} & \mathcal{D}^\dagger(\mathcal{T}) \\
\downarrow & & \downarrow & & \downarrow & & \downarrow \\
\mathrm{Hom}_L(M_w, \mathcal{D}^\dagger(L)) & \xrightarrow{\sim} & (M^\vee)_w \otimes_L \mathcal{D}^\dagger(L) & \xrightarrow[\tau_{u_w} \otimes \mathrm{Id}]{\sim} & \mathcal{T}_w \otimes_L \mathcal{D}^\dagger(L) & \xrightarrow{\sim} & \mathcal{D}^\dagger(\mathcal{T}_w)
\end{array}$$

The space $M^\vee/\mathfrak{p}_y M^\vee$ is of dimension 1 over $\mathcal{T}/\mathfrak{p}_y = L$, generated by the image u_y of u. By Proposition 8.1.3 and its proof, the space $M_w[y] = \mathrm{Symb}^{\pm}_{\Gamma}(\mathcal{D}^\dagger_w(L))[y]$ has dimension 1, and is naturally the dual of $M^\vee/\mathfrak{p}_y M^\vee$. Hence we have a commutative diagram:

$$\begin{array}{ccccccc}
\mathrm{Hom}_R(M, \mathcal{D}^\dagger(R)) & \xrightarrow{\sim} & M^\vee \otimes_R \mathcal{D}^\dagger(R) & \xrightarrow[\tau_u \otimes \mathrm{Id}]{\sim} & \mathcal{T} \otimes_R \mathcal{D}^\dagger(R) & \xrightarrow{\sim} & \mathcal{D}^\dagger(\mathcal{T}) \\
\downarrow & & \downarrow & & \downarrow & & \downarrow \\
\mathrm{Hom}_L(M_w, \mathcal{D}^\dagger(L)) & \xrightarrow{\sim} & (M^\vee)_w \otimes_L \mathcal{D}^\dagger(L) & \xrightarrow[\tau_{u_w} \otimes \mathrm{Id}]{\sim} & \mathcal{T}_w \otimes_L \mathcal{D}^\dagger(L) & \xrightarrow{\sim} & \mathcal{D}^\dagger(\mathcal{T}_w) \\
\downarrow{\scriptstyle \text{restriction}} & & \downarrow & & \downarrow & & \downarrow \\
\mathrm{Hom}_L(M_w[y], \mathcal{D}^\dagger(L)) & \xrightarrow{\sim} & M^\vee/\mathfrak{p}_y M^\vee \otimes_L \mathcal{D}^\dagger(L) & \xrightarrow[\tau_{u_y} \otimes \mathrm{Id}]{\sim} & \mathcal{T}/\mathfrak{p}_y \otimes_L \mathcal{D}^\dagger(L) & \xrightarrow{=} & \mathcal{D}^\dagger(L)
\end{array}$$

To prove (8.3.1), observe that the element $\tilde{e}$ of the upper-left corner has image $\tilde{\mu}$, by definition, in the upper right corner, hence it has image $\mathrm{ev}_y(\tilde{\mu})$ in the lower-right corner. Using the commutativity, the same element $\mathrm{ev}_y(\tilde{\mu})$ is the image by the lower line of the element "evaluation at $\{\infty\} - \{0\}$" in the lower-left corner, hence is the evaluation at $\{\infty\} - \{0\}$ of the element Φ_y in $M_w[y]$, which is the dual basis of the basis u_y of $M_w^\vee/\mathfrak{p}_y M_w^\vee$. But the latter is just the definition of μ_y. This completes the proof of (8.3.1).

Let us provide M with a norm $\| \ \|$ of R-modules, and for every $y \in U$ and w the weight of y, let us provide M_w with the quotient norm $\| \ \|_w$. We claim that there are two real constants $0 < c < C$ such that for all $y \in U$, setting $w := \kappa(y)$, the element $\Phi_y \in M_w[y]$ satisfies

$$c < \|\Phi_y\|_w < C. \tag{8.3.2}$$

To prove (8.3.2) we can clearly replace our norm $\| \ \|$ with any equivalent one. In particular, we can choose an isomorphism $M \simeq R^d$ and assume that $\| \ \|$ is the sup norm on R^d. This ensures that if we define $\| \ \|'$ as the dual norm of $\| \ \|$ on M', and $\| \ \|'_w$ as the quotient norm of $\| \ \|'$ on $M_w^\vee$, then we get the same result as if we consider the dual norm on $M_w^\vee$ of the quotient norm $\| \ \|_w$ of $\| \ \|$ on M_w.

We provide $M_w^\vee/\mathfrak{p}_y M_w^\vee$ with the quotient norm $\| \|'_y$ of the norm $\| \|_w$ on M_w. By transitivity of quotient, this is the quotient norm of the norm $\| \|$ of M. But by the theorem of Hahn-Banach $\| \|)_y$ is also the dual norm of the restriction of $\| \|_w$ on the line $M_w[y]$.

To prove the claim, recall that Φ_y is by definition the dual basis in the L-line $M_w[y]$ of the basis u_y of $M^\vee/\mathfrak{p}_y M^\vee = M_w^\vee/\mathfrak{p}_y M_w^\vee$. Thus

$$\|u_y\|'_y \|\Phi_y\|_w = 1,$$

and we are reduced to prove that there are two real constants $0 < c < C$ such that for all $y \in U$, the element $u_y \in M_w^\vee/\mathfrak{p}_y M_w^\vee = M^\vee/\mathfrak{p}_y M^\vee$ satisfies

$$c < \|u_y\|'_y < C. \tag{8.3.3}$$

Here $\| \|'_y$ is the quotient norm of the norm $\| \|'$ on $M^\vee$.

To prove (8.3.3), it is again allowable to replace $\| \|'$ by an equivalent norm on M'. Using the isomorphism $M' \simeq \mathcal{T}$, we may assume that $M' = \mathcal{T}$ and $\|\|'$ is the sup norm on the reduced affinoid $\mathcal{T}$. Then the assertion to check is that if $u \in \mathcal{T}^*$, $c < \|u_y\| < C$ for every $y \in U = \operatorname{Sp} \mathcal{T}$ and two constants c and C depending only on u, which follows from the maximum and the minimum principles.

This completes the proof of claims (8.3.2) and (8.3.3).

Now we recall that $M = \varepsilon \mathrm{Symb}_\Gamma(\mathcal{D}_K(R))^{\leq \nu}$ for some real ν, and some idempotent ε. Let us fix a positive real r such that $r < r(K)$ and $r \leq 1$. Then $\mathrm{Symb}_\Gamma(\mathcal{D}_K(R))^{\leq \nu} = \mathrm{Symb}_\Gamma(\mathcal{D}_K[r](R))^{\leq \nu}$ and thus we can provide $\mathrm{Symb}_\Gamma(\mathcal{D}_K(R))^{\leq \nu}$ with the norm $\| \|_r$. We provide M with the restriction of that norm. We provide M_w with the quotient norm, which is also the restriction of norm $\| \|_r$ on $M_w \subset \mathrm{Symb}_\Gamma(\mathcal{D}_w(L))^{\leq \nu}$. The claim shows that there are two real constants $0 < c < C$ such that for all $y \in U$, the element $\Phi_y \in M_w[y]$ satisfies

$$c < \|\Phi_y\|_r < C.$$

To finish the proof, let us assume that y corresponds to an eigenform g as in the statement of the theorem. If $\Phi_g^\pm$ is the modular symbol of Sect. 8.2.2, we have $\Phi_y = c(y)\Phi_g^\pm$. By definition, the classical modular symbol $\phi_g^\pm \in \mathrm{Symb}_\Gamma^\pm(\mathcal{V}_k(L))$ is normalized such that $\|\phi_g^\pm\|_1 = 1$. By Theorem 6.5.19(ii), it follows that we also have $\|\Phi_g^\pm\|_1 = 1$. By Proposition 6.5.12, there exists therefore some constant $D > 0$ depending of our clean neighborhood U and of the choice of r, but **independent** of the choice of the classical point y or of the corresponding modular form g such that

$$D \leq \|\Phi_g^\pm\|_r \leq 1.$$

It follows that $c \leq |c(y)| \leq CD^{-1}$, and the theorem follows. □

Corollary 8.3.2 *With U as in the above theorem, there exist a two-variable analytic function $\tilde{L}$ on $U \times \mathcal{W}$, well-defined up to an element of $\mathcal{O}(U)^*$, such that for all $y \in U$, and all $\sigma \in \mathcal{W}$*

$$\tilde{L}_p(y, \sigma) = L_p(y, \sigma)$$

where the equality is up to a scalar in L^*. Moreover, there exist two constant $0 < c < C$ such that if y corresponds to a modular eigenform g of non-critical slope, we have

$$\tilde{L}_p(y, \sigma) = c(y) L_p(g, \sigma)$$

with $c \leq |c(y)| \leq C$.

As an example of application, consider a critical Eisenstein series $E_{k,\tau,\psi,\mathrm{crit}}$, and the point x it corresponds to on the eigencurve $\mathcal{C}^{-\tau(-1)}$. Thus (cf. Theorem 8.2.5) x is a non very classical good point and $L_p(x, \sigma) = L_p^{-\tau(-1)}(E_{k,\tau\psi,\mathrm{crit}}, \sigma) = 0$ for every $\sigma \in \mathcal{W}^{-1}(\mathbb{C}_p)$. There is a sequence g_n of cuspidal newforms of level Γ and non-critical slope, whose points x_n on the eigencurve $\mathcal{C}^{-\tau(-1)}$ converge to x. Let $L_p^{\tau(-1)}(g_n, \sigma)$ denotes their p-adic L-function, well-defined up to a p-adic integer. Then

$$L_p^{-\tau(-1)}(g_n, \sigma) \longrightarrow 0, \text{ uniformly for } \sigma \in \mathcal{W}^{-\tau(-1)}(\mathbb{C}_p).$$

In particular, this provides a sequence of classical cusp forms whose p-adic L-functions have a μ-invariant which tends to ∞.

For much more about this kind of idea, see [16].

8.4 Notes and References

The first "missing" p-adic L-function, i.e. attached to a cuspidal critical-slope eigenform, was constructed by Pollack and Stevens (see [99] and [100]), in the case of a cuspidal new form f of critical slope with no companion point. The general case was treated in [8], as well as the existence of the 2-variable L-function. The (conjecturally) non-existing case of a modular forms f for which the two eigenvalues α and β are equal was omitted in [8]. Its treatment here is new. The same result has been proved independently in [21].

Most of the material of the chapter comes from [8], the unpublished paper [9], and [14]. The second part of Theorem 8.3.1 is new.

Chapter 9
The Adjoint p-Adic L-Function and the Ramification Locus of the Eigencurve

This chapter is largely inspired by Kim's 2006 Berkeley thesis [78]. Indeed the core result, namely the construction of a scalar product on the module of families of modular symbols interpolating the classical scalar product on classical modular symbols is taken directly from [78], with only minor modifications. However, we improve on Kim's treatment by proving an explicit and more general result of perfectness on this scalar product (than the one implicitly stated in [78, Theorem 11.2]), and more importantly by changing the way the adjoint p-adic L-function is constructed from the scalar product, and the way this scalar product is proved to be related to the ramification locus of the eigencurve. Actually, we propose a conceptual treatment of those two aspects together in Sect. 9.1 below. It seems to us that there is an gap in the argument of [78] concerning the descent of the adjoint p-adic L-function to the eigencurve from its normalization that our method allows to circumvent. At any rate, we obtain more precise results on the relation between the geometry of the weight map and the adjoint p-adic L-function.

Let us now describe in more details these results. As usual, we denote by $\mathcal{C}^0$ the cuspidal p-adic eigencurve on tame level N, and $\mathcal{H}_0$ the Hecke algebra generated by the operators T_l for $l \nmid Np$, U_p, and the diamond operators.

We construct, using Kim's scalar product on modular symbols, a canonical sheaf of ideals $\mathcal{L}^{\mathrm{adj}}$ of $\mathcal{O}(\mathcal{C}^0)$, and we prove (Theorem 9.4.2) that its closed locus contains all points of $\mathcal{C}^0$ where the weight map κ is not étale.

We do not know whether the ideal $\mathcal{L}^{\mathrm{adj}}$ is everywhere locally principal, but it is so in a neighborhood of most points of interest, including all classical points that are of minimal tame level N (except possibly for a few Eisenstein series of weight 2 when N is not square-free) and all their companion points when they have one. In a suitable neighborhood U of such a *good* point, we can define the adjoint p-adic

J. Bellaïche, *The Eigenbook*, Pathways in Mathematics,
https://doi.org/10.1007/978-3-030-77263-5_9

L-function L^{adj} as a generator of the ideal $\mathcal{L}^{\mathrm{adj}}$, and we can determine exactly where this function vanishes (see Theorem 9.4.7):

(i) If $x \in U$ is a point that either has non-integral weight $\kappa(x) \notin \mathbb{N}$, or is cuspidal classical[1] then $L^{\mathrm{adj}}(x) = 0$ if and only if the weight map κ is not étale at x.
(ii) Otherwise $L^{\mathrm{adj}}(x) = 0$.

We observe[2] that in case (ii) it may or may not happen that κ is étale at x.

When x is a smooth point of the eigencurve (again this includes almost all classical points), we can define two invariants at x: the degree of ramification $e(x)$ of κ at x (so κ is étale at x if and only if $e(x) = 1$), and the order of vanishing of L^{adj} at x. It is natural to ask how these invariants are related: an answer is given in Theorem 9.4.8.

Finally we justify the name of the adjoint p-adic L-function L^{adj} by proving that at cuspidal classical points of non-critical slope x corresponding to the form f_x, the value $L^{\mathrm{adj}}(x)$ is related to the value of the archimedean adjoint L-function of f_x at the near central point. Kim [78, Theorem 12.3] proves such a formula in the case f_x has wild level p^r with $r \geq 1$ and primitive nebentypus. We extend this formula to the case of a form of level p and trivial nebentypus, in particular a form f_x which is a refinement of a form of level prime to p.

9.1 The L-Ideal of a Scalar Product

In all this section, R is a Noetherian ring, and T is a finite R-algebra.

9.1.1 *The Noether Different of T/R*

Let $m : T \otimes_R T \to T$ denote the multiplication map: $m(t \otimes u) = tu$. Let I be the kernel of this map.

Lemma 9.1.1 *As an ideal of $T \otimes_R T$ the ideal I is generated by the elements of the form $t \otimes 1 - 1 \otimes t$ for $t \in T$.*

---

[1] We say that a point $x \in \mathcal{C}^0$ defined over a finite extension L of $\mathbb{Q}_p$ is *cuspidal classical* if its weight $\kappa(x)$ is an integer $k \in \mathbb{N}$ and its $\mathcal{H}_0$-system of eigenvalues appears in $S_{k+2}(\Gamma, L)$.

[2] For example, if x corresponds to an evil Eisenstein series of minimal tame level N, then it is expected that κ is étale at x, and this is known for example for the Eisenstein series E_k^{crit} (for any even $k > 4$, so for $N = 1$) when p is a regular prime. On the other hand, if x corresponds to the Eisenstein series $E_{2,\ell}^{\mathrm{crit}}$ (where $\ell \neq p$ is an auxiliary prime, and $E_{2,\ell}(z) = E_2(z) - \ell E_2(\ell z)$, and $E_{2,\ell}^{\mathrm{crit}}$ is the critical refinement at p of $E_{2,\ell}$) of an eigencurve of tame level N where N is divisible by at least two primes, then it is not hard to prove (see a work of D. Majumdar for details, [86]) that several components of the eigencurve $\mathcal{C}^0$ meet at x, and therefore that κ is not étale at x.

Proof It is clear that $t \otimes 1 - 1 \otimes t \in I$. Conversely, if $\sum_k t_k \otimes u_k \in I$ then $\sum t_k u_k = 0$, hence $\sum_k t_k \otimes u_k = \sum (1 \otimes u_k)(t_k \otimes 1 - 1 \otimes t_k)$ which shows the result.

□

A simple but important remark is in order. If M is a $T \otimes_R T$-module, it has two natural structures of T-modules, using the two natural morphisms of algebras $T \to T \otimes_R T : t \mapsto t \otimes 1, t \mapsto 1 \otimes t$. On the sub-module $M[I] = \{m \in M, Im = 0\}$, the two T-structures coincide, and in fact, $M[I]$ is the largest sub-module of M on which the two T-structures coincide. In particular, we can see $M[I]$ as a T-module unambiguously.

Definition 9.1.2 ([94]) The *Noether's different* $\mathfrak{d}_N(T/R)$ is the ideal of T which is the image of $(T \otimes_R T)[I]$ by m.

The Noether's different is also called *homological different.*

The main interest of this notion is in the following theorem (due, it seems, in this generality to Alexander-Buchsbaum). Recall that for any extension T/R of algebras, a prime $\mathfrak{P}$ of T is said *unramified* over R if for $\mathfrak{p} = \mathfrak{P} \cap R$, we have $\mathfrak{p}T_{\mathfrak{P}} = \mathfrak{P}T_{\mathfrak{P}}$, and $T_{\mathfrak{P}}/\mathfrak{P}T_{\mathfrak{P}}$ is a finite separable extension of $R_{\mathfrak{p}}/\mathfrak{p}R_{\mathfrak{p}}$. It is said *ramified* if it is not unramified.

Theorem 9.1.3 *A prime ideal* $\mathfrak{P}$ *of* S *is ramified over* R *if and only if* $\mathfrak{d}_N(T/R) \subset \mathfrak{P}$. *In other words, the closed subset of Spec T defined by the Noether's different is the ramification locus of Spec T* $\to$ *Spec R.*

For the proof, see [7, Theorem 2.7]. Their theorem is even more general, as they do not actually assume that T/R is finite, but the much weaker assumption that T is Noetherian and I is finitely generated. We shall apply this result in a situation where T/R is not only finite but also flat. In this case, $\mathfrak{P}$ is unramified over R if and only if it is étale, and if and only if it is smooth. Hence the closed set of Spec T defined by the Noether's different is also the set of points where Spec $T \to$ Spec R is not étale.

There are at least two others notions of *different*. We shall not need them, but for the sake of completeness, let us recall them quickly:

The oldest notion is *Dedekind's different* which is defined only under supplementary hypotheses. Namely one assumes that R is a domain, that T is reduced, and calling K the fraction ring of R and L the total fraction ring of T (which is finite product of fields L_i, each of them finite over K), we assume that L is étale over K (that is, each of the L_i is a separable extension of K). In Dedekind's definition as in most treatments of this notion in the literature, it is moreover assumed that R is Dedekind, but this is not actually necessary. Here is how the Dedekind different is defined: the trace map $\text{tr}_{L/K}$ defines a non-degenerate K-bilinear pairing on L: $(x, y) \mapsto \text{tr}(xy)$. Let T^* be the set of $x \in L$ such that $\text{tr}_{L/K}(xy) \in R$ for all y in T. It is clear that T^* is a T-module, and that it is actually the largest T-submodule of L such that $\text{tr}_{L/K}(T^*) \subset R$. In particular, since $\text{tr}_{L/K}(T) = R$, we have $T \subset T^*$. The *Dedekind's different* $\mathfrak{d}_D(T/R)$ is the ideal $[T : T^*]$ of T, that is the set of $x \in T$ such that $xT^* \subset T$. This definition is the one given in all textbooks on algebraic number theory (where R is invariably assumed to be Dedekind). See e.g. [109, Chapter III] for a fairly complete treatment in this case.

The most modern notion is *Kähler's different* $\mathfrak{d}_K(T/R)$ which is defined as the 0-th fitting ideal of the module of differential forms $\Omega_{T/R}$ (cf. [76]). More concretely, we can compute Kähler's different as follows: choose $x_1, \ldots, x_n$ a family of generators of T as R-algebra, and let $(P_1, \ldots, P_m)$ in $R[X_1, \ldots, X_n]$ the ideals of their algebraic relations over R. In other words $R[X_1, \ldots, X_n]/(P_1, \ldots, P_m) \simeq T$, the map sending X_i to x_i. Note that we have $m \geq n$ since T is finite over R by assumption but that it is not always possible to choose $m = n$ (when it is, T is said to be *complete intersection* over R). Form the $m \times n$-Jacobian matrix in $R[X]$ whose (i, j) coefficients is $\frac{\partial P_i}{\partial X_j}$. The Kähler's different is then the T-ideal generated by the image in T of the $(n \times n)$-minors of that matrix. This is very easy from the definitions of the module of differentials and of the Fitting ideal, both given in [59].

Recall that we always assume that R is a noetherian domain of fraction field K, and that T/R is finite. The following result summarizes the main relations between those three notions.

Theorem 9.1.4 *When T/R is flat, the ideal $\mathfrak{d}_N(T/R)$ and $\mathfrak{d}_K(T/R)$ define the same closed subset of Spec T (that is, have the same radical), namely the set of points where Spec $T \to$ Spec R is non-étale. The same is true if instead of assuming that T/R is flat, we assume that R is integrally closed, that T is reduced, and that the total fraction ring of T is étale over the fraction field of R.*

If T/R is flat and the Dedekind's different is defined, it is equal to Noether's different $\mathfrak{d}_D(T/R) = \mathfrak{d}_N(T/R)$.

If T/R is flat and locally complete intersection, then $\mathfrak{d}_N(T/R) = \mathfrak{d}_K(T/R)$.

Proof Since the 0-th Fitting ideal of a module defines the same closed subset as the annihilator of that module, the closed set defined by the Kahler's different is the support of $\Omega_{T/R}$ which is by definition the non-étale locus of Spec $T \to$ Spec R. When T/R is finite and flat, the non-étale locus is the same as the ramification locus of Spec $T \to$ Spec R since étale is "flat and unramified". The non-étale locus is also the same as the ramification locus in the case we assume that R is integrally closed that the total fraction ring of T is a product of finite separable extensions of Frac(R), for in this case, an unramified point of Spec T is flat over Spec R by [7, §4]. This completes the proof of the first paragraph of the theorem.

For the second assertion, see [94] or [7, §3] (which contain more results along this lines)

For the third assertion, see e.g. [80, Theorem 8.15]. □

Exercise 9.1.5 Let e be an integer, and let R be a DVR with uniformizer u and residual characteristic 0 or $> e$. Let $T = R[t]/(t^e - u)$. Show that $\mathfrak{d}_N(T/R) = t^{e-1}T$.

9.1.2 Duality

If M is a T-module which is finite and flat as an R-module, the module $M^\vee := \mathrm{Hom}_R(M, R)$ is also finite and flat over R, and has also a T-module structure. Moreover we have a canonical isomorphism $(M^\vee)^\vee = M$ as R-modules and even as T-modules, and for N a second T-module which is finite and flat as an R-module the transpose defines a canonical isomorphism $\mathrm{Hom}_T(M, N) = \mathrm{Hom}_T(N^\vee, M^\vee)$.

Note that the formation of $\mathrm{Hom}_R(M, N)$ and in particular of the T-module $M^\vee$ commutes with any base change $R \to R'$ in the obvious sense. (This is proved exactly as the special case $M = N$ in Sect. 2.1.) The formation of $\mathrm{Hom}_T(M, N)$ however only commutes in general with localizations of R.

If T is finite and flat over R, then the preceding apply to $M = T$ and we can define the T-module $T^\vee$, whose formation commutes with arbitrary base change. We say that the R-algebra T is *Gorenstein* if $T^\vee$ is flat of rank one over T. Locally, this is same as assuming the existence of an isomorphism of T-modules $T^\vee \simeq T$.

Proposition 9.1.6 *The natural isomorphism* $\mathrm{Hom}_R(M, N) = M^\vee \otimes_R N$ *restricts to an isomorphism* $\mathrm{Hom}_T(M, N) = (M^\vee \otimes_R N)[I]$.

Proof Recall that the natural isomorphism is the one that sends $\sum_i l_i \otimes n_i \in M^\vee \otimes_R N$ to the morphism $f : M \to N$ defined by $f(m) = \sum_i l_i(m) n_i$ (one checks locally that it is an isomorphism). It is sufficient to observe then that $\sum_i l_i \otimes n_i$ is in $(M^\vee \otimes_R N)[I]$ if and only if $\sum_i (tl_i) \otimes n_i = \sum_i (l_i \otimes tn_i)$ for all $t \in T$, if and only if f is T-equivariant. □

Corollary 9.1.7 *Assume that T/R is flat. The R-algebra T is Gorenstein if and only if $(T \otimes_R T)[I]$ is flat of rank one as T-module.*

Proof We may assume that R is local, and we have to prove in the case that T/R is Gorenstein if and only if $(T \otimes_R T)[I]$ is free of rank one as T-module.

We have $(T \otimes_R T)[I] = \mathrm{Hom}_T(T^\vee, T)$. So if T/R is Gorenstein, $T^\vee \simeq T$ and $(T \otimes_R T)[I] \simeq \mathrm{Hom}_T(T, T) = T$ as T-modules. Conversely, if $(T \otimes_R T)[I]$ is isomorphic to T, then let $f \in \mathrm{Hom}_T(T^\vee, T)$ corresponding to a generator of that module. Then for $t \in T$, $tf = 0$ implies that $t = 0$. Therefore for $t \in T$, $tf(T) = 0$ implies that $t = 0$. Since T is finite over R, it has no proper ideal having this property, and f is surjective. By equality of R-ranks, f is an isomorphism. □

Corollary 9.1.8 *If T/R is Gorenstein, $\mathfrak{d}_N(T/R)$ is a locally principal ideal.*

Proof Indeed, locally, this ideal is generated by $m(g)$ where g is a generator of $(T \otimes_R T)[I]$. □

9.1.3 The L-Ideal of a Scalar Product

Now let M and N be two T-modules, which as R-modules are finite flat. Let $b : M \times N \to R$ be an R-bilinear scalar product which is T-equivariant (that is $b(tm, n) = b(m, tn)$ for $t \in T$, $m \in M$, $n \in N$). The R-module of equivariant scalar products on M, N has a T-module structure: $(tb)(m, n) := b(tm, n) = b(m, tn)$, and is isomorphic to $(M^\vee \otimes N^\vee)[I]$.

Definition 9.1.9 We call L_b the ideal of T defined by the following construction: extend by linearity the R-linear scalar product b to a T-bilinear scalar product $b_T : (M \otimes_R T) \times (N \otimes_R T) \to T$. Set

$$L_b = b_T((M \otimes_R T)[I], (N \otimes_R T)[I]) \subset T.$$

Some clarifications may help: $(M \otimes_R T)$ and $(N \otimes_R T)$ have two T-modules structures, or what amounts to the same, a $T \otimes_R T$-module structure. The first T-structure is given by the action of T on M or N, the second by the action of T on T. When we say that b_T is T-bilinear, we are referring to the second T-structure.

We note that the formation of L_b commute with localizations on R.

Proposition 9.1.10 *We have*

$$L_b \subset \mathfrak{d}_N(T/R)$$

Proof Consider the map $B : M \otimes_R T \otimes_R N \otimes_R T \to T, m \otimes t \otimes n \otimes u \mapsto tub(m, n)$. This maps clearly factors through the quotient $(M \otimes_R T) \otimes_T (N \otimes_R T) \to T$ and induces b_T on that quotient. Hence

$$L_b = B((M \otimes_R T)[I], (N \otimes_R T)[I]).$$

There is an other way to factorize B:

$$B : M \otimes_R T \otimes_R N \otimes_R T \overset{c}{\to} (M \otimes_R N) \otimes_R (T \otimes T) \overset{b \otimes 1}{\to} (T \otimes T) \overset{m}{\to} T.$$

where the first map is just the isomorphism c given by commutativity of tensor product.

We claim that $(b \otimes 1) \circ c$ sends $(M \otimes_R T)[I] \otimes_R (N \otimes_R T)[I])$ into $(T \otimes_R T)[I]$. Indeed if $x = \sum_k m_k \otimes t_k \in (M \otimes_R T)[I]$, and $y = \sum_l n_l \otimes u_l \in (N \otimes_R T)[I]$ then $z := (b \otimes 1)(c(x, y)) = \sum_{k,l} b(m_k, n_l) t_k \otimes u_l$ and we compute, for $t \in T$

$$\begin{aligned}(1 \otimes t)z &= \sum_{k,l} b(m_k, n_l) t_k \otimes (t u_l) \\ &= \sum_{k,l} b(m_k, t n_l) t_k \otimes u_l \ \ \text{(using that } Iy = 0)\end{aligned}$$

$$= \sum_{k,l} b(tm_k, n_l)t_k \otimes u_l \quad \text{(using the } T\text{-equivariance of } b)$$

$$= \sum_{k,l} b(m_k, n_l)(tt_k) \otimes u_l \quad \text{(using that } Ix = 0)$$

$$= (t \otimes 1)z$$

Hence we see that $L_b = B((M \otimes_R T)[I] \otimes_R (N \otimes_R T)[I])) \subset m((T \otimes_R T)[I]) = \mathfrak{d}_N(T/R)$. □

Proposition 9.1.11 *Assume that $M \simeq N \simeq T^\vee$ as T-modules. Then the ideal L_b of T is principal.*

More precisely, the scalar product b can be seen as a linear map $b : T^\vee \otimes_R T^\vee \to R$. By duality, this is the same as an element $\tilde{b}$ in $T \otimes_R T$. We have $L_b = m(\tilde{b})T$.

Proof We have $(M \otimes T)[I] \simeq (T^\vee \otimes T)[I] = \mathrm{Hom}_T(T, T) = T$. Let g be a generator of this module. Similarly, let h be a generator of $(N \otimes T)[I]$. Then $b_T(g, h)$ is a generator of L_b.

For the "more precisely", let us identify M and N with $T^\vee$. Let e_i be a basis of T over R and $e_i^\vee$ a basis of $T^\vee$ over R. Take $g = h = \sum_i e_i^\vee \otimes e_i$; it is a generator of $(T^\vee \otimes T)[I]$, since it corresponds to the identity in $\mathrm{Hom}_R(T, T)$. Then by definition: $b_T(g, h) = \sum_{i,j} b(e_i^\vee, e_j^\vee)e_i e_j b(e_i^\vee, e_j^\vee)$. On the other hand, as an element of $T \otimes_R T$, $\tilde{b}$ is by definition $\sum_{i,j} b(e_i^\vee, e_j^\vee)e_i \otimes e_j$. Hence $m(\tilde{b}) = b_T(g, h)$ is a generator of L_b. □

Corollary 9.1.12 *Keep the assumptions of the preceding proposition. Then the formation of L_b commutes with arbitrary base changes. (For the sake of precision, this means the following: if R' is an R-algebra, let $M' = M \otimes_R R'$, $N' = N \otimes_R R'$ and $T' = R \otimes_R R'$. Let $b' : M' \otimes N' \to R'$ be the R'-bilinear extension of b, which is clearly T'-equivariant. Then the ideal $L_{b'}$ of T' is the ideal generated by the image of L_b by the morphism $T \to T'$).*

Proof The assumption $M \simeq N \simeq T^\vee$ as T-modules implies similar assertions over R': $M' \simeq N' \simeq T'^\vee$ as T'-modules. The element $\tilde{b}'$ in $T' \otimes_{R'} T'$ is just the natural image of $\tilde{b}$, and thus $L_{b'}$, which is the ideal generated by $m(\tilde{b}')$ is generated by the image of $m(\tilde{b})$. The result follows. □

Corollary 9.1.13 *Assume that T/R is Gorenstein, and that M and N are flat of rank one over T. Then L_b and $\mathfrak{d}_N(T/R)$ are locally principal, and $L_b = \mathfrak{d}_N(T/R)$ if and only if b is non-degenerate. More precisely, if $b = tb_0$ where $b_0 : M \times N \to R$ is a non-degenerate T-equivariant scalar product, then $L_b = t\mathfrak{d}_N(T/R)$.*

Proof Note that since T/R is Gorenstein, locally $T^\vee \simeq T$ as T-module, so we have $M \simeq N \simeq T^\vee$ and we are in a special case of the above proposition. In particular, L_b is principal (locally), and we already know that the Noether's different is principal (locally) since T/R is Gorenstein.

For the rest of the proof we assume that R is local.

Clearly b is non-degenerate if and only if for all $t \in T$, $tb = 0$ implies $t = 0$, which is equivalent to the same assertion for $\tilde{b}$. Since T is Gorenstein, $(T \otimes T)[I] = T$, and we see that b is non-degenerate if and only if $\tilde{b}$ is a generator of $(T \otimes T)[I]$.

Writing $\tilde{b} = t\tilde{b}_0$ with $t \in T$, and where $\tilde{b}_0$ is a generator of $(T \otimes T)[I]$, we have $m(\tilde{b}) = tm(\tilde{b}_0)$ hence $L_b = t\mathfrak{d}_N(T/R)$ with equality if and only if t is invertible, i.e. if and only if b is non-degenerate. □

Remark 9.1.14 Let us go back to the hypothesis of Proposition 9.1.11: $M^\vee$ and $N^\vee$ free of rank one over T. If b is non-degenerate, then b defines an isomorphism $M \simeq N^\vee$ as T-modules, so $T^\vee \simeq T$ and T is Gorenstein, and the corollary implies that $L_b = \mathfrak{d}_N(T/R)$.

9.2 Kim's Scalar Product

Let us use the same notations of Sect. 7.1: $W = \mathrm{Sp}\, R$ is an open affinoid subset of $\mathcal{W}$, which belongs to the covering $\mathfrak{C}$ (in particular R and its residue ring $\tilde{R}$ are PID, and $\mathbb{N} \cap W$ is very Zariski-dense in W), $K : \mathbb{Z}_p^* \to R^*$ the canonical character, and $\nu \geq 0$ is a real number, adapted to W. Let us fix a real r such that $0 < r < \max(r(W), p)$.

The aim of this section is to construct an R-bilinear scalar product b on $\mathrm{Symb}_\Gamma(\mathcal{D}_K[r](R))^{\leq \nu}$ that *interpolates* the so-called *corrected* scalar products on $\mathrm{Symb}_\Gamma(\mathcal{V}_k[r](\mathbb{Q}_p))^{\leq \nu}$, for $k \in \mathbb{N} \cap W(\mathbb{Q}_p)$ that we denoted by $[\,,\,]$ (See Definitions 5.3.21 and 5.2.1).

Let us precise what we mean by 'interpolate' here. For any $\kappa \in W(L)$, where L is a finite extension of $\mathbb{Q}_p$, by extension of scalars $\kappa : R \to L$ the scalar product b gives rise to a L-bilinear scalar product b_κ on $\mathrm{Symb}_\Gamma(\mathcal{D}_\kappa[r](R))^{\leq \nu} \otimes_{R,\kappa} L = \mathrm{Symb}_\Gamma(\mathcal{D}_\kappa[r](L))_g^{\leq \nu} \subset \mathrm{Symb}_\Gamma(\mathcal{D}_\kappa[r](L))^{\leq \nu}$ (cf. Theorem 7.1.16 and the definition following it). If κ is character $z \mapsto z^k$, and $k \in \mathbb{N}$, $\mathrm{Symb}_\Gamma(\mathcal{D}_\kappa[r](\mathbb{Q}_p))^{\leq \nu}$ admits the space of classical modular symbols $\mathrm{Symb}_\Gamma(\mathcal{V}_k(\mathbb{Q}_p))^{\leq \nu}$ as a quotient. By 'interpolating', we mean that the scalar product b_k factors through that quotient and is equal, on that quotient, to our old corrected scalar product.

9.2.1 A Bilinear Product on the Space of Overconvergent Modular Symbols of Weight k

Let $k \geq 0$ be an integer. Assume that L is a finite extension of $\mathbb{Q}_p$. Remember the surjective map $\rho_k : \mathcal{D}_k[1](L) \to \mathcal{V}_k(L)$, defined as the dual of the inclusion $\mathcal{P}_k(L) \to \mathcal{A}_k[1](L)$ that we shall here denote by i_k. Those maps ρ_k and i_k induce maps on modular symbols, e.g. $\rho_k : \mathrm{Symb}_\Gamma(\mathcal{D}_k[1](L)) \to \mathrm{Symb}_\Gamma(\mathcal{V}_k(L))$.

Definition 9.2.1 We define a scalar product on $\mathrm{Symb}_\Gamma(\mathcal{D}_k[1](L))$, denoted by $[\ ,\ ]_k$ by

$$[\Phi_1, \Phi_2]_k = [\rho_k(\Phi_1), \rho_k(\Phi_2)],$$

where $[\ ,\ ]$ is the corrected scalar product on classical modular symbols, defined in Definition 5.3.21. We use the same notation $[\ ,\ ]_k$ for the restriction of this scalar product to the finite-dimensional subspace $\mathrm{Symb}_\Gamma(\mathcal{D}_k[1](L))^{\leq \nu} = \mathrm{Symb}_\Gamma(\mathcal{D}_k[r](L))^{\leq \nu}$ (where r is any real number such that $0 < r < p$).

Lemma 9.2.2 *The operators in $\mathcal{H}$ (that is the T_l or U_l for every prime l, and the diamond operators) are self-adjoint for the scalar product $[\ ,\ ]_k$.*

Proof This follows from Lemma 5.3.23 and the equivariance of ρ_k. □

We shall now give an explicit description of this scalar product. Below, we consider $\mathcal{P}_k(L)$ with its **right** action of S, so that it is the contragredient of the right S-module $\mathcal{V}_k(L)$. We denote by δ the tautological product $\mathcal{P}_k(L) \times \mathcal{V}_k(L) \to L$. Similarly we shall consider $\mathcal{A}_k[r](L)$ with its natural **right** action of $S_0(p)' = \{\gamma \in S,\ p \mid c,\ p \nmid d\}$:

$$f_{|k\gamma}(z) = (\gamma' \cdot_k f)(z) = (cz+d)^k f\left(\frac{az+b}{cz+d}\right)$$

The inclusion $i_k : \mathcal{P}_k(L) \to \mathcal{A}_k[r](L)$ evidently respects our right actions. We have a natural pairing $\Delta : \mathcal{D}_k[r](L) \times \mathcal{A}_k[r](L) \to L$ defined by $\Delta(\mu, f) = \mu(f)$, which is obviously compatible with δ in the sense that

$$\Delta(\mu, i_k(P)) = \delta(\rho_k(\mu), P). \tag{9.2.1}$$

We see by an easy computation

$$\Delta(\mu_{|k\gamma}, f_{|k\gamma}) = \det(\gamma)^k \Delta(\mu, f)$$

for all $\gamma \in S_0(p) \cap S_0'(p)$, and a similar formula holds for δ. In particular, the pairings Δ and δ are pairings of $\Gamma_0(p)$-modules to L, so they define pairings, also denoted Δ and δ

$$\Delta : \mathrm{Symb}_\Gamma(\mathcal{D}_k[r](L)) \times \mathrm{Symb}_\Gamma(\mathcal{A}_k[r](L)) \to L,$$

$$\delta : \mathrm{Symb}_\Gamma(\mathcal{V}_k(L)) \times \mathrm{Symb}_\Gamma(\mathcal{P}_k(L)) \to L,$$

We now can compute:

$$\begin{aligned}[\Phi_1, \Phi_2]_k &= [\rho_k(\Phi_1), \rho_k(\Phi_2)] \text{ (by Definition 9.2.1)}\\ &= (\rho_k(\Phi_1), \rho_k(\Phi_2)_{|W_{Np}}) \text{ (by Definition 5.3.21)}\end{aligned}$$

$$
\begin{aligned}
&= \delta(\rho_k(\Phi_1), \theta_k(\rho_k(\Phi_2)_{|W_{Np}})) \text{ (by Definition 5.2.1)} \\
&= \Delta(\Phi_1, i_k\theta_k(\rho_k(\Phi_2)_{|W_{Np}})) \text{ by (9.2.1)} \\
&= \Delta(\Phi_1, V_k(\Phi_2)),
\end{aligned}
$$

where V_k is the map

$$
\begin{aligned}
V_k : \mathrm{Symb}_\Gamma(\mathcal{D}_k[r](L)) &\to \mathrm{Symb}_\Gamma(\mathcal{A}_k[r](L)) \\
\Phi &\mapsto i_k\theta_k(\rho_k(\Phi)_{|W_{Np}}).
\end{aligned}
$$

Now V_k is easily described using the definitions of its constituents: for any $D \in \Delta_0$, $\rho_k(\Phi)(D)(z^i) = \Phi(D)(z^i)$ provided that $0 \leq i \leq k$. Then $\rho_k(\Phi)_{|W_{Np}}(D)(z^i) = \Phi(W_{Np} \cdot D)(z^i_{|_k W_{Np}}) = \Phi(W_{Np} \cdot D)((-Npz)^{k-i})$. In other words,

$$
\rho_k(\Phi)_{|W_{Np}}(D) = \sum_{i=0}^{k} \Phi(W \cdot D)((-Npz)^{k-i}) l_i,
$$

where $l_i \in \mathcal{V}_k(L)$ is the linear form on $\mathcal{P}_k(L)$ sending z^j to $\delta_{i,j}$.

Using the formula for the operator θ_k given in Lemma 5.1.3, we see that

$$
\begin{aligned}
\theta_k(\rho_k(\Phi)_{|W_{Np}})(D)(z) &= \sum_{i=0}^{k} \Phi(W_{Np} \cdot D)((-Npz)^{k-i})(-1)^i \binom{k}{i} z^{k-i} \\
&= (-1)^k \sum_{i=0}^{k} (Np)^i \binom{k}{i} \Phi(W_{Np} \cdot D)(z^i)\, z^i
\end{aligned}
$$

Slightly more generally, we also have for any $e \in \mathbb{Z}$

$$
\theta_k(\rho_k(\Phi)_{|W_{Np}})(D)(z) = (-1)^k \sum_{i=0}^{k} (Np)^i \binom{k}{i} \Phi(W_{Np} \cdot D)(z^i(1 - Npez)^{k-i})\, (z-e)^i
$$

This follows by applying the above equality to Φ replaced by $\Phi_{|\gamma}$ with $\gamma = \begin{pmatrix} 1 & 0 \\ Npe & 1 \end{pmatrix} \in S$, using that γ commutes with ρ_k and θ_k and that $\gamma W_{Np} = W_{Np} \begin{pmatrix} 1 & e \\ 0 & 1 \end{pmatrix}$.

Applying i_k amounts to seeing that polynomial in z as an element of $\mathcal{A}[r](L)$. Such an element is described by its Taylor series about any $e \in \mathbb{Z}_p$, which is just

$$
\iota_k\theta_k(\rho_k(\Phi)_{|W_{Np}})(D)(z) = (-1)^k \sum_{i=0}^{k} (Np)^i \binom{k}{i} \Phi(W_{Np} \cdot D)(z^i(1 - Npez)^{k-i})\, (z-e)^i
$$

Note that in the above sum, we can replace $\sum_{i=0}^{k}$ by $\sum_{i=0}^{\infty}$ as the new terms are 0 because of their factor $\binom{k}{i}$.

To summarize,

Proposition 9.2.3 *Let k be an integer. For any $\Phi_1, \Phi_2 \in \mathrm{Symb}_\Gamma(\mathcal{D}_k[r](L))$, we have*

$$[\Phi_1, \Phi_2]_k = \Delta(\Phi_1, V_k(\Phi_2)), \tag{9.2.2}$$

where $V_k(\Phi) \in \mathrm{Symb}_\Gamma(\mathcal{A}_k[r](L))$ is defined, for any $D \in \Delta_0$, $e \in \mathbb{Z}_p$, and $z \in \mathbb{Z}_p$, by

$$V_k(\Phi)(D) = (-1)^k \sum_{i=0}^{\infty} (Np)^i \binom{k}{i} \Phi(W_{Np} \cdot D)(z^i(1 - Npez)^{k-i})\,(z-e)^i. \tag{9.2.3}$$

9.2.2 *Interpolation of Those Scalar Products*

Let L be any $\mathbb{Q}_p$-Banach algebra with a norm $|\,|$ extending the p-adic norm of $\mathbb{Q}_p$, and $\kappa \in \mathcal{W}(L)$ be a continuous character $\mathbb{Z}_p^* \to L^*$.

Definition 9.2.4

(i) If $y \in L$, $i \in \mathbb{N}$, we define $\binom{y}{i} = \frac{y(y-1)\dots(y-i+1)}{i!}$, if $i \geq 0$, and $\binom{y}{0} = 1$.
(ii) If $\kappa \in \mathcal{W}(L)$, we define $\binom{\kappa}{i} = \binom{\log_p \kappa(\gamma)/\log_p(\gamma)}{i}$, where γ is a generator of $1 + p\mathbb{Z}_p$

The first definition is standard and due to Isaac Newton. The second is easily seen to be independent of the choice of γ, and to be compatible with the first one, on the sense that $\binom{\kappa}{i} = \binom{k}{i}$ when κ is an integer k, that is to say the character $z \mapsto z^k$.

Exercise 9.2.5 Prove that $|\binom{\kappa}{i}| < p^{\frac{ip}{p-1}}$

The key observation is that formula (9.2.3) makes sense when k is replaced by an arbitrary character $\kappa \in \mathcal{W}(L)$. Indeed, let $r < r(\kappa)$; for any $\Phi \in \mathrm{Symb}_\Gamma(\mathcal{D}_k[r](L))$, $D \in \Delta_0$, $e \in \mathbb{Z}_p$, we can form the formal Taylor expansion at e:

$$T_{\kappa,\Phi,D,e}(z) := \kappa(-1) \sum_{i=0}^{\infty} (Np)^i \binom{\kappa}{i} \Phi(W \cdot D)(z^i(1 - Npez)^{-i}\kappa(1 - Npez))\,(z-e)^i. \tag{9.2.4}$$

which when $\kappa = k$ is exactly the RHS of (9.2.3). Note that the right hand side makes sense because $r < r(\kappa)$, which implies that $z^i(1-Npez)^{-i}\kappa(1-Npez) \in \mathcal{A}[r](L)$ so that this function can be integrated by the distribution $\Phi(W \cdot D) \in D[r](L)$.

Lemma 9.2.6 *Assume that $r < min(r(\kappa), p^{\frac{-1}{p-1}})$. The series in the RHS of (9.2.4) converges on the closed ball $|z-e| < r$.*

Proof We have $|(Np)^i| = p^{-i}$ since N is prime to p, $|\binom{\kappa}{i}| < p^{\frac{ip}{p-1}}$ by Exercise 9.2.5, and $|\Phi(W_{Np} \cdot D)(z^i(1-Npez)^{-i}\kappa(1-Npez))| \leq \|\Phi(W_{Np} \cdot D)\|_r$ since by Exercise 6.1.18 and $r \leq 1$,

$$\|z^i(1-Npez)^{-i}\kappa(1-Npez)\|_r \leq \sup_{z \in \mathcal{O}_{\mathbb{C}_p}} |z^i(1-eNz)^{-i}\kappa(1-Npez)| \leq 1.$$

Hence the coefficient of $(z-e)^i$ has norm at most $p^{\frac{i}{p-1}}$, which means that the series converges on any disc $|z-e| < p^{\frac{-1}{p-1}}$, hence the lemma. □

Proposition 9.2.7 *Let L be an affinoid algebra, $\kappa \in \mathcal{W}(L)$. We assume that $r <$ $\min(p^{\frac{-1}{p-1}}, r(\kappa))$.*

(i) *Fix $\Phi \in Symb_\Gamma(\mathcal{D}_\kappa[r](L))$, and $D \in \Delta_0$. Then the convergent series $T_{\kappa,\Phi,D,e}(z)$ for $e \in \mathbb{Z}_p$ define an element in $\mathcal{A}_k[r](L)$.*
(ii) *Fix $\Phi \in Symb_\Gamma(\mathcal{D}_k[r](L))$, and define $V_\kappa(\Phi) \in$ $\mathrm{Hom}(\Delta_0, \mathcal{A}_k[r](L))$ by sending $D \in \Delta_0$ to the element of $\mathcal{A}_k[r](L)$ defined in (i). Then $V_\kappa(\Phi)$ lies in $Symb_\Gamma(\mathcal{A}_\kappa[r](L))$*

Proof This can probably be shown by direct computation, but there is a more clever way to do it.

We first consider a particular case: let us assume that $L = R$ when $\mathrm{Sp}\, R = W$ is an affinoid of $\mathcal{W}$ belonging to $\mathfrak{C}$, and $\kappa = K$ is the canonical character $\mathbb{Z}_p^* \to R^*$. To prove (i), we need to prove that for e, e' in $\mathbb{Z}_p$, $T_{K,\phi,D,e}(z)$ and $T_{K,\phi,D,e'}(z)$ have the same values for any $z \in \mathbb{Z}_p$ such that $|z-e| \leq r$, $|z-e'| \leq r$. To prove that those two elements of R are equal, it suffices to prove that there images by any morphism $R \to \mathbb{Q}_p$ corresponding to an element $k \in \mathbb{N} \cap W$ are equal, since $\mathbb{N} \cap W$ is Zariski-dense in W. But those images are just $T_{k,\phi,D,e}(z)$ and $T_{k,\phi,D,e'}(z)$, and we know from the last paragraph that those two series actually define the same polynomial independent of e, namely $(-1)^k \sum_{i=0}^k (Np)^i \binom{k}{i} \Phi(W_{Np} \cdot D)(z^i)\, z^i$. Hence (i) is done in this case. For (ii), we need to prove that $V_K(\Phi)_{|K\gamma} = V_K(\Phi)$ for $\gamma \in \Gamma$. Again, this reduces to proving that $V_k(\Phi)_{|k\gamma} = V_k(\Phi)$ for all integer $k \in \mathbb{N} \cap W$, which we know since by definition $V_k(\Phi) \in \mathrm{Symb}_\Gamma(\mathcal{D}_k[r])$.

The general case (L, κ) can be reduced to the case (R, K) as follows. The character $\kappa \in \mathcal{W}(L)$ defines a morphism $\mathrm{Sp}\, L \to \mathcal{W}$ whose image is in some $W = \mathrm{Sp}\, R$ in $\mathfrak{C}$, since those W's form an admissible covering of $\mathcal{W}$. For such a $W = \mathrm{SpR}$, we have a morphism of algebra $R \to L$ and the character κ is just the composition of K with this morphism. Therefore, the case (L, κ) follows from (R, K) by base change $R \to L$ (using that the formation of $\mathcal{A}[r](L)$ commute with arbitrary base change, cf. Corollary 6.1.17). □

Definition 9.2.8 Let L be any $\mathbb{Q}_p$-affinoid algebra, $\kappa \in W(L)$, and $r < \min(p^{\frac{-1}{p-1}}, r(\kappa))$. We define a scalar product $[\ ,\ ]_\kappa$ on $\mathrm{Symb}_\Gamma(\mathcal{D}_k[r](L))$ by the formula

$$[\Phi_1, \Phi_2]_\kappa = \Delta(\Phi_1, V_\kappa \Phi_2).$$

Theorem 9.2.9

(i) *The formation of this scalar product commutes with any affinoid base change $L \mapsto L'$, in the sense that if $\Phi_1, \Phi_2 \in \mathrm{Symb}_\Gamma(\mathcal{D}_k[r](L))$, and Φ_1', Φ_2' are the image of $\Phi_1 \hat{\otimes} 1$, $\Phi_2 \hat{\otimes} 1$ by the natural morphisms (which are not always isomorphisms!) $\mathrm{Symb}_\Gamma(\mathcal{D}_k[r](L)) \hat{\otimes}_L L' \to \mathrm{Symb}_\Gamma(\mathcal{D}_k[r](L'))$, one has*

$$[\Phi_1', \Phi_2'] = [\Phi_1, \Phi_2]$$

(ii) *When $L = \mathbb{Q}_p$, $\kappa = k \in \mathbb{N}$, this scalar product coincides with the one denoted $[\ ,\ ]_k$ above.*

(iii) *All the Hecke operators in $\mathcal{H}$ are self-adjoint for $[\ ,\ ]_\kappa$.*

Proof (i) is clear since the 'Poincaré duality' pairing Δ is natural, and since the formation of V_κ commute with base change by construction. (ii) is also clear since $V_\kappa = V_k$ when $\kappa = k$ by construction. Finally (iii) is first proved for the case (R, K) where $W = \mathrm{Sp}\, R$ is an affinoid of $\mathcal{W}$ in $\mathfrak{C}$ and K the canonical weight by density from the case of integral weight (that is, Lemma 9.2.2) and then for any (L, κ) using (i). □

Using that $\mathrm{Symb}_\Gamma(\mathcal{D}_\kappa[r](L))^{\leq \nu}$ is independent of r when it is defined (that is for ν adapted to L and $r < \min(p, r(\kappa))$), the above scalar product defines by restriction a scalar product on $\mathrm{Symb}_\Gamma(\mathcal{D}_\kappa[r](L))^{\leq \nu}$. We note also that by the formal property of Δ, the scalar product $[\ ,\ ]_\kappa$ factors through the quotient $H^1_!(\Gamma, \mathcal{D}_k[r](L))$.

Theorem 9.2.10 *Let $\nu \geq 0$, and assume that L is a finite extension of $\mathbb{Q}_p$. The restriction of $[\ ,\ ]_\kappa$ to $H^1_!(\Gamma, \mathcal{D}_k[r](L))^{\leq \nu}$ is non-degenerate if $\kappa \notin \mathbb{N}$ or if $\kappa = k \in \mathbb{N}$ and $\nu < k + 1$.*

Proof Assume that $\Phi_1 \in H^1_!(\Gamma, \mathcal{D}_k[r](L))^{\leq \nu}$ is such that $[\Phi_1, \Phi_2]_\kappa = 0$ for all $\Phi_2 \in H^1_!(\Gamma, \mathcal{D}_k[r](L))^{\leq \nu}$. Then since U_p is self-adjoint, we have $[\Phi_1, \Phi_2]_\kappa = 0$ for all $\Phi_2 \in H^1_!(\Gamma, \mathcal{D}_k[r](L))$. In view of the non-degeneracy of Δ, this means that $V_\kappa(\Phi_1) = 0$. By the definition of V_κ, this means in particular that $T_{\kappa,\Phi_1,D,e}(z)$ is 0 for every $e \in \mathbb{Z}_p$, $D \in \Delta_0$. Applying this to the divisor $W_{Np}^{-1} \cdot D$ and to $e = 0$, we get that for every $D \in \Delta_0$ and every integer i:

$$\binom{\kappa}{i} \Phi_1(D)(z^i) = 0.$$

In the case $\kappa \notin \mathbb{N}$, then $\binom{\kappa}{i}$ is never 0. Therefore for all $D \in \Delta_0$, the distribution $\Phi_1(D)$ which belongs to $\mathcal{D}[r](L)$ satisfies $\Phi_1(D)(z^i) = 0$, and $\Phi_1(D)$ kills any polynomial. Since polynomials are dense in $\mathcal{A}[r](L)$ (Exercise 6.1.14), $\Phi_1(D) = 0$. Since this true for all $D \in \Delta_0$, one has $\Phi_1 = 0$.

In the case $\kappa = k \in \mathbb{N}$, then at least $\binom{k}{i} \neq 0$ for $0 \leq i \leq k$ and we have

$$\Phi_1(D)(z^i) = 0$$

for all $D \in \Delta_0$ and $0 \leq i \leq k$. This means that $\rho_k(\Phi_1) = 0$. If $\nu < k+1$, then this implies again $\Phi_1 = 0$ by Stevens's control theorem (Theorem 6.5.19). □

9.3 The Cuspidal Eigencurve, the Interior Cohomological Eigencurves, and Their Good Points

As usual p is a prime number, N an integer not divisble by p, and $\Gamma = \Gamma_1(N) \cap \Gamma_0(p)$.

We define two new eigencurves, called $\mathcal{C}_!^+$ and $\mathcal{C}_!^-$, based on the interior cohomology $(H_!^1)^{\pm}(Y_\Gamma, -)$. We use the cohomological variant of the eigenvariety machine described in Sect. 3.9. Our data for the machine (ECD1) and (ECD2) are the same as in Sect. 7.1.4: for (ECD1) we take $\mathcal{H}_0$ and for (ECD2) $\mathcal{W}$ the weight space and for $\mathfrak{C}$ the covering described *loc. cit.*. For (ECD3), for every $W = \mathrm{Sp}\,(R)$ in $\mathfrak{C}$, we choose an r such that $r < \min(r(W), p)$, and we define the complex $M_W^\bullet$ as the complex

$$(M_W^\bullet)\quad 0 \to (\mathcal{D}_\kappa(R)[r]^\Gamma)^{\pm} \to \mathrm{BSymb}_\Gamma^{\pm}(\mathcal{D}_\kappa(R)[r]) \to \mathrm{Symb}_\Gamma^{\pm}(\mathcal{D}_\kappa(R)[r]) \to 0,$$

with $(\mathcal{D}_\kappa(R)[r]^\Gamma)^{\pm}$ placed in degree -1, BSymb placed in degree 0, Symb in degree 1. It follows from Corollary 4.4.3 that this is indeed a complex, and that its cohomology is only non-zero in degree 1, with

$$H^1(M_W^\bullet) = (H_!^1)^{\pm}(Y_\Gamma, \mathcal{D}_\kappa[r](R).$$

Note that M_W^{-1}, M_W^0 and M_W^1 are all Banach modules satisfying (Pr), and that the hypotheses (ECC1) and (ECC2) are easily verified. We thus obtain the eigencurves $\mathcal{C}_!^+$ and $\mathcal{C}_!^-$.

By construction, $\mathcal{C}_!^{\pm}$ is a closed subvariety of $\mathcal{C}^{\pm}$. Indeed, $\mathcal{C}_!^{\pm}$ is constructed by glueing the Hecke algebras of modules of the form $H^1(M_W^\bullet)^{\leq \nu}$, which are quotients of the Hecke algebras of $\mathrm{Symb}_\Gamma^{\pm}(\mathcal{D}_\kappa(R)[r])^{\leq \nu}$. Hence we get a closed embedding (compatible with the map to $\mathcal{W}$ and the map from $\mathcal{H}_0$) of every local piece of $\mathcal{C}_!^{\pm}$ into a corresponding local piece of $\mathcal{C}^{\pm}$, which, by uniqueness, glue into a global closed embedding.

Proposition 9.3.1 *As subvarieties of $\mathcal{C}$, $\mathcal{C}_!^+$ and $\mathcal{C}_!^-$ are reduced, and one has $\mathcal{C}_0 = \mathcal{C}_!^+ = \mathcal{C}_!^-$.*

Proof We claim that the points of $\mathcal{C}$ of integral weight $k > 0$ and systems of eigenvalues corresponding to a non-critical slope classical cusp form of weight k (of level Γ) belongs to $\mathcal{C}_0$, $\mathcal{C}_!^+$ and $\mathcal{C}_!^-$, and are moreover Zariski-dense in $\mathcal{C}_0$, $\mathcal{C}_!^+$ and $\mathcal{C}_!^-$.

That those points belong where they should follows from Theorems 3.7.1 and 3.9.4 which shows that the systems of eigenvalues appearing in

$$S^{k+2}(\Gamma, L)^{\leq k+1} = (H_!^1)^+(Y_\Gamma, \mathcal{V}_k)^{\leq k+1} = (H_!^1)^-(Y_\Gamma, \mathcal{V}_k)^{\leq k+1}$$

(this equality following from Eichler-Simura, see Chap. 5) appear as points of weight k in the eigenvarieties $\mathcal{C}^0$, $\mathcal{C}_!^+$ and $\mathcal{C}_!^-$. Note that in each case those points are the classical points for an obvious classical structure (in the sense of Definition 3.9.5 in the two latter cases). The density of these points in the case of $\mathcal{C}_0$ results of Proposition 3.8.6, and their density in $\mathcal{C}_!^\pm$ of Proposition 3.9.6. Proposition 3.9.7 shows that both eigencurve $\mathcal{C}_!^\pm$ are reduced.

The proposition follows, since all three eigencurves appear as a reduced subspace of $\mathcal{C}$, having an identical Zariski-dense subset. □

We shall use only $\mathcal{C}^0$ below, but what we have gained from these two new constructions of $\mathcal{C}^0$ is that modules of interior cohomology are modules on the local pieces of $\mathcal{C}^0$. More precisely, If $x \in \mathcal{C}^0(L)$, and $U = \operatorname{Sp} \mathcal{T}$ is a clean neighborhood of x, let as in Sect. 7.6.1 $W = \operatorname{Sp} R$ the image $\kappa(U)$, and, for $r > 0$ sufficiently small, let $M^\pm$ be the finite free R-module $\epsilon H_!^1(\mathcal{D}_K[r](R))^\pm$ (it is independent of r) where ϵ is the idempotent corresponding to the connected component U of $\kappa^{-1}(W)$. Then $M^\pm$ has a natural structure of $\mathcal{T}$-module.

We need to define a notion of good point on the cuspidal eigencurve $\mathcal{C}^0$ analog to the one we defined on the eigencurves $\mathcal{C}^\pm$ (cf. Definition 8.1.2). The following lemma is proven exactly as Lemma 8.1.1

Lemma 9.3.2 *Let x be a point on $\mathcal{C}^0$ (of field of definition L), with $\kappa(x) = w \in \mathcal{W}$. The following statements are equivalent:*

(i) *For both choices of the sign $\pm$, the dual of the generalized eigenspace $H_!^1(Y_\Gamma, \mathcal{D}_w(L))^\pm_{(x)}$ is free of rank 1 over the algebra $\mathcal{T}_x$ of the schematic fiber of κ at x.*
(ii) *For both choices of the sign $\pm$, and for any clean neighborhood of x, the modules $(M^\pm)^\vee$ are flat of rank one over $\mathcal{T}$ at x*
(iii) *For both choices of the sign $\pm$, and for any sufficiently small clean neighborhood of x, the modules $(M^\pm)^\vee$ are free of rank one over $\mathcal{T}$ at x.*

Definition 9.3.3 A point $x \in \mathcal{C}^0$ is called *good* if it satisfies the equivalent assertions of the above proposition.

Exactly as in Theorem 8.1.5, we can prove that if x is a smooth point of $\mathcal{C}^0$ whose weight $\kappa(x)$ is of the form $z \mapsto z^k \epsilon(z)$ with ϵ of order p^t, and new away from p, then x is good.

We shall need the following lemma, which summarizes results obtained earlier:

Lemma 9.3.4 *If x is a good point, then the eigenspace $H^1_!(\mathcal{D}_w(L))^{\pm}[x]$ has dimension 1. If x is a good and of weight $w(x) = k \in \mathbb{N}$ then $H^1_!(\mathcal{V}_k(L))^{\pm}[x]$ has dimension 1 if x is very classical cuspidal, 0 otherwise, and the same is true for the generalized eigenspaces $H^1_!(\mathcal{V}_k(L))^{\pm}_{(x)}$ provided $U_p(x)^2 \neq p^{k+1}\langle p\rangle(x)$. In the former case, the map between one-dimensional spaces $\rho_k : H^1_!(\mathcal{D}_k(L))^{\pm}[x] \to H^1_!(\mathcal{V}_k(L))^{\pm}[x]$ is an isomorphism if κ is étale at x, and is 0 if κ is not. If x is a good point of weight $w(x) = k$ and $v_p(U_p(x)) < k+1$, then x is cuspidal and very classical and κ is étale at x.*

Proof The first assertion follows by duality from the definition of a good point (see Lemma 9.3.2(i)). One has $H^1_!(\mathcal{V}_k(L))^{\pm}[x]$ is non-zero if and only if x is a cuspidal very classical point by Theorem 5.3.35. If x is a cuspidal very classical point, it is new away from p since x is good and $H^1_!(\mathcal{V}_k(L))^{\pm}[x]$ has dimension 1, hence the second assertion. The fact that ρ_k is an isomorphism if and only if κ is étale at x is part of Theorem 7.6.10, noting that x, being cuspidal, is not abnormal. Finally, if x is a good point of weight $w(x) = k$ and $v_p(U_p(x)) < k+1$ by Stevens' control theorem, and it can't be Eisenstein since only critical slope Eisenstein points appear on the cuspidal eigencurve $\mathcal{C}^0$, and they satisfy $v_p(U_p(x)) = k+1$. Moreover κ is étale at x by Theorem 7.6.4. □

9.4 Construction of the Adjoint p-Adic L-Function on the Cuspidal Eigencurve

Definition 9.4.1 Let L be a finite extension of $\mathbb{Q}_p$ and $x \in \mathcal{C}^0(L)$ a point. Let $U = \operatorname{Sp} \mathcal{T}$ be a clean neighborhood of x. We define the *adjoint L-ideal* $\mathcal{L}_U^{\mathrm{adj}}$ on U as the L-ideal (cf. Definition 9.1.9) of the scalar product $[\ ,\ ]_K : M^+ \times M^- \to R$ (cf. Definition 9.2.8).

Hence $\mathcal{L}_U^{\mathrm{adj}}$ is an ideal of $\mathcal{T}$. The definition makes sense because the modules M^+ and M^- are finite flat over R and scalar product $[\ ,\ ]_K$ is $\mathcal{T}$-equivariant (cf. Theorem 9.2.9(iii)). Note that the formation of the ideal $\mathcal{L}_U^{\mathrm{adj}}$ commutes with the restriction of the clean affinoid U to a smaller clean affinoid: this follows from the commutation of the formation of $[\ ,\ ]_K$ with base change (Theorem 9.2.9(i)) and the remarks following Definition 9.1.9. In particular, since there is an admissible covering of the eigencurve $\mathcal{C}^0$ of clean affinoids, we can glue the adjoint $\mathcal{L}$-ideals

$\mathcal{L}_U^{\mathrm{adj}}$ to define a global coherent sheaf of ideal $\mathcal{L}^{\mathrm{adj}}$ on the eigencurve $\mathcal{C}^0$. The following result follows directly from Proposition 9.1.10.

Theorem 9.4.2 *The closed analytic subspace defined by the sheaf of ideal $\mathcal{L}^{adj}$ is contained in the locus of non-étaleness of the weight map κ.*

If we are willing to restrict ourselves to good points of $\mathcal{C}^0$, and to work locally, it is possible to replace the adjoint L-ideal $\mathcal{L}^{\mathrm{adj}}$ by a true p-adic adjoint L-function which satisfies an interpolation property relating it to the special value of the archimedean adjoint L-function, and to prove much more precise result on its zero locus.

Proposition 9.4.3 *Let L be a finite extension of $\mathbb{Q}_p$ and $x \in \mathcal{C}^0(L)$ be a good point. If $U = \mathrm{Sp}\,\mathcal{T}$ is a sufficiently small clean neighborhood of x, then the ideal $\mathcal{L}_U^{adj}$ of $\mathcal{T}$ is principal.*

Proof This follows from the definition of a good point and Proposition 9.1.11. □

Definition 9.4.4 If $U = \mathrm{Sp}\,\mathcal{T}$ is as in the above proposition, we define the *p-adic adjoint L-function on U*, L_U^{adj} as a generator of the ideal $\mathcal{L}_U^{\mathrm{adj}}$. Thus the p-adic adjoint L-function on U is defined up to multiplication by an element of $\mathcal{T}^*$.

Proposition 9.4.5 *Let L be a finite extension of $\mathbb{Q}_p$ and $x \in \mathcal{C}^0(L)$ be a good point. Fix $U = \mathrm{Sp}\,\mathcal{T}$ a sufficiently small clean neighborhood of x. Then for every sufficiently small real $r > 0$ there exist two real constants $0 < c < C$ such that for every point $y \in U$, with $\kappa(y) = w$ one has*

$$c\left|[\Phi_y^+, \Phi_y^-]_w\right| \leq |L^{adj}(y)| \leq C\left|[\Phi_y^+, \Phi_y^-]_w\right|,$$

where $\Phi_y^{\pm}$ are generators of the spaces $H_!^1(\Gamma, \mathcal{D}_w[r](L))^{\pm}[y]$ such that $\|\Phi_y^{\pm}\|_r = 1$. Moreover, the adjoint p-adic L-function is characterized (up to an element of $\mathcal{T}^*$) by the above property.

Proof For $0 < r < \min(p, r(\kappa))$, let us provide M^+ and M^- with the norm $\| \; \|_r$, and $M^+ \otimes_R M^-$ with the tensor norm, also denoted $\| \; \|_r$. Consider the scalar product b as an element of $(M^+ \otimes_R M^-)^\vee$. There exists real constants $0 < c < C$ such that for every $y \in \mathrm{Spec}\,\mathcal{T}(L) = U(L)$, one has $c < (\|b_y\|_r)' < C$, where b_y is the image of b in $(M^+ \otimes_R M^-) \otimes_{\mathcal{T}} \mathcal{T}/\mathfrak{p}_y = H_!^1(\Gamma, \mathcal{D}_w(L))^+[y] \otimes H_!^1(\Gamma, \mathcal{D}_w(L))^+[y]$, $\mathfrak{p}_y$ being the maximal ideal of $\mathcal{T}$ corresponding to y (and $\mathcal{T}/\mathfrak{p}_y = L$) and $\| \; \|_r'$ is the dual norm of the norm $\| \; \|_r$ on those spaces.

Fix an isomorphism $\psi^+ : (M^+)^\vee \to \mathcal{T}$ and $\psi_x : (M^-)^\vee \to \mathcal{T}$ as $\mathcal{T}$-module, so that $b \in (M^+ \otimes M^-)^\vee$ can be seen as en element of $\tilde{b} \in \mathcal{T} \otimes \mathcal{T}$, and $m(b) \in \mathcal{T}$ is by definition the p-adic L-function L_U. There exists constants $0 < c' < C'$ such that for every y in U, the generator of $\mathcal{T}/\mathfrak{p}_y$ is sent to an element $e_y^{\pm} \in H_!^1(\Gamma, \mathcal{D}_w(L))^{\pm}[y]$ of norm between c' and C' by the isomorphisms $\psi_y^{\pm}$ induced by ψ^+ and ψ^-. Since y is a good point, $H_!^1(\Gamma, \mathcal{D}_w[r](L))^{\pm}[y])$ has L-dimension 1.

Therefore $e_y^+ \otimes e_y^-$ has norm between c'^2 and c^2, and we have $\tilde{b}_y = a_y e_y^+ \otimes e_y^-$ where $a_y \in L$ is such that $c/C'^2 < |a_y| < C/c'^2$. Since $\mathcal{L}_U(y) = m(\tilde{b}_y) = a_y$, one gets the first result.

The fact that the function $\mathcal{L}_U$ is characterized by the above property up to an element of $\mathcal{T}^*$, because a meromorphic function on U (that is an element of the form f/g, for $f, g \in \mathcal{T}$, g not a zero divisor) whose norm at every closed point of $\operatorname{Spec} \mathcal{T}$ is bounded above and below away from 0 is an element of $\mathcal{T}^*$. □

Corollary 9.4.6 *Let L be a finite extension of $\mathbb{Q}_p$ and $x \in \mathcal{C}^0(L)$ be a good point. Fix $U = \operatorname{Sp} \mathcal{T}$ a sufficiently small clean neighborhood of x. Then there exists two real constant $0 < c < C$ such that for every very classical point $y \in U$ such that κ is étale at y, one has*

$$c \mid [\phi_y^+, \phi_y^-]_k \mid \leq |L^{adj}(y)| \leq C \mid [\phi_y^+, \phi_y^-]_k|,$$

where $k = \kappa(y) \in Z$ is the weight of y and $\phi_y^\pm$ are generators of the one-dimensional spaces $H_!^1(\Gamma, \mathcal{V}_k(L))^\pm[y]$ (cf. Lemma 9.3.4), normalized so that $\|\phi_y^\pm\| = 1$. Observe that the condition on y is satisfied as soon as $\kappa(y)$ is an integer $k \in \mathbb{N}$ and one has $v_p(U_p(y)) < k + 1$.

Proof By shrinking U if necessary, we can and do assume that for every $y \in U(L) - \{x\}$ such that $\kappa(y) \in \mathbb{N}$, we have $v_p(U_p(y)) < \kappa(y) + 1$. Take $0 < r < \min(p, r(\kappa))$, $r \leq 1$ and $0 < c < C$ as in the preceding proposition. For $y \in U(L) - \{x\}$ a classical point of weight $\kappa(y) = k \in \mathbb{N}$, one has that ρ_k is an isomorphism $H_!^1(\Gamma, \mathcal{D}_k[r'](L))^\pm[y]) \to H_!^1(\Gamma, \mathcal{V}_k(L))^\pm[y])$, isometric when $r' = 1$ (cf. Theorem 6.5.19) since $v_p(U_p(y)) < k + 1$. Since y is a good point, the spaces $H_!^1(\Gamma, \mathcal{V}_k(L))^\pm[y])$ are one dimensional (cf. Lemma! 9.3.4). Let $\phi_y^\pm$ be generators of those spaces of norm 1. Then by Proposition 6.5.12, there exists a real constant $0 < D \leq 1$ depending only on r, not on y, such that $D \leq \|\Phi_y^\pm\|_1 \leq 1$. It follows that $\rho_k(\Phi_y^\pm) = a_y^\pm \phi_y^\pm$ where $D \leq |a_y^\pm| \leq 1$ hence $|[\Phi_y^+, \Phi_y^-]| \leq |[\phi_y^+, \phi_y^-]| \leq D^2 |[\Phi_y^+, \Phi_y^-]|$ using Definition 9.2.1 and the corollary follows for every y except perhaps for x. Assuming x is very classical and that κ is étale at x, we see that $[\Phi_x^+, \Phi_x^-] = b[\phi_x^+, \phi_x^-]$, where b is some non-zero scalar, and by enlarging C or diminishing c, we can assume that $0 < c \leq |b| \leq C$ which concludes the proof. □

Note that if is x is a good point, then the condition $L_U^{\text{adj}}(x) = 0$ clearly does not depend on the choice of a sufficiently small clean neighborhood U of x nor of the adjoint p-adic L-function L_U^{adj}. In this case, we shall simply write $L^{\text{adj}}(x) = 0$. When x is also a smooth point of $\mathcal{C}^0$, then it makes sense to talk of the order of vanishing of L_U^{adj} at x, which again does not depend on the choice of the small clean affinoid neighborhood U of x nor of the adjoint L-function. We shall denote this integer by $\operatorname{ord}_x L_U^{\text{adj}}$.

In the following two results, we determine the zero locus, and when it makes sense, the order of vanishing, of the p-adic adjoint L-function L^{adj}.

Theorem 9.4.7 *Let L be a finite extension of $\mathbb{Q}_p$ and $x \in \mathcal{C}^0(L)$ be a good point. If either $\kappa(x) \notin \mathbb{N}$ or x is a very classical cuspidal point, then we have*

$$L^{adj}(x) = 0 \text{ if and only if } \kappa \text{ is not étale at } x.$$

Otherwise, that is when $\kappa(x) \in \mathbb{N}$ and x is not a very classical cuspidal point, we always have $L^{adj}(x) = 0$.

Proof If either $\kappa(x) \notin \mathbb{N}$ or $\kappa(x) = k \in \mathbb{N}$ and $v_p(U_p(x)) < k+1$, then the scalar product on $M^+ \times M^-$ is non-degenerate by Theorem 9.2.10 and the equivalence κ étale at $x \Leftrightarrow L^{\mathrm{adj}}(x) = 0$ follows from Corollary 9.1.13. Now assume $\kappa(x) = k \in \mathbb{N}$, and let Φ_x^+ and Φ_x^- be generators of $H^1_!(Y_\Gamma, \mathcal{D}_w[r](L))^{\pm}[x]$. Then by Proposition 9.4.5, one has up to a non-zero scalar $L^{\mathrm{adj}}(x) = [\Phi_x^+, \Phi_x^-]_k = [\rho_k(\Phi_x^-), \rho_k(\Phi_x^-)]_k$, the second equality being by definition of the scalar product. When x is a cuspidal classical point, $\rho_k(\Phi_x^-)$ and $\rho_k(\Phi_x^+)$ are non-zero if and only if κ is étale at x (Lemma 9.3.4) and when this happens $[\rho_k(\Phi_x^+), \rho_k(\Phi_x^-)]_k \neq 0$ since the scalar product $[\ ,\ ]_k$ on $H^1_!(Y_\Gamma, \mathcal{V}_k)^+[x] \times H^1_!(Y_\Gamma, \mathcal{V}_k)^-[x]$ is non-degenerate and those spaces have dimension 1. When x is not a cuspidal classical point, $\rho_k(\Phi_x^{\pm}) = 0$ by Lemma 9.3.4 and $L^{\mathrm{adj}}(x) = 0$. □

Theorem 9.4.8 *Let L be a finite extension of $\mathbb{Q}_p$ and $x \in \mathcal{C}^0(L)$ be a smooth good point. Let us call $e(x)$ the degree of the map κ at x (so $\kappa(x) = 1$ if and only if κ is étale at x):*

(i) If $\kappa(x) \neq \mathbb{N}$, or if x is a very classical point of non-critical slope (that is $v_p(U_p(x)) < k+1$ where $\kappa(x) = k \in \mathbb{N}$), then $\mathrm{ord}_x L^{adj}(x) = e(x) - 1$.
(ii) If $\kappa(x) = k \in \mathbb{N}$ and x is a very classical cuspidal point of critical slope (that is $v_p(U_p(x)) = k+1$), then $\mathrm{ord}_x L^{adj}(x) = 2e(x) - 2$
(iii) If $\kappa(x) = k \in \mathbb{N}$ and x is not a very classical cuspidal point, then $\mathrm{ord}_x L^{adj}(x) \geq 2e(x) - 1$.

The proof of this theorem will occupy the rest of this section.

Let $U = \mathrm{Sp}\, \mathcal{T}$ be a clean neighborhood of x with $\kappa(U) = W = \mathrm{Sp}\, R$, with M^+ and M^- the associated $\mathcal{T}$-modules.

Let us denote by $\mathbb{T} = \mathcal{O}_{\mathcal{C}^0,x}$ the local ring of the eigencurve $\mathcal{C}^0$ at x, and by R the local ring of the weight space at $\kappa(x)$. Let us denote by $\mathbb{M}^{\pm}$ the localization of the $\mathcal{T}$-module $M^{\pm}$ at x. By Theorem 8.1.5, A and $\mathbb{T}$ are d.v.r., and if u is an uniformizer of A, there is a uniformizer t of $\mathbb{T}$, and an isomorphism of A-algebra $\mathbb{T} = A[t]/(t^e - u)$, where e is the degree of the map κ at x. Moreover, the modules $\mathbb{M}^{\pm}$ are both free of degree e over A and of degree 1 over $\mathbb{T}$.

Let $b_0 : \mathbb{T} \times \mathbb{T} \to A$ be the bilinear A-form $b_0(t^i, t^j) = 0$ if $i + j \not\equiv e - 1 \pmod e$, and $b_0(t^i, t^j) = u^{(i+j-(e-1))/e}$ if $i + j \equiv 0 \pmod e$. One easily sees that b_0 is $\mathcal{T}$-equivariant and perfect.

The scalar product $[\ ,\ ]_K$ define by localization a scalar product $b : \mathbb{M}^+ \times \mathbb{M}^- \to A$, which is $\mathbb{T}$-equivariant, or choosing an isomorphism $\mathbb{M}^+ \simeq \mathbb{M}^- \simeq \mathbb{T}$ of $\mathbb{T}$-module, a scalar product $\tilde{b} : \mathbb{T} \times \mathbb{T} \to \mathbb{T}$ and we have $\mathcal{L}_U^{\mathrm{adj}}\mathbb{T} = \mathcal{L}_{\tilde{b}}$ since the formation of the $\mathcal{L}$-ideal commutes with localization. One has $\tilde{b} = \tau b_0$ for some $\tau \in \mathbb{T}$ since the $\mathbb{T}$-module of such scalar products is free of rank one over $\mathbb{T}$ (the module structure here is $(\tau b_0)(m, m') := b_0(\tau m, m') = b_0(m, \tau m')$). By Corollary 9.1.13, one has $\mathcal{L}_{\tilde{b}} = \tau \mathfrak{d}_N(\mathbb{T}/A)$ and the different of $\mathcal{T}$ over A is $\mathfrak{d}_N(\mathbb{T}/A) = t^{e-1}\mathbb{T}$ by Exercise 9.1.5. Hence

$$t^{\mathrm{ord}_x \mathcal{L}^{\mathrm{adj}}}\mathbb{T} = \mathcal{L}_U^*\mathbb{T} = \tau t^{e-1}\mathbb{T}.$$

To determine τ we separate the three cases of the theorem.

In case (i), b is non-degenerate so $\tau \in \mathcal{T}^*$ and we get $\mathrm{ord}_x \mathcal{L}^{\mathrm{adj}} = e - 1$.

In case (ii) and (iii) we consider the scalar product $\bar{b} \pmod u : \mathbb{M}^+/u \times \mathbb{M}^-/u \to A/u = L$ induced by b, which is the same as the scalar product $[\ ,\ ]_k : H_!^1(Y_\Gamma, \mathcal{D}_k(L))_{(x)}^+ \times H_!^1(Y_\Gamma, \mathcal{D}_k(L))_{(x)}^+ \to L$. By definition $[\Phi_1, \Phi_2]_k = [\rho_k(\Phi_1), \rho_k(\Phi_2)]_k$ where the second scalar product $H_!^1(Y_\Gamma, \mathcal{V}_k(L))_{(x)}^+ \times H_!^1(Y_\Gamma, \mathcal{V}_k(L))_{(x)}^- \to L$ is non-degenerate because it is the restriction of the non-degenerate scalar product $[\ ,\ ]$ to generalized eigenspace for a family of self-adjoint commuting operators. The spaces $H_!^1(Y_\Gamma, \mathcal{V}_k(L))_{(x)}^+$ are of dimension 1 in case (ii) and are 0 in case (iii) (cf. Lemma 9.3.4).

In case (ii), ρ_k sends a generator of $H_!^1(Y_\Gamma, \mathcal{D}_k(L))_{(x)}^\pm$ as $\mathbb{T}/u$-module on a generator of $H_!^1(Y_\Gamma, \mathcal{V}_k(L))_{(x)}^\pm$ since it is surjective, and it follows that $b(\Phi^+, \Phi^-) = [\Phi^+, \Phi^-]_k \neq 0$ in L when $\Phi^\pm$ are generators of $H_!^1(Y_\Gamma, \mathcal{V}_k(L))_{(x)}^\pm = \mathbb{M}^\pm/u$. Therefore, $b(1, 1)$ is a unit in A, that is $(\tau b_0)(1, 1) = b_0(1, \tau)$ is a unit in $\mathbb{T}$, and this implies that $\tau\mathcal{L} = t^{e-1}\mathbb{T}$. We conclude that $\mathrm{ord}_x \mathcal{L}^{\mathrm{adj}} = e - 1 + e - 1 = 2e - 2$ in this case.

In case (iii), the scalar product $[\ ,\ ]_k : H_!^1(Y_\Gamma, \mathcal{D}_k(L))_{(x)}^+ \times H_!^1(Y_\Gamma, \mathcal{D}_k(L))_{(x)}^- \to L$ is zero, so b takes values in the maximal ideal uA of A. In particular, $\tilde{b}(t^a, 1) = b_0(t^a, \tau) \in uA$. This implies $\tau\mathbb{T} \in t^e\mathbb{T}$ (for if $\tau\mathbb{T} = t^n\mathbb{T}$ with $0 \leq n \leq e - 1$, $b_0(t^{e-1-n}, \tau)$ is a unit in A). One deduces that $\mathrm{ord}_x \mathcal{L}^{\mathrm{adj}} \geq e - 1 + e = 2e - 1$ in case (iii).

This completes the proof of the theorem.

9.5 Relation Between the Adjoint p-Adic L-Function and the Classical Adjoint L-Function

9.5.1 Scalar Product and Refinements

Let $f \in S_{k+2}(\Gamma_1(N), \mathbb{C})$ be a newform. Let a_p be its coefficient at p, so that $T_f = a_p f$. According to Proposition 5.3.24 one has

$$f_{|W_N} = W(f) f,$$

where $W(f) = \pm N^{k/2}$.

Let $\Gamma = \Gamma_0(Np)$. In $S_{k_2}(\Gamma, \mathbb{C})$ the $\mathcal{H}_0^p$-eigenspace of system of eigenvalues the one of f is generated by $f(z), f(pz)$. We call it here the *old subspace*.

The Action of W_{Np} on the Old Subspace

The old subspace is preserved by the action of W_{Np}. We compute

$$\begin{aligned} f(pz)_{|W_{Np}} &= \frac{-1}{Npz^{k+2}} f\left(\frac{-p}{Npz}\right) \\ &= \frac{-1}{Npz^{k+2}} f\left(\frac{-1}{Nz}\right) \\ &= \frac{1}{p} f(z)_{|W_N} \\ &= \pm\frac{1}{p} N^{k/2} f(z) \end{aligned}$$

Thus the matrix of the action of W_{Np} from the basis $f(z), f(pz)$ to the basis $f(z), f(pz)$ has the form $w := \begin{pmatrix} x & \pm\frac{1}{p}N^{k/2} \\ y & 0 \end{pmatrix}$. To determine x and y, we note that $W_{Np}^2 = -Np$ which acts by multiplication by $(Np)^k$ on the space generated by $f(z), f(pz)$. Hence $w^2 = (Np)^k$ which gives $x = 0$ and $y = \pm p^{k+1} N^{k/2}$, so

$$w = \pm N^{k/2} \begin{pmatrix} 0 & \frac{1}{p} \\ p^{k+1} & 0 \end{pmatrix}.$$

The Adjoint of U_p on the Old Subspace

The adjoint U_p' of U_p in the space generated by $f(z), f(pz)$ is wU_pw^{-1} (cf. (5.3.19)). Since the matrix of U_p in that base is $\begin{pmatrix} a_p & 1 \\ -p^{k+1} & 0 \end{pmatrix}$ (see Sect. 2.6.4), the matrix of U_p' is $\begin{pmatrix} 0 & -p^{-1} \\ p^{k+2} & a_p \end{pmatrix}$.

The Matrix of Peterson's Product on the Old Subspace

Using this, one computes

$$\begin{aligned}(f(pz), f(pz))_\Gamma &= p^{-k-2}(f(pz), f(z)_{|U_p'})_\Gamma \\ &= p^{-k-2}(f(pz)_{|U_p}, f(z))_\Gamma \\ &= p^{-k-2}(f(z), f(z))_\Gamma\end{aligned}$$

$$\begin{aligned}(f(z), f(pz))_\Gamma &= p^{-k-2}(f(z), f(z)_{|U_p'})_\Gamma \\ &= p^{-k-2}(f(z)_{|U_p}, f(z))_\Gamma \\ &= p^{-k-2}a_p(f(z), f(z))_\Gamma - p^{-1}(f(pz), f(z))_\Gamma\end{aligned}$$

Since a_p and $(f(z), f(z))_\Gamma$ are real, and since $(f(pz), f(z))_\Gamma = \overline{(f(z), f(pz))_\Gamma}$ we get

$$(f(pz), f(z))_\Gamma = (f(z), f(pz))_\Gamma = \frac{p^{-k-1}a_p}{1+p}(f(z), f(z))_\Gamma.$$

The Corrected Scalar Product on the Old Subspace

Let $f_\alpha(z) = f(z) - \beta f(pz)$ and $f_\beta = f(z) - \alpha f(pz)$ where α and β are the roots of $X^2 - a_pX + p^{k+1}$. Note that the discriminant of this equation, $a_p^2 - 4p^{k+1}$, is negative by Hecke estimate, hence $\beta = \bar{\alpha}$.

$$\begin{aligned}[\phi_{f_\alpha}^+, \phi_{f_\beta}^-] &= (-1)^{k-1}2^{k-2}(f_\alpha, (f_\alpha)_{|W_{Np}})_\Gamma \\ &= \pm(-i)^{k-1}2^{k-2}N^{k/2}(f(z) - \beta f(pz), p^{k+1}f(pz) - \beta\frac{1}{p}f(z))_\Gamma \\ &= \pm(-i)^{k-1}2^{k-2}N^{k/2}[(-\bar{\beta}/p - \beta p/p^2)(f(z), f(z))_\Gamma \\ &\quad + p^{k+1}(f(z), f(pz))_\Gamma + \beta\bar{\beta}/p(f(pz), f(z))_\Gamma]\end{aligned}$$

$$= \pm(-i)^{k-1}2^{k-2}N^{k/2}[\frac{-a_p}{p}(f(z), f(z))_\Gamma + p^{k+1}(f(z), f(pz))_\Gamma$$
$$+p^k(f(pz), f(z))_\Gamma]$$
$$= \pm(-i)^{k-1}2^{k-2}N^{k/2}[\frac{-a_p}{p}(f(z), f(z))_\Gamma + p^{-1}a_p(f(z), f(z))_\Gamma]$$
$$= 0.$$

Note that since U_p is equivariant for [,] and $U_p f_\alpha = \alpha f_\alpha$, $U_p f_\beta = \beta f_\beta$, the above formula is obvious when $\alpha \neq \beta$.

$$[\phi^+_{f_\alpha}, \phi^-_{f_\alpha}] = (-i)^{k-1}2^{k-2}(f_\alpha, (f_\beta)_{|W_{Np}})_\Gamma$$
$$= \pm (-i)^{k-1}2^{k-2}N^{k/2}(f(z) - \beta f(pz), p^{k+1}f(pz) - \alpha\frac{1}{p}f(z))_\Gamma$$
$$= \pm (-i)^{k-1}2^{k-2}N^{k/2}[(-\beta/p - \beta p/p^2)(f(z), f(z))_\Gamma$$
$$+ p^{k+1}(f(z), f(pz))_\Gamma + \beta^2/p(f(pz), f(z))_\Gamma]$$
$$= \pm (-i)^{k-1}2^{k-2}N^{k/2}\left(\frac{-2\beta}{p} + \frac{a_p}{1+p} + \frac{\beta^2 a_p}{p^{k+2}(1+p)]}\right)(f(z), f(z))_\Gamma \tag{9.5.1}$$

Note that when $\alpha = \beta$, $2\beta = a_p$ and $\beta^2 = p^{k+1}$, so the last formula gives $[\phi^+_{f_\alpha}, \phi^-_{f_\alpha}] = 0$, consistent with the one above.

9.5.2 *Adjoint L-Function and Peterson's Product*

Definition of the Classical Adjoint L-Function

Let M be any integer. Let $f \in S_{k+2}(\Gamma_1(M), \mathbb{C})$ be a newform of nebentypus ϵ. Write $f = \sum_{n=1}^{\infty} a_n q^n$ with $a_1 = 1$. For every prime l, let α_l and β_l be the two roots of $X^2 - a_l X + l^{k+1}\epsilon(l)$. Following Shimura ([114], and also [68, §5]), we define for ψ a Dirichlet character and s a complex number of real part sufficiently large:

$$L^{\mathrm{adj}}(f, s, \psi) = \prod_{l \text{ prime}} (1 - \psi(l)\alpha_l^2 l^{-s})(1 - \psi(l)\alpha_l\beta_l l^{-s})(1 - \psi(l)\beta_l^2 l^{-s}).$$

It is easy to see that $L^{\mathrm{adj}}(f, s, \psi)$ is an holomorphic function on its half-plane of convergence. Shimura proved that it has a meromorphic continuation on $\mathbb{C}$ with possible poles at $s = k+1$ and $s = k+2$. However, there is no pole at $s = k+2$ if $\epsilon\psi$ is not a non-trivial character of order 2.

When $\psi = 1$ we write $L^{\mathrm{adj}}(f, s)$ instead of $L^{\mathrm{adj}}(f, s, 1)$.

Hida's Formula

Hida proves the following formula (see [68, Theorem 5.1]).

$$L^{\mathrm{adj}}(f, k+2, \epsilon^{-1}) = \frac{2^{2k+4}\pi^{k+3}}{(k+1)!\delta(N)MN(\epsilon)\varphi(M/N(\epsilon))}(f, f)_{\Gamma_1(M)},$$

where $N(\epsilon)$ is the conductor of ϵ, $\delta(M) = 2$ if $M = 1, 2$, $\delta(M) = 1$ otherwise, and φ is Euler's totient function.

Proposition 9.5.1 *Let N be a an integer prime to p and $f \in S_{k+2}(\Gamma_0(N), \mathbb{C})$ be a newform.*

$$[\phi^+_{f_\alpha}, \phi^-_{f_\alpha}] = \frac{(-i)^{k-1}N^{k/2+1}\delta(N)\varphi(N)}{(4\pi)^{k+3}}\left(\frac{-2\beta}{p} + \frac{a_p}{1+p} + \frac{\beta^2 a_p}{p^{k+2}(1+p)}\right)L^{adj}(f, k+2)$$

Proof Combine Hida's formula with the fact that $(f, f)_\Gamma = p(p-1)^2(f, f)_{\Gamma_1(N)}$ and Eq. (9.5.1). □

9.5.3 *p-Adic and Classical Adjoint L-Function*

Theorem 9.5.2 *L be a finite extension of $\mathbb{Q}_p$ and $x \in \mathcal{C}^0(L)$ be a good point. Fix $U = Sp\,\mathcal{T}$ a sufficiently small neighborhood of x. Then there exist two real constants $0 < c < C$ such that for every very classical point y on U corresponding to a modular form f_α of level $\Gamma_0(Np)$ of weight $k+2$ and old at p, associated with the new form f of level $\Gamma_0(N)$, one has*

$$c\left|\frac{-2\beta}{p} + \frac{a_p}{1+p} + \frac{\beta^2 a_p}{p^{k+2}(1+p)]}\right|\left|\frac{L^{adj}(f, k)}{\pi^{k+3}\Omega_f^+\Omega_f^-}\right| < |L_p^{adj}(y)| <$$
$$C\left|\frac{-2\beta}{p} + \frac{a_p}{1+p} + \frac{\beta^2 a_p}{p^{k+2}(1+p)]}\right|\left|\frac{L^{adj}(f, k)}{\pi^{k+3}\Omega_f^+\Omega_f^-}\right|$$

Proof Combine Proposition 9.5.1 with Corollary 9.4.6, noticing that the classical modular symbol denoted $\phi_y^\pm$ are $\phi_{f_\alpha}^\pm/\Omega_f^\pm$ by definition of $\Omega_f^\pm$ (cf. Remark 5.4.10). □

Chapter 10
Solutions and Hints to Exercises

Solution to Exercise 2.1.1 The result is clear when M is free. In general, there is an exact sequence $P \to Q \to M \to 0$ where P and Q are finite free. Let $P' = P \otimes_R R'$ and $Q' = Q \otimes_R R'$. By tensorizing by R' we have an exact sequence $P' \to Q' \to M' \to 0$ hence by applying the functor $\mathrm{Hom}_{R'}(-, N')$ an exact sequence

$$0 \to \mathrm{Hom}_{R'}(M', N') \to \mathrm{Hom}_{R'}(Q', N') \to \mathrm{Hom}_{R'}(P', N').$$

On the other hand by applying $\mathrm{Hom}_R(-, N)$ to $P \to Q \to M \to 0$ and then tensorizing by R', we get using that R' is R-flat, an exact sequence

$$0 \to \mathrm{Hom}_R(M, N) \otimes_R R' \to \mathrm{Hom}_R(Q, N) \otimes_R R' \to \mathrm{Hom}_R(P, N) \otimes_R R'.$$

The natural morphism $\mathrm{Hom}_R(P, N) \otimes_R R' \to \mathrm{Hom}_{R'}(P', N')$ between the last terms of the two displayed exact sequences is an isomorphism since P is finite free, and the same holds for the middle term. Hence by diagram chasing we see that $\mathrm{Hom}_R(M, N) \otimes_R R' \to \mathrm{Hom}_{R'}(M', N')$ is an isomorphism.

Solution to Exercise 2.1.2 See the solution to Exercise 2.4.2.

Hint to Exercise 2.1.3 By construction of the characteristic polynomial, this question reduces to the case where M is free, in which case it is trivial.

Solution to Exercise 2.3.2 Let $0 \neq m \in M$ be a torsion element, that is such that $xm = 0$ for some $0 \neq x \in \mathcal{T}$. Since $\mathcal{T}$ is finite over R, there is a monic polynomial $P(X) = X^d + a_{d-1}X^{d-1} + \cdots + a_0 \in R[X]$ such that $P(x) = 0$. Choosing P of minimal degree d, we may assume that $a_0 \neq 0$, for otherwise $Q(x)x = 0$ with $Q(X) := X^{d-1} + a_{d-1}X^{d-2} + \cdots + a_1 \in R[X]$, and thus since $\mathcal{T}$ is a domain $Q(x) = 0$, contradicting the minimality of d. But then $0 = P(x)m = a_0 m$

J. Bellaïche, *The Eigenbook*, Pathways in Mathematics,
https://doi.org/10.1007/978-3-030-77263-5_10

since $xm = 0$, which shows that m is R-torsion, in contradiction with our running assertion that M is locally free over R.

Solution to Exercise 2.3.6 The surjective tautological map $\mathcal{H} \otimes R \to \mathcal{T}_A$ factors through $\mathcal{T}$ since any element $h \in \mathcal{H} \otimes R$ such that $\psi(h) = 0$ obviously acts by 0 on A. Hence we have a natural surjective map $\mathcal{T} \to \mathcal{T}_A$. Its kernel is by construction the ideal generated by the elements $\psi(h)$ of $\mathrm{End}_R(M)$ that vanish on A. If M/A is torsion, such an endomorphism $\psi(h)$ is torsion in $\mathrm{End}_R(M)$, hence is 0 since M is a faithful R-module.

For 2. and an example of non-injectivity, takes $A = B = 0$. For an example of non-surjectivity, take $A = B = R$ and $M = A \oplus B$ with $\mathcal{H}$ generated by one element T which acts as the identity on M. Then the map $\mathcal{T} \to \mathcal{T}_A \times \mathcal{T}_B$ is the diagonal embedding $R \to R^2$ which is not surjective.

3. is proved exactly as the second part of 1.

For 4. let $e = (T - a)/(b - a)$ in $\mathcal{T}$, so e acts by 0 on A and 1 and B, that is the image of $e \in \mathcal{T}$ is 0 in $\mathcal{T}_A$ and 1 in $\mathcal{T}_B$. If $t_A \in \mathcal{T}_A$ and $t_B \in \mathcal{T}_B$, there exists $u_A \in \mathcal{T}$ that maps on t_A in $\mathcal{T}_A$ (since the map $\mathcal{T} \to \mathcal{T}_A$ is surjective by 1.) and $u_B \in \mathcal{T}$ that maps on t_B in $\mathcal{T}_B$ (same reason). Then $t := (1 - e)t_A + et_B \in \mathcal{T}$ maps to (t_A, t_B) in $\mathcal{T}_A \times \mathcal{T}_B$.

Hint to Exercise 2.3.7 For 1. consider the map $\mathcal{T} \to \mathcal{T}^\vee$ that sends an endomorphism of M to its transpose.

2. is just a reformulation of the classical result that over any field a square matrix is conjugate to its transpose.

For 3. and R not a field, check that one just need to find a matrix in $M_n(R)$ which is not conjugate (in $\mathrm{GL}_n(R)$) to its transpose. Then check that for $R = \mathbb{Z}$, the matrix $T = \begin{pmatrix} 1 & -5 \\ 3 & -1 \end{pmatrix}$ is not similar to its transpose over $\mathbb{Z}$.

For 3. and $R = \mathbb{C}$, take $M = \mathbb{C}^3$ and $\mathcal{H}$ generated by two elements X and Y acting on M by the matrices $X = \begin{pmatrix} 0 & 0 & 0 \\ 1 & 0 & 0 \\ 0 & 0 & 0 \end{pmatrix}$ and $Y = \begin{pmatrix} 0 & 0 & 0 \\ 0 & 0 & 0 \\ 1 & 0 & 0 \end{pmatrix}$.

Solution to Exercise 2.3.7 1. and 2. should be clear after the hint.

For 3. and the case of a ring, checking that the matrix $T = \begin{pmatrix} 1 & -5 \\ 3 & -1 \end{pmatrix} \in M_2(\mathbb{Z})$ is not conjugate to its transpose is a straightforward but tedious computation. For a conceptual proof, and an explanation of how this example was found and how to find many other ones, see [49], especially Example 15.

For 3. and the case of a field, considering the example given in the hint we observe that $\mathrm{Im}\, Y \oplus \mathrm{Im}\, X$ has dimension 2 while $\mathrm{Im}\, {}^tY \oplus \mathrm{Im}\, {}^tX$ has dimension 1. There can therefore be no matrices $P \in \mathrm{GL}_3(\mathbb{C})$ such that $PXP^{-1} = {}^tX$ and $PYP^{-1} = {}^tY$, which shows that as an $\mathcal{H}$-module, M is not isomorphic to $M^\vee$.

Solution to Exercise 2.3.9 For 1. let $M = R^2$, and $\psi : a + b\epsilon \mapsto \begin{pmatrix} a & b \\ 0 & a \end{pmatrix}$ for $a \in R, b \in I$. For 2. take $R = \mathbb{C}[X, Y]$. Since R is regular of dimension 2, any finite torsion-free R-algebra is an eigenalgebra, and there are plenty of them that are non-flat.

Solution to Exercise 2.4.2 The eigenalgebra $\mathcal{T}$ is the quotient of $\mathbb{Z}_p[T]$ by the ideal of operators that act trivially on M. Let $P(X) \in \mathbb{Z}_p[X]$. If $P(T) = 0$ as an operator on M, then P is divisible by the minimal polynomial $(X-1)^2$ of T in $\mathbb{Q}_p[X]$, so is divisible by $(X-1)^2$ in $\mathbb{Z}_p[X]$ (since the algorithm of euclidean division by a monic polynomial like $(X-1)^2$ is identical in $\mathbb{Z}_p[X]$ and $\mathbb{Q}_p[X]$). It follows that $\mathcal{T} \simeq \mathbb{Z}_p[X]/(X-1)^2$ with the isomorphism sending T on X.

The eigenalgebra $\mathcal{T}'$ is simply $\mathbb{F}_p$ since T acts as the identity on $M \otimes \mathbb{F}_p = \mathbb{F}_p^2$. The map $\mathcal{T} \otimes_{\mathbb{Z}_p} \mathbb{F}_p = \mathbb{F}_p[X]/(X-1)^2 \to \mathcal{T}' = \mathbb{F}_p$ is the map that sends X on 1. Of course this map is not an isomorphism.

Solution to Exercise 2.5.4 This is standard: by induction on the number of generators of $\psi(\mathcal{H})$, one reduces to the case of one generator T of $\psi(H)$. In this case let $P(X)$ be the characteristic (or minimal) polynomial of T and write $P(X) = (X - a_1)^{n_1} \dots (X - a_r)^{n_r}$ where the a_i are distinct. Let $P_i(X)$ be the same polynomial with the factor $(X - a_i)^{n_i}$ removed. Then the P_i's are relatively prime, and a Bézout relation between them shows that $M = \sum_i \ker(T - a_i)^{n_i}$, a sum which is obviously direct.

Solution to Exercise 2.5.5 We obviously have $M[\chi] \otimes k' \subset M'[\chi']$ in M' and the equality follows from the equality of dimensions, since the rank of a system of linear equations over k does not change when we extend the scalar to k'. Same argument for $M_{(\chi)}$.

Solution to Exercise 2.5.6 By Exercise 2.5.5, we may assume that k is algebraically closed. The operators $\psi(T)$, $T \in \mathcal{H}$, on the non-zero space $M_{(\chi)}$ commute so they have a common non-zero eigenvector v, whose system of eigenvalue is χ. So $M[\chi] \neq 0$.

Solution to Exercise 2.5.7 The only non-trivial part is the surjectivity of $M_{(\chi)} \to N_{(\chi)}$ but it can be checked when the field is assumed to be algebraically closed by Exercise 2.5.5 and then it follows easily from Exercise 2.5.4.

Solution to Exercise 2.5.8 We may assume that k is algebraically closed. Since $M = \oplus_\chi M_{(\chi)}$ by Exercise 2.5.4, we get a decomposition $M^\vee = \oplus_\chi M^\vee_{(\chi)}$, and by applying directly Exercise 2.5.4 an other decomposition $M^\vee = \oplus_\chi (M^\vee)_{(\chi)}$. Since $M^\vee_{(\chi)} \subset (M^\vee)_{(\chi)}$ the two decompositions are the same.

For an example where $\dim M[\chi] = \dim(M^\vee)[\chi]$, consider the case $M = \mathbb{C}^3$ with the action of $\mathcal{H} = \mathcal{Z}[X, Y]$ introduced in the solution of Exercise 2.3.7.

Solution to Exercise 2.5.15 Setting $d = \dim_k M$, the algebra of matrices that are upper unipotent by blocks $(d/2, d/2)$ when d is even, or $((d-1)/2, (d+1)/2)$ when d is odd, is commutative and has the required dimension.

Solution to Exercise 2.6.3 If we have two writings $\Gamma s\Gamma = \coprod_{i=1}^{n} \Gamma s_i = \coprod_{j=1}^{m} \Gamma s'_j$ then obviously $n = m$ and there is a a permutation σ of $\{1, \dots, n\}$ such that $s_i = \gamma_i s'_{\sigma(i)}$ for $i = 1, \dots, n$, where γ_i is some element of Γ. Thus $\sum_{i=1}^{n} v_{|s_i} = \sum_{i=1}^{n} v_{|\gamma_i s'_\sigma(i)} = \sum_{i=1}^{n} v_{|s'_\sigma(i)} = \sum_{j=1}^{n} v_{|s'_j}$, which shows that the formula is independent of the choice of the s_i.

If $\gamma \in \Gamma$, since

$$\coprod_{i=1}^{n} \Gamma(s_i\gamma) = (\coprod_{i=1}^{n} \Gamma s_i)\gamma = \Gamma s\Gamma\gamma = \Gamma s\Gamma = \coprod_{i=1}^{n} \Gamma s_i,$$

the result of the preceding paragraph shows that $\sum_{i=1}^{n} v_{|s_i\gamma} = \sum_{i=1}^{n} v_{|s_i}$, which proves that $v_{|[\Gamma s\Gamma]}$ is invariant by Γ.

Solution to Exercise 2.6.4

$$\begin{aligned}
U_l f(z) &= \sum_{a=0}^{l-1} f_{|k\begin{pmatrix} 1 & a \\ 0 & l \end{pmatrix}}(z) \\
&= \sum_{a=0}^{l-1} l^{k-1} l^{-k} f((z+a)/l) \text{ (by (2.6.2))} \\
&= \sum_{a=0}^{l-1} l^{-1} f((z+a)/l) \\
&= \sum_{a=0}^{l-1} l^{-1} \sum_{n=0}^{\infty} a_n e^{2i\pi n(z+a)/l} \\
&= \sum_{n=0}^{\infty} a_n q^n l^{-1} \sum_{a=0}^{l-1} e^{2i\pi na/l} \\
&= \sum_{n=0}^{\infty} a_n q^n 1_{l\mathbb{N}}(n) \\
&= \sum_{n=0}^{\infty} a_{ln}(f) q^n
\end{aligned}$$

Solution to Exercise 2.6.15

1. By Theorem 2.6.14, a basis of $\mathcal{E}_k(\Gamma_1(N) \cap \Gamma_0(p))$ consists of the forms $E_{k,\chi,\psi,t'}(z)$ with $LRt' \mid Np$ and $\chi\psi$ is trivial on $(\mathbb{Z}/p\mathbb{Z})^* \subset (\mathbb{Z}/pN\mathbb{Z})^*$ (and $LRt' \neq 1$ if $k = 2$). These conditions implies that L and R are both prime to p, because if one of them, say L were not, χ would be non-trivial on $(\mathbb{Z}/p\mathbb{Z})^*$, so ψ would also be non-trivial on $(\mathbb{Z}/p\mathbb{Z})^*$, and R would also be divisible by p, contradicting $LR \mid Np$ since N is not divisible by p. Thus the family of form $E_{k,\chi,\psi,t}$ with $LR \mid N$ and $LRt \mid Np$ form a basis of $\mathcal{E}_k(\Gamma_1(N) \cap \Gamma_0(p))$. All these forms are p-old since $LR|(Np)/p$.
   Recall that the $E_{k,\chi,\psi,t}$ where $LRt \mid N$ are a basis of $\mathcal{E}_k(\Gamma_1(N))$. The basis of $\mathcal{E}_k(\Gamma_1(N) \cap \Gamma_0(p))$ describes above is the family of the forms $E_{k,\chi,\psi,t}$ and $E_{k,\chi,\psi,pt}$ for $LRt \mid N$. Thus we see that the dimension of $\mathcal{E}_k(\Gamma_1(N) \cap \Gamma_0(p))$ is twice the dimension of $\mathcal{E}_k(\Gamma_1(N))$.
2. This follows immediately from question 1.
3. To see that these vectors are U_p-eigenvectors and that their U_p-eigenvalues are as given is an easy computation using (2.6.13). (A similar computation is done in Lemma 6.7.2 below).
4. is trivial from question 2, and question 3.
5. First observe that the form $E_{k,\chi,\psi,t,\mathrm{crit}} = E_{k,\chi,\psi,t}(z) - \chi(p)E_{k,\chi,\psi,tp}(z)$ vanishes at the cusp ∞. Indeed, note that we have by definition $E_{k,\chi,\psi,t}(\infty) = E_{k,\chi,\psi,tp}(\infty) = E_{k,\chi,\psi}(\infty)$. Now either $L \neq 1$, in which case the $E_{k,\chi,\psi}(\infty) = 0$ by definition, or else $L = 1$, in which case $\chi = 1$, hence $\chi(p) = 1$ and $E_{k,\chi,\psi,t}(\infty) - \chi(p)E_{k,\chi,\psi,tp}(\infty) = E_{k,\chi,\psi,t}(\infty) - E_{k,\chi,\psi,tp}(\infty) = 0$.
   Thus by linearity any form $g \in \mathcal{E}_k(\Gamma_1(N) \cap \Gamma_0(p))_{\mathrm{crit}}$ vanishes at ∞. Since U_p commutes with action of $\Gamma_1(N)$, $\mathcal{E}_k(\Gamma_1(N) \cap \Gamma_0(p))_{\mathrm{crit}}$ is stable by $\Gamma_1(N)$, and it follows that every $g \in \mathcal{E}_k(\Gamma_1(N) \cap \Gamma_0(p))_{\mathrm{crit}}$ vanishes at every cusp in the $\Gamma_1(N)$-equivalence class of ∞.
   The converse assertion follows by dimension counting.
6. In the case $k = 2$, the form $E_{2,1,1,1}$, which does not exist, should not be considered in the family of forms descibred in question 1. This family will still be a basis of $\mathcal{E}_2(\Gamma_1(N) \cap \Gamma_0(p))$, and the dimension of that space will be twice the dimension of $\mathcal{E}_2(\Gamma_1(N)$ plus one.

In question 2 we should keep the same definition of $E_{2,\chi,\psi,t,\mathrm{ord}}$ and $E_{2,\chi,\psi,t,\mathrm{crit}}$ when $LRT > 1$, but in in the case $L = R = t = 1$, we should define $E_{2,1,1,1,\mathrm{ord}}$ as $E_{2,p}$ and refrain from defining $E_{2,1,1,\mathrm{crit}}$. Then the statement of questions 2, 3, 4 and 5 would still hold.

Solution to Exercise 2.6.21 For 1., $M[\lambda]$ has a basis $f(z), f(lz), \dots, f(l^{m_l}z)$. Assume $l \nmid N_0$ first. For $m = 1, 2, 3$ respectively, the matrix of U_l in this basis is

$$\begin{pmatrix} a_l & 1 \\ p^{l-1}\epsilon_l & 0 \end{pmatrix}, \quad \begin{pmatrix} a_l & 1\,0 \\ p^{l-1}\epsilon_l & 0\,1 \\ 0 & 0\,0 \end{pmatrix}, \quad \begin{pmatrix} a_l & 1\,0\,0 \\ p^{l-1}\epsilon_l & 0\,1\,0 \\ 0 & 0\,0\,1 \\ 0 & 0\,0\,0 \end{pmatrix}$$

For $m > 3$, the matrix is the same as in the case $m = 3$, extended with entries just above the diagonal equal to 1 and 0 elsewhere.

If $m \geq 3$, this matrix is not semi-simple since 0 is a root of multiplicity $m - 1$ of its characteristic polynomial, but the kernel of this matrix has dimension 1. If $m = 1$ (resp. $m = 2$), the characteristic polynomial is $(X^2 - a_l X + l^{k-1}\epsilon_l)$ (resp. $X(X^2 - a_l X + l^{k-1}\epsilon_l)$). If $X^2 - a_l X + l^{k-1}$ has two distinct roots (necessarily non-zero), then the characteristic polynomial has simple roots and U_l is diagonalizable. If this polynomial has a double root, then we see for $m = 1$ that U_l is not diagonalizable as it is not scalar. For $m = 2$, U_l is not diagonalizable either, because by looking at the restriction of U_l on the stable space generated by $f(z), f(lz)$ we are reduced to the case $m = 1$.

If $l \mid N_0$, then the matrix of U_l is for $m = 3$ (for instance)

$$\begin{pmatrix} u_l & 1 & 0 & 0 \\ 0 & 0 & 1 & 0 \\ 0 & 0 & 0 & 1 \\ 0 & 0 & 0 & 0 \end{pmatrix}.$$

We leave the analysis of this case to the reader.

For 2., note that $\mathcal{H}$ acts semi-simply on $M[\lambda]$ if and only if each of the U_l, for $l \mid N/N_0$ acts semi-simply (since those operators commute and the others act by scalars). So fix an $l \mid N/N_0$. For d a divisor of N/N_0 not divisible by l, call $M_d \subset M[\lambda]$ the subspace generated by $f(dz), f(dlz), \dots, f(dl^{m_l}z)$. Then M_d is stable by U_l and the matrix of U_l in the above basis is the same as written above (in particular is independent on d). Since $M[\lambda]$ is the sum of the M_d for d as above, U_l acts semi-simply on $M[\lambda]$ under the same conditions as in 1, and the result follows.

Solution to Exercise 2.7.7 Point 1. is a direct application of the commutative diagrams of sets we have written.

Let K' be a finite Galois extension such that $G_{K'}$ acts trivially on the set of points of non-closed points of Spec $\mathcal{T}$ (this is possible sine $\mathcal{T} \otimes K$ is étale over K), and define R', $\mathfrak{m}'$, k' as usual. Recall that $\mathcal{T}_{R'} = \mathcal{T} \otimes_R R'$ is étale over R' if and only if $\mathcal{T}$ is étale over R, since R'/R is faithfully flat. Observe that every irreducible component of Spec $\mathcal{T}_{R'}$ has generic degree 1 over R' since its generic point is defined over K'. Thus every irreducible component of Spec $\mathcal{T}_{R'}$ is isomorphic (through the structural map) to Spec R'. We thus see that Spec $\mathcal{T}_{R'}$ is non-étale if and only if it has more irreducible component that connected component, and 2. follows from 1. applied to $\mathcal{T}_{R'}$.

Solution to Exercise 2.7.8 $\mathcal{T} = \mathbb{Z}_p[X]/(X^2 - p^{a+b})$ where the X corresponds to $\psi(T)$. Let $n = a + b$. Then $\operatorname{Spec} \mathcal{T}$ is connected iff $n \geq 1$, irreducible iff n is odd, regular iff $n \leq 1$, and étale over R iff $n = 0$. The module M is free over $\mathcal{T}$ iff $ab = 0$.

Hint to Exercise 2.7.11 Take $R = k[[X]]$ (k any field) and $\mathcal{T} = \{(P, Q) \in R^2,\ P(0) = Q(0)\}$. Note that $\mathcal{T}$ is a finite torsion-free R-algebra over R which is a PID, so is really an eigenalgebra. Observe that $\mathcal{T}_K \simeq K^2$, where $K = \operatorname{Frac}(R) = K((X))$. Consider the map $\chi : \mathcal{T} \mapsto k[[X]]/(X^2)$, $(P, Q) \mapsto P(0) + (P'(0) + Q'(0))X$. Check that this is a morphism of algebra which is not liftable.

Solution to Exercise 2.7.12 $C = \mathbb{Z}/2\mathbb{Z}$.

Solution to Exercise 2.7.13 The inclusion between finite R-modules $(M \cap A) \oplus (M \cap B) \hookrightarrow M$ becomes after tensorizing by K the isomorphism $A \oplus B \xrightarrow{\simeq} M$. This implies that the cokernel C of this inclusion is torsion. It is finite as a quotient of M. This proves question 1.

The map p_A restricted to $M \cap A$ has kernel $M \cap A \cap B = 0$, so this map identifies $M \cap A$ with a submodule of M_A. Now consider the composition $M \xrightarrow{p_A} M_A \to M_A/p_A(M \cap A) = M_A/(M \cap A)$, which is surjective as a composition of surjections. Its kernel is the set of $m \in M$ such that $p_A(m) \in p_A(M \cap A)$ that is such that m differs from an element of $M \cap A$ by an element of $\ker p_A = M \cap B$. In other words, the kernel of this map is $(M \cap A) \oplus (M \cap B)$, and this map realizes an isomorphism $M/((M \cap A) \oplus (M \cap B)) \to M_A/(M \cap A)$, which proves question 2.

The kernel of the map (p_A, p_B) is $(M \cap B) \cap (M \cap A) = 0$ so this map may be used to identify M to a submodule of $M_A \oplus M_B$. The composition $M_A \hookrightarrow M_A \oplus M_B \to (M_A \oplus M_B)/M$ has for kernel the set of $m \in M_A$ such that there exists $m' \in M$ satisfying $p_A(m') = m$, $p_B(m') = 0$. The latter condition on m' is equivalent to $m' \in M \cap A$, so the condition on m is equivalent to $m \in p_A(M \cap A)$ and our map realizes an injection $C = M_A/(M \cap A) \to (M_A \oplus M_B)/M$. This map is easily seen to be surjective, which proves question 3.

Solution to Exercise 2.7.17 Let $A = Kf$, B the orthogonal of A for the bilinear product. The non-vanishing of $\langle f, f \rangle$ ensures that $B \cap A = 0$, and since $\dim B + \dim A = \dim M_K$ by the non-degeneracy of the bilinear product, $M_K = A \oplus B$. So we can apply Proposition 2.7.16 which gives the result.

Hint to Exercise 2.7.20 Let $R = k[[X]]$, k a field. Let $\mathcal{T} = \{(p, q, r) \in R^3,\ p(0) = q(0) = r(0),\ p'(0) = q'(0) + r'(0)\} \subset R^3$. Check that $\mathcal{T}$ is a torsion-free subalgebra of R^3. Let $\mathcal{T}_A = \{p \in R\} = R$ and $\mathcal{T}_B = \{(q, r) \in R^2,\ q(0) = r(0)\} \subset R^2$, which are also R-algebras. The obvious maps $\mathcal{T} \to \mathcal{T}_A$ and $\mathcal{T} \to \mathcal{T}_B$ are clearly surjective, while their product $\mathcal{T} \to \mathcal{T}_A \times \mathcal{T}_B$ is injective. Check that there exists a module M of rank 3 over R, with a action of an operator T, and a T-stable decomposition $M_K = A \oplus B$ such that $\mathcal{T}, \mathcal{T}_A, \mathcal{T}_B$ are the eigenalgebras of M, $M \cap A$, $M \cap B$ respectively, and the natural maps $\mathcal{T} \to \mathcal{T}_A$ and $\mathcal{T} \to \mathcal{T}_B$ are the ones we define. Also check that $\mathcal{T}_K = K^3$, so every point of Spec $\mathcal{T}_K$ is defined over K.

Consider the character $\chi : \mathcal{T} \to R/\mathfrak{m}^2$ sending (p, q, r) to $p \mod \mathfrak{m}^2$. This characters obviously factors through $\mathcal{T}_A$. Show that this characters also factors through $\mathcal{T}_B$. So it is an *eigencongruence* between A and B modulo $\mathfrak{m}^2$. However, show that there is no congruence modulo $\mathfrak{m}^2$ between system of eigenvalues appearing in A and B.

Hint to Exercise 2.7.21 Apply the variant of the Deligne-Serre lemma.

Solution to Exercise 2.7.22 By Prop 2.7.15 there is $f \in A$, $g \in B$ such that $f \equiv g \pmod{\mathfrak{m}}$ but $f \not\equiv 0 \pmod{\mathfrak{m}}$. Since $\dim A = 1$, f is an eigenvector for $\mathcal{T}$, defining a character $\phi_f : \mathcal{T} \to R$, which obviously factors through $\mathcal{T}_A$. Since $f \equiv g$, we have $Tg \equiv \phi_f(T)g \pmod{m}$ for $T \in \mathcal{T}$, which shows that the character $\phi_f \pmod{m} : \mathcal{T} \to R/\mathfrak{m}$ factors through $\mathcal{T}_B$. Lifting this character into a character $\phi' : \mathcal{T}_B \to R$ (using the variant of Deligne-Serre's lemma) gives the congruence $\phi' \equiv \phi_f \pmod{\mathfrak{m}}$ we were looking for.

A counter-example to the converse is provided by Example 2.7.19.

Hint to Exercise 2.7.27 Prove that $\mathcal{O}_M/\mathcal{T}$ is killed by π^v.

Hint to Exercise 2.8.2 Consider the case $w = 12, k = 1, R = \mathbb{Z}_2$.

Solution to Exercise 2.8.2 $M_{12}(\mathrm{SL}_2(\mathbb{Z}), \mathbb{Z})$ has basis (E_{12}, Δ). The reduction modulo 2 of the q-expansions of those forms are $\tilde{E}_{12} = 1$ and $\tilde{\Delta} = \sum_{n \text{ odd}} q^{n^2}$. In particular, all Hecke operators T_p sends both forms to 0, on the Hecke algebra $\mathcal{T}_{\mathbb{F}_2}$ in this case is just $\mathbb{F}_2$. However, the rank of $\mathcal{T}$, that is the dimension of $\mathcal{T}_{\mathbb{Q}}$ is 2 because the eigenform E_{12} and Δ have distinct eigenvalues (or because of Proposition 2.6.16). This shows that the map $\mathcal{T} \otimes_R k \to \mathcal{T}_k$ is not an isomorphism.

Hint to Exercise 2.8.3 By looking on the formula defining Hecke operator on q-expansion, show that for $f \in M_k(\mathrm{SL}_2(\mathbb{Z}), \mathbb{F}_p)$, one has $T_p f^p = f$ and for m an integer not divisible by p, $T_m f^p = (T_m f)^p$.

Solution to Exercise 2.8.3 We know from the proof of the proposition above that the rank of $\mathcal{T}$ and the dimension of $\mathcal{T}_k$ are equal, and equal to $n = \dim S_w(\mathrm{SL}_2(\mathbb{Z}), \mathbb{Q})$, which is the number of system of eigenvalues appearing in $S_w(\mathrm{SL}_2(\mathbb{Z}), \bar{\mathbb{Q}})$. The rank of $\mathcal{T}^p$, that is the dimension of $\mathcal{T}^p$ is also the number of systems of eigenvalues for all Hecke operators excepted T_p, in $S_w(\mathrm{SL}_2(\mathbb{Z}), \bar{\mathbb{Q}})$, but by the strong multiplicity one theorem, that is the same as the number n of systems of eigenvalues for all Hecke operators. Hence $\dim \mathcal{T}^p \otimes_R k = n$. Therefore it suffices to prove that $\dim \mathcal{T}_k^p < n$, or equivalently, that the inclusion $\mathcal{T}_k^p \subset \mathcal{T}_k$ is strict.

Suppose the contrary. Then as an operator of $S_w(\mathrm{SL}_2(\mathbb{Z}), \mathbb{F}_p)$, T_p is a linear combination $\sum a_i T_{m_i}$ where the a_i are in $k = \mathbb{F}_p$, and the m_i are integers not divisible by p. Thus $\tilde{\Delta} = T_p \tilde{\Delta}^p = \sum a_i T_{m_i} \Delta^p = (\sum a_i T_{m_i} \tilde{\Delta})^p$ using the hint (which is an easy computation), and this is a contradiction since $\tilde{\Delta}$ is not a p-power since its q-expansions begins with q.

Solution to Exercise 2.8.8 This maps sends the trace of the universal representation at T_l to $T_l \in \mathcal{T}_{0,\mathfrak{m}_i}$.

Solution to Exercise 3.1.1 If x is multiplicative, we can compute for $z \in R$: $|z| = |xx^{-1}z| \leq |x||x^{-1}z| \leq |x||x^{-1}||z| = |z|$, and all these inequalities are thus equalities. In particular $|z| = |x||x^{-1}z|$. Setting $z = xy$, we get $|xy| = |x||y|$. The converse and the second assertions are clear.

Solution to Exercise 3.1.2 If two norms are equivalent, it is trivial that they define the same topology. Conversely, if $|\ |_1$ and $|\ |_2$ define the same topology, then $\{m \in M, |m|_1 < 1\}$ is open, hence contain a set of the form $\{m \in M, |m|_2 < \eta\}$ for some η. For any non-zero $m \in M$, there exists an integer n such that $1/p \leq |p^n m|_1 < 1$, so that $p^{-n}|m|_2 = |p^n m|_2 < \eta$, or $|m|_2 < \eta p^n \leq \eta p |m|_1$. By symmetry, the two norms are equivalent.

Hint to Exercise 3.1.10 Can you find first a (commutative) $\mathbb{C}^*$-algebra with a non-closed principal ideal? Such algebras are isomorphic to $\mathcal{C}(X, \mathbb{C})$, the algebra of continuous functions from a compact space X to $\mathbb{C}$ (Gelfand-Mazur). What are their closed ideals?

Solution to Exercise 3.1.10 Let A be the $\mathbb{Q}_p$-Banach algebra of all continuous functions $f : \mathbb{Z}_p \to \mathbb{Q}_p$, with norm $|f| = \sup_{x \in \mathbb{Z}_p} |f(x)|$.

Let us define two functions g and h on $\mathbb{Z}_p$ by $g(x) = |x|^{-1}$ and $h(x) = |x|^{-2}$ for $x \in \mathbb{Z}_p - \{0\}$, and $g(0) = h(0) = 0$ (here $|x|$ is a real number which is in fact rational hence is seen as an element of $\mathbb{Q}_p$).

Then it is easy to see that g and h are p-adically continuous at 0, and even locally constant everywhere else. Hence they belong to A.

The ideal Ah does not contain g, because if it did, $g \in Ah$, one would have $g = fh$ with $f \in A$. On $\mathbb{Z}_p - \{0\}$, one thus would have $f(x) = g(x)/h(x) = |x|$ so $f(x)$ has no p-adic limit at 0, a contradiction since f is continuous on $\mathbb{Z}_p$.

But the closure of Ah contains g. Indeed, for $n \in \mathbb{N}$, let f_n be defined by $f_n(x) = |x|$ if $|x| > p^{-n}$, and $f(x) = p^{-n}$ if $|x| \leq p^{-n}$. It is clear that $f_n \in A$. Moreover $f_n h$ agrees with g on $|x| > p^{-n}$, and if $|x| \leq p^{-n}$, $|f_n(x)h(x)| = |p^{-n}||x|^2 \leq p^n p^{-2n} = p^{-n}$, and $|g(x)| \leq p^{-n}$, so $|f_n g - g| \leq p^{-n}$. This makes clear that $f_n h$ converges uniformly to g in A.

Hence Ah is not closed, and property (3.1.8) is not satisfied for A.

Solution to Exercise 3.1.11 Let M be an orthonormalizable A-module, with orthonormal basis $(e_i)_{i\in I}$, and N a submodule generated by r vectors $f_j = \sum_{i\in I} a_{i,j}e_i$. As there are finitely many j's, and as for each j, $\lim_{i\in I} a_{i,j} = 0$, there is a term a_{i_0,j_0} which has the largest absolute value of all $a_{i,j}$'s. Up to changing the order of the f_j's and the e_i's, we may assume that $(i_0, j_0) = (1, 1)$, so that, after our hypothesis on A, all $a_{i,j}$ belong to $Aa_{1,1}$, say $a_{i,j} = b_{i,j}a_{1,1}$. Changing f_j into $f_j - b_{1,j}f_1$ for $j = 2, \dots, r$ ensures that the $a_{1,j}$ are all 0 for $j > 1$. Then changing every e_i for $i \in I - \{1\}$ into $e_i - b_{i,1}e_1$ we get a new orthonormal basis of M in which the $a_{i,1}$ are 0 for $i \neq 1$. In other words, in our new basis, $f_1 = a_{1,1}e_1$.

We can repeat the process with the submodule N' of N generated by the vectors $f_2, \dots, f_r$ in the orthonormalizable sub-module M' of M generated by the vectors e_i for $i \in I - \{1\}$, ensuring that, after changing the generating family (f_i) of N' and the orthonormal basis (e_i) of M', $f_2 = a_{2,2}e_2$. Repeating again we can assume that the module N is generated by $a_{i,i}e_i$. (We have just proved a version of the elementary divisors theorem in orthonormalizable modules using the usual pivot algorithm).

But it is clear that such a module is closed: it is the module of elements $\sum_i x_ie_i$ satisfying the closed conditions $|x_i| \leq |a_{i,i}|$ for $i \in \{1, \dots, r\}$, $x_i = 0$ for $i \in I - \{1, \dots, r\}$.

Hint to Exercise 3.1.23 First do the case where P is orthonormalizable and use the open mapping theorem.

Solution to Exercise 3.1.23 First assume that P is orthonormalizable with basis $(e_i)_{i\in I}$. One has $|\alpha(e_i)| \leq |\alpha|$ for every i, hence by the Open Mapping theorem one can choose bounded elements m_i of M such that $f(m_i) = \alpha(e_i)$. There is therefore a continuous morphism $\beta : P \to M$ sending e_i to m_i, and one has $f\beta(e_i) = \alpha(e_i)$ for every i, hence $f\beta = \alpha$.

When P is potentially orthonormalizable, just change the norm.

In the general case, assume $P \oplus Q$ is potentially orthonormalizable. Then for every surjective continuous $f : M \to N$, and $\alpha : P \to N$, the map $\tilde{\alpha} : P \oplus Q \to N$ equal to α on P and to zero on Q defines by the preceding case a map $\tilde{\beta} : P \oplus Q \to N$ such that $f\tilde{\beta} = \tilde{\alpha}$, and upon defining $\beta = \tilde{\beta}_{|P}$ one has $f\beta = \alpha$.

Solution to Exercise 3.2.1 Assume $\nu' > \nu$. There exists a sequence (u_n) of integers such that $u_n - n\nu \to +\infty$ but $u_n - n\nu' \to -\infty$. Then the sequence $F_n := p^{u_n}T^n$ converges to 0 for $v(\cdot, \nu)$ but not for $v(\cdot, \nu')$.

Solution to Exercise 3.2.2 Write $F_n = \sum_{m=0}^{\infty} a_{m,n}q^m$. Saying that $(F_n)_{n\in\mathbb{N}}$ is Cauchy for $v(\cdot, \nu)$ amounts to saying that for every A, there exists $N > 0$ such that for integers $n, n' \geq N$ and every m, $v_p(a_{m,n} - a_{m,n'}) - m\nu > A$. This implies in particular that for m fixed, the sequence $(a_{m,n})_{n\in\mathbb{N}}$ is Cauchy. Let us call a_m its limit in R, which is independent of ν, and set $F = \sum_{m=0}^{\infty} a_mT^n \in R[[T]]$. Next, we show that for all ν, $v(F_n - F, \nu) \to \infty$. It clearly suffices to show that for all ν, $v(F_n - F, \nu)$ is bounded below (by a bound depending on ν). But this follows from F_n being Cauchy for $v(-, \nu)$.

Solution to Exercise 3.2.6 Let $F = \sum_n a_n T^n$, $G = \sum_n b_n T^n$, $FG = \sum_n c_n T^n$, $N = N(F, \nu)$, $M = N(G, \nu)$. One has

$$c_n = \sum_{i+j=n} a_i b_j$$

hence $v_p(c_n) - n\nu \geq \max_{i+j=n} v_p(a_i) - i\nu + v_p(b_j) - j\nu$. By assumption, one has $v_p(a_i) - i\nu \geq v_p(a_N) - N\nu$, $v_p(b_j) - j\nu \geq v_p(b_M) - M\nu$ and if $i > M$ or $j > N$, the first or the second if these inequalities is strict. We deduce that $v_p(c_{N+M}) = v_p(a_N) + v_p(b_M)$, and that, for any n, $v_p(c_n) \geq v_p(a_N b_M) - (N + M)\mu = v_p(c_{N+M}) - (N + M)\nu$ with a strict inequality when $n \geq N + M$.

Solution to Exercise 3.2.12 Write $F(T) = 1 + \sum_{n=1}^{\infty} a_n T^n$. Then $a(\sum_{n=1}^{\infty} a_n a^{n-1}) = -1$.

Solution to Exercise 3.3.3 Since a_d is invertible, by the maximum modulus principle applied to a_d^{-1} we have $s := \sup_{x \in \mathrm{Sp}\, R} v_p(a_d(x)) < \infty$. Let $\nu_0 = \max(0, s - v_p(a_{d-1}), \ldots, s - v_p(a_0))$. Then if $\nu \geq \nu_0$, $x \in \mathrm{Sp}\, R$ and $0 \leq i \leq d-1$, one has $v_p(a_d(x)) - d\nu \leq s - d\nu = s - (d - i)\nu - i\nu \leq s - (d - i)\nu - \nu_0 \leq s - (d-i)\nu - (s - v_p(a_{d-i})) = v_p(a_{d-i}) - (d-i)\nu$ which shows that Q is strongly ν-dominant.

Solution to Exercise 3.3.4 Write $N = N(Q, \nu)$ for simplicity. Since a_N is multiplicative, $v_p(a_N) = v_p(a_N(x))$ for all x. For all $n \leq N$, one therefore has $v_p(a_n(x)) - n\nu \geq v_p(a_n) - n\nu \geq v_p(a_N) - N\nu = v_p(a_N(x)) - N\nu$, which shows, since Q_x is of degree N, that $N(Q_x, \nu) = N$.

For the converse, any polynomial $Q = a_0$ of degree 0 is strongly ν-dominant provided that $a_0(x) \neq 0$ for all x, that is provided a_0 is invertible in R. If $v_p(a_0)$ is not constant, however, a_0 is not multiplicative.

Solution to Exercise 3.3.5 Question 1 is trivial. For question 2, consider $x \in \mathrm{Sp}\, R$. Since Q is ν-dominant, its dominant term is invertible so Q_x has also degree N, and since Q is strongly ν-dominant, Q_x is ν-dominant. So $(aQ)_x = a(x)Q_x$ is ν-dominant because every element is multiplicative in $L(x)$ and question 1 applies. It follows that aQ is strongly ν-dominant.

For question 3 choose e.g. $R = \mathbb{Q}_p\langle X\rangle$ and two elements a, b in R^* such that $v_p(a) = v_p(b) = 0$ but $v_p(ab) = 2$. Then $bT + p$ is 0-dominant but not $a(bT + p)$.

Solution to Exercise 3.3.11 This is clear by criterion (i) of Prop. 3.3.9, since Z_F amd Z_G are closed subvarieties of Z_{FG}.

Hint to Exercise 3.6.4 1. is already clear at the level of local pieces. 2. is easy. For 3., observe that a ν which is adapted to M'_W is also adapted for M_W, using that the characteristic power series of U_p on M_W divides the one of M'_W. This reduces 3. to proving the existence of the closed immersion for the local pieces $\mathcal{E}_{W,\nu} \hookrightarrow \mathcal{E}'_{W,\nu}$, which is easy. For 4., apply 3. twice and observe that the eigenvariety for M_W^2 is the same as for M_W.

Solution to Exercise 3.7.2 A system of eigenvalues appearing in M''_W appears either in M_W or in M'_W. Hence the exercise follows from Theorem 3.7.1.

Solution to Exercise 3.8.2 In an irreducible one-dimensional rigid analytic space, any set with an accumulation point is Zariski-dense. If $x \in a + b\mathbb{N}$, and V is an affinoid closed ball of radius $r > 0$ around x, then for n big enough $x + p^n b\mathbb{N} \in V \cap X$ and this set has x as accumulation point.

Solution to Exercise 3.8.3 The set $p^{\mathbb{N}}$ of powers of p is Zariski-dense as it has 0 as accumulation point. But its intersection with a small ball of center $p^0 = 1$ is reduced to $\{1\}$, hence is not Zariski-dense.

Solution to Exercise 3.8.4 Take $X = \cup_{n\in\mathbb{Z}}\{(x, y),\ n < |x| < n+1,\ y = nx\}$.

Solution to Exercise 4.1.1

1.(a) Let A, B, C be such a triangle. Let D be the point such that $(ABDC)$ is a parallelogram. Then there is no points of $\mathbb{Z}^2$ in that parallelogram, except A, B, C and D since any point in $(ABDC)$ either lies on the triangle (ABC) or on the triangle (BCD) in which case the symmetric of P with respect to the middle of (BC) would be an integral point in (ABC).

We can assume that A is the origin $(0, 0)$, by translation. Let Λ be the lattice in $\mathbb{R}^2$ generated by the vectors $\vec{AB}$, $\vec{AC}$. Clearly $\Lambda \subset \mathbb{Z}^2$. We claim that this inclusion is an equality. Indeed, if $p \in \mathbb{Z}^2$, there is $v \in \Lambda$ such that $p - v$ lies in the parallelogram $(ABCD)$. Since $p - v$ is in $\mathbb{Z}^2$, it is by the above either A, B, C or D. In any case $p - v \in \Lambda$ so $p \in \Lambda$.

It follows that the vectors $\vec{AB}$, $\vec{AC}$ generates $\mathbb{Z}^2$. Hence their determinant is ± 1, and the area of the triangle (ABC) is $1/2$,

1.(b) Let $A = (0, 0)$, $B = (a, c)$, $C = (b, d)$. Assume that the angle $\vec{AB}$, $\vec{AC}$ is less than 180° by exchanging B and C if necessary. If the triangle (A, B, C) has area $1/2$, the matrix $\gamma = \begin{pmatrix} a & b \\ c & d \end{pmatrix}$ has determinant ± 1, and actually 1 by our angle hypothesis. Thus $\gamma \in \mathrm{SL}_2(\mathbb{Z})$ sends $\{\infty\} - \{0\}$ to $\{a/c\} - \{b/d\}$

1.(c) Let $A = (0, 0)$, $B = (a, c)$, $C = (b, d)$. Assume that the angle $\vec{AB}$, $\vec{AC}$ is less than 180° by exchanging B and C if necessary. We prove by induction on the number of points of $\mathbb{Z}^2$ in the triangle (ABC) that $\{a/c\} - \{b/d\}$ lies in the $\mathbb{Z}[\mathrm{SL}_2(\mathbb{Z})]$-module generated by $\{\infty\} - \{0\}$. Since those divisors clearly generate Δ_0 as $\mathbb{Z}$-module, this would be sufficient to prove Manin's lemma.
To start the induction, note that when (ABC) has only three points in $\mathbb{Z}^2$, we know that it has area $1/2$ by 1.(a). So by 1.(b) we are done.
In general, if (ABC) has more than three points in $\mathbb{Z}^2$, let $D = (e, f)$ be a fourth point. By replacing D by an other point of $\mathbb{Z}^2$ in the segment $[AD]$ (which lies inside the triangle (ABC)), we can assume that e and f are

relatively prime integers. Note that D is not on $[AB]$ (resp. nor on $[AC]$) since we have assumed that a and c (resp. b and d) are relatively prime. Therefore the triangles (ABD) and (ADC) are non-flat, and have strictly less integral points that (ABC). By induction hypothesis, we know that $\{a/c\} - \{e/f\}$ and $\{e/f\} - \{b/d\}$ both belong to the $\mathbb{Z}[\mathrm{SL}_2(\mathbb{Z})]$-module generated by $\{\infty\} - \{0\}$. Therefore their sum also does, which proves the induction step.

2. is an easy computation: $(1+\sigma)(\{\infty\} - \{0\}) = \{\infty\} - \{0\} + \{0\} - \{\infty\} = 0$, and

$$(1+\tau+\tau^2)(\{\infty\} - \{0\}) = \{\infty\} - \{0\} + \{0\} - \{1\} + \{1\} - \{\infty\} = 0.$$

Solution to Exercise 4.1.2 If $\Gamma = \mathrm{SL}_2(\mathbb{Z})$ then Exercise 4.1.1 tells us that there is an exact sequence

$$\mathbb{Z}[\mathrm{SL}_2(\mathbb{Z})]^2 \overset{\mu}{\to} \mathbb{Z}[\mathrm{SL}_2(\mathbb{Z})] \overset{M}{\to} \Delta_0 \to 0$$

where μ sends $(1, 0)$ to $1 + \tau + \tau^2$ and $(0, 1)$ to $1 + \sigma$. This is a finite presentation of Δ_0 as left $\mathbb{Z}[\mathrm{SL}_2(\mathbb{Z})]$-module.

If Γ has index n in $\mathrm{SL}_2(\mathbb{Z})$-module, then $\mathbb{Z}[\mathrm{SL}_2(\mathbb{Z})]$ is free of rank n as $\mathbb{Z}[\Gamma]$-modules, hence a finite presentation

$$\mathbb{Z}[\Gamma]^{2n} \overset{\mu}{\to} \mathbb{Z}[\Gamma]^n \overset{M}{\to} \Delta_0$$

as left $\mathbb{Z}[\Gamma]$-modules.

Solution to Exercise 4.1.3 By induction on the number of integral points in a polygon, any convex polygon can be covered by triangles whose vertices are in $\mathbb{Z}^2$ and that contain no other point of $\mathbb{Z}^2$. Since we know by (a) that Pick's theorem is true for any of those triangles, it is sufficient to check that Pick's theorem is additive, that is holds for any polygon with vertices in $\mathbb{Z}^2$ that is the union of two smaller polygon with vertices in $\mathbb{Z}^2$, disjoint interior, and an edge in common, provided Pick's theorem is true for those two smaller polygons. This is an easy computation. We leave the details to the reader.

For non-convex polygons, the answer depends on what you call a *polygon*. Read [82].

Solution to Exercise 4.1.7 Since $\mathrm{Symb}_\Gamma(V) = \mathrm{Hom}_{\mathbb{Z}[\Gamma]}(\Delta_0, V)$, this follows from the fact that Δ_0 is a finitely generated $\mathbb{Z}[\Gamma]$-module, which follows from the first question of Exercise 4.1.1.

Solution to Exercise 4.2.1 We have

$$\Gamma\begin{pmatrix}1&0\\0&l\end{pmatrix}\Gamma=\coprod_{a=0}^{l-1}\begin{pmatrix}1&a\\0&l\end{pmatrix}\coprod\begin{pmatrix}l&0\\0&1\end{pmatrix}$$

for $l \nmid N$ and the same without the last matrix if $l \mid N$. Since all those matrices have determinant l, the result follows from the definition of the Hecke operators.

Solution to Exercise 4.3.1 Replacing G by f^*G, we may assume $Y = X$, $f =$ Id, that is we need to prove that $G(X)$ is a sub-module of M if X is connected. If X is empty, $G(X) = 0$ and the result is true. Assume X is not empty and let $x \in X$. We claim that the natural map $G(X) \to G_x \simeq M$ is injective. If s, s' are two sections of G, both the sets where they are equal and different are open, because this is true for section of a constant sheaf, and this assertion is local. Thus if $s, s' \in G(X)$ have same image in G_x, they are equal on some open subset U of X containing x, and the locus where they are equal, being a non-empty open and closed subset of X, is X.

Hint to Exercise 4.3.2 Cover $[0, 1]$ by finitely many open sub-intervals $[a_1, b_1)$, $(a_2, b_2), \ldots, (a_n, b_n]$ with $0 = a_1 < b_1 < a_2 < b_2 < \cdots < a_n < b_n = 1$ such that G is constant on each interval. Correct the isomorphisms of G to a constant sheaf on each of these intervals so that they glue together.

Hint to Exercise 4.3.3 For 1., consider a path $\gamma : [0, 1] \to X$ with $\gamma(0) = x$, $\gamma(1) = y$ and set $G' = \gamma^*G$. By Exercise 4.3.2, G' is the constant sheaf, hence a natural isomorphism $G'_0 \simeq G'([0, 1]) \simeq G'_1$, and since $G'_0 = G_x$, $G'_1 = G_y$, we get the isomorphism i_γ.

Solution to Exercise 5.1.1

1. One has $Q_{|\gamma}(z, 1) = Q(az + b, cz + d) = (cz + d)^k Q(\frac{az+b}{cz+d}) = Q(z, 1)_{|\gamma}$, using that Q is homogeneous of degree k. Thus the map $Q(X, Y) \mapsto Q(z, 1)$ is S-equivariant, and it is obviously an isomorphism.
2. Clear from 1.

Solution to Exercise 5.1.2

1. The expression $\sum_{\sigma\in S_k}\prod_{i=1}^k \lambda_i(P_{\sigma(i)})$ is obviously symmetric in the P_i's so it defines an element of $(\mathrm{Sym}^k\mathcal{P}_1(R))^\vee$ and s_k is well-defined. It is clearly a morphism of S-modules.
2. Let $j_1, \ldots, j_k$ be a family of elements of $\{0, 1\}$. Then $s_k(l_0^{\odot m} \odot l_1^{\odot k-m})(z^{j_1} \odot \cdots \odot z^{j_k})$ is equal to the number of $\sigma \in S_k$ such that $j_{\sigma(1)} = \cdots = j_{\sigma(m)} = 0$ and $j_{\sigma(m+1)} = \cdots = j_{\sigma(k)} = 1$. The number of such σ is $\binom{k}{m}$ if the number of j_i equal to zero is k, and is 0 otherwise. Therefore $t_k^\vee \circ s_k(l_0^{\odot m} \odot l_1^{\odot k-m})$ is the linear form on $\mathcal{P}_k(R)$ that sends z^m to $\binom{k}{m}$ and z^i to 0 if $i \neq m$. In other words, it is the

element $\binom{k}{m} l_m$ of $\mathcal{V}_k(R)$. Since $(l_0^{\odot m} \odot l_1^{\odot k-m})_{m=0,\ldots,k}$ is a basis of $\mathrm{Sym}^k V_1(R)$ and $l_0, \ldots, l_k$ is a basis of $\mathcal{V}_k(R)$, we see that $t_k^\vee \circ s_k$ is an isomorphism if $k!$ is invertible in R.

Solution to Exercise 5.1.4

1. An element in $\mathcal{P}_k(R)^{\Gamma_\infty}$ is a polynomial $P(z)$ of degree at most k such that $P(z+1) = P(z)$. This implies that $P(z) - P(0)$ has at least p zeros, so $P(z) = P(0)$.
2. The surjectivity is standard. The action of Γ on $\mathcal{P}_k(\mathbb{F}_p)$ factors through its quotient $\mathrm{SL}_2(\mathbb{F}_p)$. By question 1, an element of $\mathcal{P}_k(\mathbb{F}_p)^\Gamma$ is a constant polynomial r , and its invariant by $\begin{pmatrix} 0 & -1 \\ 1 & 0 \end{pmatrix}$ means that $r = rz^k$, hence $r = 0$ if $k > 0$.
3. Left to the reader.

Solution to Exercise 5.3.3

$$\begin{aligned}(\phi_f, \phi_{i^{k-1}g}) &= (2i)^{k-1}[(f, i^{k-1}g)_\Gamma + (-1)^{k-1}(i^{k-1}g, f)_\Gamma] \\ &= (2i)^{k-1}(-i)^k[(f, g)_\Gamma + (g, f)_\Gamma] \\ &= 2^k i^{k-1}(-i)^{k-1}\mathrm{Re}(f, g)_\Gamma.\end{aligned}$$

Solution to Exercise 5.3.7 No, we have $\dim_{\mathbb{C}} W[\chi] = \dim_{\mathbb{C}} \bar{W}[\bar{\chi}]$.

Solution to Exercise 5.3.8 The map c is clearly $\mathbb{R}$-linear, so we only need to check that $c(iw) = ic(w)$. If $w = w_1 + iw_2$ with $w_1, w_2 \in W_{\mathbb{R}}$, then $iw = -w_2 + iw_1$, so $c(iw) = \overline{-w_2} + i\overline{w_1} = i(\overline{w_1} + i\overline{w_2}) = ic(w)$.

Hint to Exercise 5.3.28 $\Gamma_0(N)$ has 2^r cusps, which are parametrized by the subsets $a \subset \{l_1, \ldots, l_r\}$; if a is such a subset, the set c_a of rational numbers which are l-integral (resp. not l-integral) if $l \in a$ (resp. $l \notin a$) is a $\Gamma_0(N)$-class in $\mathbb{P}^1(\mathbb{Q})$, and all these classes have the form c_a for a unique $a \subset \{l_1, \ldots, l_r\}$.

Solution to Exercise 5.3.28 A basis of $\mathrm{Hom}(\mathbb{P}^1(\mathbb{Q}), \mathbb{C})^{\Gamma_0(N)}$ is given by the maps $\delta_a : \mathbb{P}^1(\mathbb{Q}) \to \mathbb{C}$, sending x to 1 or 0 according to whether $x \in c_a$ or not. Here a runs among the subset of $\{l_1, \ldots, l_r\}$.

For $l \nmid N$, $(T_l\delta_a)(x) = \delta_a(lx) + \sum_{i=0}^{l} \delta_a((x+i)/l)$. But if l_j is a prime dividing N then $(x+i)/l$ is l_j-integral if and if x is, and the same holds for lx. Thus $T_l\delta_a = (l+1)\delta_a$.

If $l \nmid N$, then we have $U_l\delta_a(x) = \sum_{i=0}^{l} \delta_a((x+i)/l)$. If x is l-integral, then $(x+i)/l$ is l-integral for exactly one i. If x is not l-integral, then neither is $(x+i)/l$. At other primes l', the l'-integrality of $(x+i)/l$ is the same as that of x.

Therefore,

$$U_l\delta_a = (l-1)\delta_{a\cup\{l\}} + l\delta_a \text{ if } l \notin a, \text{ and } U_l\delta_a = \delta_a \text{ if } l \in a.$$

Thus a basis of $\mathcal{H}$-eigenvectors in $\mathrm{Hom}(\mathbb{P}^1(\mathbb{Q}), \mathbb{C})^{\Gamma_0(N)}$ is given by the

$$f_a := \sum_{a\subset a'\subset\{l_1,\ldots,l_r\}} \delta_{a'}$$

for a running among subsets of $\{l_1, \ldots, l_r\}$. One has $T_l f_a = (1+l)f_a$ if $l \nmid N$, $U_l f_a = f_a$ if $l \in A$, and $U_l f_a = l f_a$ if $l \notin A$. One sees that each generalized $\mathcal{H}$-eigenspace has dimension 1.

Finally, remember that $\mathrm{BSymb}(\Gamma_0(N), \mathbb{C})$ is $\mathrm{Hom}(\mathbb{P}^1(\mathbb{Q}), \mathbb{C})^{\Gamma_0(N)}/\mathbb{C}$ where $\mathbb{C}$ is the subspace of constants, that is $\mathbb{C} f_\emptyset$ in our basis. Thus the f_a with a non-empty form a basis of eigenvectors in $\mathrm{BSymb}(\Gamma_0(N), \mathbb{C})$.

Hint to Exercise 5.3.30 Look up the description of Eisenstein series of weight 2 and square-free level N in Sect. 2.6.4. Show that with the notation thereof, a basis of $\mathcal{E}_2(\Gamma_0(N), \mathbb{C})$ is given by the $E_{2,1,1,t}$ for $t \nmid N$, $t \neq 1$. Show that $T_l E_{2,1,1,t} = (1+l)E_{2,1,1,t}$. Compute the action of the U_l's on this basis and conclude.

Solution to Exercise 5.3.32 The map lV_l is the right action of the matrix $\begin{pmatrix} l & 0 \\ 0 & 1 \end{pmatrix}$ on $\mathrm{BSymb}_{\Gamma_0(N)}(\mathbb{C})$. On an element δ_a of the basis given in the solution of Exercise 5.3.28, one has $lV_l(\delta_a)(x) = \delta_a(lx) = \delta_a(x)$ for $x \in \mathbb{P}^1(\mathbb{Q})$, hence lV_l is the natural inclusion for boundary modular symbols, and the image of μV_l is the subspace $\mathrm{BSymb}_{\Gamma_0(N)}(\mathbb{C})$ in $\mathrm{BSymb}_{\Gamma_0(Nl)}(\mathbb{C})$. On the other hand, for modular forms, the image by the map $f \mapsto V_l(f) = f(lz)$ of $M_2(\Gamma_0(N), \mathbb{C})$ has trivial intersection with $M_2(\Gamma_0(N), \mathbb{C})$ in $M_2(\Gamma_0(Nl), \mathbb{C})$ by the theory of old forms. Thus the diagram cannot be commutative.

Solution to Exercise 5.3.37

1. On may rewrite (5.3.20) as

$$\phi_{k,u,v}\left(\gamma \cdot \frac{u}{v}\right)(P(z)) = P_{|\gamma}\left(\frac{u}{v}\right)v^k.$$

We must first prove that the LHS is well-defined, that is that the right-hand side depends only on P and $\gamma \cdot \frac{u}{v}$, not on P and γ. In other word we need to prove that if γ is replaced by $\gamma\gamma_1$, with $\gamma_1 \in \Gamma_1(N)$ fixing u/v, the the RHS is left unchanged. That is, we need to check for all P that

$$P_{|\gamma}\left(\frac{u}{v}\right)v^k = P_{|\gamma\gamma_1}\left(\frac{u}{v}\right)v^k.$$

Replacing $P_{|\gamma_1}$ by P, we need to check

$$P\left(\frac{u}{v}\right)v^k = P_{|\gamma_1}\left(\frac{u}{v}\right)v^k$$

which amounts to $v^k = (c_1 u + v_1 d)^k$ if $\gamma_1 = \begin{pmatrix} a_1 & b_1 \\ c_1 & d_1 \end{pmatrix}$. But since γ_1 fixes u/v, then $\frac{a_1 u + b_1 v}{c_1 u + d_1 v} = \frac{u}{v}$, and since $a_1 u + b_1 v$ and $c_1 u + d_1 v$ are two relatively prime integers, one has $a_1 u + b_1 v = \pm u$ and $c_1 u + d_1 v = \pm v$, hence $(c_1 u + d_1 v)^k = v^k$. This shows that $\phi_{k,u,v}$ is well-defined.
Then we compute for $\gamma_1 \in \Gamma$,

$$\begin{aligned}(\phi_{k,u,v})_{|\gamma_1}\left(\gamma\frac{u}{v}\right)(P(z)) &= \phi_{k,u,v}\left(\gamma_1\gamma\frac{u}{v}\right)(P_{|\gamma_1^{-1}}(z)) \text{ (by definition)}\\ &= P_{|\gamma_1^{-1}\gamma_1\gamma}(z)v^k\\ &= P_{|\gamma}(z)v^k\\ &= \phi_{k,u,v}\left(\gamma\cdot\frac{u}{v}\right)(P(z))\end{aligned}$$

This shows that $\phi_{k,u,v} \in \mathrm{Hom}(\Delta, \mathcal{V}_k)^{\Gamma_1(N)}$.
2. is easy and 3. is a computation left to the reader.

Solution to Exercise 5.4.4 In the definition of $f_\chi(z)$, introducing the easy equality $\chi(n) = \frac{\sum_{a \,(\mathrm{mod}\, m)} \bar{\chi}(a) e^{2i\pi na/m}}{\tau(\bar{\chi})}$ gives the formula for $f_\chi(z)$. For the fact that $f_\chi(z)$ is a modular form, see [113, Prop. 3.64].

Solution to Exercise 5.4.6 if ψ is any continuous character $\mathbb{A}_{\mathbb{Q}}^*/\mathbb{Q}^* \to \mathbb{C}^*$, then $\psi_{|\mathbb{R}_+^*}$ as the form $x \mapsto x^s$ for some unique $s \in \mathbb{C}$. Then $\psi|\ |^{-s}$ is a continuous character on $\mathbb{A}_{\mathbb{Q}}^*/\mathbb{Q}^*\mathbb{R}_*^+$ and we conclude easily.

Solution to Exercise 5.4.12 When m is prime to $\mathfrak{p}$, the polynomials $(z - a/m)^j$ are $\mathfrak{p}$-integral, so the right hand side of (5.4.6) is clearly $\mathfrak{p}$-integral. Let us assume that $k = 0$. Then since $\phi_f^\pm$ are $\mathfrak{p}$-normalized, there is an irreducible fraction a/m such that $\phi_f(\{\infty\} - \{a/m\})/\Omega_f^\pm$ has $\mathfrak{p}$-valuation 0. Thus by linear independence of characters, there is a character χ of conductor m such that $\sum_{a \,\mathrm{mod}\, m)} \bar{\chi}(a) \frac{\phi_f^\pm(\{\infty\}-\{a/m\})}{\Omega_f^\pm}$ also has $\mathfrak{p}$-valuation zero.

Solution to Exercise 6.1.4 The image of the norm on this space contains $p^{1/2}$ which is not in the image of the norm on $c_0(\mathbb{Q}_p)$. Therefore these spaces can't be isometric.

Solution to Exercise 6.1.13 Write $x = 1 + y$ with $y \in R$, $|y| < 1$.

Write $x = 1 + y$ with $y \in R$, $|y| < 1$. For k a natural integer, let $\binom{z}{k}$ the polynomial $z(z-1)\dots(z-k+1)/k! = \sum_{n=1}^{k} a_{n,k} z^n$ in $R[z]$.

Since $k!a_{n,k}$ is an integer, $|a_{n,k}| \leq |k!|^{-1} \leq p^{k/(p-1)}$.

The coefficient of z^n is ± 1 times the reciprocal of the product of $n+1$ integers between 1 and k. Each of these integers has reciprocal of norm at most k hence $|a_{n,k}| < k^{n+1}$.

Now consider the formal powers series

$$g(y,z) = \sum_{k=0}^{\infty} \binom{z}{k} y^k = \sum_{\substack{k=0,\dots,\infty \\ n=1,\dots,\infty}} a_{n,k} y^n z^m$$

in $R[[y,z]]$. For m fixed and $y \in R$, $|y| < 1$, the real numbers $|a_{n,m} y^n| \leq n^{m+1}|y|^n$ goes to zero, and are bounded by $c/(1-|y|)^{m+1}$ with c some absolute constant. Hence the series $\sum_{n=0}^{\infty} a_{n,m} y^n$ converges to an element b_m and $|b_m z^m| < c|z|^m/(1-|y|)^{m+1}$. This shows that for $y \in R$, $|y-1| < 1$ the series $g(y,z) \in R[[z]]$ converges for $|z| < 1 - |y| = 1 - |1-x|$. On the other hand, for z a positive integer in that domain, one clearly has $g(y,z) = (1+y)^z = x^z$. Hence the result.

Solution to Exercise 6.1.14 Choose a finite set E as in the above proposition and $f \in \mathcal{A}[r](R)$. For every integer $N \geq 0$, there exists by Lagrange's theorem a polynomial P_N in $R(z)$ such that for every $e \in E$ and every n such that $0 \leq n \leq N$, $P_N^{(n)}(e)/n! = a_n(e)$. Thus $\|P_N - f\|_r = \sup_{n>N} \max_{e\in E} |a_n(e)|r^n$ which goes to zero as N goes to infinity.

Solution to Exercise 6.1.18 $B[\mathbb{Z}_p.r]$ is the disjoint union of open balls $B(e,r) \subset \mathbb{C}_p$ for e running in E (defined as above). On each of this ball the power series $\sum a_n(e)(z-e)^n$ defining f converges to a function of f_e by definition of $\mathcal{A}[r]$, and one has $\sup_{B(e,r)} |f_e(z)| = \sup |a_n(e)|r^n$. The collection of the f_e's defines a function on $B[\mathbb{Z}_p, r]$ that extends f, and whose r-norm is $\|f\|_r$, hence 1. and 2. The point 3. is clear.

Solution to Exercise 6.2.1 For question 1, if d is an ultrametric distance that is invariant by translation, then $|x| := d(x,0)$ is a norm: the first property is obvious, and we have $|x-y| = d(x-y,0) = d(x,y) \leq \max(d(x,0), d(0,y)) = \max(|x|,|y|)$. Conversely, if $|\ |$ is a norm, then $|-x| = |0-x| \leq |x|$ and by symmetry $|x| = |-x|$ for all x. If then we set $d(x,y) := |x-y|$, then clearly one has $d(x,y) = 0$ if and only if $x = y$; $d(x,y) = d(y,x)$; $d(x+z, y+z) = d(x,y)$; and $d(x,z) = |x-z| = |((x-y)-(z-y)| \leq \max(|x-y|, |z-y|) = \max(d(x,y), d(z,y)) = \max(d(x,y), d(y,z))$.

For question 2, assume that the topology of M is defined by the semi-norms $(p_n)_{n\in\mathbb{N}}$. Then define $|m| = \sum_{n\in\mathbb{N}} \frac{\max(p_n(m)),1}{2^n}$. This norm makes $(M,+)$ a normed group.

Hint to Exercise 6.2.2 See [105, Cor. 9.4].

Solution to Exercise 6.2.3 Questions 1 and 3 are clear from the definition. For question 2, recall that the canonical surjection $\pi : X \to X/\ker f$ is open; indeed, if U is open, we need to show that $\pi(U)$ is open which means that $\pi^{-1}(\pi(U)) = U + \ker f$ is open, which is clear. Hence $f = \bar{f} \circ \pi$ is open if and only if $\bar{f}$ is open. Since $\bar{f}$ is a continuous isomorphism, it is open if and only if it is an homeomorphism. (After [27, §1.1.9].)

For question 4, cf [27, §1.1.9].

For question 5, note that both X and Y are normable groups by the above exercise. If f is strict, $f(X) \simeq X/\ker f$ is complete, hence closed in X. Conversely, if $f(X)$ is closed, it inherits a structure of Fréchet, and the isomorphism $\bar{f} : X/\ker f \to f(X)$ is thus an homeomorphism by the open mapping theorem ([105, Prop. 8.6]).

For question 6, recall that f strict means that the continuous isomorphism $\bar{f} : X/\ker f \to f(X)$ is an homeomorphism. The map $\overline{f'} : X'/\ker f' \to f'(X')$ is just $(\bar{f}) \otimes 1$ by flatness of R' over R. Hence we are reduced to the case where f is an isomorphism. In this case, that f is an homeomorphism means that for every semi-norm q in our family of semi-norms on Y, one has two semi-norms p_1 and p_2 in our family on X and two positive constant C_1 and C_2 such that $C_1 p_1 \leq f^* q \leq C_2 p_2$. But then clearly $C_1 p_1' \leq f'^* q' \leq C_2 p_2'$ and it follows that f' is an homeomorphism.

For question 7, note that we obviously have $\ker f \subset \ker \hat{f}$, hence $\widehat{\ker f} \subset \ker \hat{f}$. Since f is strict, for every $\epsilon > 0$, there exists $\delta > 0$ such that if $|f(x)| < \delta$, $|x + k| < \epsilon$ for some $k \in \ker f$. Let $x \in \hat{X}$ such that $|f(x)| < \delta/2$. Let x_n be a sequence of elements of X converging to x. For n large enough, one has $|x_n - x| < \epsilon$ and $|\hat{f}(x_n) - \hat{f}(x)| < \delta/2$, so $|f(x_n)| = |\hat{f}(x_n)| < \delta$ and $|x_n + k_n| < \epsilon$ for some $k_n \in \ker f$. Thus $|x + k_n| < 2\epsilon$ which proves that $\hat{f}$ is strict. To prove that $\ker \hat{f} = \widehat{\ker f}$, let x in $\ker \hat{f}$. The reasoning above show that for every $\epsilon > 0$, there is a $k_n \in \ker f$ such that $|x + k_n| < 2\epsilon$. Thus $x \in \widehat{\ker f}$ and $\widehat{\ker f} = \ker \hat{f}$. The assertion for $\operatorname{Im} f$ is similar and left to the reader.

Solution to Exercise 6.2.5 For question 1, let (m_n) be a sequence in $c(M)$. For every open set U of $c(M)$ containing (m_n), there exists an $\epsilon > 0$ and a finite subset S of I such that U contains all sequences (m_n') such that $q_i((m_n - m_n')) < \epsilon$ for $i \in J$. Since (m_n) goes to 0, we can find $N \in \mathbb{N}$ such that for $n > N$, and for $i \in S$, $p_i(m_n) < \epsilon$. If (m_n') is the sequence (m_n) for $n < N$ and 0 afterwards, then we see that $(m_n') \in U$. Hence the result.

For question 2 the map $c(R) \times M \to c(M)$, $((r_n)_{n\in\mathbb{N}}, m) \mapsto (r_n m)_{n \in M}$ is bilinear, hence define a natural continuous map $c(R) \otimes M \to c(M)$, whose image clearly contains all eventually zero sequences. Such sequences are dense in $c(M)$ by question 1, hence the result.

Solution to Exercise 6.3.1 Such a morphism κ has image in the maximal compact subgroup $\mathcal{O}_{\mathbb{C}_p}^*$. If $\mathfrak{m}$ is the maximal ideal of $\mathcal{O}_{\mathbb{C}_p}$, then $U := \kappa^{-1}(1 + \mathfrak{m})$ is open in the compact group $\prod_l \mathbb{Z}_l^*$, hence of finite index. The group U can be written as $T \times \prod_{l \neq p} \mathbb{Z}_l$ with T torsion, but $\kappa(T) = 1$ since $1+\mathfrak{m}$ has no torsion, and $\kappa(\mathbb{Z}_l) = 1$, since if $x \in \mathbb{Z}_l$, $\kappa(l^n x) = \kappa(x)^{l^n} \to 1$ while since $\kappa(x) \in 1 + \mathfrak{m}$, $\kappa(x)^{p^n} \to 1$, and choosing for every n a Bezout relation $a_n l^n + b_n p^n = 1$ we deduce that $\kappa(x) = 1$. Hence $\kappa(U) = 1$.

Solution to Exercise 6.3.6 The character κ_p is constant on the ball $1 + qp^n\mathbb{Z}_p$ (and each of its multiplicative translation) but not on the ball $1 + qp^{n-1}\mathbb{Z}_p$. Thus $\kappa_p(1 + pz)$ extends to a function constant on the ball $B(0, r)$ in $\mathbb{C}_p$ for any $r < |q||p|^{n-2}$ (and its additive translations) but not for $r = |q||p|^{n-2}$. Hence one has $r_n = 1/p^{n-1}$ if $p > 2$, $r_n = 1/p^n$ if $p = 2$. Question 2 follows immediately.

Solution to Exercise 6.3.9

1. Another generator has the form γ^a with $a \in \mathbb{Z}_p^*$. But

$$\frac{\prod_{i=0}^{k-1} \log_p(\gamma^{-aj}\sigma(\gamma)^a)}{(\log_p \gamma^a)^k} = \frac{\prod_{i=0}^{k-1} a \log_p(\gamma^{-j}\sigma(\gamma))}{a^k(\log_p \gamma^a)^k}$$
$$= \frac{\prod_{i=0}^{k-1} \log_p(\gamma^{-j}\sigma(\gamma))}{(\log_p \gamma)^k}.$$

2. The function $\log_p$ vanishes at order 1 at roots of unity and nowhere else. The result follows easily.
3. Clear from the definition.
4. Left to the reader.

Solution to Exercise 6.3.10 By Scholium 6.3.3, we have, fixing a generator γ of $1 + q\mathbb{Z}_p$ an isomorphism

$$\mathcal{O}(\mathcal{W}) = \prod_{\kappa_f} \{f = \sum_{n=0}^{\infty} a_n T^n \in \mathbb{Q}_p[[T]], |a_n| r^n \to 0 \text{ for every positive real } r < 1\}$$

compatible with the isomorphism $\Lambda = \prod_{\kappa_f} \mathbb{Z}_p[[T]]$ described above. Thus if $|f(w)| \leq 1$ for all w in $\mathcal{W}(\mathbb{C}_p)$, we have by the maximum modulus principle that for every n and every $r \in p^{\mathbb{Q}}$, $r < 1$, $|a_n| r^n \leq 1$. Letting $r \to 1$, we get $|a_n| \to 1$, hence $a_n \in \mathbb{Z}_p$, and $f \in \mathbb{Z}_p[[T]]$.

Solution to Exercise 6.4.10 This is trivial.

Solution to Exercise 6.4.12 Let $a_n = p^m = n + 1$ if $n = p^m - 1$ for an integer m, $a_n = 0$ otherwise. Then $\lim a_n = 0$ so $f(z) = \sum a_n z^n \in \mathcal{A}[1](\mathbb{Q}_p)$. However, the antiderivative $g(z) = \sum b_n z^n$ of f has $b_{p^m} = 1$ for all m so $g \notin \mathcal{A}[1](\mathbb{Q}_p)$.

Solution to Exercise 6.4.15 Why should it?

Solution to Exercise 6.4.16 This map is a morphism of Fréchet spaces and has a closed image by Theorem 6.4.14. Therefore it is strict by question 5 of Exercise 6.2.3.

Solution to Exercise 6.5.7 Assume that $\Phi_{|U_p} = 0$. Then for every $D \in \Delta_0$, and every $f \in \mathcal{A}^\dagger(L)$ we have

$$
\begin{aligned}
0 = \Phi_{|U_p}(D)(f) &= \sum_{a=0}^{p-1} \Phi(\beta(a, p) \cdot D)_{|_\kappa \beta(a,p)}(f) \\
&= \sum_{a=0}^{p-1} \Phi(\beta(a, p) \cdot D)\,(f(pz - a))
\end{aligned}
$$

Fix $a \in \{0, \dots, p-1\}$. Let h be any function on $\mathbb{Z}_p$ in $\mathcal{A}^\dagger(L)$ with support in $a + p\mathbb{Z}_p$. For some $r > 0$, $h \in \mathcal{A}[r](L)$. Let f be the function on $\mathbb{Z}_p$ defined by $f(z) := h((z + a)/p)$, so that $h(z) = f(pz - a)$. Then $f \in \mathcal{A}[r/p](L) \subset \mathcal{A}^\dagger(L)$, and the equation above gives

$$0 = \Phi(\beta(a, p) \cdot D)(h).$$

Since every $D \in \Delta_0$ has the form $\beta(a, p)D'$ for some $D' \in \Delta_0$, we see that for every $D \in \Delta_0$, and every $h \in \mathcal{A}^\dagger(L)$ with support in $a + p\mathbb{Z}_p$,

$$\Phi(D)(h) = 0.$$

Now letting a varies and using linearity, we see that $\Phi(D)(h) = 0$ for every $h \in \mathcal{A}^\dagger(L)$, Hence $\Phi(D) = 0$.

Solution to Exercise 6.5.11 Consider the following example: $V = \mathbb{Q}_p^{(\mathbb{N})}$ and U_p is the operator sending $(u_0, u_1, u_2, u_3, u_4, \dots)$ to $(u_0, u_0, u_1, u_2, u_3, \dots)$. Let $W = \mathbb{Q}_p$ with U_p acting on it as the identity. Then the map $V \to W$, $(u_0, u_1, u_2, \dots) \mapsto u_0$ is U_p-equivariant and surjective, but $V^{\leq 0} \to W^{\leq 0}$ is not surjective.

Solution to Exercise 6.5.20 Note that $f\left(\gamma \cdot \frac{u}{v}\right)$ makes sense since f is a function on $\mathbb{Z}_p$ and $\gamma \cdot \frac{u}{v}$ is a p-integer. Checking that $\Phi_{k,u,v}$ is a modular symbol is exactly the same computation as in question a of Exercise 5.3.37, since $\Phi_{k,u,v}$ is defined by the same formula as $\phi_{k,u,v}$ with the polynomial $P \in \mathcal{P}_l(L)$ replaced by a function $f \in \mathcal{A}_k^\dagger(L)$, and the action of Γ on these spaces are defined by the same formulas. It is then clear that $\Theta_k(\Phi_{k,u,v}) = \phi_{k,u,v}$.

Hint to Exercise 6.6.1 See [43].

Hint to Exercise 6.6.4 See [42].

Hint to Exercise 6.6.6 This follows from the main results of [83, §3].

Hint to Exercise 6.6.8 See [98].

Hint to Exercise 6.6.9 The functions $\log_p^+(1+x)$ and $\log_p^-(1+x)$ defined in [98] are good examples.

Solution to Exercise 6.7.1 One has $f(pz) = p^{-1-k} f_{|\begin{pmatrix} p & 0 \\ 0 & 1 \end{pmatrix}}(z)$, and the results follows from the fact that $\begin{pmatrix} p & 0 \\ 0 & 1 \end{pmatrix} \Gamma \begin{pmatrix} p & 0 \\ 0 & 1 \end{pmatrix}^{-1} \subset \Gamma_1(N)$.

Solution to Exercise 6.7.5 Everything, except the last assertion follows easily from Atkin-Lehner's theory, see Sect. 2.6.4. For the last assertion, we note that the matrix of U_p in the basis $f, V_p f$ is $\begin{pmatrix} a_p & 1 \\ -\epsilon(p)p^{k+1} & 0 \end{pmatrix}$, with characteristic polynomial $(X-\alpha)(X-\beta)$. In the case $\alpha = \beta$, if this matrix was diagonalizable, it would be scalar, which is clearly not the case.

Solution to Exercise 6.7.8 To get from the first to the second line, we need $z^j \in \mathcal{P}_k$.

Solution to Exercise 7.1.6 There is no solution: I don't know the answer. I would say no.

Solution to Exercise 7.1.14 Let $m = \sum_{i=1}^{\infty} a_i e_i$ be an element of M, with $a_i \in R$, $\lim a_i = 0$. One compute $u(m) = \sum (a_1 p^i T^i + p^i a_{i+1}) e_i$. If $u(m) = 0$, then $a_{i+1} = -a_1 T^i$, but then if $a_1 \neq 0$, the sequence a_i doesn't converge to 0, contradiction. Hence $a_1 = 0$, $m = 0$ and u is injective. After tensorization by $R' = \{\sum c_n T^n, |c_n| r^n \to 0\}$ provided with the norm $|\sum c_n T^n| = \sup_m |c_n| r^n$, $u \otimes 1$ is still injective because R' is R-flat, but the kernel of $u \hat{\otimes} 1$ contains the element $e_1 - \sum_{i=2}^{\infty} T^{i-1} e_i$.

Hint to Exercise 7.4.11 Look up the appendix of [33].

Solution to Exercise 7.4.11 We have $\rho_x = 1 \oplus \omega_p$. Since $U_p(x) = p$, in a neighborhood of x all classical point y are such that ρ_y is irreducible (cf. the appendix of [33].) Hence by Ribet's lemma (cf. e.g. [10]) we can construct a non trivial extension, in the category of $G_{\mathbb{Q},lp}$-representations, of 1 by ω_p, which is crystalline at p by Lemma 7.4.10. If there was infinitely many classical points y in the given neighborhood of minimal tame level 1 (instead of l), then that extension would also be unramified at l: but there is non-trivial extension of 1 by ω_p in the category of $G_{\mathbb{Q}}$-representations that is unramified everywhere and crystalline at p, because $\mathbb{Q}$ has only finitely many units (cf. the appendix of [33]).

Solution to Exercise 7.6.15

1. By assumption $(\rho_f)_{|G_K}$ is the sum of two continuous characters χ and χ'. Those characters are distinct because there restriction to a decomposition group $G_{\mathbb{Q}_p} \subset G_K \subset G_{\mathbb{Q}}$ have distinct Hodge-Tate weights 0 and $k-1>1$.
2. Since ρ_g is absolutely irreducible, the pseudo-representation $(\tau_{\mathbb{T}}, \delta_{\mathbb{T}})$ is attached to a true continuous representation $\rho : G_{\mathbb{Q}} \to \mathrm{GL}_2(\mathbb{T})$. Let us prove that $\rho_{|G_K} \pmod{\mathfrak{m}}^2 : G_K \to \mathrm{GL}_2(\mathbb{T}/m^2)$is the sum of two continuous characters. Let us choose a basis of $\mathcal{T}^2$ where $\rho(g) \equiv \begin{pmatrix} \chi(g) & 0 \\ 0 & \chi^c(g) \end{pmatrix} \pmod{m}$. In that basis, we can write $\rho(g) = \begin{pmatrix} a(g) & b(g) \\ c(g) & d(g) \end{pmatrix}$ for a general $g \in G_K$, and we have $b(g) \equiv c(g) \equiv 0 \pmod{\mathfrak{m}}$. Multiplicativity implies that $a(gg') = a(g)a(g') + b(g)c(g')$ hence $a(gg') \equiv a(g)a(g') \pmod{\mathfrak{m}}^2$. Hence a is a continuous character $G_K \to (\mathcal{T}/\mathfrak{m}^2)^*$, and so is d, and the pseudo-representation $(\tau_{\mathbb{T}}, \delta_{\mathbb{T}}) \pmod{\mathfrak{m}^2}$ restricted to G_K is the sum of the two characters a and d.
3. It follows from question 2 that the reducibility ideal of the pseudo-representation $(\tau_{\mathbb{T}}, \delta_{\mathbb{T}})$ restricted to the decomposition group $G_{\mathbb{Q}_p}$ is contained in $\mathfrak{m}^2$. By [12, §5.4], this reducibility ideal contains the ideal of the schematic fiber of κ at x_g (actually, is equal, see [12, §7.1]). This implies that κ is not étale at x_g.

Solution to Exercise 8.1.4 This follows from Lemma 7.6.2.

Solution to Exercise 9.1.5 Since T is flat and complete intersection over R, one has $\mathfrak{d}_N(T/R) = \mathfrak{d}_K(T/R)$. By definition, $\mathfrak{d}_K(T/R) = \frac{\partial t^e - u}{\partial t} = et^{e-1}T = t^{e-1}T$.

Solution to Exercise 9.2.5 One has $|\log_p \kappa(\gamma)| < 1$ and $|\log_p(\gamma)| = 1/p$, so $|\log_p \kappa(\gamma)/\log_p(\gamma)| < p$. Hence $\left|\binom{\kappa}{i}\right| < p^i |i!|^{-1}$. Since $v_p(i!) < i/(p-1)$, $|i!|^{-1} < p^{i/(p-1)}$ and the result follows.

Bibliography

1. Y. Amice, J. Vélu, Distributions p-adiques associées aux séries de Hecke, in *Journées Arithmétiques de Bordeaux (Conference, University of Bordeaux, Bordeaux, 1974)*. Astérisque, No. 24–25 (Société mathématique de France, Paris, 1975)
2. F. Andreatta, A. Iovita, G. Stevens, Overconvergent modular sheaves and modular forms for GL_2/F. Israel J. Math. **201**(1), 299–359 (2014)
3. F. Andreatta, A. Iovita, V. Pilloni, Le halo spectral. Ann. Sci. Éc. Norm. Supér. **51**(3), 603–655 (2018)
4. A. Ash, G. Stevens, Modular forms in characteristic l and special values of their L-functions. Duke Math. J **53**(3), 849–868 (1986)
5. A. Ash, G. Stevens, p-adic deformations of arithmetic cohomology (Preprint, 2008)
6. A.O.L. Atkin, J. Lehner, Hecke operators on $\Gamma_0(m)$. Math. Ann. **185**, 134–160 (1970)
7. M. Auslander, D.A. Buchsbaum, On ramification theory in noetherian rings. Amer. J. Math. **81**, 749–765 (1959)
8. J. Bellaïche, Critical p-adic L-function. Invent. Math. **189**(1), 1–60 (2012)
9. J. Bellaïche, p-adic L-functions of critical CM forms. http://people.brandeis.edu/~jbellaic/
10. J. Bellaïche, Ribet's lemma, generalizations, and pseudo-characters, notes from Hawaii CMI summer school. http://people.brandeis.edu/~jbellaic/
11. J. Bellaïche, Introduction to the conjecture of Bloch-Kato, notes from Hawaii CMI summer school. http://people.brandeis.edu/~jbellaic/
12. J. Bellaïche, G. Chenevier, On the eigencurve at classical weight 1 points. J. Inst. Math. Jussieu **5**(2), 333–349 (2006)
13. J. Bellaïche, G. Chenevier, Families of Galois representations and Selmer groups. Soc. Math. France. Astérisque **324**, 1–314 (2009)
14. J. Bellaïche, S. Dasgupta, The p-adic L-functions of evil Eisenstein series. Compos. Math. **151**(6), 999–1040 (2015)
15. J. Bellaïche, M. Dimitrov, On the eigencurve at classical weight 1 points. Duke Math. J. **165**(2), 245–266 (2016)
16. J. Bellaïche, R. Pollack, Congruences with Eisenstein series and μ-invariants. Compos. Math. **155**(5), 863–901 (2019)
17. J. Bergdall, Ordinary modular forms and companion points on the eigencurve. J. Number Theory **134**, 226–239 (2014)
18. V.G. Berkovich, Étale cohomology for non-Archimedean analytic spaces. Publ. Math. I.H.E.S. **78**, 5–161 (1993)
19. A. Betina, Les variétés de Hecke-Hilbert aux points classiques de poids paralléle 1. J. Théor. Nombres Bordeaux **30**(2), 57–607 (2018)

J. Bellaïche, *The Eigenbook*, Pathways in Mathematics,
https://doi.org/10.1007/978-3-030-77263-5

20. A. Betina, M. Dimitrov, Geometry of the eigencurve at CM points and trivial zeros of Katz p-adic L-functions (2021). arxiv:1907.09422
21. A. Betina, C. Williams, Arithmetic of p-irregular modular forms: families and p-adic L-functions (2020). arXiv:2011.02331
22. A. Betina, M. Dimitrov, A. Pozzi, On the failure of Gorensteinness at weight 1 Eisenstein points of the eigencurve (2021). arxiv: 1804.00648
23. C. Breuil, M. Emerton, Représentations p-adiques ordinaires de $GL_2(\mathbb{Q}_p)$ et compatibilité local-global. Astérisque **331**, 255–315 (2010)
24. O. Brinon, C. Conrad, CMI Summer School Notes on p-adic Hodge Theory, Available at http://math.stanford.edu/~conrad/papers/notes.pdf
25. A. Borel, Some finiteness properties of adèle groups over number fields. IHES Publ. Math. **16**, 5–30 (1963)
26. S. Bosch, *Lectures on Formal and Rigid Geometry.* Lecture Notes in Mathematics, vol. 2105 (Springer, Cham, 2014)
27. S. Bosch, U. Guntzer, R. Remmert, *Non-archimedian Analysis.* Grundlehren der mathematischen Wissenschaften, vol. 261 (Springer, Berlin, 1984)
28. N. Bourbaki, *Espaces vectoriels topologiques. Chapitres 1 à 5*, (nouvelle édition) (Masson, Paris, 1981)
29. K. Buzzard, Eigenvarieties, in *Proceedings of the LMS Durham Conference on L-Functions and Arithmetic* (2007)
30. K. Buzzard, F. Calegari, The 2-adic eigencurve is proper. Doc. Math. Extra Vol. 211–232 (2006)
31. H. Carayol, Sur les représentations galoisiennes modulo *l attachées aux formes modulaires.* Duke Math. J. **59**(3), 785–801 (1989)
32. P. Cartier, *Representations of p-adic groups: a survey,* in *Automorphic Forms, Representations and L-functions (Proceedings of Symposia in Pure Mathematics, Oregon State University, Corvallis, Ore., 1977)*, Part 1 (Proceedings of Symposia in Pure Mathematics, XXXIII) (American Mathematical Society, Providence, 1979), pp. 111–155
33. G. Chenevier, Familles p-adiques de formes automorphes et applications aux conjectures de Bloch-Kato. Ph. D. Thesis, defended on June 13th, (2003) (Paris VII, under the supervision of M.Harris), 192 p. available on http://gaetan.chenevier.perso.math.cnrs.fr/pub.html
34. G. Chenevier, Familles p-adiques de formes automorphes pour GL_n. J. Reine Angew. Math. **570**, 143–217 (2004)
35. G. Chenevier, une correspondance de Jacquet-Langlands p-adique. Duke Math. J. **126**(1), 161–194 (2005)
36. G. Chenevier, The p-adic analytic space of pseudocharacters of a profinite groups and pseudorepresentations over arbitrary rings, in *Automorphic Forms and Galois Representations: Volume 1.* London Mathematical Society London Mathematical Society, vol. 414 (Cambridge University Press, Cambridge, 2014), pp. 221– 285
37. R. Coleman, Classical and overconvergent modular forms. Invent. Math. **124**(1–3), 215–241 (1996)
38. R. Coleman, p-adic Banach spaces and families of modular forms. Invent. Math. **127**(3), 417–479 (1997)
39. R. Coleman, B. Edixhoven, On the semi-simplicty of the U_p-operator on modular forms. Math. Annalen **310**, 119–127 (1998)
40. R. Coleman, B. Mazur, The eigencurve, in *Proceedings of the Durham, 1996.* Lecture Notes Series, vol. 254 (London Mathematical Society, London, 1998)
41. R. Coleman, F. Gouvêa, N. Jochnowitz, E_2, Θ, and overconvergence. Internat. Math. Res. Notices **1995**(1), 23–41 (1995)
42. P. Colmez, Notes du cours de M2, Distributions p-adiques. https://webusers.imj-prg.fr/~pierre.colmez/M2.html
43. P. Colmez, Notes du cours de M2, La fonction Zeta p-adique. https://webusers.imj-prg.fr/~pierre.colmez/M2.html

44. P. Colmez, Fonctions L p-adiques., Séminaire Bourbaki, Vol. 1998/99. Astérisque **266**, Exp. No. 851, 3, 21–58 (2000)
45. P. Colmez, Les conjectures de monodromie p-adiques, in Sém. Bourbaki 2001–02, exp. 897. Astérisque **290**, 53–101 (2003)
46. B. Conrad, Irreducible components of rigid analytic spaces. Annales de l'institut Fourier **49**, 905–919 (1999)
47. B. Conrad, Modular curves and rigid-analytic spaces. Pure Appl. Math. Q. **2**(1), 29–110 (2006)
48. B. Conrad, Several Approaches to Non-Archimedean Geometry, in *p-Adic Geometry*. University Lecture Series, vol. 45 (American Mathematical Society, Providence, 2008), pp. 9–63
49. K. Conrad, Ideal classes and matrix conjugation over $\mathbb{Z}$, Available on http://www.math.uconn.edu/~kconrad/blurbs/gradnumthy/matrixconj.pdf
50. H. Darmon, F. Diamond, R. Taylor, Fermat's Last Theorem in ed. by J. Coates, S.-T. Yau, *Elliptic Curves, Modular Forms and Fermat's Last Theorem* (International Press, Vienna, 1997)
51. H. Darmon, A. Lauder, V. Rotger, Overconvergent generalised eigenforms of weight one and class fields of real quadratic fields. Adv. Math. **283**, 130–142 (2015)
52. P. Deligne, *Formes modulaires et représentations l-adiques*, Séminaire Bourbaki vol. 1968/69 Exposés 347–363. Lecture Notes in Mathematics, vol. 179 (Berlin, New York, 1971)
53. P. Deligne, M. Rapoport, Les schémas de modules de courbes elliptiques., in *Modular Functions of One Variable, II (Proceedings International Summer School, University of Antwerp, Antwerp, 1972)*. Lecture Notes in Mathematics, vol. 349 (Springer, Berlin, 1973), pp. 143–316
54. P. Deligne, J.-P. Serre, Formes modulaires de poids 1. Ann. Sci. École Norm. Sup. **7**(4), 507–530 (1974)
55. S. Deo, On the eigenvariety of Hilbert modular forms at classical parallel weight one points with dihedral projective image. Trans. Amer. Math. Soc. **370**(6), 3885–3912 (2018)
56. F. Diamond, J. Shurman, *A First Course in Modular Forms*. Graduate Texts in Mathematics, vol. 228 (Springer, New York, 2005)
57. H. Diao, R. Liu, The eigencurve is proper. Duke Math. J. **165**(7), 1381–1395 (2016)
58. M. Eichler, Eine Verallgemeinerung der Abelschen integrale. Math. Zeitschr. **67**, 267–298 (1957)
59. D. Eisenbud, *Commutative Algebra with a View Toward Algebraic Geometry*. Graduate Texts in Mathematics, vol. 150 (Springer, Berlin, 1995)
60. E. Ghate, An introduction to congruences between modular forms, in *Currents Trends in Number Theory* (Allahabad, 2000) (Hindustan Book Agency, New Delhi, 2002), pp. 39–58
61. E. Ghate, On the local behavior of ordinary modular Galois representation, in *Modular Curves and Abelian Varieties*. Progress in Math, vol. 224 (Birkhäuser, Basel, 2004), pp. 105–124
62. R. Godement, Topologie algébrique et théorie des faisceaux. Hermann, Paris (new edition, 1997)
63. B. Gross, On the factorization of p-adic L-series. Invent. Math. **57**, 83–95 (1980)
64. A. Grothendieck, Sur quelques points d'algèbre homologique. in Modular curves and Abelian varieties. Progress Math. Tohoku Math. J. **9**, 119–221 (1957).
65. D. Hansen, Universal eigenvarieties, trianguline Galois representations, and p-adic Langlands functoriality. With an appendix by James Newton. J. Reine Angew. Math. **730**, 1–64 (2017)
66. A. Hatcher, *Algebraic Topology* (Cambridge University Press, Cambridge, 2002)
67. S. Hattori, J. Newton, Irreducible components of the eigencurve of finite degree are finite over the weight space. J. Reine Angew. Math. **763**, 251–269 (2020)
68. H. Hida, Congruences of Cusp forms and special values of their zeta functions. Invent. Math. **63**, 225–261 (1981)
69. H. Hida, On congruence divisors of cusp forms as factors of the special values of their zeta functions. Invent. Math. **64**(2), 221–262 (1981)
70. H. Hida, Galois representations into $\mathrm{GL}_2(\mathbb{Z}_p[[X]])$ attached to ordinary cusp forms. Invent. Math. **85**(3), 545–613 (1986)

71. H. Hida, Iwasawa modules attached to congruences of cusp forms. Ann. Sci. École Norm. Sup. **19**(2), 231–273 (1986)
72. H. Hida, *Elementary Theory of L-Functions and Eisenstein Series*. London Mathematical Society Student Texts, vol. 26 (Cambridge University Press, Cambridge, 1993)
73. N. Jacobson, *Schur's result on commutative matrices*. Bull. Amer. Math. Soc. **50**, 431-436 (1944)
74. U. Jannsen, On the l-adic cohomology of varieties over number fields and its Galois cohomology, in *Galois Groups Over* $\mathbb{Q}$ (Berkeley, CA, 1987). Mathematical Sciences Research Institute Publications, vol. 16, (Springer, New York, 1989), pp. 315–360
75. C. Johansson, J. Newton, Extended eigenvarieties for overconvergent cohomology. Algebra Number Theory **13**(1), 93–158 (2019)
76. J. Glynn, *Algebra und Differentialrechnung*. (German) Bericht über die Mathematiker-Tagung in Berlin, Januar, 1953 (Deutscher Verlag der Wissenschaften, Berlin, 1953), 58–163
77. N. Katz, p-adic L-functions for CM fields. Invent. Math **49**, 199–297 (1978)
78. W. Kim, Ramifications points on the eigencurve, PhD Thesis, Berkeley (2006)
79. M. Kisin, Overconvergent modular forms and the Fontaine-Mazur conjecture. Inv. Math. **153**(2), 373–454 (2003)
80. E. Kunz, *Residues and Duality for Projective Algebraic Varieties.* With the assistance of and contributions by David A. Cox and Alicia Dickenstein. University Lecture Series, vol. 47 (American Mathematical Society, Providence, 2008)
81. S. Lang, *Algebra*, 3rd edn. (Addison-Wesley, Boston, 1993)
82. I. Lakatos, *Proofs and refutations. The Logic of Mathematical Discovery.* (Paperback edition. Originally published in 1976.) Cambridge Philosophy Classics (Cambridge University Press, Cambridge, 2015)
83. M. Lazard, Les zéros des fonctions analytiques d'une variable sur un corps valué complet (French) Inst. Hautes Études Sci. Publ. Math. **14**, 47–75 (1962)
84. R. Liu, D. Wan, L. Xiao, The eigencurve over the boundary of weight space. Duke Math. J. **166**(9), 1739–1787 (2017)
85. R. Livné, On the conductors of mod l Galois representations coming from modular forms. J. Number Theory **31**(2), 133–141 (1989)
86. D. Majumdar, Geometry of the eigencurve at critical Eisenstein series of weight 2. J. Théor. Nombres Bordeaux **27**(1), 183–197 (2015)
87. J. Manin, Parabolic points and Zeta-functions of modular curves. Math. USSR Izv. **6**, 19 (1972)
88. J. Manin, Periods of cusp forms, and p-adic Hecke series. Mat. Sb. (N.S.) **92**(134), 378–401 (1973)
89. H. Matsumura, *Commutative Ring Theory*, vol. 8 of Cambridge Studies in Advanced Mathematics, 2nd edn. (Cambridge University Press, Cambridge, 1989)
90. B. Mazur, Modular curves and the Eisenstein ideal. Inst. Hautes Études Sci. Publ. Math. **47**, 33–186 (1977)
91. B. Mazur, A. Wiles, The class field of abelian extensions of $\mathbb{Q}$. Invent. Math. **76**(2), 179–330 (1984)
92. B. Mazur, J. Tate, J. Teitelbaum, On p-adic analogues of the conjectures of Birch and Swinnerton-Dyer. Invent. Math. **84**(1), 1–48 (1986)
93. T. Miyake, *Modular Forms*. Springer Monographs in Mathematics (Springer, Berlin, 2006). Translation of the 1976 Edition
94. E. Noether, Idealdifferentiation und differente. J. Reine Angew. Math. **188**, 1–21 (1950)
95. L. Nyssen, Pseudo-representations. Math. Annalen **306**, 257–283 (1996)
96. J. Park, p-adic family of half-integral weight modular forms via overconvergent Shintani lifting. Manuscripta Math. **131**(3–4), 355–384 (2010)
97. V. Pilloni, Overconvergent modular forms. Ann. Inst. Fourier **63**(1), 219–239 (2013)
98. R. Pollack, On the p-adic L-function of a modular form at a supersingular prime. Duke Math. J. **118**(3), 523–558 (2003)

99. R. Pollack, G. Stevens, Overconvergent modular symbols and p-adic L-functions. Ann. Sci. Éc. Norm. Supér. **44**(1), 1–42 (2011)
100. R. Pollack, G. Stevens, Critical slope p-adic L-functions. J. Lond. Math. Soc. **87**(2), 428–452 (2013)
101. S. Ramanan, *Global Calculus*. Graduate Studies in Mathematics, vol. 65 (American Mathematical Society, Providence, 2005)
102. K. Ribet, Endomorphisms of semi-stable abelian varieties over number fields. Ann. Math. Second Ser. **101**(3), 555–562 (1975)
103. K. Ribet, Congruence relations between modular forms, in *Proceedings of the International Congress of Mathematicians*, vol. 1, 2 (Warsaw, 1983) (PWN, Warsaw, 1984), pp. 503–514
104. R. Rouquier, Caractérisation des caractères et pseudo-caractères. J. Algebra **180**(2), 571–586 (1996)
105. P. Schneider, *Nonarchimedean Functional Analysis*. Springer Monographs in Mathematics (Springer, Berlin, 2002)
106. I. Schur, Zur Theorie vertauschbaren Matrizen. J. Reine Angew. Math. **130**, 66–76 (1905)
107. Y. Sella, Comparison of sheaf cohomology and singular cohomology (2016). arXiv:1602.06674
108. J.-P. Serre, Endomorphismes complètement continus des espaces de Banach p-adiques. Inst. Hautes Études Sci. Publ. Math. **12**, 69–85 (1962)
109. J.-P. Serre, *Local Fields*. Translated from the French by Marvin Jay Greenberg. Graduate Texts in Mathematics, vol. 67 (Springer, New York, 1979)
110. J.-P. Serre, *Classes des corps cyclotomiques (d'après K. Iwasawa)*, in *Séminaire Bourbaki*, vol. 5, Exp. No. 174, 83–93 (Société mathématique de France, Paris, 1995)
111. G. Shimura, Sur les intégrales attachées aux formes automorphes. J. Math. Soc. Japan **11**, 291–311 (1959)
112. G. Shimura, An ℓ-adic method in the theory of automorphic forms, unpublished (1968)
113. G. Shimura, *Introduction to the Arithmetic Theory of Automorphic Functions* (Princeton University Press, Princeton, 1971)
114. G. Shimura, On the holomorphy of certain Dirichlet series. Proc. London Math. Soc. **31**, 79–98 (1975)
115. V. Shokurov, *Shimura integrals of cusp forms*. Izv. Akad. Nauk SSSR Ser. Mat. **44**(3), 670–718, 720 (1980)
116. E. Spanier, Tautness for Alexander-Spanier cohomology. Pacific J. Math. **75**(2), 561–563 (1978)
117. E. Spanier, Singular homology and cohomology with local coefficients and duality for manifolds. Pacific J. Math. **160**(1), 165–200 (1993)
118. N.E. Steenrod, Homology with local coefficients. Ann. Math. Second Ser. **44**(4), 610–627 (1943)
119. W. Stein, *Modular Forms, A Computational Approach* (American Mathematical Society, Providence, 2007). Available on https://wstein.org/books/modform/modform/eisenstein.html#explicit-basis-for-the-eisenstein-subspace
120. G. Stevens, Rigid analytic modular symbols. Preprint, available on http://math.bu.edu/people/ghs/research.d
121. R. Taylor, Galois representations associated to Siegel modular forms of low weight. Duke Math. J. **63**, 281–332 (1991)
122. R. Taylor, A. Wiles, Ring-theoretic properties of certain Hecke algebras. Ann. Math. **141**(3), 553–572 (1995)
123. E. Urban, Eigenvarieties for reductive groups. Ann. Math. **174**(3), 1685–1784 (2011)
124. M. Višik, Nonarchimedean measures connected with Dirichlet series, Math USSR Sb. **28**, 216–218 (1976)
125. A. Wiles, On ordinary λ-adic representations associated to modular forms. Invent. Math. **94**(3), 529–573 (1988)
126. A. Wiles, The Iwasawa conjecture for totally real fields. Ann. Math. **131**(2), 493–540 (1990)

Printed in the United States
by Baker & Taylor Publisher Services